Quantity

FOOD PRODUCTION, PLANNING, AND MANAGEMENT

Quantity
FOOD PRODUCTION, PLANNING, AND MANAGEMENT

■

THIRD EDITION

John B. Knight
Indiana University Purdue University Fort Wayne

Lendal H. Kotschevar
Florida International University

John Wiley & Sons, Inc.
New York • Chichester • Weinheim • Brisbane • Singapore • Toronto

Library of Congress Cataloging-in-Publication Data:

0-471-33347-6

Printed in the United States of America.

10 9 8 7 6 5 4 3 2 1

For his love, dedication, and service to God, country, and mankind, and for his contributions to the hospitality industry and its students, the authors dedicate this book to Dr. Lewis J. Minor, Professor Emeritus, Michigan State University, and founder of the L. J. Minor Corporation, Cleveland, Ohio.

And to my loving wife, Whitney, and our three children, Brad, David, and Jacqueline, as well as to my parents, Herb and Eleanor Knight, and my in-laws, Leonard and Joan Jones.

—*John B. Knight*

Contents

■

■ APPENDICES 423

Preface

———————————— ■ ————————————

The planning necessary to run a quantity foodservice operation, the preparation of the food within that operation, and the management of the entire facility may seem overwhelming, even to longtime professionals. This book presents the details necessary for successful quantity food production. It is intended primarily for students within hospitality management programs, but those studying culinary arts, club management, nutrition and dietetics, and food operations, as well as industry personnel wishing to learn more about food and beverage management, will find the material presented critical to improving professionalism within operations.

ORGANIZATION

Quantity Food Production, Planning, and Management systematically presents all phases of the foodservice operation. Part I begins with a new challenge to food and beverage students and operators alike in the chapter titled, "Raising the Standard." Management issues dealing with trends in nutrition and health are considered next, at the beginning of the book, in light of the importance of these issues in today's society. Menu planning, as the initial step in designing or operating any foodservice facility, is discussed in the following chapter. The reader then learns about the foodservice equipment needed to prepare a menu successfully. Controls, including the purchasing, receiving, storing, and issuing of food, are discussed next, followed by a new chapter on service and dining etiquette. Management principles dealing with human resources, product

and profits, and property and promotions are each covered in detail in the next chapters.

Quantity food production is the subject matter of Part II. In this revised edition, numerous chapters have been updated to discuss computer applications, and the material is presented in a fashion that enables both experienced and inexperienced readers to benefit. Cooking principles are examined first, followed by sanitation and safety; pantry products; stocks, soups, and sauces; fruits, vegetables, and cereals; meats, fish, and poultry; bakeshop production; and dairy products and eggs. Each chapter covers a different important aspect of the multifaceted job of managing a foodservice in light of the many products available on the market today.

FEATURES

Vocabulary terms are clearly defined as they appear, a new glossary is provided at the end of the appendices, and text material is continually highlighted with illustrations. At the beginning of each chapter are learning objectives and selected Web site references, and at the end of each chapter, problems and questions, including case studies, are posed to test the reader's ability to apply the information covered and to ensure comprehension of the important subjects discussed.

New in the third edition are the following sections:

- The Foodservice Industry and Its Mission
- Cultural Diversity

- Employee Productivity
- Career Opportunities
- Total Quality Management
- The "Heart-Healthy" Environment
- Genetic Engineering and the Food Supply
- The New Food Label
- Smoking vs. the Nonsmoker
- Dining Etiquette
- The Legal Environment

- Wages, Taxation, and Government Intervention
- New Industry Growth Segments
- Innovative Cooking Methods
- Cooking Trends
- The Americans with Disabilities Act
- Hazard Analysis Critical Control Points

An **Instructor's Manual** (ISBN: 0-471-37677-9) to accompany the textbook is available from the publisher.

Acknowledgments

■

Special thanks are given to the following people at Indiana University Purdue University–Fort Wayne, who offered their suggestions and assistance throughout the preparation of the book:

Louise Pruse
Secretary

Ken Jordan
Computer Expert

Elmer Denman
Photographer

Bobbi Shadle
Graphic Design Artist

Linda Lolkus
Registered Dietitian
Visiting Assistant Professor

In addition, the following people gave of their time and expertise in the completion of the book:

Jeremy Corson
Chef
Oakwood Inn and Conference Center
Syracuse, Indiana

Pam Klopfenstein
Culinary Instructor
Anthis Career Center
Fort Wayne, Indiana

Finally, three hospitality management students must be recognized for their contributions in the reading and editing of the final manuscript. Their contributions are greatly appreciated:

Jason Buchheit

Travis Dale

Wei Zhou

The authors appreciate the industry support provided by all who contributed pictures, essays, and other material incorporated into this book.

John B. Knight
November 1999

PART I

Quantity Food Planning and Management

CHAPTER 1

Raising the Standard

■

LEARNING OBJECTIVES

By the end of this chapter, the reader will:

1. Understand the foodservice industry, its mission, and organization.

2. Recognize the career opportunities available and the levels of responsibility required.

3. Be able to distinguish between such terms as *management, cultural diversity, employee productivity,* and *total quality management.*

4. Know the power the computer plays within the modern foodservice operation and appreciate its applications, from menu planning and forecasting to customer ordering and delivery and payment.

■

SELECTED WEB SITE REFERENCES

American Culinary Federation
www.acfchefs.org

American Hotel & Motel Association
www.ahma.com

Club Managers of America
www.cmaa.org

Education Institute of the American
 Hotel & Motel Association
www.ei-ahma.org

Educational Foundation of the National
 Restaurant Association
www.edfound.org/

National Restaurant Association
www.restaurant.org

NCR Corporation
www.ncr.com

Remanco International Inc.
www.remanco.com

■

THE FOODSERVICE INDUSTRY AND ITS MISSION

Foodservice is one of the largest industries, projecting nearly $354 billion in sales in 1999. Foodservices contribute substantially to the mobility of individuals and the flexibility of one's social life.

The foodservice industry employs more people than does any other industry. Of every $2.50 spent for food, $1 is spent in foodservices. The industry has enjoyed steady growth since before World War II, with an average increase of about 5 percent per year. Its growth has been the most rapid of that of all retail industries.

The 750,000 or more foodservices that provide this food are widely dispersed and vary from small snack or take-out shops, drive-ins, and coffee shops to moderate-sized restaurants and hotel operations, to huge foodservices such as those serving the military, schools, and large hospital complexes. Airline foodservices alone do millions of dollars in business each year.

Because the foodservice industry is labor intensive and because it is divided into many small units that must produce in small amounts, it has never been able to obtain the benefits of mass production. The industry is characterized by low productivity, low profit margins, low capital investment, and high labor cost per dollar of sales. The industry also has a high labor turnover rate: workers move from job to

job with ease, and each year many move in and many others move out.

The steady flow of workers in and out, plus job-hopping, reduces operating efficiency and is costly. It has been estimated that training a new worker to replace an experienced one can cost the operation as much as 8 percent of a worker's annual salary for the first year. The industry has moved to reduce its labor turnover by trying to provide stable, financially attractive positions for workers. Although pay is low in some categories, it is good in others, especially at the supervisory and management levels. Graduates from four-year colleges in the hospitality industry find plenty of jobs available at salaries equal to those offered to beginning engineers.

Many foodservice operations are family-owned or individual-owned enterprises, but this characteristic has been constantly eroding over the last thirty years. Today, massive foodservice chains do most of the dollar business, and the small, single-outlet enterprise's share of the market is shrinking. The large corporate organization is the leader in the industry.

Within the last thirty years, the type of food used in foodservices has gradually changed. More processed foods have found their way into use, some of which are practically ready to serve. Many operations still want to produce their own food, but even they now use more processed foods to do it. Another development is the central commissary, where foods are mass-produced and then shipped to satellite units to be served. The problem of maintaining quality while obtaining desirable economies has slowed the growth of central commissary production, however, and some operations that had instituted central commissaries have now discontinued them.

The industry has had to change to meet a changing patron market, too. People eat less, have clearer concerns for their health, are eating out more, and are more discriminating shoppers. The industry has become highly competitive, and its management is now largely composed of people who have had training in business or in specialized schools that teach foodservice management.

■

TRAINING THE LABOR FORCE

The foodservice industry employs millions of people, and each year several hundred thousand new workers must be found to fill newly created or newly vacated jobs. Until recently, much of this new labor was recruited from "off the streets." These people simply walked in and were assigned jobs whose duties they learned by experience. Their teachers were other workers who may or may not have known how to do the job; as a result, the new worker often learned to do the job incorrectly or inefficiently. Upper-echelon jobs were often filled by promotion from within the operation's existing ranks. Not infrequently, the manager or even the president of a large corporation was once a pot washer or bus person.

Within the last forty years, a sizable educational system for training employees at all levels of the industry has emerged. Unions, too, have concerned themselves with educational courses—in part so that they can certify the job abilities of members better. Unions often work with public schools, but in some cases they have established their own training and certification centers. The government has also been active in promoting short educational programs to train workers for the foodservice industry.

Professional associations have set up programs to give educational opportunities to their members. The Club Managers Association has developed an educational program that must be taken by any member who wishes to use the title CM (Certified Manager). The American Dietetic Association requires all dietitians to be certified by examination and to continue to accumulate educational and professional points during their professional lives if they intend to use the title RD (Registered Dietitian) after their names. The School Foodservice Association and some of its state organizations now require all workers to gain educational credits in order to be certified. The Educational Foundation of the National Restaurant Association has its Food Professional Manager Program, offering certification to members, and the Dietary Managers Association (DMA) offers professional development to its membership as well. Comparable programs exist in other professional groups in the industry. In addition, the industry sponsors a great many educational scholarships for interested persons.

Home study courses are popular, too. Many universities and other schools sponsor training programs for workers, supervisors, and managers. At some institutions, 5000 or more people per year register for correspondence courses. Many high schools offer courses for waiters, waitresses, cooks, and other service positions.

Many community colleges and junior colleges have excellent programs for training workers. The emphasis varies from vocational courses for those wanting to work as chefs or maître d's to economics and management training for people interested in becoming food and beverage managers, sales managers, or catering or banquet managers. Four-year hospitality programs commonly accept credits from these schools.

Nearly 200 universities and colleges have four-year programs in the area, and some offer master's degrees and a few the doctorate. Over 2000 graduates from four-year schools are thought to enter the foodservice field each year—and the industry could use twice that number. Each year companies make recruiting visits to educational institutions, offering graduates a three month or longer training program before being moved into supervisory or managerial positions. Because of the rapid growth of many businesses, some graduates hold important positions within four or five years after leaving college. The growth of graduate programs to develop top executives and teachers for the industry has also been significant.

Most four-year programs are offered in the College of Business, but others may be found in the College of Home Economics or the College of Food

Technology, and still others exist as separate schools. The discipline of foodservice management is today well recognized in academic circles. Most academic programs require that students spend many hours working in the industry; internships have been established whereby students gain on-the-job experience and receive academic credit. Such an approach works best because—although many basic facts can be learned in class—the student must be exposed to situations that are related to the real work world in order to develop into a successful manager.

Our workforce in the foodservice industry is rapidly changing. Native Americans, Latinos, Asians, and African-Americans now make up a significant part of the labor force, requiring management to understand cultural differences among employees. Also, the number of older people and those with disabilities in the foodservice industry has increased significantly.

Cultural Diversity

Workforce diversity is no longer just a gender or religion management consideration. While it originally referred simply to a person's gender or ethnic group, today, workforce diversity encompasses differences in age, affectional orientation or sexual preference, economic status, educational background, ethnicity, gender, geographic location, income, marital status, lifestyle, and religion, to name a few characteristics.

Foodservices that manage workforce diversity efficiently benefit from more effective recruiting, improved management quality, increased market share and productivity, and reduced labor costs through lower employee turnover. Although developing and maintaining programs that enhance diversity can be difficult to implement, the benefits far outweigh the disadvantages. A comprehensive and carefully planned approach for implementation is required.

One foodservice company, McDonald's, has been successful through its *Changing Workforce Programs*, whereby different courses are offered to help employees work through gender and cultural differences. These courses, such as Managing the Changing Workforce, Women's Career Development, Black Career Development, Hispanic Career Development, and Managing Cultural Differences and Managing Diversity, encourage open communications.

The end result is a raising of the standard of professionalism within the foodservice industry. As labor is tight in many markets, workforce diversity must play an even greater role in the success of foodservice operations tomorrow. At the present time, Latinos and African-Americans make up 25 percent of our total population, and this percentage is expected to grow.

Employee Productivity

Another concept that raises the standard within the industry is employee productivity. People who enjoy coming to work generally produce more, as will workers who believe that their work will be valued or lead to advancement. The challenge is to clearly define what good work is for the employee and how that can be measured. Reward systems are the key to recognizing individual contributions in order to encourage greater productivity.

In one operation, ability to work in any station within the kitchen was looked upon as excellent work. That way, employees could assist others in accomplishing tasks at hand when a rush would come. Accordingly, employees were encouraged to learn each other's job and were recognized as having done so by receiving a pin to wear for each station learned. With the knowledge of sauté, broiler, saucier, garde-manger, and other positions under one's belt, that employee became highly productive for the kitchen as a whole. Morale improved and turnover was lowered.

With improved productivity, any foodservice operation will attract and retain employees better. A positive atmosphere in which employees are recognized for their contributions, rather than being criticized for their mistakes, will do much to improve productivity. Supervisors should look for employees to do the job approximately correctly, at which point recognition should be given.

■

MANAGEMENT AND FOODSERVICES

This book is designed to move the student in a logical, sequential order through the various management steps involved in quantity foodservice (we shall define a *quantity foodservice facility* as one preparing foods in quantity to serve to people within a designated period of time). Management is given first priority in this introductory chapter, although it presents only a brief summary of management's scope and functions. The ensuing chapters consider the basics of planning quantity food production and service, and then take up specifics of food production. Food production issues are viewed from the standpoint of what a manager ought to know to be able to manage these operations well and to ensure that adequate standards in food production and service are satisfied. The management skills needed to operate a foodservice are presented in such a way as to prepare the reader for the responsibilities of an entry-level position in the industry. By integrating planning, preparation, and management in an orderly fashion, this book is designed to help the student, manager, owner/operator, or other interested person build the skills necessary to operate a successful quantity foodservice facility.

Management Defined

Management is sometimes defined as the element in an enterprise that plans, staffs, organizes, directs, and controls an enterprise so that it reaches its goals. It is management's job to establish goals, plan how to reach them, and then organize, staff, lead, and control so as to reach them. Often, these actions are called the *function of management*. Sometimes, *management* is defined as reaching goals through people. This is a good definition because it underscores that managers alone cannot make an operation a success; it takes others working with management to accomplish this.

Some workers see management as the boss giving the orders, as the one getting the most pay for doing the least work, or as the sector of the company that knows less about the operation than anyone else. Such opinions are often based on a lack of knowledge about what the manager is or does. Management's failure to inform members of the operation about its essential role can be costly to its image.

Good management often consists of the ability to plan operations intelligently and then to select and direct people so that the plans come to a successful conclusion. The more effectively managers prevail upon others to do a good job, the better management

succeeds in its job. Many people are involved in achieving an enterprise's goals, and management is responsible for getting people's energy going in the right direction.

Levels of Responsibility

The hierarchy in a foodservice has three levels: management; mid-management or supervisory; and workers. To maintain a smooth-functioning facility, employees at each level must adequately perform the essential functions allocated to that level.

Management comprises the chairman and members of the board in a corporation, the president, the manager, and even the assistant managers. Mid-management may reach as high in the organization as the assistant managers and may extend as low as supervisors; sales managers, food and beverage managers, and purchasing department heads are typically considered mid-management, as are chefs, chef stewards, the maître d'hôtel, the catering manager, a food production manager, and a dietitian in charge of a hospital's dietary department. The workers are everybody else. In a traditional continental-cuisine organization, the executive chef is a mid-management person under a food and beverage manager or other manager. The *sous chef* is a supervisor under the executive chef, and the *chefs des partis* (heads of production, such as the *chef de saucier* in charge of the sauce, gravy, and soup department, and the garde-manger, who heads the cold meat section) are sub-supervisors under the *sous chef*. Under these people come workers such as cooks, helpers, and others working in the organization. In most foodservices, the number of people at the worker level far exceeds the number of people at the other two levels.

The management level plans action, the mid-management initiates and supervises action, and the workers carry it out. It is important that all employees in the organization know what type of action they are responsible for. When people at a lower level try to assume the functions of others at a higher level (or vice versa), conflict can occur. The intruding person is resented as encroaching on someone else's turf. A person promoted from one level to a higher one sometimes encounters problems in this regard because he or she finds it difficult to break away from the action patterns learned in the lower level. This is seen when a supervisor promoted from worker ranks fails to supervise without constantly pitching in and doing jobs others should be doing.

Career Opportunities

The employment outlook in foodservice is excellent. The reason for this is due to the unprecedented growth in the industry over the past two decades, resulting in exceptional opportunities, especially for the graduates of management programs.

The managerial job pool is growing at a rate better than the rest of the economy. In some segments, such as contract foodservice and quick-service restaurants, the need for entry-level management is acute. New opportunities will arise in management for those willing to serve the "baby boomers" as they age in assisted-care living facilities.

With the trend of foodservices being increasingly dominated by large national and multinational corporations, the demand for managers with a formal

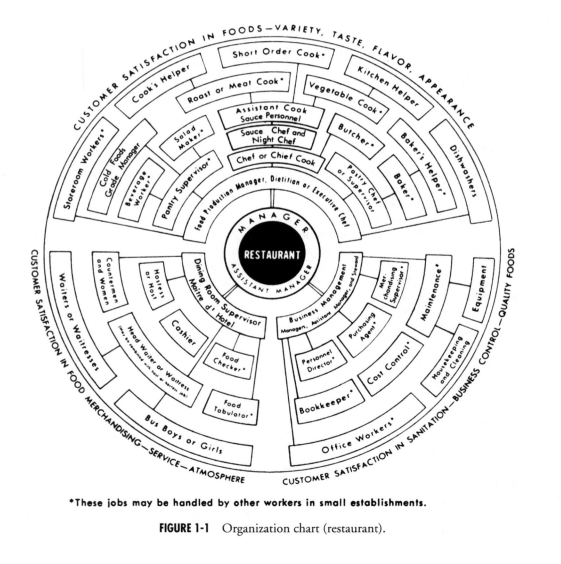

*These jobs may be handled by other workers in small establishments.

FIGURE 1-1   Organization chart (restaurant).

education will continue to increase. At the same time, the industry will provide a vast array of opportunities for those people who possess a strong entrepreneurial spirit. For innovative, hard-working, service-oriented people who want to own their own business, foodservice offers almost unlimited potential.

Graduates of management programs typically begin their careers as management trainees, assistant managers, or supervisors. Starting salaries are competitive with those in other industries, but the potential for advancement significantly exceeds those in other industries for the capable person who is willing to work hard. Opportunities also exist for graduates in the areas of accounting, marketing and sales, finance, and human resource management. Of course, foodservice consulting as a career for the seasoned industry professional can be very rewarding.

In deciding if a career in foodservice is for them, individuals must be people- and service-oriented. Hours to be worked are often during hours when others are relaxing or at play. Relocating for the right job may take a foodservice manager across the country. Finally, to ensure a smooth-running operation, the foodservice manager often works more than a 40-hour week. Despite the hard work, the rewards can be great for those seeking opportunities for professional growth, good salaries, responsibility, and high levels of personal satisfaction.

■
ORGANIZATION

Getting people into the right place so they do the right job is called "setting up an organization." A *business organization* may be defined as an arrangement of people in jobs designed to accomplish the goals of the operation. Figures 1-1 to 1-3 offer examples of the many ways organization charts can be set up.

Organizational structures vary according to the needs of the enterprise, the jobs done, and the people in the organization. Organizations should be dynamic, not static. They must be changed as needed, and organizational charts should reflect such changes. Obviously, just setting up an organizational chart does not build an organization. Creating the living organization—by selecting and organizing people on the job—is one of the most important responsibilities of management.

Types of Organizations

Different organizations are needed to do different things for operations. Basically, there are three kinds of organizations: line, staff, and functional. Most organizations are combinations of these. For example, even though a line organization may be set up, parts of the operation may be strictly based on functional tasks or on staff work that is required to support the line operation.

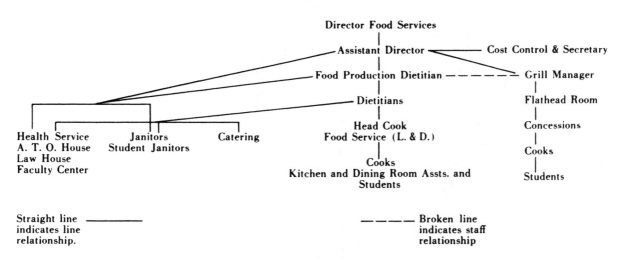

FIGURE 1-2 Organization chart (foodservice at a university).

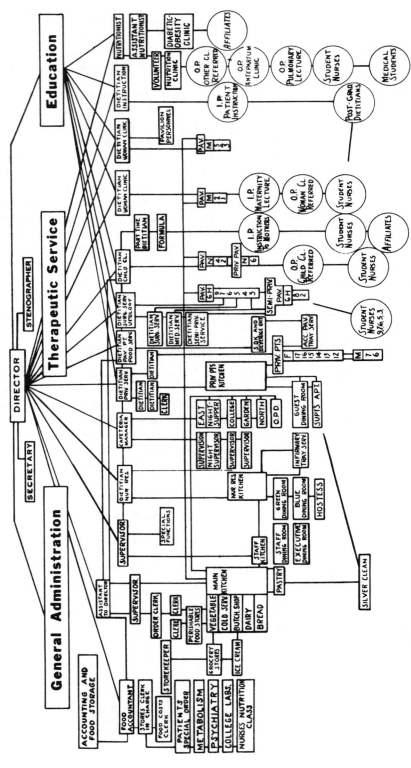

FIGURE 1-3 Organization chart (nutrition department of a hospital).

Line Organization

Military organizations usually are considered the best examples of line organizations. The lines of authority and responsibility flow from the head down through the various mid-management layers to the workers. In a line organization, a manager directs an assistant manager to do something; that person in turn directs a supervisor, who in turn directs a worker or workers. In reverse order, the report of work done or problems encountered flows back up through the same line of authority until it reaches the manager. No bypassing of layers is permitted either on the way down or on the way up. Sometimes this flow of authority and responsibility is called the *chain of command.*

In a line organization, the top position bears full responsibility for everything that happens in the organization. Such responsibility cannot be delegated to others. The power to make a decision or do a job can be vested in a subordinate, but the way this duty is discharged by the subordinate remains the responsibility of the chief officer. A good rule is "you can delegate power but you can't delegate responsibility."

Staff Organization

Staff organizations are usually support organizations established to render a service to the line organization. The staff does not take steps to see that things are done, nor does it have authority to implement any decisions it reaches or to act on any research it gathers. These are tasks for the line organization to perform. Every organization has staff positions—such as a secretary who works for the manager—that carry no line authority. It is improper for such a person to give orders to others in the line. Another staff organization is the accounting department, which renders a service to management by keeping records and furnishing essential financial and accounting data. A consulting dietitian serves in a staff position in relation to the superintendent of a small hospital or nursing home. Much trouble can occur when staff members step out of their staff roles and assume an unwarranted line role and give orders.

Functional Organization

If a large corporation were to set up its organizational structure on a functional basis, it might set up a division that was in charge of sales, promotion, and advertising; another that was in charge of planning; another that was in charge of procurement; and so on—seeing that all functions needed to bring the operation to its goals were established. The various functional divisions of a corporation coordinate their activities and (usually) report to a high-level line organization. It is possible for line organizations to exist under a functional one. Thus, individual operating foodservices that receive instructions from various functional divisions are set up as line organizations. Functional organizations are best suited to large corporate units.

Management's Responsibility

In any foodservice, daily events raise the need for action by management. Management must be prepared to see that the proper decisions are made so that the operation functions efficiently. Poorly devised action or inaction by management can ultimately cause an operation to fail. Proper procedures and records must be maintained. Employees must be selected with care and given adequate training. Personnel records must be established and maintained, complete with job descriptions. Supervisors must be properly instructed as to plans and actions and must be given adequate authority and backing to ensure that the work progresses satisfactorily. Planning, production, and service must flow smoothly so that patrons are satisfied and functions occur when they should. Making the operation run like clockwork requires considerable management skill. Sound decision making is the product of good judgment, practical knowledge, and astute thinking. The foodservice business is a dynamic, fast-changing one, and management must be equal to its demands.

■

COMPUTERS AND MANAGEMENT

The job of managing has been considerably assisted by the computer, which has changed many clerical tasks from manual to mechanical processes, giving management more accurate and more reliable information in a shorter time. This permits better decision making and faster action in response to changing conditions or unexpected developments. The overall result of the computer revolution has been better planning, improved organizational control, and more up-to-date and more nearly complete information. Many tasks are simplified, and some are performed almost automatically. Consequently, management has more time to devote to planning and to direct supervision.

Although major computer application came relatively late to the foodservice industry, it is now being extended rapidly into many functions. Some of the more common tasks are discussed in the subsections that follow. In some cases, a more complete explanation is given in a later chapter. Although various computer applications are presented, few (if any) operations have yet put together computer programs that include all of the things a computer can do that are discussed here. The use of computers and the coverage they provide in foodservice operations will undoubtedly continue to be extended in the future.

Menu Planning

Early use of the computer in foodservice operations came in the areas of payroll, inventory, accounting, and point-of-sale transactions—largely because systems covering these areas could easily be adapted from other industry computer applications. The first original programs unique to the foodservice industry were programs set up in the 1950s and 1960s for computer planning of menus. These early programs had as their goal the production of menus planned to meet nutritional standards within certain cost constraints; they were largely developed for health-related foodservices.

Today, menu-planning programs are used to produce menus that meet specific patron and operational needs within certain cost limitations. The program can be set to offer a desirable range of individual menu items, such as the following four soups: one a cream soup; another clear; another heavy with vegetables and other items; and the fourth a purée. The menu programming can include instructions to avoid too much repetition of ingredients or flavors, such as having tomatoes appear a limited number of times in dishes on the menu.

Different forms of preparation can also be specified, to ensure that different cooking methods are found in the new items offered, such as a braised meat item, a roast, a broiled meat, and a deep-fried meat

item. The computer can also be programmed to present a balance in meat, fish, and poultry offerings. Special combinations can be established, such as having Yorkshire pudding always appear automatically whenever roast ribs of beef are on the menu, having a horseradish sauce always appear with boiled tongue, or having deep-fried onion rings always appear with sautéed calf's liver. The frequency with which menu items reappear from week to week can also be set. If a retirement home wants chicken to appear on the menu for every Sunday dinner, the computer can be instructed to make this happen. The computer can give a printout of nutritional values for meals so that management can monitor them to ensure that the proper nutrients are being served.

Recipe size can be expanded or reduced automatically. Thus, a recipe can be exploded (increased) from 25 portions to 150, 250, or 1000 just by giving the computer the right instructions. Other constraints or needs can be input to the computer and it will comply with them in generating its menus. Quite flexible programs are possible, and often management can override the computer by adding, deleting, or otherwise changing what it does. The list of menu items that is entered into the computer to use in menu planning is called the *menu file*. Normally, each menu item is identified in the computer by its code number, but some computers can identify items by name as well.

Standardized Recipes

Many menu items in the menu file require special preparation before service, and so must be backed up by one or more recipes. The recipes listed in this manner are kept on what is called a *recipe file*. Thus, an item titled Baked Cod Mornay may have two recipes: one for baking the cod in a mornay sauce; and one for making the mornay sauce, which is a béchamel (cream) sauce to which cheese has been added. (Actually, the mornay sauce might be listed as a variation under the béchamel sauce recipe.) Thus the mornay sauce in the Baked Cod Mornay recipe actually becomes an ingredient in the baked cod recipe. These other recipes for sauces and other items supporting the main recipe are sometimes called *subrecipes*.

Recipes should be standardized to give a known quantity and a known quality at a known cost. Standardization is accomplished by pretesting recipes and precalculating costs.

Recipes should specify the code number or code name, the total amount of food produced and/or the number of portions, the portion size, the ingredients, the amount of each ingredient used, and the method of handling ingredients to produce the recipe. Cooking times, cooking temperatures, panning instructions, and other relevant information should also be given. Variations and substitutions can be listed at the bottom of the recipe. Total cost of the recipe and cost per portion are usually listed, and sometimes menu selling price is, too. In some cases it may be desirable to identify other items that are served with the item produced by that recipe or to include plating or serving instructions. A picture of how the item should appear is sometimes added, and some computers possess the necessary graphics capabilities for this.

Ingredient Files

An ingredient file consists of all ingredients listed for recipes. Ingredients are identified in the computer either by their code numbers or their code names. The ingredient code number should either be the same as that used in purchasing or be closely related to it. Thus, a 6/10 case of tomato catsup might have a purchasing and receiving number of P2148, but on the inventory and issue records it is identified as 12148 because the shipping unit (the case) is opened and broken down in the storeroom and issued in individual No. 10 cans. Pricing also differs. The price of P2148 (a case composed of six cans) is $24.60, and the price of 12148 (a single can) is $4.10. The ingredient price and code may require a third set of numbers, since the catsup used in recipes is often measured by volume. Thus, in the recipe, the ingredient code may be R2148 and the cost $0.035 per fluid ounce. If a recipe called for 1½ pints of catsup, the computer would convert 1½ pints into 24 fluid ounces and multiply 24 × $0.035, for a total ingredient cost of $0.84. Some computers cannot easily convert pints into fluid ounces; in such situations the recipe must give the percent of the can to be used. Similarly, in bar work, making conversions from liter bottles into ounce serving portions is difficult for some computers to do, in which case the bar recipes must express quantities of ingredients in milliliter portions.

Changes in recipes or ingredient costs must be input as they occur so that the recipes can be updated and so that the information compiled by the computer reflects what is actually the case. Often a change in the purchase price of one ingredient automatically changes prices in other files.

Recipes often specify the condition in which an ingredient is to be added to others during preparation of the item. For example, a recipe may call for the ingredient to be sifted, beaten, blanched, melted, diced, or strained.

Forecasting

To obtain an estimate of costs and/or profitability of a specific menu mix (group of menu items), management must develop a projection of the number of menu items that will be sold. This projection is called a *forecast*.

Forecasting is still a rather inexact science. The many variables involved in future demand are often difficult to detect and evaluate. Historical data—such as amount sold for the same meal on the same day last week, last month, or last year—can be used. A popularity index can be kept to record how many portions were sold each time the item appeared on the menu. Another technique is to evaluate the other items competing against a particular menu item and estimate how much they add or detract from its sales. An item that might sell well when grouped with some items might not do as well against others.

The effect of weather, events such as conventions, athletic contests, and other independent factors may have to be taken into account. Food-eating trends and other long-term factors deserve consideration, as well. In some foodservices, fairly good estimates of sales can be developed by noting reservations or bookings (such as with hotels or airlines), occupancy projections for dormitories over weekends, normal attendance for sporting events, or other information that indicates what quantity of patrons to expect. Hospitals find that their patient count usually goes down before and during a holiday season, but hospitals have an advantage in forecasting that many do not, in that many have patients mark menus a day ahead, indicating the foods they want served to them.

Forecasting, if done with some care, is a better guide to future demand than just plain guessing or playing hunches, although these should not be ruled out entirely. With some practice and experience, prognosticators will find that their forecasts become increasingly precise and increasingly accurate.

Once the forecast is made and input into the computer, the computer can generate cost, profit, and other financial data. The computer will read the menu item, select its proper recipe, calculate from the recipe the ingredient cost and the portion food cost, and multiply the latter cost by the number of portions given in the sales forecast to establish a projected material cost for the menu item. Repeating this process for each item on the menu yields a total food cost figure for the period covered in the projection. If a selling price for each menu item is input into the computer, the forecast number of sales times the selling price gives the expected income from the item; again, when this is done for all menu items, the sum of the income figures will equal the total sales. These computations enable management to estimate how well the menu is going to do. If other cost projections are programmed, the computer can generate a net profit estimate and make other projections that are of great help to management.

Production Planning

An important part of managing a foodservice is production planning. There is a need to know when to produce foods, how much, when, where it will be needed, and other information. Unless such information is given, successful operation cannot be achieved. Such planning is the responsibility of top management, but it is often delegated to others.

General Considerations

The first step in planning production is to decide what is to be produced and when. This requires that a menu be planned and the recipe to be used be selected along with the date, meal, and sometimes, time in hours and minutes when it will be needed. The cost must also be determined to see if it fits into budget or menu price restraints. Sometime in this preliminary planning one must decide if adequate labor, equipment, and other facilities are available to produce what is needed. Also, is the amount of labor sufficient

and sufficiently skilled to do the job? After all this, the decision of how much must come, and this requires a forecast. Next, a check must be made to see if the materials for production are on hand, and, if not, are ordered in the proper quality and quantity adequate to serve production needs. Proper receiving and storage must be given and the requisitions for supplies withdrawn from inventory must be made and delivered. Production must occur at the proper time. This is frequently an important consideration in planning. Menu items take various times to be ready for service. An aspic must be made at least 4 hours before it is needed, to be set sufficiently for service. A tenderloin steak may be done to order. Bread and many bakery items are prepared hours ahead. Often, production requirements are such that the equipment used for production is pressed to produce what is needed and those doing the production must make sufficient adjustment to accomplish what is desired. (Menu planners should watch to see that planning is done to give a balance to the equipment on hand and not overload. A menu requiring all oven preparation can often result in a production failure or, if not, in a lot of frustration.) Sometimes, items can be prepared in advance, such as soups and sauces, and placed into a bain marie or steam table to await production. Planning is such that all units of a production department must at the time of a meal come together in unison with production complete. The start of a meal is like that of the start of a symphony orchestra in which every player is ready when the maestro gives the first stroke of the baton. It must proceed as an orchestra working in unity together.

Although service is usually not a part of production planning, it must be considered. Successful production can succeed or fail with proper or improper service. Proper dishup is needed. Portions must be sized properly and displayed attractively. The size of the dish holding the order or orders must be right, not too large nor too small. The portion should not come over the rim. It is said, "The rim belongs to management, and the rest to the server." Attractive dishware must be used. Dishware color and decoration should fit the environment of the facility. Thus, Chinese food should be served in Chinese-type dishware. Tableware should fit into the decor of the operation. The proper ware should be used. If an upper-scale restaurant wants to feature Snails Bourguignon, they should be served on an indented snail plate with the right holding tongs and extractor. Otherwise, the dish loses much interest and may result in guest dissatisfaction.

Food should be properly garnished. Although it is not possible for all garnishes to be edible, some might not be—paper frills used to cover the rib ends of lamb chops or of a pork crown roast, for instance. The garnish should not overwhelm the food but should serve as an attractive accompaniment to it. Contrasting complementary colors between the garnish and food should be sought. The garnish should be fresh appearing, not "tired." Certain garnishes are typical of certain foods, and a garnish common to one type of ethnic food must be used with care with other types of goods. Unusualness should be sought. Thus, a bread or cream horseradish sauce is often an accompaniment to roast beef, but an unusual and interesting difference is to present the roast beef with several horseradish curls on the plate, omitting the sauce.

Three Types of Production Planning

Three methods of planning production are commonly used by foodservices, although there are others. These are the laissez-faire method, the production sheet method, and the ingredient room method. The first is common in many restaurants and other commercial foodservices. The second is common in institutional feeding and also in many commercial operations, while the third is common in institutions and in large-scale production units such as central commissaries.*

The *laissez-faire method* basically leaves most of production planning to those who will be doing it. Management or sub-management may indicate roughly some of the basic guidelines, such as menu items to be produced and time, but leave the remainder to employees. In many operations this works well. However, in some, management may not even plan

---

*The ingredient room method was developed by the U.S. Navy Supply Research and Development Facility in Bayonne, New Jersey, in the 1950s. It was adapted from a production method observed in the mass production of food at the Horn and Hardart central commissary. Katherine Flack, director of foodservices for New York State institutions, saw the plan in action and adapted it to her operations, calling the assembly area in the storerooms the ingredient room. From there it was adopted into many state foodservice, commercial, and other foodservice operations.

the menu. In a hotel, management usually leaves production planning to a food and beverage director or even to a chef. In a hospital the director seldom takes any action in production planning. This is left to dietitians or foodservice directors, or even at times those lower down in the organization.

Where those actually doing the production also have the responsibility for production planning delegated to them, the lead employees with the responsibility usually have been employed in the operation for some time and know the needs or have sufficient experience in other foodservices to know what is needed if they are a new employee. Often, the person doing the purchasing confers with those doing the production to ascertain items needed, quality, amounts, and so on. They know from experience, for example, the times when planning must be done so that foods are on hand when needed. Besides setting up for direct deliveries to supply the necessary items, some system of inventory withdrawal will be worked out either by management or the accounting division. Times for starting production are known from experience, and allocation of the work to be done is almost automatic. At times, those requiring items from another section may remind those in the section of this need, such as notifying the bakery section that so many vol-au-vent shells are needed for Tuesday's lunch. For this type of production planning to work, it is necessary that the entire operation work together as a team.

The *production sheet method* usually requires that management or submanagement plan production, and usually the latter procedure is used. This requires that previous to production the menu and other steps be planned and then a sheet is drawn up detailing the production need, the amount to be produced, the time needed, and if food is delivered to different service areas, where such food is to go, when, and how much. In a health facility, special diet needs may be noted, and production needs grouped under these categories. Often, in a hospital general menu is planned, and from this the special diets are taken with the required modifications being made. The use of a special diet kitchen to prepare some diets is used today for only very special diets, and many health care operations have dispensed with them entirely. Usually, the production sheet also informs purchasing how much of specific items will be needed, although in some cases, this may be left to those producing the foods. Variations in this method are possible where experienced production employees are on the staff and certain decisions such as amounts to order, and so on, are left to them.

The *ingredient room method* requires that a file of standardized recipes be available, usually computerized. Mid-management staff usually have delegated to them the planning of the menu along with other preliminary planning decisions. This person often decides on the menu and the amounts to be produced. Once these decisions are made, the items needed along with amounts are fed into the computer and a printout of the quantities and quality of foods is obtained to send to the storeroom if needed from inventory, to purchasing if they must be ordered, to the vegetable preparation section, or to other basic supply units. These needs sent to these areas are divided by section or production area along with the time required. (All this must be hand-computed if computerization is not used.)

The storeroom usually has a separate area or room where it weighs out or otherwise measures the amounts required for each recipe. This must be done with extreme accuracy in *exact* quantities to produce the desired product in the quantity required. Recipe amounts are then segregated by section following which the storeroom delivers the items to the areas where needed. Preparers then process the items according to the production sheet definitions. One requirement that must be impressed on preparers and enforced is that all ingredients delivered *must* be remeasured. This is to prevent the preparer from blaming the storeroom for failing to deliver the correct amount in the event of a recipe failure.

The ingredient room method has many advantages. It saves worker time. If a worker has to leave the station in which the person works and travel to the storeroom to get items, a high-priced worker is often spending time just walking and waiting to get the items. If a lower-paid storeroom clerk can do the job and deliver the items, the worker then stays in the work area and produces. It also gives good organization to production. It is obvious that such a system requires proper time planning. Menus and preliminary planning must be prepared sufficiently in advance for storerooms, vegetable preparation, or purchasing to

have the items delivered when needed. However, once the system is worked out and put into motion, production usually proceeds smoothly, without a hitch.

Quality Control

Critical Quality Control Points

A consistent level in the quality of menu items is achieved by establishing a standard for the product and then setting up checks at critical control points to see that the desired quality is achieved. These points should include correct ingredients and their measurement; preparation techniques; times and temperatures; correct tools, equipment, and utensils; condition of product during various stages of preparation; sanitation; portion size; and other factors required in special products. Each product will have its own standard, so there will be as many standards as there are menu items produced.

It is important that the checks and procedures established should be in as much detail as the standard. Thus, recipes should detail the critical quality control points. Again, pictures of the product showing the right procedure or the results of the wrong one are helpful. If rather inexperienced personnel are to prepare a product, it might be well to go over the critical points before starting preparation.

Formulation of Standards

A *standard* is something established as a rule or basis of comparison in measuring quality, quantity, value, or other factor of a product. In foods it covers factors such as color, volume, shape, crumb, crust, density or thickness, tenderness, juiciness, and so on—in fact, any attribute that affects the quality of a product. As noted, each product will have its own standards. Often, standards must differ according to the type of operation. Also, standards may even have to differ within a specific operation; thus, portion sizes may be varied between adults and small children. The number of standards required will depend on the number of items served in a facility. A drive-in selling only carbonated beverages, milk shakes and malts, hamburgers, and french fries will have only a few, whereas an upscale restaurant that changes menus for every meal, every day, and season to season will have many.

Often, assistance in setting up standards for foods can be obtained. Recipes often list many factors that one needs to cover in a standard, such as type, quality, and amount of ingredient; techniques to use; times and temperatures; utensils, equipment, and tools to use; condition of product during preparation, cooking, or baking; and as a finished product. Anyone with a good foundation in food preparation can thus take a recipe and formulate much of a standard.

Often, food preparation texts give standards for products, and these can be used as reminders of factors to put into one's own standard for the product. Trade associations, purveyors of food products, and others often have standards for their products. Advertisements of foods also give pictures of products and product descriptions that can be helpful.

The federal government has many quality standards for foods. Thus, one can say in a standard for roast beef: "The quality and type of product should be a Range C, No. 109, upper Choice grade roast, aged in Cryovac 21 days." One then knows that it will be a beef roast from 19 to 22 pounds in weight, cut and tied ready to be roasted, of a specific quality to be satisfactory in flavor, juiciness, and tenderness to give good eating—these three factors being decided largely by the grade and aging specified—and sized to give a generous portion. Also, the federal government has many grade standards for foods that can be helpful in establishing factors to cover. Thus, a U.S. Department of Agriculture (USDA) score sheet for judging the quality of canned asparagus lists color, defects, character, and condition of the canning liquor as factors to judge. Using this, one would then proceed to add other critical quality factors required. In some of these federal standards for foods, one will find helpful information such as the syrup density or density of a product such as tomato puree.

In many operations personnel have the knowledge and skills necessary to produce the type of product desired. Seeking their help can often result in a good standard. One can stand and observe such a person preparing an item, noting the actions taken, and from time to time asking a question to explain why something must be done in a certain way.

In judging products to decide on the standard to use, a panel of judges can be assembled and products scored. This is often a technique used to decide on the quality of liquors to serve as well as products. The panel should know the standard desired and what will be suitable to satisfy the patrons served.

This brings us to the final determinant in what the standard should be—the patron. A standard that satisfies the trade is essential, and no matter what management or anyone else in the operation believes, standards that do not satisfy will not keep a facility in business long. Thus, getting patrons to indicate what they like and do not like or might like or might not like can be extremely helpful and necessary in establishing a standard, and one must start from this premise. Otherwise, formulation of the standard may be a futile exercise.

Menu Analysis

Menus can be analyzed in various ways, ranging from highly qualitative methods to complex mathematical ones. Subjective menu evaluation, menu counts (popularity indexes), Hurst's menu scoring, menu factor analysis, the matrix analyses of Miller, Smith–Kasavana, Pavesic, and Kotschevar, goal analysis, and break-even analysis are ten common approaches. Some involve radically different approaches and give different information; and even those that use similar methods vary in the information they give.

A detailed discussion of menu analysis is beyond the scope of this book at this point. However, such a discussion is included in Appendix A for those wishing to delve deeper into this area. Some facets of menu analysis are also discussed in Chapter 8 under "Menu Engineering" and "Break-Even Analysis and Its Calculation."

Customer Ordering

One of the more recent developments in computer use within quantity food production has been seen in customer ordering. While the fast-food and quick-service segments have led the way, other dining establishments are not far behind in their implementation of this latest hardware and software.

Touch-sensitive screens are now used in operations where service counter personnel once took the order. Customers are moved through a series of graphics in which they literally touch the screen to select the food item desired. In the case of play areas where a mother wishes to watch her children without having to leave them to order and pick up food, touch-sensitive screens offer a method by which the mother identifies her table number and the food is brought to her.

Merchandising is enhanced through a voice that encourages the customer to order the largest drink since it is the best value. Since a tally is kept on the screen as to how much the customer has spent in the ordering process, customers are more likely to spend more in that they know how much money they have and how much they have spent in ordering. Thus, an extra apple pie or order of french fries are or are not purchased.

Devices are now also available that allow a server at the table to take orders, punch them into the device, and have the orders filled by the kitchen. When the order is ready for pickup, the device beeps; thus, the server is spared taking trips to the kitchen to place orders and to see if they are ready for pickup.

Delivery and Payment

Just as the delivery of food to the mother watching her children in a restaurant's play area has been made possible through computers, so is the payment for food ordered at a touch-sensitive screen. By having stored-value cards similar to those used to prepay a phone call through a calling card, the customer is able to pay for his or her meal electronically via computer.

The system then allows the customer to order food without assistance, pay for it without interaction with another person, and pick up the food or have it delivered, as is the case in the play area. Ordering at drive-up windows is being revolutionized, as the "loud speaker order taker," who never seems to get the order right, is being replaced by a touch-sensitive screen that merchandises products through graphics, keeps a tally of the total dollar spent, and ensures that the correct items are ordered. Another drive-up window system has a "scoreboard" installed which lights up each item the customer has ordered. This innovation is linked to the point-of-sale personal computer to help ensure against communication glitches between the customer and order taker.

Computers will continue to affect the heart of the house delivery and payment systems as well. Purchasing agents will order on-line through the Internet. The future of the computer on food production, planning, and management will greatly affect the efficiency of operations, as has been seen in recent years.

Complementary Computer Applications

The quantity of items to purchase or to issue is determinable when the quantities of ingredients needed to produce the menu are calculated by the computer. These quantities can be further distinguished as quantities needed for immediate use or as quantities to have on inventory. Timely compilation of such information can help operations avoid stock shortfalls and other frustrating food production problems.

When an ordered item is received, the computer can be informed of the date, the item's code number, purveyor, amount, cost, a description of the goods, whether the item went directly to production (directs) or to stores (inventory), and so on. If programmed to do so, the computer adds the new amount to inventory and calculates a new inventory amount. Directs are charged out as, in effect, issues on the receiving date. By having the computer add directs to storeroom issues, management can obtain a food cost for the day; put into ratio with sales, this gives the percentage food cost. Thus, one might see the following:

Storeroom issues	$ 865.74
Directs	46.83
Total food cost	$ 912.57
Sales	$2,944.23
Percentage food cost	31%

Deducting issue values from inventory and adding receipts enables the computer to generate an updated inventory called a *perpetual inventory*. All these programmed tasks are rapidly and accurately performed by the computer, eliminating much slow and tedious manual manipulation and calculation.

Much useful information for making management decisions can be obtained at the cash register, often called point of sale (POS). With a POS system, management at any time can ascertain what checks are out, how long checks have been out, what orders remain unfilled, how many items of a particular kind have been sold, what their dollar value is, what volume of sales has been generated by various servers (this is called a *productivity report*), what amounts credit card charges, room charges, and tips come to, and many other things. At the end of a recording period, total sales, total item or food group sales, and other desirable data are at hand. It is now possible for the computer to run an analysis of how well planning projections made prior to production anticipated actual results. This alone is of tremendous value to management as a tool for refining the processes of planning and decision making. In the days when manual compilation of such data was necessary, the results often were unavailable until too long after an actual occurrence for management to take corrective action.

In addition, after all results are in, the computer can transfer essential data to the general ledger file so that important financial reports such as cash flow, balance sheet, and profit-and-loss report can be compiled. Thus, almost from cradle to grave, the computer can be effective in food production management. It has changed clerical, decision-making, and other time-consuming tasks into almost automatic operations and has extended management's control and base of information so that better management is achieved.

■

TOTAL QUALITY MANAGEMENT

Although the term *total quality management* has been used extensively, there has been some disagreement as to the nomenclature in use. Some refer to the subject matter as just quality management; others refer to it as continuous quality improvement. There is one aspect that has generated very little disagreement and that is the fact that management should focus on eliminating "quality" problems before they occur rather than attempting to address them after they have happened. This is especially true in the area of customer service but is also applicable in the area of food production. Both are important and must be given careful consideration.

Customer Service

The key to quality in service is to have a systematic, planned process for returning dissatisfied customers to a state of satisfaction after service has failed to live up to expectations. Since customer turnover can have a tremendous financial impact on an operation's bottom line, customer satisfaction is a must for total quality management.

For example, the loss of a loyal customer is one thing, but the alienation of potential customers due to negative word-of-mouth referrals can be even worse. Dissatisfied customers are likely to complain to nine or ten others about service imperfections, while satisfied customers typically express contentment to only four or five others. Furthermore, research indicates that it costs five times more to replace a customer than to retain one. Hospitals sometimes find that doctors and patients decide on which hospital to patronize because of high satisfaction with the food at that hospital.

The management issue, then, is to empower service providers to engage in techniques to ensure customer satisfaction when, in fact, the customer is not satisfied. The fiscal and informational resources must be at hand for the server to see that the customer departs the operation satisfied. Recovery actions often involve "doing it right the second time" but, more specifically, correcting what was done wrong immediately, so that when there is a second time, the customer notices that matters have been corrected.

Recovery actions may take the form of compensation, apology, reparation, and empathy. Each must be understood by servers and applied correctly depending on the situation. Total quality management in the area of customer service is ongoing and must be given top priority.

Food Production

The same holds true in food production. When the product produced or served is not up to the standard, the professional must recognize that fact and take corrective action. Checkpoints come from the cook, expediter (one who checks plated food prior to it leaving the kitchen), server, and customer. If at any point, one of those people says that the item produced is not what it should be, quality is affected and change must take place.

Satisfaction can be improved by replacing the item immediately or offering the customer a different choice. When it comes to food production, the key is to correct the system such that another poor item is not served in the future. This may range from retraining the cook to regarnishing the plate. If it is the server's fault, such as serving cold food due to delay in pickup or taking down and serving the wrong order, the standards in the service area must be rechecked.

Total quality management must be constant in any operation in order to raise the standard. Both customer service and food production are affected. Management must be aware of what is being done incorrectly and make every effort to improve in those areas.

■

SUMMARY

The foodservice industry is the fifth largest industry in the United States. It employs more than 10 million people and accounts for 4 percent of our gross national product. The industry is highly labor intensive and has many problems, because of its wide dispersal of operating units, its low productivity, and its need to produce on demand. It is a healthy industry, however, enjoying one of the best and most stable growths of any industry in the nation. It performs important services in our economic, social, and recreational programs. The industry is characterized by family-owned or single-ownership operations, but this is changing rapidly as chains and multiple-type corporations take over.

Training the large number of employees needed to operate the foodservice industry was at one time quite haphazard and unprogrammed. Today, various formal training programs are springing up, bringing a better-trained employee into the industry. Unions as well as professional associations are beginning to develop training programs. Many people are also taking foodservice courses in high schools, community or junior colleges, or four-year universities. A number of graduate programs have been started, and correspondence courses are used heavily. Foodservices that manage workforce diversity efficiently, benefit

from more effective recruiting, improved management quality, increased market share and productivity, and reduced labor costs through lower employee turnover.

One of the best definitions of management is: *Management is the achievement of goals through people.* Three levels of action can be identified in any foodservice organization: management; mid-management or supervision; and workers. It is the job of the first to plan action, of the second to start action, and of the third to act. Proper activity by personnel in each area of action promotes efficiency and smoothness in the foodservice operation.

Management's responsibility is to provide the planning, organizing, staffing, controlling, and leading necessary to bring the enterprise to its goals. Mid-management's responsibility is usually to receive direction from management and see that the required steps are taken to move the enterprise to its goals. Workers' responsibilities are job-related; because they are specific, these duties are best outlined in job classifications.

Foodservices can be organized in three ways: line, staff, and functional. The last two are frequently combined with the first. A line organization is one in which authority and responsibility flow down from the top through various positions. When reversed, the flow proceeds by the same step-by-step process to the top. Staff organizations usually exist to supply information to the line. Functional organizations serve specific functions within the line organization.

Every person holding a position in an organization should have the proper authority and responsibility to discharge the position's requirements. A failure to establish both power and accountability can cause problems in the organization. Delegation is a procedure by which the duty to perform a job is assigned to a subordinate to carry out. In such cases, the proper authority to do the job must be given; but it is not possible for the person delegating the job to escape responsibility for whether and how well the job is done.

The computer has significantly changed the planning, management, and implementation of quantity food production. Its effectiveness in almost all phases of foodservice operations has helped management refine traditional methods of planning, control, and decision making.

If properly programmed, the computer can plan a menu within specified constraints, print out standardized recipes for any number of portions desired, set up a file of ingredients so that recipes can be priced out, and calculate portion costs. If a forecast of number of portions to be sold is made, the computer can generate a food cost figure; if given the selling price, it can compute the total dollar sales an item will bring in. Other arrangements allow the computer to estimate what total sales will be.

The computer is also useful in performing various menu analyses (which are of special benefit to management for decision making and planning), in calculating quantities to purchase, in totaling inventory, in keeping track of issues and directs, in setting up receiving reports, in maintaining the payroll, and in making many point-of-sale calculations.

Total quality management must be a constant endeavor in any operation in order to raise the standard. Both customer service and food production are affected. Management must be aware of what is being done incorrectly and make every effort to improve in those areas.

CHAPTER REVIEW QUESTIONS

1. R. A. Kroc developed McDonald's from a tiny group of hamburger drive-ins into one of the world's largest foodservice corporations. He was an engineer and had no foodservice experience. One of the reasons for his success was that he approached management from an engineer's standpoint. Just what is management, how does it work, and why would a person with a strong engineering or scientific background have attributes that might translate into excellent management abilities?

2. A large foodservice operation finds that its menu is in need of improvement: it lacks good profitability, and management feels that it is deficient in patron appeal. You are a shift manager and know that some of the items on the menu do not make a good gross profit because their food cost is too high and some of them are unappealing. You feel you can help the manager by explaining what you feel is wrong with the menu, but you want some proof. What analytical methods would you use to analyze the menu and to present evidence of your results to the manager?

3. How big is the foodservice industry in dollar sales and as an employer?

4. What are some of the problems in the industry?

5. What educational facilities are available to those who would like to work in the industry? Identify the opportunities for all levels of work and management.

6. What is *management?*

7. What are the functions of management?

8. Give an example in which a staff organization and a functional organization are both part of a line operation. What would these organizations do for the line operation?

9. What jobs can a computer do for management? How does it perform these jobs?

CASE STUDIES

1. Merril Len Manson is thinking about getting out of the rock entertainment business and into foodservice. He is interested in trying to decide which type of restaurant to open and has decided to do research on dining trends. Rocker Manson has contacted you as a bright student of management and has asked you to use the Web site *www.restaurant.org* to assist him. What are the

trends that might affect his new operation? What kind of operation should he open? Where should he locate that operation? Support your decision with information from the Web site.

2. In looking around for new computer products to use in her many restaurants, Ms. Arrow Smith has found the Web site *www.culinarysoftware.com* and is very excited. She has called you to summarize the products available from the site and make

recommendations to her as to which products you feel are best and why.

3. A career in hospitality management can be exciting for anyone. Using the information at *www.ei-ahma.org, www.restaurant.org,* and in this chapter, overview the career opportunities in the hospitality industry. What career interests you most and why? Support your answer with information found on-line.

■

VOCABULARY

Foodservice industry

Processed foods

Cultural diversity

Functions of management

Maître d'hôtel

Chefs des partis

Chef de saucier

Garde-manger

Organization

Line organization

Staff organization

Functional organization

Control points

Decision making

Menu planning

Standardized recipe

Recipe file

Ingredient file

Forecast

Menu analysis

Directs

Issues

Perpetual inventory

Total quality management

Ingredient room

■

ANNOTATED REFERENCES

Bartlett, M. 1994. *The Best of Restaurant and Institutions Winning Food Service Ideas.* New York: John Wiley & Sons, Inc., pp. 1–104. (Food trends are considered in one section.)

Cornyn, J., and J. Coons-Fasano, with M. Schechter. 1995. *Noncommercial Foodservice: An Administrators Handbook.* New York, John Wiley & Sons, Inc. (The management of noncommercial foodservices is presented.)

Cullen, N. C. 1996. *The World of Culinary Supervision, Training, and Management.* Upper Saddle River, N.J.:

Prentice Hall. (Written from the chef's prospective, this book covers all management topics.)

Dornenburg, A., and K. Page. 1998. *Dining Out.* New York: John Wiley & Sons, Inc. (The restaurant critic's job is considered, as are other matters of interest to the foodservice manager.)

Egerton-Thomas, C. 1995. *How to Open and Run a Successful Restaurant.* New York: John Wiley & Sons, Inc. (From financial considerations to human resources, all management aspects are considered.)

Marvin, B. 1992. *Restaurant Basics: Why Guests Don't Come Back and What You Can Do About It.* New York: John Wiley & Sons, Inc. (Recognizing what annoys customers and correcting bad situations in restaurants are presented in this text.)

McCool, A., and F. Smith, with D. Tucker. 1994. *Dimensions of Noncommercial Foodservice Management.* New York: John Wiley & Sons, Inc. (Contract foodservice management is reviewed in this book.)

Powers, T. 1995. *Introduction to the Hospitality Industry.* New York: John Wiley & Sons, Inc. (The foodservice industry and its careers are discussed over many chapters.)

Rande, W. L. 1996. *Introduction to Professional Food Service.* New York: John Wiley & Sons, Inc. (The industry overviewed with management considerations is the basis for this text.)

Riegel, C., and M. Dallas. 1998. *Hospitality and Tourism Careers.* Upper Saddle River, N.J.: Prentice Hall. (From first jobs to professional development, this book covers the topic of careers in total.)

Starr, N. 1997. *Viewpoint: An Introduction to Travel, Tourism, and Hospitality,* 2nd ed. Upper Saddle River, N.J.: Prentice-Hall, pp. 323–340. (This chapter is on careers in the industry.)

CHAPTER 2

Understanding Trends in Nutrition and Health

■

LEARNING OBJECTIVES

By the end of this chapter, the reader will:

1. Appreciate the role that nutrition plays in one's health and sense of well-being.

2. Be able to distinguish between various types of diets.

3. Learn of current trends in nutrition and be able to better understand why people eat the way they do.

4. Be able to identify the various nutrients and the role they play in bodily functions.

5. Recognize the new food label and understand the information and its importance.

6. Know the impact that genetic engineering is having on our food supply and be able to appreciate its impact on the foodservice industry.

7. Understand the concerns of both the smoker and nonsmoker as issues arise regarding smoking within foodservice operations.

■

SELECTED WEB SITE REFERENCES

AIDS and Young People
www.avert.org/young.htm

American Dietetic Association
www.eatright.org

Dietary Managers Association (DMA)
www.dmaonline.org

Healthfinder
www.healthfinder.org

International Food Service Executives Association
www.ifsea.org

National Association of College & University Food Services (NACUFS)
www.nacufs.org

Food sustains us and enables us to grow. If we select our food well, we take one important step toward having good health. Good nutrition is important to us from the earliest weeks after conception until our death. A good diet is needed not only to promote essential living functions but also to protect against some of our biggest killers: heart attacks, strokes, cancer, diabetes, and others. Seven of the top causes of deaths in this country are diet-related.

■

HEART-HEALTHY ENVIRONMENT

Despite our high standard of living, the United States is not the healthiest nation in the world. Part of the cause could be our diets. Many people in this nation do not eat well, and many eat foods that lead to impaired health.

Studies have shown that poor nutrition is not restricted to the underprivileged (who may not be able to buy the right kind and amount of food); it also afflicts many people who can afford a good diet but either do not know how to eat properly or do not care.

Within the last decade or so, people have been hearing more about food and how it affects our health. As a result, many people have become concerned about what they eat, and their dietary patterns have changed. The convincing evidence that obesity leads to serious health problems has been a stimulus to reduce weight, along with the fact that people feel better and look better when they are thinner. Reducing salt intake is a major step in combating or avoiding hypertension (high blood pressure). The danger of heart problems is lessened by efforts to cut down on saturated fat and cholesterol intake. We are on a health-diet binge that shows no sign of abating. People today want to eat well to be well, and they are aggressively pursuing the dietary way. A study by the National Restaurant Association (NRA) found that about one-third of all customers were concerned about their diets and sought healthful menu items. About the same number were interested in a healthful diet but not aggressively so. Another third said they do not care and eat what they like.

The latter group was composed of more younger people than the other groups.

One way in which dietary patterns are changing is that people are not consuming as much red meat (beef, lamb, and pork). These meats contain saturated fats that may be harmful in too large amounts. More white meats, such as chicken and fish, are being consumed because their fat is less saturated. Along with eating less red meat, people are reducing their consumption of eggs, butter, and other fatty foods to lower their saturated fat, cholesterol, and calorie intake. There is a distinct trend toward eating more fresh fruits and vegetables, although the total quantity of processed and fresh vegetables and fruits consumed has not increased significantly. People are also eating more high-fiber foods such as whole-grain cereals, bran, vegetables, and fruits. Table 2-1 gives the saturated fat, unsaturated fat, cholesterol, and calorie content of some foods.

Public concern has challenged foodservices to become more knowledgeable about nutrition and more careful in planning menus that meet the needs of patrons. Foodservice operations today must have a basic knowledge of how to select and prepare foods in a manner that preserves the nutrients they contain and how to avoid foods that are not desirable for some patrons. It has been found that selling nutrition pays, and many foodservices are becoming more expert in nutrition because it improves the bottom line to do so. Educational institutions are also finding it advisable to include instruction on nutrition. Thus, in the future we should see even more attention being paid to nutritional matters. An added impetus is public and governmental pressure. Already the fast-food industry is being pressured to publish the nutritional value of the foods it serves. This requirement could spread throughout the industry.

Unfortunately, there is so much widely circulated misinformation about nutrition and diets in the public mind that foodservices find it difficult to meet all the ideas patrons have about diet and about the foods they should eat. If a foodservice decides to embark on a nutritional program, it should know how to do it, it should set it up carefully, and it should execute a carefully planned strategy to see that it is done right.

How much responsibility do foodservices have in providing nutrition for patrons? Are they responsible for seeing an adequate diet is selected and consumed? There is no easy answer. It depends upon the person selecting and eating the food, the foodservice, and the situation.

People today eat less because they are less active. Alcohol consumption is dropping. Instead of hard

TABLE 2-1 Fats, Cholesterol, and Calories in Some Foods

Food	Fat (%)		Cholesterol[a] (%)	Calories (per 100g)
	Saturated	Unsaturated		
Beef, choice	12	11	0.07	301
Chicken	2	3	0.06	124
Eggs	4	5	0.50	163
Fish, lean	2	4	0.07	79
Lamb	12	9	0.07	276
Milk, whole	2	1	0.01	65
Pork, lean	4	4	0.07	185
Salmon	5	5	0.04	222
Veal	5	4	0.09	207

[a] 0.07% means that a person receives 70 milligrams of cholesterol in 100 grams (about 3 ounces) of product.

spirits, more wine and lighter alcoholic beverages are selected. A glass of white wine or a glass of orange juice is a common request at a cocktail party. On average, the consumption of rich desserts is down, but many restaurants find that patrons eating out may want to splurge and eat a luscious, rich, high-calorie dessert as a special treat. Many seek foods that for a given number of calories offer a high nutrient yield. Often, these foods are called *nutrient-dense*, in contrast to *empty-calorie* or *junk* foods that give calories but little or no other nutrients. Calories are not bad in themselves, but eating only calories with little or no accompanying nutrients creates nutritional problems.

The public concern for better nutrition has been felt by foodservices, as is evidenced by changes in patron's food-selecting habits and by increased demand for specific kinds of foods and preparation. Many hotels and other operations feature low-calorie and low-fat foods. Terms such as "Lean 'n' Light" and "Heart Food" are often at the column head. Some foodservices ask the American Heart Association or American Dietetic Association to approve the recipe for a menu item and then indicate such approval when this food appears on the menu. Some put a small heart after each entrée approved by the American Heart Association. This change in eating patterns has gone beyond being a fad; it is a permanent change.

In fact, restaurants that make a nutrition claim such as "low fat," "light," or "heart healthy" on a menu item must be able to back up the claim, according to regulations from the Food and Drug Administration (FDA). For instance, if a foodservice operation labels its tuna salad as "low fat," it must meet the FDA definition of low fat, which is 3 grams of fat or less per serving.

Although a nutrient analysis of the healthier food item is not needed on the menu, the operation must be able to provide that information if it is requested by a diner. The bigger responsibility comes as the cook or chef is producing the food and whether or not the formula written in the recipe is being followed. This is discussed further at the end of the chapter.

Major Nutrient Groups

Some forty nutrients are needed by the body. These fall into categories such as carbohydrates, proteins, fats, vitamins, and minerals. Water is not a nutrient, but it is essential because it allows the body to carry on its functions. Fiber is not a nutrient, either, but it too is essential. The diet should be designed to allow these items to be consumed in adequate amounts for good health.

More and more evidence is being accumulated about the *specific* effect that certain foods have on certain diseases. Thus, orange juice is now recommended for consumption to help avoid cancer or to help people with malignancies. Many other foods are also known to have an influence in avoiding or in helping to combat certain other diseases. Thus, it may be quite possible in the future that physicians will be prescribing certain foods as medicines. Foodservices may find patrons ordering certain types of food or meals because they have certain types of illnesses. It is a new avenue and one that will be of high interest to many, increasing customer interest in healthy menus and the need for foodservices to pay attention to this customer want.

Calorie Nutrients

We get energy from three nutrients—four, if we include alcohol. The three are carbohydrates, proteins, and fats. The body can turn these into calories for energy. Carbohydrates are used almost exclusively for energy, but protein and fat play additional important energy roles.

Protein. The major material used in making muscle, blood, skin, hair, and the internal organs is protein. Bones and teeth start out as protein substances; then, as minerals replace the protein, they become hard tissues. Proteins are needed to make growth possible, to repair body tissues, and to form hormones, antibodies, vitamins, and other essential substances. Animal products furnish good-quality protein; cereals combined with legumes also furnish fairly good protein; nuts and seeds supply some protein in the diet. Even fruits and vegetables contain some protein, but not in significant amounts.

Protein is made up of units called *amino acids*. There are twenty-two known amino acids, eight of which the body cannot manufacture and which (as a result) we must get from our diets. We call these eight *essential* amino acids. A *complete* protein is one that contains all eight essential amino acids. An *incomplete* protein is one that lacks one or more essential amino acids or has them in poor balance. When we combine cereals and legumes, we get complete proteins, but each separately has only an incomplete protein makeup. Soybean protein is nearly complete.

We need about 15 percent of our total calories to come from protein; about 25 percent of this should be animal protein. Vegetarians and others are able to live without animal proteins because they combine incomplete plant proteins to produce effectively complete ones.

If the diet lacks calories from fats or carbohydrates, the body breaks down body proteins and uses them for energy. This is an expensive way to get energy; even in weight-loss diets, carbohydrates should be eaten in sufficient amount to spare protein from metabolic attack.

Carbohydrates. Carbohydrates, our main source of energy, come from foods such as cereals, beans, syrups, and sugars. These foods are our cheapest source of energy. Our bodies break down carbohydrates into a simple sugar called *glucose;* this is what we burn for energy and heat. The liver can change glucose into a body starch called *glycogen.* This and the small amount of sugar in our muscles and blood constitute our energy reserve. If we use up the reserve and have no carbohydrates, we have to get the energy we need from fat or protein. About 55 to 65 percent of our total calories should come from carbohydrates. As has been noted, some carbohydrate must be in the diet (about 100 grams per day) to help burn fat in a reducing diet or serious problems can occur.

Carbohydrate is also our most efficient source of energy. An athlete whose diet is high in carbohydrate and not too high in protein has improved stamina and strength. Athletes should not have too much fiber in their diets. For this reason, they do better on carbohydrates from cereals and potatoes than on carbohydrates from fruits and vegetables.

Fats (Lipids). Fats and fatty substances belong to the family called *lipids.* They are rich sources of energy, but only a few are known to be essential in the diet. In their disfavor, they give a lot of calories and also are associated with cardiovascular disease. The fat-soluble vitamins A, D, E, and K are found in fats. Fatty foods stay in the stomach longest, while carbohydrates stay the shortest time and proteins fall in between (but closer to fats). Thus, we say a meal that is high in fat and protein "sticks to the ribs." Food mixtures of fat, protein, and carbohydrate stay in the stomach just about as long as fatty foods do.

Fats are often reduced in diets to reduce calories or to reduce blood cholesterol. Although all fats, regardless of type, yield 9 calories per gram, certain fats, such as highly saturated fats, contribute more blood cholesterol than others. The monounsaturated and polyunsaturated fats actually tend to reduce blood cholesterol and for this reason are recommended when fats are to be in the diet. A *saturated fat* is one that has all the hydrogen attached to it that it can hold. An *unsaturated fat* can hold more hydrogen. Solid, firm fats such as beef or lamb fats tend to be more saturated than soft, greasy fats or oils such as chicken fat, fish oils, olive oil, or canola oil. Coconut oil is, however, highly saturated and corn oil slightly less so. There are two types of unsaturated fats: monounsaturated and polyunsaturated—*mono* indicating that only one place is open for hydrogen to join, and *poly*, that there are many places. They tend to pick up cholesterol at their unsaturated points and carry it to the liver, where the liver breaks down the cholesterol. Saturated fats tend to increase blood cholesterol.

Body Regulators

Vitamins and minerals are important because they help carry on essential bodily functions. Minerals also have importance because they form a part of hard tissues and constitute important substance-forming materials in many body tissues.

Vitamins. If the body were a machine, the energizing spark provided by vitamins could be compared to the flash of a spark plug that gets the combustion engine of the machine going. If the spark were not there, the machine would not run. Even though vitamins are needed in only minute amounts (like the flash of a spark plug), this amount is vital to good health.

The body cannot make many vitamins; most of them must be gleaned from our food. We know that various vitamins are essential, and we even know how much of some vitamins is needed; on the other hand, we do not know how much of some others is, and as a result, scientists are unsure whether all vitamins have yet been identified.

Because fat-soluble vitamins can be stored in excess in the body, consuming too much of them can cause health problems and even death. Vitamin A is required for tissue health, especially tissues of the nose, ear, throat, and alimentary canal. It is essential to good eyesight, can help fight off infections, helps regulate cell growth, and can delay senility and increase longevity. It is found in yellow and orange fruits and vegetables and in leafy green vegetables as a precursor called *carotene*. Butter and eggs also contain it. It appears in a different form and to a lesser extent in some other foods. Vitamin A is relatively stable in foods and in cooking. We need about 5000 international units (IU) of this vitamin per day. A green leafy, yellow, or orange vegetable every other day is a good safeguard. Vitamin D is found in milkfat, fish, fish oils, and organ meats. It is often added to margarine and other foods as well. It prevents rickets, helps build bones and teeth, keeps up muscle tone, and reduces the danger of osteomalacia (a ricketslike problem experienced by some adults). Ultraviolet rays from sunlight or lamps striking the skin change egosterol or dehydrocholesterol into vitamin D. Vitamin E, found in many foods but most plentifully in seed and nut oils, functions largely as an antioxidant in the body. Contrary to some people's belief, it does not help prevent infertility in humans. It may, however, help prevent damage from smog and other harmful gases in the atmosphere. Vitamin E is an antioxidant and helps to prevent undesirable oxidations in the body. Vitamin K is required for blood clotting and may have something to do with blood manufacture. It is found in many vegetables, especially the leafy green ones. Of all these fat-soluble vitamins, vitamin D is the most toxic in excess. About 10 micrograms of vitamins D and E per day is sufficient for an adult male. No requirement has been set for vitamin K.

The B-vitamin group is made up of about ten vitamins. They were once identified as a single vitamin, vitamin B, but today we have many B vitamins. The chemical names are largely used today. Thiamine (B_1) is found in meats, whole-grain or enriched cereals, and legumes. It is needed for muscle action, including heart, nerve, and brain functions, good appetite, normal digestion, proper growth, fertility, lactation, the ability to utilize carbohydrates, and avoidance of beriberi (a muscle and nerve disease). Riboflavin (B_2) is found in leafy vegetables, milk products, meats, and cereals. It helps maintain skin, nerves, and eyes; it works with pyridoxine (vitamin B_6) to change tryptophan (an amino acid) into a vitamin (niacin). It is essential in oxidation reactions in the body. Niacin is found in organ meats, flesh, nuts, and cereals. A diet high in tryptophan provides plenty of niacin. It cures pellagra, works as a coenzyme in removing hydrogen from substances so that the body can use them, and helps maintain skin and the nervous system. Pyridoxine (B_6) is widely distributed in plant and animal foods; flesh, liver, vegetables, and whole-grain cereals are the best sources. It maintains skin, helps prevent anemia, and acts as a coenzyme for many bodily reactions. A lack of pyridoxine causes convulsions. Folacin is widely found in the green part of plants. It works as a coenzyme to further body reactions, helps in the use of amino acids, helps make chole (a vitamin), helps make purines and pyrimidines, and works with vitamin B_{12} and vitamin C to prevent anemia. Vitamin B_{12} (cobalamin) is found only in animal proteins. It is mainly associated with the prevention of pernicious anemia. It is not found in plants and so some vegetarians may lack it. The recommended amount of B vitamins per day in an adult male's diet is as follows: thiamine, 1.4 milligrams (mg); riboflavin, 1.7 mg; niacin, 18 mg equivalent; pyridoxine, 2.2 mg; folacin, 400 micrograms (μg); and B_{12}, 3 μg.

There are no recommended amounts for the other B vitamins. Pantothenic acid is plentiful in most animal and plant foods. It helps to produce energy in the body. Biotin, which is found in organ meats, chicken, eggs, milk, most fresh vegetables, and some fruits, is needed to metabolize carbohydrates, fats, and proteins. Choline is found in brains, organ meats, eggs, yeast, wheat germ, milk products, meats, and vegetables. It functions in protein and fat metabolism. Myoinositol is found in organ meats, brains, whole-grain cereals, and yeast. It probably performs metabolic functions.

Most of the B vitamins are fragile and can be destroyed if cooked too long. Thiamine, pyridoxine, and some others are destroyed if placed in an alkaline medium, such as when soda is added in cooking vegetables. An acidic medium helps preserve them.

Riboflavin and some other B vitamins are harmed by the ultraviolet rays of sunlight. Other B vitamins oxidize easily. All are quite soluble in water; for this reason soaking, cooking in water, or otherwise exposing foods to water can lead to a leaching out of the B vitamins. Taking care in the storage, preparation, cooking, and holding of foods rich in B vitamins can do much to preserve the vitamins. Following the *food guide pyramid* (see Fig. 2-1) gives an adequate amount of them all, especially if whole-grain cereals are included in the diet.

Vitamins often join with minerals, enzymes, or other vitamins to do their jobs. When they join an enzyme, the substance is called a *coenzyme*. Many coenzymes perform bodily functions. Some vitamins and minerals work only when other substances are present. Thus, eating just a few of the essential nutrients does not work: a full, balanced group must be consumed to get the job done. The presence of one substance often demands another. Thus, the quantity of thiamine regulates the quantity of energy produced in the body, since thiamine must be present for energy production to occur.

Vitamin C is often called "the sailors' vitamin" because in past centuries sailors who stayed out at sea for long periods of time without vitamin C developed scurvy, "the sailors' disease," evidenced by hemorrhaging under the skin, swollen and bleeding gums, and great weakness. Magellan and a number of his men died of scurvy during the expedition's famous circling of the earth. The disease was caused by the prolonged reliance on a salt meat and biscuit diet with no fresh vegetables. Later, a British physician discovered that the citrus fruit family contained a substance that would cure scurvy, and the British navy ordered all its ships to carry lemons or limes on cruises in order to prevent it. This is why the British sailor is called a *limey*.

Vitamin C is found in good supply in citrus fruits, tomatoes, leafy vegetables of the cabbage family, tropical fruits, and potatoes. It is essential for the production of collagen, the substance that initiates the formation of bones and teeth. It also aids in wound healing and is related to the metabolism of some amino acids and to other bodily functions. It helps the body produce some hormones and helps maintain the bones and teeth.

Vitamin C (also known as ascorbic acid) is very fragile. It can be destroyed by oxidation and cooking (especially in alkaline mediums), and it is so water-soluble that it can easily be leached out of foods. In quantity cooking, much of the ascorbic acid in food tends to be destroyed. Therefore, foodservice operations usually do not plan on having a liberal supply of

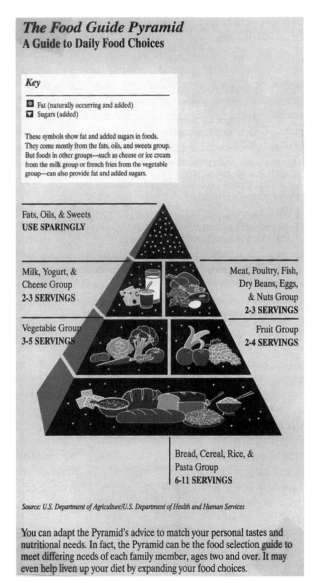

FIGURE 2-1 Food guide pyramid.

The Food Guide Pyramid
A Guide to Daily Food Choices

Key

■ Fat (naturally occurring and added)
▽ Sugars (added)

These symbols show fat and added sugars in foods. They come mostly from the fats, oils, and sweets group. But foods in other groups—such as cheese or ice cream from the milk group or french fries from the vegetable group—can also provide fat and added sugars.

Fats, Oils, & Sweets
USE SPARINGLY

Milk, Yogurt, & Cheese Group
2-3 SERVINGS

Meat, Poultry, Fish, Dry Beans, Eggs, & Nuts Group
2-3 SERVINGS

Vegetable Group
3-5 SERVINGS

Fruit Group
2-4 SERVINGS

Bread, Cereal, Rice, & Pasta Group
6-11 SERVINGS

Source: U.S. Department of Agriculture/U.S. Department of Health and Human Services

You can adapt the Pyramid's advice to match your personal tastes and nutritional needs. In fact, the Pyramid can be the food selection guide to meet differing needs of each family member, ages two and over. It may even help liven up your diet by expanding your food choices.

the vitamin in the foods served; they often advise people to obtain a good source of vitamin C in the morning for breakfast and then not worry about getting it the rest of the day. By eating plenty of fresh fruits and vegetables, however, a person's diet will usually have an adequate amount of vitamin C without such a morning precaution.

Minerals. Minerals participate in many bodily functions. They make body tissues such as bones and teeth. They carry out essential reactions such as the transfer of electrical nerve impulses. Some form parts of vitamins, as cobalt does in vitamin B_{12}; others help make proteins, as sulfur does in the proteins that make up the hair, skin, and nails. Similarly, iron is used in making blood, and iodine is an element in thyroxine, which regulates the body's cell functions. Our muscles contract and relax because of minerals. They are also vital in maintaining a proper acid-alkali balance in the body. Some fifteen minerals are used by the body. We think we know how much iron, zinc, iodine, calcium, and phosphorus are needed, but not how much magnesium, sodium, potassium, copper, cobalt, manganese, sulfur, chlorine, fluorine, molybdenum, chromium, and selenium.

Because of the number of minerals we need, our diet must contain a variety of foods to get all in adequate amounts. Milk and milk products, meat or its equivalent, fruits, vegetables, and cereals together provide the needed variety. Iron is well supplied in dried fruits and meats, especially organ meats. Milk provides calcium and phosphorus; phosphorus can also be supplied by meat, fruits, and vegetables. Iodine is best found in iodized salt, seafoods, and fruits and vegetables grown in soil of good iodine content. If the pyramid plan is followed, the diet usually includes enough minerals. The sources of various minerals and the body's use of each mineral are given in Table 2-2.

Water. About two-thirds of the body is water. Water is an essential part of cells and forms a major part of body fluids in which many essential processes of the body take place. Water helps regulate temperature through evaporation from the skin. Another important use of water is in digesting foods, after which it must be reabsorbed in the intestines. Water also acts as a lubricant to reduce friction between moving body parts, thus preventing fatigue.

Water is taken into the body in fluids and foods. It is produced when food is converted into energy. The amount of water in the body can vary. Salt and some other minerals increase the amount held; lowering their intake causes the body to lose water and thus weight. In some health problems, a swelling of the body called *edema* results when water is retained, making the body appear puffy in parts. A finger pressed against the flesh at such a place leaves a depression that goes away slowly. Prolonged protein deficiency or kidney disease can cause edema.

The body eliminates water through urine, feces, sweat, skin evaporation, and breath. Harm to the kidneys or the buildup of harmful substances in the blood may occur if the body lacks water (a condition called *dehydration*.) Excessive perspiration may cause dehydration and, if severe enough, may cause heat exhaustion, heat stroke, or even death. The average individual should have the equivalent of six to eight glasses of water daily in his or her food.

Individual's Responsibility and Dieting

Patrons eating in foodservices often have primary responsibility for making nutritious choices. Those who freely select where and what they eat should select a foodservice that can provide what they want and need and then should select foods properly. If a patron is not free to select the place and what is consumed, however, the foodservice has some responsibility. Thus, if a person elects to eat at a drive-in every day and always buys a hamburger, a carbonated beverage, and fries, the foodservice should not be blamed for the person's faulty diet; but if the patron is in a hospital or other institution where all food is provided, the foodservice is obliged to see that the diet adequately meets the patron's needs. In short, depending on circumstances, patrons have a varying responsibility in selecting a diet that is nutritious.

Most commercial operations are subject to free selection of place and of what is consumed. Such operations' responsibility is to ensure that the food offered contains all the nutrients that it should contain, whether or not the food offers a complete and balanced diet. If the operation offers a complete menu from which patrons can order a complete meal, it has

TABLE 2.2 Sources and Uses of Minerals

Mineral	Good Source	Use in Body
Calcium	Milk products, shellfish, egg yolk, some fruits and vegetables	Element in bones and teeth; helps muscles function and maintains bones so osteoporosis and osteomalacia do not occur; helps blood coagulate and helps maintain body's acid–base balance; pregnant or lactating women and rapidly growing children need more than do others
Chlorine	Meats, cheese, table salt	Important part of gastric juices; combines with many of the body's essential substances
Chromium	Brewer's yeast, flesh, nuts, whole grains, American cheese	Thought to be useful in changing carbohydrate to energy
Cobalt	Animal products only	Important part of vitamin B_{12}
Copper	Cocoa, tea, coffee, nuts, beef liver, pork liver	Helps make blood
Fluorine	Fluoridated water, flesh, plant foods	Helps make and maintain strong teeth and bones
Iodine	Seafood, iodized salt, fruits, vegetables	Governs reactions that control the body's speed of metabolism; lack may lead to swelling of the thyroid gland, causing a goiter; lack can cause a pregnant woman to give birth to a mentally deficient child
Iron	Egg yolk, dried legumes, meat, enriched or whole-grain cereals, molasses	Element in blood; found in muscle tissues
Magnesium	Leafy green vegetables, nuts, soybeans, snails	Functions in muscle contraction and nerve reactions
Manganese	Nuts, tea, cocoa, coffee, bran, canned pineapple, whole-grain cereals	Helps form blood and some coenzymes
Molybdenum	Organ meats, dark green vegetables, cereals, legumes	Probably involved in the catabolism of some substances in the body
Phosphorus	Milk, flesh, egg yolk, cereals, legumes, nuts	Works with calcium to make hard tissues; important in the development of energy; helps maintain body's acid–base balance
Potassium	Bran, molasses, potatoes, tea, coffee, legumes, spices, yeast, cocoa	Important in muscle, brain, and nerve cell functioning
Selenium	Animal products, whole-grain cereals	Thought to be related to antioxidant reactions
Sodium	Table salt, root vegetables, flesh, processed foods	Maintains fluid and acid–base balances
Sulfur	Whole grains, legumes, cheese, lean flesh, nuts	Important part of some essential amino acids
Zinc	Eggs, milk products, flesh, peanut butter, whole-grain cereals	Element in insulin and some enzymes; important in the sexual maturation of males and in successful pregnancies

an obligation to offer items that together constitute a nutritious meal; but if the operation offers only partial meals, no such responsibility attaches. Patrons eating partial meals should ensure on their own that they complete their dietary requirements at some other time during the day. Nonetheless, foodservices must recognize patrons' concerns about nutrition and must try to meet the industry's responsibilities. More than 25 percent of our nutrition is provided outside the home, and foodservices thus have a very real responsibility to the nation's welfare and health.

From time to time in the discussion above reference has been made to the normal requirement for an individual. These all have been given for an adult male but it can vary because of age, gender, or other factors. A revision of the nutrient recommendations is currently under way. Table 2-3 lists the latest recommendations for nutrient intakes. It is evident from this table that people's nutritional needs vary. The data given are for the average person; people with special circumstances will need more or less than the tabulated amount of some nutrients.

People today are interested in weight control, reduced intake of sodium (salt), low-cholesterol or low-fat diets, high-fiber food, special diets for geriatric eating problems, and alcohol restriction. These and some others are covered in the sections that follow.

Weight Control

The desire to lose weight is perhaps the most widespread nutritional concern in the United States today. The way for a person to lose weight is to lower caloric intakes below what is needed for the body to function. Exercise increases the body's caloric need. Drugs can help, but this is dangerous and should only be done under medical advice. A person's weight usually increases slowly, and the best way to get rid of excess weight is to embark on a program of losing weight at about the same rate at which it was originally gained. Crash diets are unhealthful and do not usually lead to permanent weight loss. There is also a danger in going below a food intake level of 1200 calories per day for an adult because below this level it is difficult to get a sufficient supply of all necessary nutrients. Permanent weight control occurs when a person

changes from the old diet to a lower-calorie diet and stays on the new one.

Lean meats, fruits, vegetables, low-fat dairy products, and other low-calorie foods should be consumed in place of high-fat foods to reduce calories. A good daily diet might include the following items:

- Two 8-oz glasses of nonfat milk (or equivalent)
- Two 2-oz portions of lean meat, fish, or poultry (or other good protein equivalent)
- Four 3-oz portions of fruits and vegetables
- Four 1-oz portions of grains (or carbohydrate equivalent), preferably whole grains

It is important to lower the fat intake in low-calorie diets. A gram of fat has 9 calories whereas a gram of protein or carbohydrate has 4 calories. A gram of pure alcohol has 7 calories. To increase calories, select appropriate additional meats, grains, desserts, and other rich foods.

Although some think carbohydrates are fattening, they are not. Fat and alcohol are the big fatteners. Three ounces of lean chicken breast without skin— one of the lowest-calorie meats—has three times more calories than 3 oz of plain baked potato. Some carbohydrate is essential to dieters when dieting, to burn up fat. If about 500 calories (5 oz) of carbohydrate is not consumed, dangerous by-products of fat breakdown can induce a ketonic coma and then death.

Some foodservices are misinformed on what foods are low-calorie. They will include a hamburger patty in a low-calorie meal, thinking they have chosen a low-calorie product. But the hamburger contains 325 calories because of its fat. Selecting a lean fish such as cod or sole, a chicken breast, or a lean piece of meat increases the likelihood of keeping a modest meal below 500 calories. A vegetable plate with a poached egg (80 calories) on a thin square of whole-wheat toast, ½ cup each of four vegetables (such as carrots, green peas, baked tomato, and broccoli), ¼ head of lettuce with a low-calorie dressing, and a glass of low-fat milk equals about 350 calories. Topping it off with a low-calorie gelatin dessert with fresh fruit, low-calorie canned fruit, or a low-calorie sherbet or ice milk keeps the entire meal under 500 calories.

TABLE 2-3 Nutrient Recommendations[a]

I. 1997–1998 Dietary Reference Intakes

Age (yr)	Recommended Dietary Allowance (RDA)								Adequate Intake (AI)					
	Thiamine (mg)	Riboflavin (mg)	Niacin (mg NE)	Vitamin B_6 (mg)	Folate (µg DFE)	Vitamin B_{12} (µg)	Phosphorus (mg)	Magnesium (mg)	Vitamin D (µg)	Pantothenic Acid (mg)	Biotin (µg)	Choline (mg)	Calcium (mg)	Fluoride (mg)
Infants[b]														
0.0–0.5	0.2	0.3	2[c]	0.1	65	0.4	100	30	5	1.7	5	125	210	0.01
0.5–1.0	0.3	0.4	4	0.3	80	0.5	275	75	5	1.8	6	150	270	0.5
Children														
1–3	0.5	0.5	6	0.5	150	0.9	460	80	5	2.0	8	200	500	0.7
4–8	0.6	0.6	8	0.6	200	1.2	500	130	5	3.0	12	250	800	1.1
Males														
9–13	0.9	0.9	12	1.0	300	1.8	1250	240	5	4.0	20	375	1300	2.0
14–18	1.2	1.3	16	1.3	400	2.4	1250	410	5	5.0	25	550	1300	3.2
19–30	1.2	1.3	16	1.3	400	2.4	700	400	5	5.0	30	550	1000	3.8
31–50	1.2	1.3	16	1.3	400	2.4	700	420	5	5.0	30	550	1000	3.8
51–70	1.2	1.3	16	1.7	400	2.4	700	420	10	5.0	30	550	1200	3.8
>70	1.2	1.3	16	1.7	400	2.4	700	420	15	5.0	30	550	1200	3.8
Females														
9–13	0.9	0.9	12	1.0	300	1.8	1250	240	5	4.0	20	375	1300	2.0
14–18	1.0	1.0	14	1.2	400	2.4	1250	360	5	5.0	25	400	1300	2.9
19–30	1.1	1.1	14	1.3	400	2.4	700	310	5	5.0	30	425	1000	3.1
31–50	1.1	1.1	14	1.3	400	2.4	700	320	5	5.0	30	425	1000	3.1
51–70	1.1	1.1	14	1.5	400	2.4	700	320	10	5.0	30	425	1200	3.1
>70	1.1	1.1	14	1.5	400	2.4	700	320	15	5.0	30	425	1200	3.1
Pregnancy	1.4	1.4	18	1.9	600	2.6	d	+40	d	6.0	30	450	d	d
Lactation	1.5	1.6	17	2.0	500	2.8	d	d	d	7.0	35	550	d	d

II. 1989 Recommended Dietary Allowances

Age (yr)	Energy (kcal)	Protein (g)	Vitamin A (µg RE)	Vitamin E (mg α-TE)	Vitamin K (µg)	Vitamin C (mg)	Iron (mg)	Zinc (mg)	Iodine (µg)	Selenium (µg)
Infants										
0.0–0.5	650	13	375	3	5	30	6	5	40	10
0.5–1.0	850	14	375	4	10	35	10	5	50	15
Children										
1–3	1300	16	400	6	15	40	10	10	70	20
4–6	1800	24	500	7	20	45	10	10	90	20
7–10	2000	28	700	7	30	45	10	10	120	30
Males										
11–14	2500	45	1000	10	45	50	12	15	150	40
15–18	3000	59	1000	10	65	60	12	15	150	50
19–24	2900	58	1000	10	70	60	10	15	150	70

25–50	2900	63	1000	10	80	60	10	15	150	70
51+	2300	63	1000	10	80	60	10	15	150	70
Females										
11–14	2200	46	800	8	45	50	15	12	150	45
15–18	2200	44	800	8	55	60	15	12	150	50
19–24	2200	46	800	8	60	60	15	12	150	55
25–50	2200	50	800	8	65	60	15	12	150	55
51+	1900	50	800	8	65	60	15	12	150	55
Pregnancy	+300	60	800	10	65	70	30	15	175	65
Lactation										
First 6 months	+500	65	1300	12	65	95	15	19	200	75
Second 6 months	+500	62	1200	11	65	90	15	16	200	75

III. Tolerable Upper Intake Levels for Selected Nutrients (per Day)[e]

Age (yr)	Vitamin D (μg)[f]	Niacin (mg)	Vitamin B_6 (mg)	Folate (μg)[f]	Choline (mg)	Calcium (mg)	Phosphorus (mg)	Magnesium (mg)	Fluoride (mg)
Infants[g]									
0.0–0.5	25	—	—	—	—	—	—	—	0.7
0.5–1.0	25	—	—	—	—	—	—	—	0.9
Children									
1–3	50	10	30	300	1000	2500	3000	65	1.3
4–8	50	15	40	400	1000	2500	3000	110	2.2
9–13	50	20	60	600	2000	2500	4000	350	10.0
14–18	50	30	80	800	3000	2500	4000	350	10.0
Adults									
19–70	50	35	100	1000	3500	2500	4000	350	10.0
>70	50	35	100	1000	3500	2500	3000	350	10.0
Pregnancy	50	35	100	1000	3500	2500	3500	350	10.0
Lactation	50	35	100	1000	3500	2500	4000	350	10.0

[a] The nutrient and energy standards known as the Recommended Dietary Allowance (RDAs) are currently being revised. The new recommendations are called Dietary Reference Intakes (DRIs) and include two sets of values that serve as goals for nutrient intake—Recommended Dietary Allowances (RDAs) and Adequate Intakes (AIs). In part I are presented the new RDAs and AIs for the 14 nutrients revised to date; in part II are presented the 1989 RDAs for the remaining nutrients and energy, which will serve until new values can be established. In addition to the values that serve as goals for nutrient intakes (presented in parts I and II), the DRIs include a set of values called tolerable upper intake levels—the maximum amount of a nutrient that appears safe for most healthy people to consume on a regular basis (part III).

[b] For all nutrients, an AI was established instead of an RDA as the goal for infants; for the eight vitamins and choline, the age groupings are 0 through 5 months and 6 through 11 months.

[c] The AI for niacin for this age group only is stated as milligrams of preformed niacin instead of niacin equivalents.

[d] Values for these nutrients do not change with pregnancy or lactation. Use the value listed for women of comparable age.

[e] An upper level was not established for thiamine, riboflavin, vitamin B_{12}, pantothenic acid, and biotin because of a lack of data, not because these nutrients are safe to consume at any level of intake; all nutrients can have adverse effects when intakes are excessive.

[f] To convert μg to IU, multiply by 40: for example, 50 μg × 40 = 2000 IU.

[g] Upper levels were not established for many nutrients in the infant category because of a lack of data.

Food should not be cooked in fat, and butter or margarine should be used sparingly as a seasoning. Often, if care is taken to hold down fat, the calories in a meal drop considerably.

Low-Salt (Low-Sodium) Diet

Too much sodium can cause hypertension (high blood pressure) or other health problems in some people. Hypertension can increase the risk of a stroke or heart attack. Plant and animal foods contain some natural sodium, but most of our sodium comes from table salt (sodium chloride). About a third of our sodium comes from our own salting of foods, another third comes from processed food we buy, and the final third is present naturally in foods. A normal person can consume 3 to 8.5 grams (g) of salt a day without experiencing any problems. A teaspoon of salt weighs 7 g. Most normal people have no problem handling a slight sodium excess. The Salt Institute estimates that the average daily intake is 3.5 g. Some drinking waters are high in sodium and must be avoided by people observing low-sodium diets. A sodium-exchange water softener can add sodium to tap water. Except in special dietary cases, the salt added to food should contain iodine, an essential nutrient.

Sodium is an essential nutrient. It helps us to maintain our body fluid balance and supplies the electrical charge that makes possible cell metabolism. Our nervous and muscular systems would not work without it. Reducing salt does not work to reduce hypertension in many people; only a minority are salt-sensitive and benefit from it. Today, doctors usually prescribe a drug to control hypertension, with moderate to no restriction of high-sodium-containing foods. In 1990 the *American Journal of Hypertension* advised against a low intake of sodium for the general population because in many it could actually increase hypertension, disturb sleep patterns, and decrease resistance against some diseases. In 1995 the American Heart Association published a study in which over 1900 adult males participated; it found that those who had a low excretion of sodium had four times as many heart attacks as those with a normal or high excretion. As a result of these and other studies, most authorities now say that the matter is an individual problem and

should be handled as such.* Foodservices should undersalt all foods and let patrons add an additional amount to suit their tastes. The menu or a table tent may state: "We purposely undersalt our foods and provide extra salt at the table for those who wish to add more." Another nice gesture is to put a good salt substitute on the table. The NRA feels that leaving the salt out of food is not the answer. Salt is a substance that brings out the flavor of food and adds to its palatability and that foodservices should attempt to cover their low salting of foods with spices and other flavor heighteners. Customers can also be enticed away from salted foods with menus that additionally feature highly flavorful and attractive fruit and vegetable combinations, such as a small rib lamb chop in the center of a colorful circle of vegetables, or a chicken breast sandwich with a bowl of vegetable soup, or a colorful fruit plate with a bit of low-sodium, low-fat cheese and bran muffins. There are many successful foodservices today cashing in on healthful, attractive novel food combinations. The Hyatt hotel chain finds that 38 percent of its customers are attracted to its healthful food offerings because they find them appetizing as well as healthful.

Most people on low-sodium diets should be conversant with the sodium content of foods and know which to select. They can also request that there be no salt on foods. Foodservices, too, should be informed as to the sodium content of foods. Many substances contain sodium, including baking powder, baking soda, monosodium glutamate, and sodium sulfite. With normal precautions on both sides, there should not be a problem in satisfying a low-sodium diet in an ordinary foodservice.

Low-Fat or Low-Cholesterol Diet

It is not difficult to serve a low-fat meal if certain rules are observed. No fried or fatty foods should be served. Butter and margarine should be used sparingly. Foods rich in fat such as pastries and other desserts should be avoided. Foods such as mayonnaise, avocado (about

---

*The source of much of the information given in this paragraph comes from an article in the National Restaurant Association's (NRA's) *Restaurants USA*, **19**(3), 24–29, March 1999.

one-quarter fat), and peanut butter contain the same amount of fat per ½ oz as 4 oz of roast beef. It is often desirable to reduce saturated fat and increase unsaturated fat in the diet. A foodservice can meet this need by consulting a good table that indicates the fat content of foods.

It is also wise to refer to a table of food values in planning a low-cholesterol diet. Saturated fats should be restricted, and so (probably) should calories. Unsaturated fats help lower the cholesterol content of the blood. When a person loses weight, a frequent complementary occurrence is the lowering of blood cholesterol levels. The American Heart Association has stated that cholesterol in foods should be limited to less than 300 mg per day. An egg contains 274 mg of cholesterol. Our bodies need about 900 mg of cholesterol per day to make bile, which helps to emulsify fats in the intestines and thus make them absorbable, to produce sex and energy hormones, to keep the skin lubricated, to make vitamin D, to use in nerves and brain cells, and to perform other functions. We could not exist without cholesterol. A number of people find that reducing cholesterol in the diet does little good, because the body generates its own to make up for the deficiency. In one test, over 1800 men who had high blood cholesterol levels were for seven years put on a low-cholesterol diet; the men averaged a 3.5 percent lowering of serum cholesterol. A person is considered at risk for cardiovascular problems if the serum cholesterol level is higher than 220 mg per deciliter of blood. Thus, if a person has a blood cholesterol of level of 300 mg and reduces it by 3.5 percent, the blood cholesterol level is lowered by 10.5 mg but remains at a still-dangerous level of 289.5 mg. Exercise, weight reduction, no smoking, spicy foods, garlic, fibers from oatmeal and other grains, fruits and vegetables, hard water, and other food items and dietary changes can help reduce cholesterol. The problem is largely genetic—less than 5 percent of the families in the United States account for around 50 percent of heart-related deaths—and a low-cholesterol diet together with other good living patterns seems to be about the best way to try to lower serum cholesterol. Some people at risk can take drugs, but others cannot because they are vulnerable to bad side effects of the drugs.

Cholesterol is blamed for causing cardiovascular problems because it joins with other substances to form a plaque that accumulates in the artery walls and causes them to harden and become inflexible. Deposits of plaque can develop to such an extent that they severely constrict the interior diameter of an artery or even plug it. All such problems with arteries make the heart work harder. People differ in the kind of cholesterol they carry in the blood. Some carry a lot of high-density lipoproteins (HDLs), which are high in protein but low in cholesterol and for this reason are more soluble in the blood and less apt to form plaque deposits. Other people carry a lot of low-density lipoproteins (LDLs), which encourage deposits. Eating saturated fats seem to build up LDLs, while eating polyunsaturated fats builds up HDLs. Most vegetable oils contain polyunsaturated fats.

Fats are essential for the maintenance of life. They place protective padding around organs and are important components of the brain and nervous system. They are needed especially by growing children for the good development of these systems. Fats are necessary for the absorption of the fat-soluble vitamins A, D, E, and K. If a person replaces the calories lost from fat reduction by consuming too many refined carbohydrates, he or she can develop insulin problems and develop diabetes. Added problems can develop with such a high-carbohydrate diet if vitamin B is not increased above ordinary levels, because vitamin B is needed to metabolize carbohydrates.

The dietary guidelines recommend that we consume no more than 30 percent of our total calories in fat, with no more than 10 percent of our total calories being saturated fats. This would be about 2½ oz of fat in a 2000-calorie diet. The American Dietetic Association estimates that the average daily intake is 34 percent for total fat calories, of which 12 percent is saturated fat. The American Heart Association and other authorities have indicated that ultra-low-fat diets are not beneficial for most people. A diet of 15 percent fat calories can increase health risks. An extremely low fat diet can impose serious health risks for children and pregnant women.

As with salt, fats are helpful in increasing the palatability of foods by giving them richness, moisture, and often flavor. Menu planners and chefs can do much, however, to make reduced fat foods acceptable, and even desired. Some of the items mentioned above under menu items low in salt are also low in fact. Tasty pasta dishes can be low in fat. The use of many low-cholesterol or low-fat products, such as Egg Beaters, or low-fat cheese or frozen desserts can be used without serious loss of palatability. If one has to fry or deep-fry to please customers, using frying oils that can withstand the frying rigors but are low in saturated fats can be beneficial; using cottonseed oil instead of corn oil, for instance, can result in a lower-saturated-fat product. Gravies and sauces are often rich in fats, but many operations today have found that even low-fat gravies and sauces can win high praise from diners. Stir-fried items are popular and often are low in fat. Novelty with good food judgment can do much to overcome a reduction in fat content.

High-Fiber Diet

Fiber is largely composed of a complex carbohydrate called *cellulose*. It occurs in bran, and it forms the skeletal structure of fruits and vegetables, but it is not found in flesh foods. A diet high in fiber can reduce a person's chance of developing intestinal cancer, diverticulosis, constipation, and appendicitis; it can even promote weight loss by displacing other, higher-calorie foods in a normally filling diet. Fiber reduces cholesterol in the digestive tract, thus lowering its chances of being absorbed into the body. There are two ways of identifying fiber amounts. One is in terms of *crude fiber*, which is the fiber that remains in food after treatment with alkalis and acids. The other is in terms of *dietary fiber*, which is the fiber actually present in food in the diet. A person needs about 6 g of crude fiber (about 15 g of dietary fiber) per day. A serving (1 oz) of bran flakes contains 6 oz of dietary fiber. A 4-in. piece of corn-on-the-cob contains 6 to 8 oz of dietary fiber. Eating the amount of fruits and vegetables indicated in the pyramid plan (discussed previously) supplies enough fiber, especially if whole-grain cereal products are selected.

It is possible to get too much fiber, since fiber can cause digestive disturbances and can even cause dehydration by pulling too much water out of the body and into the feces. It may also cause foods to pass too quickly through the digestive system, resulting in poor nutrient absorption.

Water Supplies, Filtration, and Safety

Despite progress, U.S. water supplies are monitored erratically and may contain contaminants with long-term health effects. Millions drink water that violates federal health guidelines. The question for many is: Is my water safe?

The truth about U.S. drinking water is simultaneously reassuring and disturbing. Our water has never been safer. Age-old waterborne diseases such as cholera, dysentery, and polio, which ravage poor countries, are rare here. But new threats specific to the late twentieth century are surfacing and must be faced squarely by government at all levels.

Unfortunately, it hardly matters where one gets water, since any source can become tainted. The biggest threats to the nation's water supply include three broad categories of contaminants: industrial and agricultural chemicals, lead, and biological organisms. Although one might assume that someone is checking water supplies 24 hours a day, in actuality there are no requirements for daily, or even weekly, monitoring for chemicals, lead, or pesticides. Some communities check for contaminants once every four years, if at all, and drawing samples is on the honor system. So how does one know how safe water supplies are?

By law, one may ask local water officials, but the responses may not be easily understood. Even when water meets government standards, it may not be safe for everyone to drink. Results may look fine, but some take no chances and have a filter installed. Yet the first step should be to have water tested.

Most experts recommend testing as the first line of defense, especially for common but dangerous contaminants such as lead. If tests confirm that water is contaminated, bottled water can be a good solution. On the other hand, if the test shows no problem, the good news is that there is no need to buy bottled water or to install a filter. If a problem is discovered, the water may be retested after corrective measures have been taken.

Remember that the tricks of disreputable marketers of water filters are limited only by the imagination. The critical thing is knowing what the problem is and knowing what technology will do. Foodservice operators are encouraged to take appropriate steps to have water tested and to monitor technology to provide the safest water possible to patrons.

GENETIC ENGINEERING AND THE FOOD SUPPLY

Worldwide, almost 90 percent of the human food supply is provided by only fifteen crop species and eight livestock species, small numbers when compared with the estimated 10 to 30 million species inhabiting our biosphere. Introducing genes from various organisms into crops and livestock has long been regarded as a promising way to ensure the continued productivity of agriculture and forestry.

New techniques permit scientists to remove individual genes from one living organism and transplant them into another. The result is the transfer of genetic traits between plants and animals that could not occur by natural reproductive techniques. Such genetic engineering is the best hope for increasing food production to sustain the world's growing population. In fact, the twenty-first century may prove to be the Age of the Gene.

Beef, Pork, and Produce

Through genetic engineering, healthier and safer foods can be produced in a much more environmentally sensitive way. For example, a vaccine is being introduced that will use genetic alteration techniques to end the mechanical castration of animals, a process that retards growth. The vaccine will eliminate the traits in male animals that cause them to fight and that make their meat taste strong, while retaining the potential for faster and larger growth.

In the earliest stages of development, genetic engineering was used on breeding strains of wheat that would withstand the rigors of harsh climate and voracious pests. Canola was made in this way to be a valuable oilseed plant that produces a cooking oil that is lower than most other oils in unhealthy saturated fats.

Tomorrow, designer animals, the result of genetic programs, could be customized to increase growth, shorten gestation, and enhance nutritional value. Farmers will be able to order the genes they want from gene banks for transmission to local biofactories, where animals with the desired characteristics will then be produced and shipped.

Genetics should allow future farmers to produce plants with higher yields that are more resistant to disease, frost, drought, and stress. These products will have higher protein, lower oil, and more efficient photosynthesis rates than ever before. Natural processes such as ripening will be enhanced and controlled.

If variety is what one wants, imagine a tossed salad of bright orange tomatoes, chocolate-colored bell pepper rings, and wedges of blue lettuce garnished with maroon carrot curls. The selection of produce in size, shape, and color will only get broader. The bottom line is a product improved in both quality and safety.

New miniature iceberg lettuce heads the size of tennis balls are true dwarfs, with all of the characteristics of iceberg lettuce. Imagine serving a salad without the usual cutting and chopping of lettuce. Other new produce items could include "popbeans," called *nuñas*, which burst and expand when heated rapidly and have a nutty taste and high fiber content; "plumcots," a new fruit with an apricot's flavor and a plum's firmness; and strawberries high in ellagic acid, which medical studies have shown inhibited the start of cancer.

Extending Shelf Life

One of the principal goals of genetic engineering involving agricultural biotechnology is to increase the value and yields of crops and livestock production. One biotech product to reach supermarkets has been a tomato that tastes good long after it has been picked. It has been genetically altered to soften or ripen more slowly than conventional tomatoes. By isolating the gene that causes softening, copying it, and placing it in the tomato backward, scientists slowed down the softening process, allowing farmers to leave the tomatoes on the vine longer to develop more flavor.

From the introduction of the new tomato, three camps emerged: purists who consider the "slow-rot" tomato a dangerous step toward "Frankenstein food"; fence-sitters who support genetic engineering in some cases; and those who view the tomatoes as scientific progress that will help control the food supply, reduce food waste, and offer consumers a better product.

Fruits give off an ethylene gas which helps them to ripen, and fruit ripeners are available that capture this gas and cause the fruit stored in them to ripen faster. Fresh vegetables deteriorate rapidly when there is even a small amount of ethylene present, so when fresh fruits and vegetables are stored together, the shelf life of the vegetables is shortened. In most operations it is not possible to have separate storage areas for these products, so the undesirable effect of ethylene must be borne. However, a device is available on the market that captures the ethylene from fruits and prevents it from causing the ripening of fruits and the deterioration of vegetables. Thus, the shelf life of both is extended.

Flavor Enhancement

Genetics will allow farmers to customize and fine-tune crops, building in flavor, sweeteners, and preservatives while increasing nutritional value. The first step in this process, called *agrogenetics*, is to identify disease-resistant genes, and the second step is to put them into the plants.

Some items on tomorrow's menu might include: "protrout," a superprotein fish; "pig-no-more," ultralean bacon; "octo-squid," a combo of two underwater species; "beefison," a meat with venison's flavor and beef's bulk; "shrimpsters," a less-squishy oyster produced with shrimp genes; "puck," a duck-flavored pork; "sworduna," a swordfish-flavored tuna; "quicken," a quail–chicken transgenic fowl; and a seaweed dip seasoned with spring onion genes.

Ultimately, genetic engineering may produce crossover transgenics such as plants with animal genes and animals with plant genes to provide new meaning to "chocolate milk" and "duck à l'orange." Genetic engineering and the food supply is one topic that foodservice professionals need to watch carefully.

■ THE NEW FOOD LABEL

One school of thought maintains that information is power and that consumers, including patrons at foodservice facilities, can never know too much about the food they buy and consume, and the more information the better. Accordingly, in 1990, the U.S. Congress passed the Nutrition Labeling and Education Act (NLEA), and new food labels were introduced in May 1994. New regulations now limit health claims, and new labels use standardized portion sizes and focus on nutrients associated with chronic diseases.

The purpose of the food label reform was simple: to clear up confusion that has prevailed on supermarket shelves for years, to help consumers choose more healthful diets, and to offer an incentive to food companies to improve the nutritional qualities of their products. Key features include:

- Nutrition labeling for almost all foods. Consumers can now learn about the nutritional qualities of almost all the products they buy.

- A new, distinctive, easy-to-read format (see Fig. 2-2) that enables consumers to find the label more quickly and the information they need to make healthful food choices.

- Information on the amount per serving of saturated fat, cholesterol, dietary fiber, and other nutrients that are of major health concern to today's consumers.

- Nutrient reference values, expressed as percent daily values, that help consumers see how a food fits into an overall daily diet.

- Uniform definitions for terms that describe a food's nutrient content—such as "light," "low-fat," and "high-fiber"—to ensure that such terms mean the same for all products on which they appear. These descriptions are particularly helpful for consumers trying to moderate their intake of calories or fat and other nutrients, or for those trying to increase their intake of certain nutrients, such as fiber.

- Claims about the relationship between a nutrient or food and a disease or health-related condition,

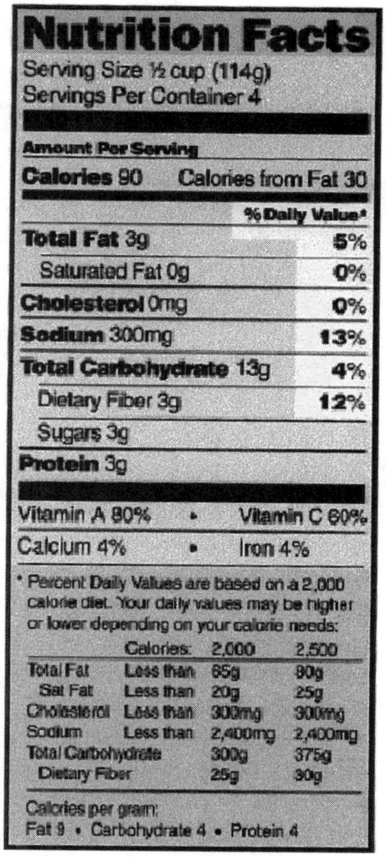

Nutrition Facts

Serving Size ½ cup (114g)
Servings Per Container 4

Amount Per Serving

Calories 90   Calories from Fat 30

                                    % Daily Value*

Total Fat 3g   **5%**

   Saturated Fat 0g   **0%**

Cholesterol 0mg   **0%**

Sodium 300mg   **13%**

Total Carbohydrate 13g   **4%**

   Dietary Fiber 3g   **12%**

   Sugars 3g

Protein 3g

Vitamin A 80%   •   Vitamin C 60%

Calcium 4%   •   Iron 4%

* Percent Daily Values are based on a 2,000 calorie diet. Your daily values may be higher or lower depending on your calorie needs:

	Calories:	2,000	2,500
Total Fat	Less than	65g	80g
Sat Fat	Less than	20g	25g
Cholesterol	Less than	300mg	300mg
Sodium	Less than	2,400mg	2,400mg
Total Carbohydrate		300g	375g
Dietary Fiber		25g	30g

Calories per gram:
Fat 9 • Carbohydrate 4 • Protein 4

FIGURE 2-2 New food label.

such as calcium and osteoporosis, and fat and cancer. These are helpful for people who are concerned about eating foods that may help keep them healthier longer.

- Standardized serving sizes that make nutritional comparisons of similar products easier.
- Declaration of total percentage of juice in juice drinks. This enables consumers to know exactly how much juice is in a product.
- Voluntary nutrition information for many raw foods.

The Reference Daily Intake (RDI) list of the FDA is used as the basis for judging what a day's proper allowance is (see Table 2-4). The values on this list differ from those given in Table 2-3.

There is a new set of dietary components on the nutrition panel. The mandatory (boldface) and voluntary components and the order in which they must appear are:

- **Total calories**
- **Calories from fat**
- Calories from saturated fat
- **Total fat**
- **Saturated fat**
- Polyunsaturated fat
- Monounsaturated fat
- **Cholesterol**
- **Sodium**
- Potassium
- **Total carbohydrate**
- **Dietary fiber**
- Soluble fiber
- Insoluble fiber
- **Sugars**
- Sugar alcohol (e.g., the sugar substitutes xylitol, mannitol, and sorbitol)
- Other carbohydrate (the difference between total carbohydrate and the sum of dietary fiber, sugars, and sugar alcohol, if declared)
- **Protein**
- **Vitamin A**
- Percent of vitamin A present as beta-carotene
- **Vitamin C**
- **Calcium**
- **Iron**
- Other essential vitamins and minerals

If a claim is made about any of the optional components, or if a food is fortified or enriched with any of them, nutrition information for these components becomes mandatory.

These mandatory and voluntary components are the only ones allowed on the nutrition panel. The list

TABLE 2-4   FDA's Reference Daily Intake Values for Use in Food Labeling

	Adults and Children Over 4 Years	Children Under 4 Years	Infants Under 13 Months	Pregnant or Lactating Women
Biotin	0.3 mg	0.15 mg	0.15 mg	0.3 mg
Calcium	1.0 g	0.8 g	0.8 g	1.3 g
Copper	2 mg	1 mg	1 mg	2 mg
Folacin	0.4 mg	0.2 mg	0.2 mg	0.8 mg
Iodine	150 μg	70 μg	70 μg	150 μg
Iron	18 mg	10 mg	10 mg	18 mg
Magnesium	400 mg	200 mg	200 mg	450 mg
Niacin	20 mg	9.0 mg	9.0 mg	20 mg
Pantothenic acid	10 mg	5 mg	5 mg	10 mg
Phosphorus	1.0 g	0.8 g	0.8 g	1.3 g
Protein[a]	65 g	28 g	25 g	65 g
Riboflavin	1.7 mg	0.8 mg	0.8 mg	2.0 mg
Thiamine	1.5 mg	0.7 mg	0.7 mg	1.7 mg
Vitamin A	5000 IU	2500 IU	2500 IU	8000 IU
Vitamin B_6	2.0 mg	0.7 mg	0.7 mg	2.5 mg
Vitamin B_{12}	6 μg	3 μg	3 μg	8 μg
Vitamin C	60 mg	40 mg	40 mg	60 mg
Vitamin D	400 IU	400 IU	400 IU	400 IU
Vitamin E	30 IU	10 IU	10 IU	30 IU
Zinc	15 mg	8 mg	8 mg	15 mg

[a] If the protein efficiency ratio of protein is equal to or better than that of casein, the RDI is 45 g for adults and pregnant or lactating women, 20 g for children under 4 years of age, and 18 g for infants.

of single amino acids, maltodextrin, calories from polyunsaturated fat, and calories from carbohydrates, for example, may not appear as part of the nutrition facts on the label. The required nutrients were selected because they address today's health concerns. The order in which they must appear reflects the priority of current daily recommendations. Thiamin, riboflavin, and niacin are no longer required in nutrition labeling because deficiencies of each are no longer considered of public health significance. However, they may be listed voluntarily.

The regulations also spell out what terms may be used to describe the level of a nutrient in a food and how they can be used. These are the principal terms:

- *Free.* This term means that a product contains no amount of, or only trivial or "physiologically inconsequential" amounts of, one or more of these components: fat, saturated fat, cholesterol, sodium, sugars, and calories. For example, *calorie-free* means fewer than 5 calories per serving and *sugar-free* and *fat-free* both mean less than 0.5 g per serving. Synonyms for "free" include "without," "no," and "zero."
- *Low.* This term can be used on foods that can be eaten frequently without exceeding dietary guidelines for one or more of these components: fat, saturated fat, cholesterol, sodium, and calories. Thus, descriptors are defined as follows:

- *Low-fat:* 3 g or less per serving.
- *Low-saturated fat:* g or less per serving
- *Low-sodium:* 140 mg or less per serving
- *Very low sodium:* 35 mg or less per serving
- Low-cholesterol: 20 mg or less and 2 g or less of saturated fat per serving
- *Low-calorie:* 40 calories or less per serving

Synonyms for "low" include "little," "few," and "low source of."

- **Lean and extra lean.** These terms can be used to describe the fat content of meat, poultry, seafood, and game meats.
  - *Lean:* less than 10 g fat, 4.5 g or less saturated fat, and less than 95 mg cholesterol per serving and per 100 g
  - *Extra Lean:* less than 5 g fat, less than 2 g saturated fat, and less than 95 mg cholesterol per serving and per 100 g
- *High.* This term can be used if the food contains 20 percent or more of the daily value for a particular nutrient in a serving.
- *Good source.* This term means that one serving of a food contains 10 to 19 percent of the daily value for a particular nutrient.
- *Reduced.* This term means that a nutritionally altered product contains at least 25 percent less of a nutrient or of calories than the regular, or reference, product. However, a reduced claim cannot be made on a product if its reference food already meets the requirement for a "low" claim.
- *Less.* This term means that a food, whether altered or not, contains 25 percent less of a nutrient or of calories than the reference food. For example, pretzels that have 25 percent less fat than potato chips could carry a "less" claim. "Fewer" is an acceptable synonym.
- *Light.* This descriptor can mean two things:
  1. That a nutritionally altered product contains one-third fewer calories or half the fat of the reference food. If the food derives 50 percent or more of its calories from fat, the reduction must be 50 percent of the fat.
  2. That the sodium content of a low-calorie, low-fat food has been reduced by 50 percent. In addition, "light in sodium" may be used on food in which the sodium content has been reduced by at least 50 percent.

The term *light* can still be used to describe such properties as texture and color, as long as the label explains the intent—for example, "light brown sugar" and "light and fluffy."

- *More.* This term means that a serving of food, whether altered or not, contains a nutrient that is at least 10 percent of the daily value more than the reference food. The 10 percent of daily value also applies to "fortified," "enriched," and "added" claims, but in those cases, the food must be altered.

Alternative spelling of these descriptive terms and their synonyms are allowed—for example, "hi" and "lo"—as long as the alternatives are not misleading.

As far as foodservice facilities are concerned, the FDA wants to make information available to the dining public, especially in those cases in which nutritional claims are made. FDA's regulations permit restaurants to promote their healthier menu fare using the following:

- *Specific claims about a menu item's nutrient content.* For example, low fat or high fiber. These are known as *nutrient claims.*
- *Claims about the relationship between a nutrient or food and a disease or health condition.* For example, a dish that is low in fat, saturated fat, and cholesterol might be able to carry a claim about how diets low in saturated fat and cholesterol may reduce the risk of heart disease. These are known as *health claims,* and they may initially appear on the menu in simple terms, such as *heart healthy.* Further information about the claim should be available somewhere on the menu or in other labeling—for example, with the accompanying nutrition information that must be provided on request.

Consumers can use these claims to spot foods that may be more healthful for them. They also can look for statements giving what FDA considers general dietary guidance. For example, the salad section may start with the message "Eating five fruits and vegetables a day is an important part of a healthy diet." This statement would refer to the National Cancer Institute's recommendation that Americans eat more fruits and vegetables to help reduce their risk of cancer and heart disease.

Restaurants do not have to provide nutrition information about foods that do not bear nutrient content or health claims or that are referred to in general dietary guidance messages. However, restaurateurs need to be careful that the general guidance they provide on the menu doesn't turn into a claim, such as "Fruits and vegetables can help reduce the risk of cancer." This, then, would require the item to meet FDA's nutrition information and claims' requirements.

Claims that promote a nutrient or health benefit must meet certain criteria established by the FDA and U.S. Department of Agriculture; for example, the food must provide a requisite amount of the nutrient or nutrients referred to in the claim. In addition, a menu item carrying a health claim must provide significant amounts of one or more of six key nutrients, such as Vitamin C, iron, or fiber, and cannot contain a food substance at a level that increases the risk of a disease or health condition. For example, a restaurant meal that contains 26 g of fat (40 percent of the daily value for fat) or 960 mg of sodium (40 percent of the daily value for sodium) is disqualified from making a heart-healthy claim. These rules also apply to claims used in the labeling of commercial food products. But the requirements for further information differ between restaurant and commercially manufactured foods.

To meet FDA's criteria, food manufacturers may choose to do chemical analyses to determine the nutritional value of their products. But the criteria for menu items are more flexible, and under FDA's requirements, restaurants may back up their claims with any "reasonable" base, such as databases, cookbooks, or other secondhand sources that provide nutrition information.

Also, restaurants do not have to provide the standard nutrition information profile and more exacting nutrient content values required in the nutrition facts panel of packaged foods. Instead, restaurants can present the information in any format desired, and they have to provide only information about the nutrient or nutrients to which the claim is referring. They can say simply that the amount of the nutrient in question does not exceed the limit imposed by FDA—for example, "This low-fat restaurant dish provides no more than 5 grams of fat per serving."

Although nutrition information is not required to appear on the menu, it must be made available to consumers when they request it. Restaurants can present it in a printed format—such as a notebook—or by having the staff recite it.

FDA is granting restaurants more flexibility with the following terms being used:

- *Low sodium, low fat, low cholesterol.* These claims mean that the item contains low amounts of these nutrients.

- *Light.* This means that the item has fewer calories and less fat than the food to which it is being compared. (Restaurants may continue to use the term *light* for reasons other than as a nutrient content claim—for example, "lighter fare" to mean that the dishes contain smaller portions. However, its meaning must be clarified on the menu.)

- *Healthy.* This means that the item is low in fat and saturated fat, has limited amounts of cholesterol and sodium, and provides significant amounts of one or more of the key nutrients vitamins A and C, iron, calcium, protein, or fiber.

- *Heart healthy.* This has two possible meanings:

  1. The item is low in saturated fat, cholesterol, and fat and without fortification provides significant amounts of one or more of six key nutrients. This claim will indicate that a diet low in saturated fat and cholesterol may reduce the risk of heart disease.

  2. The item is low in saturated fat, cholesterol, and fat, provides without fortification significant amounts of one or more of six key nutrients, and is a significant source of soluble fiber (found in fruits, vegetables, and grain products). This claim will indicate that a diet low in saturated fat and cholesterol and rich in fruits, vegetables, and grain products that contain some types of fiber (particularly soluble fiber) may reduce the risk of heart disease.

Foodservice menu regulation has taken the following route:

- *1990:* Congress passes the Nutrition Labeling and Education Act (NLEA), which makes nutrition information mandatory for most foods. Among the few foods exempted were restaurant items, unless they carried a nutrient or health claim.

- *January 1993:* FDA issues regulations under NLEA that requires restaurants to comply with regulations for nutrient and health claims that appear on signs and placards. Menu claims are exempt.

- *March 1993:* Two consumer advocacy groups, Public Citizen and Center for Science in the Public Interest, file suit against the Department of Health and Human Services and FDA charging that the menu exemption violates the NLEA and Administrative Procedure Act.

- *June 1993:* FDA proposes to require that menu items about which claims are made be subject to the nutrient and health claims' regulations.

- *June 1996:* Because FDA failed to finalize its June 1993 proposal, the U.S. District Court in Washington, D.C., rules that Congress intended restaurant menus to be covered by NLEA and orders FDA to amend its nutrition labeling and claims regulations to include menu items about which claims are made.

- *August 1996:* FDA issues a final rule removing the restaurant menu exemption and establishing criteria under which restaurants must provide nutrition information for menu items.

- *May 2, 1997:* FDA's regulations for nutrition labeling of restaurant menu items that bear a claim take effect.

■

SMOKERS VERSUS NONSMOKERS

As government dictates the smoking policies of foodservice facilities, a war has emerged in many communities. California restaurants and bars have become smoke-free. Other states, cities, and communities are looking to do the same. Lawsuits are being filed, restaurateurs are fighting back, but the writing is clear in society today. Those who smoke will have to do it elsewhere.

While Australia, Canada, Israel, the United Kingdom, and the United States are leading the world with laws banning smoking in public places, Brazil, China, Egypt, France, Germany, Italy, and Russia may have desires to limit smoking, but the smokers are not complying.

Workplaces and restaurants are the public venues that expose the greatest number of people to environmental tobacco smoke. Because of their prolonged exposure, staff in bars and dining rooms are most at risk. The statistic is that there is a 50 percent increase in lung cancer risk among foodservice workers.

A study in California following the ban of smoking reports good news for bar workers, who state that congestion has cleared, throats are no longer sore, and health is much improved following the "no smoking" signs being lit.

The debate will continue, no doubt, but foodservices are under attack and the future appears to be that smokers will be lighting up only in enclosed, separate rooms designated for smoking, if at all. Time will tell, but the hospitality industry does have a responsibility to accommodate the nonsmoker as well as the smoker, as has been seen in hotels with both smoking and nonsmoking rooms.

■

FOODSERVICE MANAGER'S RESPONSIBILITY

The responsibility for nutrition varies among foodservices. A commercial foodservice in which the customer has free choice of patronizing and selecting what he or she wants has a different responsibility from a noncommercial operation in which there is no full selection of food by the people eating there. The first should make it possible to select a balanced meal if the menu is extensive enough. All foods served should contain the expected nutrients. An assisted-care living facility serving a "captive audience" should see that the food is nutritionally adequate in all respects. A health facility serving special diets should be able to assure that the food is nutritionally adequate and meets the person's special dietary needs.

How much responsibility should a foodservice have to inform the people who eat there about nutrition values? Should a foodservice be responsible for educating patrons in good nutrition principles? Should it serve foods that meet some of the more common, simple dietary needs?

Obviously, some of the questions have been answered by the government, as stated earlier. If nutrition claims are being made by the operation, the responsibility to communicate those claims to the patron is required by law.

Preparing Food

Because of the interest in nutrition, menus of many foodservices should be constructed more carefully to conform to dietary principles and meet dietary needs. An operator planning a menu should keep in mind the following questions:

- Are menus planned and portions controlled to assure that customers get the type and the amount of food preferred?

- Are procedures and recipes used which assure that each portion served contains the nutrients it should and that these are not lost in preparation procedures or in the use of poor recipes?

- Are proper procedures used in storing and handling food?

- Is the food attractively merchandised and served in an appealing way to help build good appetites?

- Is there a display of the nutritive value of foods served for which nutritional claims are made?

- Do menus provide foods that meet nutritional requirements?

- Can a person select a balanced meal if a complete menu, with a full range of courses, is provided?

- How can the menu help to educate nutritionally?

Training Employees

Even though most commercial foodservices do not have people on their staff who are capable of planning diets, the foods served should contain the appropriate nutrients. These should not be lost in handling, preparing, cooking, or in holding for service, as noted above. It is desirable to attempt to train service personnel to know enough about the menu and diets to suggest foods that might meet the needs of a patron on a simple diet such as a low-salt, low-fat, low-cholesterol, low-saturated fat, low-calorie, or diabetic diet.

As people eat more and more away from home, the need for more nutritional guidance and better nutritional planning will become clear. Foodservices should recognize that they are an important part of the national nutritional picture and should assume greater responsibility in this area.

Using Products to Reduce Fat, Sugar, and Salt Intake

In making claims regarding nutrition, the foodservice operator must now be able to back up those claims according to FDA regulations. Accordingly, using products to reduce fat, sugar, salt intake, and the like is appropriate, but if claims are made, the facts need to be available for the patron to review.

These claims may be based on nutrient analysis from a cookbook recipe or computer program. Entrées need not be sent into a laboratory for analysis. Of course, some patron may order a plate of food to go, run it down to the nearest laboratory, and if it doesn't come out right, report the foodservice operation. This is not likely to happen, but it does suggest that there is a responsibility for using products to reduce fat, sugar, and salt intake and making claims about it. The industry will continue to introduce new products that are heart healthy. Americans like to talk light and eat heavy. That is, a person who orders a small entree, light in calories, is sometimes seen overcompensating by eating a gooey fudge dessert. The bottom line is that the patron has the responsibility for eating correctly, but the foodservice operation must provide the opportunity for that to occur.

■ SUMMARY

Nutrition involves the relationship between the food a person eats and how well this food enables the body to maintain good health. Despite our high standard of living, the United States is not the healthiest nation in the world. Americans are concerned about the foods they get in foodservices, feeling that the foods offered often fail to provide proper nutrients. This is especially true in the heart-healthy environment in which the nation finds itself today.

Foodservices have responded to public concern by offering patrons what they want. Menus reflect the changing patterns of selections by patrons and the demand for different foods and methods of preparation. Red meats are falling out of favor because many people believe that they contain undesirable levels of cholesterol and saturated fats. Fewer calories are being consumed. Alcohol consumption is also dropping, in part because it contributes significant calories and in part because problems with alcoholism are being assessed realistically.

The nutrients we need are classified as proteins, carbohydrates, fats, vitamins, and minerals. Water and fiber are not nutrients but are essential in a good dietary program. The energy-producing nutrients are carbohydrates, proteins, and fats. Carbohydrates, primarily serve to provide energy; proteins and fats do other jobs. Proteins make up muscle and provide the building units for antibodies, hormones, vitamins, and other essential bodily products. For a protein to be completely beneficial, it should contain all of the essential amino acids (the building blocks in proteins that the body cannot manufacture). Proteins in flesh foods and dairy products are largely complete. A complete protein can also be made by combining incomplete proteins—such as those in legumes and cereals—in a single meal. Only one or two fats in their form as lipoproteins are needed by the body. Other fats interchangeably perform the general duties required of fats in the body. For complete metabolization of fats in the body, a person must use some carbohydrate as well.

Vitamins are body regulators that do many things to make bodily functions occur. Minerals also help bodily functions to occur and provide building materials for the body.

Water is needed in the amount of about 1½ quarts per day. It is obtained from fluids, foods, and the breakdown of foods in the body. It is eliminated from the body through exhaled air, evaporation from the skin, sweat, urine, and feces. Salts and other regulators determine the quantity of water needed in the body.

The major concerns of patrons today are weight control, low-salt or low-sodium diets, low-fat or low-cholesterol diets, and high-fiber diets. It is possible for foodservices to satisfy all of these special concerns.

Despite progress, U.S. water supplies can be erratically monitored and may contain contaminants with long-term health effects. Foodservice operators are encouraged to take appropriate steps to have water tested and to monitor technology to provide the safest water possible to patrons.

Worldwide, almost 90 percent of the human food supply is provided by only fifteen crop species and eight livestock species, small numbers when compared with the estimated 10 to 30 million species inhabiting our biosphere. Introducing genes from various organisms into crops and livestock has long been regarded as a promising way to ensure the continued productivity of agriculture and forestry. Through genetic engineering, healthier and safer foods can be produced in a much more environmentally sensitive way. Genetic engineering and the food supply is one topic that foodservice professionals will wish to watch carefully.

In 1990 the U.S. Congress passed the Nutrition Labeling and Education Act (NLEA), and new food labels were introduced in May 1994. The purpose of the food label reform was simple: to clear up confusion that has prevailed on supermarket shelves for years, to help consumers choose more healthful diets, and to offer an incentive to food companies to improve the nutritional qualities of their products.

The Recommended Daily Dietary Allowances of the National Academy of Sciences is the primary guideline for the desirable nutrient intakes in our diets according to gender, age, and life conditions. The Reference Daily Intake (RDI) list offers recommendations about the amounts of nutrients that should be in the American diet. The RDI quantities are used as a measure of the nutrients provided in many packaged foods.

As government dictates the smoking policies of foodservice facilities, a war has emerged in many communities. California restaurants and bars have become smoke-free. Other states, cities, and communities are looking to do the same. The hospitality industry has a responsibility to accommodate the nonsmoker as well as the smoker, but the debate continues as to how best to provide that accommodation.

The foodservice's responsibility for providing nutrition varies, depending on whether or not patrons have free choice to select where they eat and what they eat. If nutrition claims are being made by the operation, the responsibility to communicate those claims to the patron is required by Food and Drug Administration regulations.

Foodservice operations should ensure that the nutrients that should be in food at service are there. If an operation does not serve complete meals, it has no obligation to see that the foods served are nutritionally complete; if it does offer a complete menu, the operation should make sure that it is possible to select from the menu a completely nutritious meal. Health-related operations are obligated to see that patrons get all the nutrients they need in their food.

■

CHAPTER REVIEW QUESTIONS

1. You are the manager of a restaurant serving yuppies who want healthful but very good-tasting food. What sort of menu would you offer? How would most of the foods be cooked? Would you use any promotions? Would you try to offer foods for any modified diets?

2. What kind of nutritional information is communicated through newspapers, magazines, TV, and other media? Is it informative? slanted? How well does it agree with known scientific information?

3. What kind of nutritional information can you get from your local American Dietetic Association, American Heart Association, and others? Do they provide information on special diets? What are the basic points in the American Heart Association's publications? How helpful do you think such material is to those who need such information?

4. How well does your own diet agree with the food guide pyramid? Keep a record for several days to find out.

5. How much responsibility do you think patrons have in deciding their own diets when eating in a commercial operation at which free choice is made as to where and what to eat? Do you agree that fast-food outlets serving only partial meals are not responsible for seeing that patrons eat a complete meal at the operation? How much responsibility does a college dormitory have in this regard when it serves students most of their food? How much responsibility does a retirement home have when it serves all of the food consumed?

6. Do the public health concerns mentioned in this chapter agree with any of yours? Do you agree with the ideas expressed about them?

■

CASE STUDIES

1. Ima Piggy, being well rounded and the owner of Ima's Big and Tall Shops, has decided to go into the restaurant business. She is excited because the menu will feature all of her favorite specials and desserts. In attempting to make the operation attractive to all markets, though, she has decided to include a few heart healthy items, such as "low-fat pudding," "low-salt (low-sodium) French onion soup," and "low-cholesterol tuna salad." She has learned that you are willing to consult with her and wishes to know how she must comply with regulations to inform her patrons of the

heart healthy claims she is making about her few menu items. By using the Internet at *www.fda.gov*, research for Ms. Piggy what she must do to comply with FDA regulations and present that research on a sheet of paper that she will be able to hand to her customers. Does the Web site *www.culinarysoftware.com* offer any assistance to the foodservice operator wishing to present nutritional information to the dining public? If so, how, and what does it cost?

2. In an attempt to assist Ima Piggy further, you boldly suggest that she would do well to go on a diet to reduce her weight. Although at first she is disgusted with the idea, you decide to help her by providing a training manual on how to read and use the new food labels found on packaged food. Using the Web site *www.fda.gov*, prepare and present your training manual based on what you have found to assist Ms. Piggy in her weight reduction program.

■

VOCABULARY

Nutrition

Cholesterol

Saturated fat

Unsaturated fat

Fiber

Glucose

Vitamin

Vitamins A, D, E, K

Vitamin B_{12}

Choline

Pantothenic acid

Molybdenum

Lipid

Essential amino acid

Complete protein

Ethylene

Thiamine

Riboflavin

Niacin

Pyridoxine

Biotin

Myoinositol

Folacin

Fluorine

Genetic engineering

Coenzyme

Carbohydrate

RDA

RDI

Calcium

Iodine

Manganese

Chlorine

Vitamin C

Cobalt

Potassium

Chromium

Edema

Protein

Fat-soluble vitamin

Water-soluble vitamin

Phosphorus

Sulfur

Magnesium

Sodium

Zinc

Copper

Iron

Selenium

■

ANNOTATED REFERENCES

Basketts, M., and E. Mainella. 1999. *The Art of Nutritional Cooking*. Upper Saddle River, N.J.: Prentice Hall. (Nutrition in cooking is highlighted in this text.)

Drummond, K. E. 1998. *Nutrition for the Foodservice Professional*, 3rd ed. New York: John Wiley & Sons, Inc. (Nutrition as it affects the hospitality industry is presented in this book.)

Gielisse, V., M. E. Kimbrough, and K. G. Gielisse. 1999. *In Good Taste*. Upper Saddle River, N.J.: Prentice Hall, pp. 51–68. (Section on nutritional cooking presents ingredients, flavor enhancers, and marinades, as well as contemporary cooking methods.)

Kapoor, S. 1995. *Professional Healthy Cooking*. New York: John Wiley & Sons, Inc. (From recipe modification to cutting the fat, this book tells how to cook more nutritionally.)

Kapoor, S. 1996. *Healthy and Delicious: 400 Professional Recipes*. New York: John Wiley & Sons, Inc. (This book provides heart healthy recipes.)

Khan, M. A. 1998. *Nutrition for Foodservice Managers*. New York: John Wiley & Sons, Inc. (From concepts to management applications, this text presents nutrition in total.)

Knight, J. B. 1997. *Managing Foodservice Operations*. Dubuque, Iowa: Kendall/Hunt Publishing Company. (A system approach for health care and institutions is provided in this text by the Dietary Managers Association.)

Producing a Menu

■

LEARNING OBJECTIVES

By the end of this chapter, the reader will:

1. Know the various components of a feasibility study in order to research the market for whom a menu will be produced.

2. Be able to distinguish between the various types of menus.

3. Understand how to design a menu from sequencing menu items to writing descriptive copy and printing it.

4. Recognize how to make a profit through "heart of the house" considerations such as managing labor, raw materials, and facilities and equipment.

5. Learn of the important role the computer can play in producing a menu.

■

SELECTED WEB SITE REFERENCES

American School Food Service Association
www.asfsa.org

International Association of Culinary Professionals
www.iacp-online.org

International Food Service Executives Association
www.ifsea.org

International Hotel & Restaurant Association
www.ih-ra.com

Meeting and Event Planning Center
www.eventplanner.com

MenuPro
www.soft-café.com

Professional Convention Management Association
www.pcma.org

Society for Foodservice Management
www.sfm-online.org

The menu in a modern foodservice operation functions as the operation's primary sales medium. Therefore, it should be written to sell and inform. The content of the menu also defines the operation and establishes a direction for management. It determines the items purchased, their cost, the operation's personnel and facility requirements, and the types of service that will be offered. The operation's decor, atmosphere, theme, and service style revolve around the menu as well. At the front of the house and in the heart of the house, the menu establishes the style and content of the operation.

The plan and design of the menu must attract customers to the front of the house and must encourage them to return. Menus that attempt to attract too broad a market are less effective than those that zero in on (or *target*) a specific market. A foodservice cannot be all things to all people: it must be suited to a specific known group or segment of the population. Consequently, the first step in planning a menu is to determine the target market. This clientele's needs and desires must be taken into consideration in constructing the type of menu that will give the operation its best chance for success.

The *front of the house* is the area of the operation in which direct contact with the customer is made; the focal point here is clearly the dining room. Because the front of the house is so vital to an operation's success, management must consider certain front-of-the-house needs in planning a menu. Effective merchandising and selling are the primary objectives of menu planning, and these have an effect on style of

service and the variety of menu offerings. Front-of-the-house considerations must be in balance with heart-of-the-house concerns such as food and labor costs, equipment, space and storage, skill of personnel, and availability of foods. Neglecting to establish such a balance can result in a menu failure and an operation that does not achieve desired goals. Therefore, weighing front- and heart-of-the-house considerations is an essential step in the menu-planning process.

■

ATTRACTING THE CUSTOMER

Every operation should attempt to build an image that it believes will best appeal to the target market. The image is built from a combination of things: food items offered, style of service, pricing, atmosphere, location, type of customer attracted, and personality and philosophy of management. The menu provides a means of conveying these image-building factors to the customer.

The menu must be designed to communicate the operation's ability to satisfy the needs of a specific market. In many cases the wants and needs of a specific market can be identified on the basis of socio-economic variables such as religious background, age, family size, amount of disposable income, ethnic background, and level of education. These are only a few of the variables that may be considered in defining a target market. The menu planner has a responsibility to study and know the market so the menu can be designed to meet that market's needs.

In the original planning of a foodservice, the concepts on which a design is based are defined in a *feasibility study*. The feasibility study covers a host of factors, including site selection, type and location of competition, size, cost, financing, budgeting, payback of capital and interest within a given period, and the number and type of employees, as well as presenting a detailed analysis of the target market and a preliminary menu designed to satisfy the market's needs.

In hospitals and educational institutions, for example, menus are designed to meet a customer's need for three meals per day, seven days a week. A menu in a first-class restaurant may be written for patrons who eat out only on special occasions, such as couples celebrating their anniversaries. One market may be drawn to a menu featuring hamburgers and milkshakes, while another may have special dietary needs. The following questions need to be addressed in targeting a market and designing a menu:

- Who are the customers?
- Where do they come from?
- How much disposable income do they have?
- Where do they live?
- How do they get here?
- How old are they?
- What is their ethnic background?
- What types of jobs do they have?
- How do they spend their leisure time?
- How often do they eat out?
- In what type of community am I located?
- Are there legal or political restraints?
- Where else can these customers eat?
- Who are the direct competitors?
- What does the competition offer?
- Should I imitate the competition or be different?
- What other types of food or service might be successful?

Any menu that does not meet the needs of the market is unlikely to prove successful and should be changed. A precise analysis of the market offers guidelines to management for determining the theme or logo, location, menu structure, equipment needs, and overall image. Recent studies have identified shifts in customer preferences, as well as an increase in the number of meals eaten away from home. The menu planner should be aware of these trends and make adjustments accordingly.

There may be certain limits within which an operator must work in planning a menu. Some facilities are required to follow specific nutritional guidelines. For example, in a school foodservice covered by federal regulations, a type A lunch for elementary school students must supply one-third of a child's daily nutritional requirements. Hospitals must provide medically prescribed meals rather than meet personal preferences. This frequently causes problems for a hospital, because people would rather eat what they want instead of what is good for their health.

Restaurants and commercial operations may also face certain constraints. Seasonal availability of some products must be considered—especially in the case of seafood or fresh produce. Many items that were only available on a seasonal basis in the past are now on the market year round. The price of such items (lobster, for example) can still vary considerably, however, and may force certain operations to drop otherwise desirable selections from their menus. Other constraints may be encountered in designing a menu for a pre-existing facility. A limited level of skill among workers or a lack of equipment can prevent certain changes from taking place. A new menu must conform to the existing operation; if it does not, confusion and erratic or low-quality performance are likely results.

■ TYPES OF MENUS

Once a target market has been determined and its needs identified, menu planning begins. There are many different menu types, and each is written for the needs of the particular market it serves. Menus may be à la carte, table d'hôte, du jour, limited, cyclical, or other. Children's menus can be important, too. Some markets may require a combination of menu types.

A La Carte Menu

Food items are priced separately in an à la carte menu, allowing a person to select precisely the items desired

SONNY'S RIBS
115 S. State St.

512-7681

ONION RINGS

Half Loaf	1.95	**Full Loaf**	2.95

APPETIZERS

Fried Cheese 3.75
 Golden browned mozzarella sticks,
 seasoned and served with marinara sauce.

Regular Order	2.25	**Large Order**	3.25

Potato Skins
 Filled with real bacon and
 cheddar cheese.

Regular Order	2.25	**Large Order**	3.75

AND OTHER THINGS

Baked Potato with Butter & Sour Cream	.95
French Fries	.95
Corn on the Cob	.95
Cole Slaw	.95

SALADS

Cobb Salad	4.25
Tossed Salad	1.25
Sliced Tomatoes with Bermuda Onion	1.15

RIBS

All rib entrees are served with cole slaw
and french fries.

Baby Back Ribs	10.95

 House specialty. The choicest lean U.S. ribs
 cut from the tenderloin.

St. Louis Style Ribs	8.95

 Select cut U.S. spare ribs. Tender and meaty.

Bountiful Beef Ribs	8.95

 Big, juicy ribs. Hearty and flavorful.

JUMBO PACKS

Each order serves approximately four people.
Does not include cole slaw or french fries.
Barbecue sauce is included.

Baby Back Ribs (40)	38.40
St. Louis Style Ribs (40)	30.40
Bountiful Beef Ribs (16)	27.80
Baby Back Chicken Combo	
(20 ribs, 2 chicken halves)	27.80
St. Louis Chicken Combo	
(20 ribs, 2 chicken halves)	25.90
Chicken (4 halves)	17.80
White Chicken (8 breasts)	19.80
Barbecue Sauce	
Pint	1.95
Quart	3.50
Gallon	15.20

BARBECUE

All entrees are served with cole slaw and
french fries.

Barbecued Shrimp & Baby Back Ribs	11.95

 A skewer of large grilled barbecued shrimp,
 served with our Baby Back Ribs.

Barbecued Chicken & Ribs	9.95

 with our Baby Back Ribs (.75 extra).

One Half Barbecued Chicken	6.45

 with White Meat Only (.50 extra).

GRILLED ENTREES

All entrees are served with cole slaw and
french fries.

Marinated Boneless Breasts of Chicken	7.45

 Two boneless breasts of chicken, marinated
 then grilled. Served on a bed of rice.

Charbroiled Ribeye Steak	11.95

 U.S.D.A. Choice

Filet Mignon on a Skewer	7.95

 Marinated Kabob with mushrooms, peppers,
 onions and tomatoes.

BEVERAGES	DESSERTS
Soft drinks	Chocolate Beast
Iced Tea	Key Lime Pie
Coffee	Cheese Cake
Juices / Milk / Tea	
.85	2.50

$15.00 Minimum Required on all Delivery Orders.

FIGURE 3-1 Sample à la carte menu.

(see Fig. 3-1). It is popular in commercial foodservices, since selling separate items brings in more revenue than selling items at a group price. A semi-à la carte menu may offer an entrée (main dinner offering) with salad, vegetable, and beverage at one price; customers who want an appetizer, a dessert, or other additional food must pay extra.

Another approach is to use à la carte pricing for the entrée as well as semi-à la carte pricing for it in combination with other foods. This is done either by listing entrées separately on an à la carte page or by indicating two prices beside each entrée: one for the entrée alone, and one for the entrée plus the extras. The items included as extras should be clearly indicated, in order not to confuse the customer about what is being offered at à la carte prices. Other menus present two prices: one for an item selected à la carte, and the other for the table d'hôte offering.

Table d'Hôte Menu

A table d'hôte menu is the opposite of an à la carte menu and had its beginning at old roadside inns. The literal translation is "table of the host." An innkeeper would offer "room and board," with the sleeping accommodations being in the room and the board being the table of the host (see Fig. 3-2). It offers a complete meal at a fixed price, although a choice of some items such as salad, entrée, or dessert may be offered. The table d'hôte offering might include appetizer, salad, soup, entrée, beverage, and perhaps dessert. However, spiraling food and labor costs have caused complete menus of this sort to become quite expensive, and many people want a meal with far fewer calories than they would get with such a repast. Offering a more limited selection on the table d'hôte menu makes it possible to reduce the price and yet give enough food to satisfy the customer's appetite. A table d'hôte menu is desirable in an operation that knows what its customers want and can forecast closely what these needs will be. The ability to predict which items on the menu will be selected is essential. If the number of guests is guaranteed, the prediction may be quite precise. Institutional and industrial operations may find limited menus of this type suitable, whereas elegant restaurants may use a traditional table d'hôte or an à la carte menu.

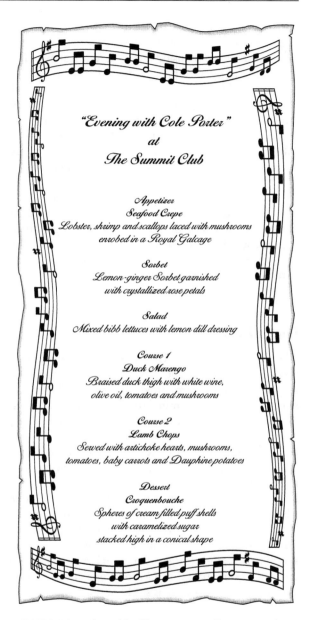

FIGURE 3-2 The table d'hôte menu offers a complete meal, such as the one seen here, at an all-inclusive price.

Du Jour Menu

The term *du jour menu* means "menu of the day." Such a menu must be planned and written daily, since it will reflect only the foods produced for that day.

This offers a convenient way to make use of carry-overs and food bargains; but if it neglects front- and heart-of-the-house considerations, it may not meet the needs of the operation. The du jour menu has the advantage of flexibility not only with respect to food items but with respect to the skill of the employee who is to produce them. Usually, only a few foods or their combinations are offered. Often, an operation offering a du jour menu is a specialty one that serves original and special foods to a limited clientele.

Nonetheless, other operations may use such a menu. For example, special foods may be offered on special days to suit market needs. This provides variety along with known foods. On the whole, though, other menus could be more suitable if the only rationale for such a menu hinges on the use of carryovers and food bargains. The normal rule in a foodservice is that good planning leaves no carryovers.

Limited Menu

The limited menu evolved from more traditional types and may in fact be a hybrid of them. It may or may not offer the same foods every day. The entrée selections are usually limited to between six and twelve items, and some menus may offer only two or three choices. A representative but small selection of appetizers, salads, entrées, and desserts is also offered.

Because of rising costs, limited menus have become increasingly popular as a way to reduce labor costs and food waste so as to improve production control and labor productivity. Planning problems and production complexity are simplified as the menus become more limited. Limited menus must be pointed to relatively specific markets, since they offer fewer items. They have met with great success in fast-food operations and in specialty restaurants such as steakhouses.

Cyclical Menu

The cyclical menu schedules foods for days within certain time periods, repeating the selections every two to six weeks. It is often wise to change cycles with the seasons and to provide special offerings on certain days. Thus, an operation may have four different cyclical menus to supply meals over the course of four seasons. Sundays and Fridays are days on which people especially remember food, so a slower-running cycle may be used on these days than on other days within the cycle. For example, a menu could be planned with an eighteen-day regular cycle, to which a four-Sunday cycle and a four-Friday cycle are built in (see Table 3-1). The eighteen-week or four-season-a-year cyclical menu is the most usual, but cycles within this period may repeat. Flexibility is extremely desirable. A cyclical menu that does not allow for changes or for the use of carryover foods can become costly and may fail to meet the operation's needs. Cyclical menus do, however, give personnel a greater knowledge of production and quality requirements and more opportunity for better planning. This is because workers are faced with day-to-day changes but at the same time can see a repetition that helps in standardization and reduction of costs. Management forecasts of requirements are often improved as well.

Cyclical menus reduce menu-making time as well as labor time. Most foodservices can adapt cyclical menus, even in a limited fashion. Some hotels operate on a strict seven-day cycle, which makes food production coincide with employees' time on the job. This helps simplify production requirements, since employees prepare the same foods every week and gain proficiency and efficiency with repetition. Management can then more precisely identify costs and quantities needed.

Children's Menu

A child's menu is not a necessity for all operations but can be good for those wishing to bring in adults by ensuring that the kids have a good time. If the children are excited about their dining experience, they will be the ones to recommend where to go the next time the question is raised about where to dine.

Although there are no set rules about children's menus, they should be exciting and different from the regular adult menu. Allowing the child to participate in some activity while considering the menu is also good. For example, mazes to follow, coloring to do, games to play, or puzzles to complete can all be fun activities allowing the adults time to talk while the kids are enjoying themselves (see Fig. 3-3).

TABLE 3-1   Complete Menu Cycle for 13 Weeks or 18 Weeks[a]

Week	Sunday	Monday	Tuesday	Wednesday	Thursday	Friday	Saturday
1	S-1	D-1	D-2	D-3	D-4	F-1	D-5
2	S-2	D-6	D-7	D-8	D-9	F-2	D-10
3	S-3	D-11	D-12	D-13	D-14	F-3	D-15
4	S-4	D-16	D-17	D-18	D-1	F-4	D-2
5	S-1	D-3	D-4	D-5	D-6	F-1	D-7
6	S-2	D-8	D-9	D-10	D-11	F-2	D-12
7	S-3	D-13	D-14	D-15	D-16	F-3	D-17
8	S-4	D-18	D-1	D-2	D-3	F-4	D-4
9	S-1	D-5	D-6	D-7	D-8	F-1	D-9
10	S-2	D-10	D-11	D-12	D-13	F-2	D-14
11	S-3	D-15	D-16	D-17	D-18	F-3	D-1
12	S-4	D-2	D-3	D-4	D-5	F-4	D-6
13	S-1	D-7	D-8	D-9	D-10	F-1	D-11
14	S-2	D-12	D-13	D-14	D-15	F-2	D-16
15	S-3	D-17	D-18	D-1	D-2	F-3	D-3
16	S-4	D-4	D-5	D-6	D-7	F-4	D-8
17	S-1	D-9	D-10	D-11	D-12	F-1	D-13
18	S-2	D-14	D-15	D-16	D-17	F-2	D-18

[a] This cyclical menu offers no two of the same meals on the same day throughout its 13- or 18-week cycle. Eighteen daily menus (D) have been specified, and four Friday (F) and four Sunday (S) menus have been written to meet the wants and needs of the market that this particular operation serves.

FIGURE 3-3   This children's menu from Chi-Chi's actually allows the child to wear the menu as a sombrero, due to its unique cut and design. (Printed with permission of Chi-Chi's, Inc.)

Menu Variety

To capture a market, more than one menu may be required. For instance, there may be a need for separate breakfast, luncheon, and dinner menus or for different menus for different food operations. The menu a coffee shop needs is different from those needed by a bar, a shopping center during tea hours, or a late supper dining room. Suiting the menu to these needs is an art, and clever merchandising—both with food and with names—can do much to promote customer interest. Suiting the menu to the amount that patrons want to spend is very important. Attention to special nutritional needs may attract customers.

■

DESIGNING THE MENU

One of the first things the customer sees when handed a menu is its design. If the lighting is low, the design of the menu may need to be bolder. In a fast-food operation, the design may be relatively simple and plain. Four factors are especially important in menu construction: menu sequence; the copy, layout, and printing style; the cover; and the flexibility of the menu.

Menu Sequence

The sequence in which foods are placed on the menu usually mirrors the order in which foods are eaten (see Table 3-2). Menus are read from the outside pages to the inside, from top to bottom, and from left to right. Many people agree that the best menu position for gaining the reader's attention is on the inside center or inside right-hand page. Entrées should be given the best positions, since they bring in the largest amount of money, but alcoholic beverages should not be neglected in such positioning, because they can earn the largest profit per dollar of sale.

		Dinner		
Breakfast	Lunch	Limited Menu Du Jour Menu	A la Carte Menu Table d'Hôte Menu	French Cuisine Menu
Juices	Appetizers	Appetizers	Cocktails	Hors d'oeuvres
Fruits	Soups	Soups	Appetizers	Soups
Cereals	Entrées	Entrées	Soups	Eggs
Eggs	Vegetables	Desserts	Entrées	Fish
Omelets	Salads[a]	Accompaniments[b]	Vegetables	Entrées
Griddle items	Breads		Desserts	Roasted and broiled items
Meats	Desserts		After-dinner drinks	Vegetables and pastas
Vegetables	Beverages		Accompaniments[b]	Cold buffet
Breads				Salads
Beverages				Cheese
				Desserts
				Fruits
				Beverages

TABLE 3-2   Common Menu Sequences for Various Menu Types

[a] Salads as accompaniments to entrées are usually listed after the entrées on the menu. Salads served alone are usually listed before the entrée selection.

[b] Side orders, salads, sandwiches, beverages, and similar accompaniments should be located on the menu as customer demand dictates.

Entrée order is important. The most popular entrées or group of entrées—or the entrées that the operation most wants to sell—should be listed first. The first and last entrée items listed are those most frequently read. Prices should not be arranged in order from least to greatest or from greatest to least. Mixed prices are best. Unless meat, fish, and poultry are offered in separately headed groups, it is not desirable to arrange entrées by animal type or other similar category in an undifferentiated list. Again, mixing is best.

Alcoholic beverages should be listed on the menu in the sequence in which they would be consumed in the course of a meal. Thus, cocktails appear at the beginning, wines in the middle, and after-dinner drinks at the end. Wines identified on a separate wine list card or booklet will not be noticed by customers as often as wines shown on the menu with the food items, unless special attention is brought to it by the server. Some menus list wines on the back of the menu.

The back of the menu is usually used, however, to indicate the operation's address, phone number, days and hours of operation, credit cards honored, and facilities offered. This information informs patrons of important facts relating to the operation. An interesting bit of history or legend about the foodservice or geographical area can also appear there.

Copy, Layout, and Printing Style

The copy on a menu names and describes each item being offered. The layout of the menu sets individual menu items or groups of items apart from others. The printing style (font) selected can make reading the menu easy or difficult. Appropriate print size and style, effective use of headings, descriptions, and space, and attractive arrangement of the copy can make a menu easy to comprehend.

Items or groups of items should bear names people recognize and understand. If a name is unclear or does not give the right connotation, additional descriptive copy is desirable. Such a descriptive explanation can help sell the item; it should be brief and should describe the item realistically. Not everyone is good at writing interesting descriptive copy. Usually, a menu writer/designer familiar with menu

terms, culinary words, and ways to enhance a menu's interest should do it. Foreign words should be avoided unless patrons can be expected to understand them clearly. Simplicity helps improve understanding and broaden communication. Some menus are improved by general descriptive copy built around a special feature, character, or service, or around the history of the establishment or community in which the operation is located. In an operation selling Maryland crab dishes, for example, a story about the Maryland blue crab, how it lives, how it is caught fresh each day by professional crabbers, and how skilled cooks swiftly turn the day's catch into the many detectable dishes on the menu can heighten the customers' interest and build an image of freshness and quality about the product. Using colorful, evocative words is an art; overdoing it can produce garish results.

The advice of a professional menu planner should be used if no one in the operation can put good copy together. Some printing companies make a specialty of planning and setting up menus.

Menu copy should be set in a style of type that is easily readable. Plain Bodoni Roman type is one of the best. Type size for regular copy should be 10-point to 14-point, and type size for headings should be as much as 18-point (about ¼ in. high). Adequate spacing between lines should be maintained, since copy can be confusing if lines are too close together. Mixing type among regular, bold, script, and italics may be done to gain emphasis, but too much mixing causes confusion and clutters the menu. Wide margins should be allowed; altogether, they can account for nearly 50 percent of the entire page. In the space used for copy, from one-third to one-half of the area should be left blank. One of the most common faults in menu writing is to try to squeeze too much onto the page. Studies have shown that this defeats the purpose of the menu, which is to quickly convey a message that sells. Too much information on the page confuses patrons, who consequently are not drawn to the menu items the operation particularly wishes to sell.

Emphasis can be added by using borders and boxes. Prices should be clearly presented so that customers know how much an item or a group of foods will cost.

The color of the menu paper and the color of print can also help to make things stand out. Dark print on light paper is the easiest to read and thus creates the least confusion. Light yellow or light green print on light paper is hard to read. The contrast between paper and ink must be strong to make reading easy and comprehension good. Cafeteria menus and others that are placed on the wall or in places where all customers can see them should be planned so that they quickly give the message the operation wants them to give.

Among the more common mistakes in menu planning are the following:

- Descriptive copy is left out when it is needed.
- The wrong emphasis is given. For instance, a cup of tea may receive as much attention as a prime rib entrée.
- Emphasis is lost because print size and style are not used properly.
- The menu lacks originality.
- The menu is designed for the wrong market.
- Much needed information (such as a listing of certain food or drink items) is omitted.
- Pricing is not clear.
- The menu is wrong for the operation.
- Patrons do not see valuable copy because clip-ons or other added material cover up portions of the menu.

Cover

The cover of the menu should bear a symbol of the operation. A reproduction of a photograph, old print, woodcut, or coat-of-arms can be used for this purpose. This symbol is often called a *logo* and should represent a theme carried through in the decoration of the establishment.

The cover is often planned last because its size depends on the menu's size. Using one color of print on a white background or using two complementary colors is not expensive. Four-color photography produces all colors of the spectrum but is expensive. The cost of using color photography includes obtaining the right color picture to begin with. A professional photographer's fee can add significantly to the expense.

The paper (or *cover stock*) chosen should be heavy, durable, and grease-resistant. Paper that produces too much glare should be avoided, however, since copy is not easily read from it. The cover stock should be stiff enough to remain upright in the hand without bending. The paper used most often for this purpose is called heavy cover, Bristol, or tag stock and is at least 0.006 in. thick. Paper used inside the cover for additional copy can be lighter in weight; strong, heavy book paper is usually good enough.

Flexibility

The flexibility of the menu—its ability to withstand changes without having to be discarded—is important in inflationary times (when prices rise rapidly), as well as in an environment in which the preferences of patrons change relatively rapidly. Food fads rise and fall. If an operation tries to keep up fads for merchandising reasons, its menus can become obsolete quickly; and it is expensive to throw away menus.

Some operations resort to the du jour menu or even a wall board to obtain needed flexibility. This was a custom in the first restaurants in Paris. A neatly written board told patrons as they entered what was offered that day. Later, wooden boards were hung from the belts of waiters so that they could read to guests what was available at that meal. Printed menus did not come into being until much later.

Today, there is a return to the old plan, but for a different reason—to gain flexibility not only in items but in pricing. Some operations like to change parts of the menu daily or weekly, while leaving other parts the same. A change can be effected by adding a paper insert or attachment to the firmer, harder, printed menu. Appetizers, side dishes, desserts, and beverages do not change frequently, so these can be in print on the harder material; meanwhile, entrées and other offerings that may change more frequently are typed or printed on light paper and can be changed each day. Clip-ons or tip-ins may be used if they are styled to match the menu. Inexpensive inserts can also be used. In many fine dining establishments, it has become customary for waiters to repeat the list of the daily specials to the customer from memory, providing the least costly way of making daily changes.

Presentations other than the traditional printed menu are also becoming more and more prevalent. Alternative approaches include listing menu items on pieces of wood, meat cleavers, or wine bottles. The table tent has been a popular device for merchandising alcoholic beverages and is now being used to promote desserts and other foods. In some establishments that feature buffet menus or salad bars, the operator allows the food to do the selling. Attractive displays and presentations can be a very convincing promotional tool.

Nothing becomes staler faster to the frequent customer than the same menu. Normally, menus should be planned to last three months, since after that point customers tire of them. The design should be flexible if new menus are to be created this often.

■

COMPUTERS AND PRODUCING A MENU

Computers can assist the menu planner by maintaining various files on possible menu items and by offering advanced help in all stages of publishing a menu. Files can be created listing menu items, their composition, and any special considerations relating to a specific item. A cross-referenced database can list hundreds or thousands of potential menu items, and the planner has immediate access to this information through a computer terminal. Seasonality of various ingredients can be noted, as well as previous problems encountered in producing an item. By predicting sales, the computer can give an estimate of a menu's profitability and indicate purchase needs.

Operations stressing personalized service sometimes maintain files on their customers. Survey responses, birthdays and anniversaries, and names of family members can be kept on file. The survey results can be used to determine shifting preferences, which may lead to menu changes. Customers can be mailed birthday and anniversary greetings, possibly resulting in a return visit to the foodservice for the impending celebration. Greeting the customer and family by name can give the customer a feeling of being appreciated, leading to long-term patronage.

The other front-of-the-house area in which computers are having a major impact is in the publishing of menus. Traditionally, a menu is planned, designed, turned over to a publisher, and finally received for use. This process can take anywhere from several weeks to several months. However, with the introduction of sophisticated word processing and publishing computer packages, laser and color printers, and low-priced color copiers, this process can be completed in a day. As a result, the menu can be changed on a daily basis (or as often as desired), while maintaining high quality in the presentation of the menu.

■

MAKING A PROFIT

A properly designed and planned menu can increase profits by increasing the size of the average check, minimizing waste, speeding service (to increase dining room turnover), and improving overall operating efficiency. Understanding these factors and their relationship to the menu will help a manager improve the operation's performance and increase profits. In a nonprofit operation, benefits can be measured in terms of reduced costs.

A menu must emphasize popular foods to achieve a high volume of sales. It must also feature high-profit items. The popularity and profitability of each item put on the menu needs to be examined and the item's contribution to total sales must be weighed. Controlling food and labor costs is especially important, since together these consume roughly two-thirds of the operation's income. With the escalation of energy and other costs, however, the need to look at all costs becomes imperative.

The basic cost of food, adjusted to take into account preparation and cooking losses, yields the actual cost of the food. Arriving at a low food cost is not always desirable, since an inexpensive product may require so much additional preparation labor that the final menu item has a higher overall cost and a lower profit. For example, the cabbage used in cole slaw may be cheap, but once the costs of cleaning, shredding, and making the salad are added in, cole slaw may cost more than a salad made from hearts of lettuce, which have a higher food cost but can be prepared more quickly. The selling price of a food is often established on the basis of the food cost. If menu items carry too high a food cost, the result may be a selling price that customers are unwilling to pay. There are many factors to consider in arriving at food costs. Continued careful analysis must be practiced in selecting items that meet both the foodservice's cost constraints and the price range the customer is willing to pay.

Many operators are turning to value-added foods in an effort to control labor costs. Historically, a change from fresh food to processed (canned, dried, or frozen) food occurred because of the need to improve profits; such a change was not always accompanied by improved or even steadily maintained quality standards. Often, difficult decisions must be made by management when profit margins become too narrow. On the other hand, many of the new value-added foods are of excellent quality.

Since a value-added food incorporates labor, its base price is higher than that of the regular food item the foodservice would use to make it. Consequently, basing the selling price on the cost of the value-added item might make the selling price too high. (If the cost of an item is $1.00 as a raw item but $1.50 as a value-added item, fixing the selling price at 35 percent of food cost would make the same item sell for about $3.00 in the first case and for about $4.50 in the second.) Therefore, some adjustment in setting the selling price must occur when one of these new kinds of foods is used. When these value-added foods raise food cost, management should see that the cost of labor drops equally or more than this rise.

Planning for profit requires that a menu be designed to achieve a proper mix of high- and low-profit items. Highly popular, low-profit items on a menu can outperform other items and reduce the operation's profitability. Even though such a popular item may be desirable on the menu, it must be positioned so as not to bury or overwhelm the other offerings. One alternative is to raise the price of the popular item.

■

PLANNING THE MENU

A menu cannot be planned solely to satisfy front-of-the-house interests. It must also satisfy the needs of the heart-of-the-house. Instead of being concerned with attracting customers, heart-of-the-house considerations are directed toward producing the product and toward accommodating constraints imposed by limits on food availability, equipment, personnel, and cost. Heart-of-the-house activities revolve around purchasing, storing, preparing, holding, and serving food. These constraints and processes must be considered in planning a menu.

Preparation and Service Personnel

The responsibility for planning the menu belongs to management, but the views of other personnel should be sought before final decisions are made. A menu can be set up roughly and then given to the chef or head cook for review and suggestions. Others in the preparation team may also be consulted. An item may not be feasible because of a lack of equipment or because of some other factor that management may not know about. For instance, to make good, hard-crusted bread, there must be steam in the bake oven. Without knowing this (and without having the proper type of oven), management might put this bread on the menu, expecting the bakers to produce it. In this case, the bakers can help by pointing out the hidden factor.

The practicability of producing some item may be open to question for other reasons as well. Sometimes an item cannot be produced properly under the stress of heavy orders during peak periods. The chef or head cook will know this and should advise against its inclusion. Making menu planning a team effort also allows people outside management to feel responsible for the menu and may therefore increase its chances of performing well.

Besides the skill and ability of employees, the efficient distribution of tasks must be considered. A menu can easily overload one employee or one division of the cooking or preparation area, while leaving another section with little or no work to do. The workloads, as dictated by the menu, must be balanced. Employees who are overworked or required to exert a lot of effort to accomplish a particular task tend to do poor work, and the quality of the menu items involved suffers as a result. Management should anticipate changing some production demands seasonally or even by day of the week. Warm weather increases demand for such items as cold plates and complete salad dishes. On a Saturday night, the grill may be overwhelmed with fry orders. Shaping the menu to assist in spreading out the work can do much to win the confidence and respect of workers and so to improve work performance and product quality.

Raw Materials

Various food components and raw materials can require different preparation and cooking times. Such variations in time can be critical in deciding whether an item can or cannot be placed on the menu. Customers may not be willing to wait the length of time required for production, or the preparation time may interfere with the production of other items in the kitchen. Quality and cost must also be considered. At certain times of the year, foods are out of season. Although customers may find it pleasing to see them on the menu, either the cost or the quality may make including them impractical.

A menu should present a variety of choices for its customers. People may come to a restaurant for its beef, but if one person in the dining party only eats seafood, the menu should be able to accommodate that person. In establishing a varied menu, availability of the product must be considered. Management must ask, if I cannot get this product, is there a suitable substitute? For instance, a restaurant may offer a halibut mousse with shrimp newburg sauce that is prepared using fresh halibut. If the fresh product were unavailable, a high-quality frozen product could be substituted without a detectable difference in flavor. It should be noted, however, that if the restaurant's menu states that the operation uses fresh halibut, *only* fresh halibut can be used. Truth in menus has become a topic of discussion and legislation in the last few years, as a result of controversy over misleading menu descriptions used by some restaurateurs. If a menu states that a product is fresh, it cannot have been frozen; if an item is described as sautéed in butter, it cannot be sautéed in margarine; if sour cream is specified, an imitation cannot be used. The planner of the menu must have a good knowledge of food products and preparation and must make use of a wide repertoire of dishes to satisfy customer needs and at the same time satisfy the heart of the house.

The form, shape, color, temperature, texture, and overall presentation of a particular dish must be considered when deciding whether to include it in a menu. A dish should be judged both on its own and in relation to other menu offerings. The components that make up a dish should exhibit variety in size, shape, form, and texture, uniting in a harmonious display. Forms and shapes should be definite and varied. An overabundance of cubed items or items cut into hard angles should be avoided. The food on a plate should not be uniformly flat and low; combining various heights produces a more pleasing effect. Proper use of colors can add tremendously to plate appeal. Light-colored items presented together can look bland, but when mixed with brightly colored, complementary vegetables or fruits, they help create an exciting and appetizing display. Contrasts of temperature can bring variety to a dish and enhance flavor as well (hot soups and cold salads, for example). Textures are also significant. A meal consisting of soft foods offers no contrast of textures and, consequently, is unappealing. A meal of soft meat loaf, mashed potatoes, and boiled carrots, for example, does not hold much textured appeal. But replacing the carrots with a crisp salad quickly achieves a nice mix. Adding diced celery to tuna fish salad is another example of mixing textures to enhance interest. Ultimately, all of these elements combine to create a single, overall impression. Judicious use of garnishes can enliven a dull-looking menu item and bring it to life, adding texture, flavor, and color (see Table 3-3 and Fig. 3-4).

The flavor of the food item and the resulting blend of flavors in the dish as a whole must be given high priority. An excess of strong flavors will clash and leave an equally strong (and unfavorable) impression. By the same token, a group of bland flavors may not create any impression at all. A combination of complementary flavors—strong and bland—however, can result in a high-quality product. The flavor of roast beef is complemented by a good sharp horseradish sauce. A fruit salad can be improved with the addition of a slightly sweet honey French dressing. A tart, flavorful fruit blend containing strawberries or cherries combines beautifully with cheesecake. Creating interesting flavor combinations is part of good menu planning. Many young American chefs are gaining recognition with imaginative new combinations and

TABLE 3-3　Garnishes and Their Yields

Item	Portion	Yield
Apples, ring	1 ring	5 rings per 113-size apple
sliced	2 thin wedges	12 wedges per 113-size apple
Apricots	½ or 1 fruit	20 halves or 40 wedges per lb
Avocado	1 slice	30 slices per avocado, 24 per crate size
Banana, split, 1-in. slice	1 or 3 round slices	12 bananas or 2½ c slices per lb AP; 1 banana (3 to lb AP) yields 30 1-in. slices, split
Blackberries	3 berries	1 qt yields 100 berries
Blueberries	3 to 5	1 qt (1½ lb) 360 to 800 berries
Cantaloupe	3 balls or small wedges	30 balls or 45 wedges per 45-size fruit per crate
Capers	1 t	10 to 15 capers per t
Cheese, cottage	1 No. 20 scoop (1½ oz)	10½ scoops per lb
cream	2 T (1 oz)	for stuffing celery, 16 portions per lb
shredded, dry	1 T (¼ oz)	64 portions per lb
shredded, moist	1 T (½ oz)	32 portions per lb
Cherries, maraschino	½ or 1	640 per gal
sweet, fresh	1 fruit	40 per lb
Chocolate tidbits	1 T	40 portions per lb
Coconut, long shred	1 T rounded	1 lb equals 6½ c or 60 portions
Currants	3 fruits	1 lb equals 150 currants
Dates	1 fruit	60 dates per lb
Decorettes	1 t	160 portions per lb
Endive, curly	1 leaf	45 per head
Figs	1 fruit	48 per box, 6 lb
Grapefruit	1 to 2 sections	12 sections per grapefruit
Grapes	3 fruits	50 medium-sized grapes per lb
Kumquats	1 fruit	1 lb equals 24 kumquats
Lemons, wedge	⅙ to ⅛ fruit	1 doz lemons yields 144 rind twists
Limes, wedge	¼ to ⅙ fruit	1 doz limes yields 62 twists or rind
Mint	2 to 3 leaves	300 leaves per bunch
Mushrooms, cap	1 cap	15 to 20 caps per lb AP
Nuts, chopped	1 T	1 lb chopped is 4 c
salted for tea	1 T	1 lb nuts is about 4 c; use 3 lb for 100 people
Olives, green	1 or 2 fruits	1 qt (1¼ lb) equals 100 extra-large olives
stuffed, sliced	1 or 2 fruits	1 medium-sized olive yields 6 slices
ripe	1 or 2 fruits	1 qt small size yields 120 olives
Oranges, sections	3	8 to 9 sections per 82-size orange; 1 doz orange rinds yields 164 rind twists
Parsley, curly	1 sprig	80 sprigs per bunch
Peach	1 wedge	8 wedges per medium peach
Pear	1 wedge or slice	12 wedges per 5-oz pear
Pepper, ring	1 ring	10 rings per medium-sized pepper

TABLE 3-3   Continued

Item	Portion	Yield
Pickles, sweet, medium (3 in.) [a]	½ pickle	24 pickles per qt
Pineapple	1 wedge or 2 to 3 diced pieces	60 wedges or 150 diced pieces per 18-size pineapple
Plums, Santa Rosa	1 medium	70 per till (5 × 5 size)
Pomegranate	5 seeds	25 garnishes per pomegranate
Potato chips or shoestring potatoes	¾ oz	1 c; 8 ounces is about 2½ qt
Prunes, dried	1 fruit	30 to 40 per lb AP
Radishes	1 or 2	15 to 20 per bunch; 1 bunch 10 oz; 1½ c topped and tailed equals 8 oz or about 25 radishes
Raspberries	5 berries	1 qt yields 300 berries
Rhubarb	1 or 2 curls	1 lb yields 100 curls
Sardines	1 3 in. long	1 lb yields 48 sardines
Strawberries	1 berry	1 qt yields 60 medium-sized berries
Tangerines	3 to 4 sections	10 sections per tangerine
Tomatoes	1 wedge	8 wedges per medium tomato
Walnuts, whole	½ nut	8 oz equals 2 c or about 150 calories
Watercress	1 sprig	30 sprigs per bunch

[a] Pickles sized per gallon are frequently used: gherkins, 200; pickle rings or slices, 400; small sweets (3 in.), 80 to 100; large dills (4½ in.), 25.

presentations, resulting in distinctive and outstanding menus.

The key to good menu planning is knowing how to plan a sequence of flavors from the beginning to the end of a meal. Acidic or spicy foods can help whet the taste buds at the start of a meal. A flavorful consommé or bouillon can do the same. Meat flavors stimulate the taste buds in the mouth and start the flow of gastric juices in the stomach, which helps to spur the appetite. Alcohol can also accomplish this, but it destroys the sense of taste, especially if consumed to excess. Sweet or bland foods often are served at the end of the meal to close off the appetite and give a feeling of satisfaction.

Many foods need special flavor combinations. Roast pork is served with applesauce because the tart, sweet sauce helps mask the oiliness of the pork. Cranberries help to modify the sulfury flavor of turkey. Mint sauce masks the pungent flavor of lamb and mutton.

The seasonality of foods needs to be considered in menu planning. Modern methods have gone a long way in extending seasons and in shipping foods from

FIGURE 3-4   Simple garnishes have transformed this braised lamb shank into an elegant meal.

other growing areas during times when they are not available from local supplies. But much variability remains in the market offerings, and some foods are not on the market—either at all or in sufficient supply—at certain times of the year. If foods are purchased out of season, they will cost considerably more and often their quality will be inferior to what it is when the item is at the peak of its season. Many seasonality charts are available that show when foods are apt to be scarce. Seasonality even extends to fish, meat, and poultry. Market reports can also advise menu planners on foods that will be in good supply and foods that will not be. The USDA maintains a market report system that can advise buyers.

Featuring foods when they are in season can do much to brighten up a menu. Thus, a broiled grapefruit in January or February (when this fruit is at its peak) can be a taste treat as well as an unusual dish. A dessert of fresh strawberries with sour cream and light brown sugar in May or June is apt to win customer approval and requires little preparatory labor. Certain fish come onto the market at special times and should be featured during these periods. Eggs are highest in quality and lowest in price in the spring and early summer. A warm-weather omelet festival offering a wide choice of different kinds of omelets can lend a lot of interest to a menu. Promoting certain kinds of foods when people are thinking about them is both smart menu planning and smart merchandising. Pumpkin pie around Halloween takes on a special meaning. A peach shortcake in August with an appropriate descriptive line—"Georgia peaches are here again; try our famous Georgia Gold Shortcake made with mellow, rich, sliced Georgia peaches and our own delicate sponge cake, topped with whipped cream"—may encourage a dessert sale that otherwise would not be made.

As discussed in Chapter 2, nutritional balance is another increasingly important consideration in menu planning. Not only can a menu win friends by offering nutritionally balanced combinations (and even some special diets), it can also draw the attention of customers to principles of good nutrition and thus contribute to increased knowledge on their part.

Facilities and Equipment

A facility should be planned and equipment selected after the menu is written; the menu should not be planned after the facility has been built and the equipment is already in place. If the menu is allowed to determine the design of the production facility and the equipment requirements, lower operating costs and a better product will result. Many foodservices have more space and equipment than they need. As much as 35 percent of some operations' equipment is unnecessary, and many operations allocate space for activities that never take place. Operation and capital costs are far less when a facility is built to meet special menu needs. If proper analysis of the menu occurs, a foodservice in an elementary school may eliminate a broiler and a deep fryer from its plans, since they will rarely be used in producing meals for the children. Some hospital kitchens in Texas were planned with only steamers, steam-jacketed kettles, mixers, ovens, sinks, and tables, since these were the only pieces of equipment needed. A dishwasher was used in the dishroom. Productivity of meals ran well over twenty meals per labor hour, in contrast to the average productivity of four meals per labor hour in many hospitals.

Achieving a proper production flow is necessary if efficient production of food is to occur. Menu items can be selected to ensure a workload that is evenly distributed among different work stations and results in an orderly and timely flow of production. Materials should flow smoothly from storage to preparation and from preparation to service. A well-organized flow between processes will minimize the number of steps that must be walked between work areas. The hazards of poor sanitation may be increased when workflow is poorly organized. A poorly planned menu can result in crowded work stations and a backlog of work. Efficiency is lost; and once this happens, sanitation is usually the first thing overlooked. For example, a cook may cut up chickens on a meat block and, being rushed, not clean it properly, but proceed to chop onions and celery which are used to make a bread dressing. The onions and celery pick up *Salmonella* bacteria from the block, which contaminate the dressing, which then poisons some customers and gives rise to a costly lawsuit for the foodservice. If the flow of work is so disorganized that the time lag between broiler and service is excessive, an item such as a broiled steak should not be on the menu. Attempting to send cooked steaks up to a twenty-fifth-floor dining room from a basement kitchen creates an impossible situation—and yet it is done. Needless to say, customers are rarely happy with the results.

Directing Food Selections

The menu planner can do much to influence customers to select certain foods. Often this can be done by emphasizing the food. Descriptions can help gain attention, as can placing items in boxed areas or giving them special spacing. The items that appear first and last in a group list are the ones most often seen, while those in the center are the ones most often missed. An operation that does not wish to sell steaks, but feels compelled to have a steak listed on the menu, can bury it at the center of the list of entrée items, while placing the higher-gross-profit items at the beginning or end of the column. Special headings or bold type can make something stand out, and clip-ons or tip-ins also have built-in prominence.

Menus should be planned so that foods are balanced between lower- and higher-cost items. It is often wise to invite guests to help themselves to liberal portions of salad materials at a salad bar. This helps the "big eater" to fill up, after which a normal or even slightly smaller portion of the more expensive items will suffice to complete the meal later. Serving a good, substantial soup or salad before a meal can accomplish the same result. In planning a buffet, the lower-cost foods should come first, so customers fill their plates with them and have less room for the more expensive ones (such as roast beef) at the end of the buffet.

Lower costs can be realized if sufficient volume is built around items. The cost of materials and labor can be lowered, since it costs just about as much to cook 20 gallons of a product as to cook 40 gallons. Similarly, mixing 30 portions in a mixer takes almost the same amount of energy and time as mixing 300 portions. By planning a menu with these costs in mind and putting items on it that can develop sufficient volume, operational costs can be lowered. Some-

times menu prices may have to be manipulated to achieve better volume. That is, it may be better to sell 500 hamburgers with a gross profit of 12 cents each (giving a $60 margin) than to sell 300 hamburgers with a gross profit of 16 cents each (giving a $48 margin). Management needs good cost information to make informed judgments in these areas. Precosting methods are discussed in detail in Chapter 8.

■

SUMMARY

The menu is central to the entire foodservice operation. It is the starting point in the planning process and establishes a direction for further planning. The front of the house is designed to meet the needs and wishes of the customer, and the menu must reflect this intention. At the same time, heart-of-the-house needs must also be considered and balanced with those of the front of the house. The goal of a menu is to attract customers. To achieve this, a menu should be designed to appeal to a target market whose demographic characteristics must be identified and whose needs must be determined. This is the first step in the menu planning process, and a good market assessment can contribute a great deal to ultimate success.

The next step in the process is to decide which type of menu to use. A number of different approaches have been used in the past. An à la carte menu prices each food separately. A table d'hôte menu sells foods grouped together for a set price; the group sold as a unit may or may not constitute a complete meal. The du jour menu covers only one day's offerings; the next day, a different set of menu items is offered. The limited menu offers between six and twelve entrée or main dinner items, which reduces planning, production, and service problems and strengthens oversight control. Cyclical menus provide daily changes in meals offered over a set period, after which the menu is repeated; the repetition simplifies planning and improves production performance. Although there are no set rules about children's menus, they should be exciting and different from the regular adult menu. Allowing the child to participate in some activity while considering the menu is best.

Front-of-the-house needs must be addressed in designing the menu. First, the sequence in which menu items are listed on the menu should follow the progression in which foods are presented during a meal. Positioning and location are important in directing patrons' attention to specific items that the operation is most eager to sell. Second, the copy of the menu is important because it can improve communication between patrons and the salespeople as well as enhancing sales. The kind and size of type, its spacing and layout, and other typographic factors can do much to improve menus and help patrons. The cover should be practical and cleanable, and should bear the logo of the operation; it should be completed last, to ensure its compatibility with the menu as planned. Finally, a flexible menu will enable the operator to adjust to changes in customer preferences and eating habits.

Heart-of-the-house considerations involve the purchasing, storage, and preparation of food, as well as the service to the customer. These factors must be considered in the menu planning process.

Computers are becoming a useful tool for the menu planner. Information about the type and range of menu item choices can be kept on file. Market information can easily be stored and retrieved, allowing the operation to track customer preferences and to address customers on a personal level. Computers give planners more flexibility, since planning and publishing can be completed in a single day, on the premises—and a new menu can be in the customers' hands that night. Making a profit in a profit-oriented operation (or not exceeding budgeted allowances in a nonprofit one) is a primary consideration. Each entrée's popularity must be analyzed against its profitability. The costs of using convenience foods—as opposed to foods requiring full preparation—must be considered.

Planning a menu with due consideration for the needs of the heart of the house requires that a number of factors be taken into account. First, preparation and service personnel should assist in the planning of the menu to ensure that items to be prepared are within the competence of the staff and are not inconsistent with the workload desired. Time factors must also be considered. Issues of high cost, unavailability, or fluctuating supply of raw materials may force menu planners to offer a different set of selections. Facilities and equipment must be rationally related to the number of patrons to be served, the type of service, the availability of equipment, the balancing of menu production requirements and equipment capabilities, and many other factors.

Management must strongly consider the popularity of menu items in deciding which ones to include. Costs can fluctuate with the volume of business and with the volume of materials, machines, and labor used in a foodservice. Management can control these costs by placing specific items on the menu or by keeping them off. The influence of management on profitability by selecting items according to popularity and cost can thus be substantial.

Such factors as reducing waste, increasing productivity, and using food efficiently should be weighed in planning the menu. Standardized recipes that give a known quantity of food of a known quality can be used to maintain standards and control production. Using such recipes also gives an operation greater flexibility. Yield and quantity tests help define the yield that can be expected from quantities purchased and thereby contribute significantly to the process of identifying food costs with precision. Quality of items can also be checked to see if the standards set by the operation are met.

■

CHAPTER REVIEW QUESTIONS

1. What are the important questions to answer when doing a feasibility study? Design a questionnaire that could be used by management in determining who its customers are. Assume that the operation is on the main street of a shopping

district, and that management will use this questionnaire to request information from patrons finishing their meals.

2. Compare the characteristics of the six menu types listed in this chapter. Give advantages and dis-

advantages of each. Identify the kind of foodservice in which each would be found. Try to find examples of each type and show them to the class.

3. Analyze the design of a menu by critiquing it with reference to the factors for menu construction: sequence, copy, cover, and flexibility. Is this menu a more effective sales tool because these factors are present or is it lacking because of failures to use them properly? What common menu mistakes does it have? How could its design be improved?

4. Investigate the use of convenience foods as opposed to the use of foods requiring full preparation to see whether the costs and the quality of both are equal. List some of the more popular convenience foods and explain how they can contribute to making a profit in an operation.

5. Discuss planning the menu in light of how it may contribute to making a profit in a specific type of foodservice. Be sure to include such factors as preparation and service personnel, raw materials, and facilities and equipment.

■

CASE STUDIES

1. Jennie Hendricks, assistant to the regional manager of OFU Restaurant Group, has just been assigned a project by her boss, Nola Redding. Nola has not been entirely happy with the process by which unit managers have been designing menus for their operations. Some menus have worked better than others, and some operations seem to make the transition to new menus much more easily than others. Nola feels that certain guidelines should be followed in designing a menu, and she is not sure that all of the unit managers are following them. Consequently, she has given Jennie the job of establishing guidelines for designing menus.

   Jennie began work on this project by first analyzing the types of restaurants that the company operates. The OFU Restaurant Group operates a number of different types of properties in different locales, ranging from family-style restaurants located in suburban shopping centers to fine dining establishments in the heart of a city's financial district. Each operation is very different from the next. Taking this into account, Jennie has decided that the guidelines must be very general, so as to be applicable to a wide range of situations. She sees this project developing into a manual that will outline the steps a manager should follow and will identify key points to take into consideration. This is the first project that Jennie has been assigned since she joined the company (after graduating from a restaurant management program) and she hopes to make a strong positive impression on her new boss. Nola has told Jennie that she expects a two-page report on her desk in the next 72 hours detailing the goals, objectives, and major points to be addressed by the finished manual. If you were in Jennie's position, what would be in that report?

2. Mike Jorden's new sports restaurant and bar had opened successfully. He used pictures, winning basketballs, and some crusty uniforms from championship games to decorate the walls. Fans were excited to dine amidst the paraphernalia. Unfortunately, they were not excited about the informality of his menu. Comment cards suggested that a menu cover would be appropriate and well received to give the menu a professional look. Learning of your expertise in the area of menu design, he has called for suggestions as to what cover he might use in his mid-price-range foodservice operation. Using the Web site *www.menusasap.com* for your research, suggest to Mike which covers he might purchase, and why. What would the covers cost him? Which would you recommend?

■

VOCABULARY

Market	Children's menu
Logo	Feasibility study
Cycle	A la carte
Semi-à la carte	Du jour
Entrée	Limited menu
Copy	Target market
Image	Food cost
Sequence	Table d'hôte
Heart of the house	Front of the house
Convenience foods	Selling price
Seasonality	Complementary flavors

■

ANNOTATED REFERENCES

Khan, M. A. 1991. *Concepts of Foodservice Operations and Management.* New York: John Wiley & Sons, Inc., pp. 39–84. (A chapter on menus and menu planning is presented in this book.)

Kotschevar, L. H. 1987. *Management by Menu,* 2nd ed. Chicago: Educational Foundations of the USA. (Menus are considered in total in this classic text.)

Mill, R. C. 1998. *Restaurant Management: Customers, Operations, and Employees.* Upper Saddle River, N.J.: Prentice Hall, pp. 107–127. (Pricing and designing the menu are discussed in this chapter.)

Scanlon, N. L. 1992. *Catering Menu Management.* New York: John Wiley & Sons, Inc. (Catering menus and their management are the basis for this text.)

Scanlon, N. L. 1993. *Restaurant Management.* New York: John Wiley & Sons, Inc., pp. 55–96. (Menu development and pricing are covered in two chapters.)

Wigger, G. U. 1997. *Themes, Dreams, and Schemes.* New York: John Wiley & Sons, Inc. (Banquet themes, menus, and management are discussed in this book.)

Implementing Equipment

■

LEARNING OBJECTIVES

By the end of this chapter, the reader will:

1. Recognize how equipment is implemented into the various subsystems of a foodservice facility, from receiving to cleanup.
2. Be able to identify foodservice equipment, including utensils, pots and pans, and small tools.
3. Learn of new innovations in equipment as well as

how the computer has affected equipment design and operations.
4. Know how to evaluate foodservice equipment and distinguish between the advantages and disadvantages of similar pieces of equipment.

■

SELECTED WEB SITE REFERENCES

Alcatel
www.alcatel.com/

Blanco Gmbh + Co. KG
www.blanco.com/

Cambro
www.cambro.com/

Hatco Corporation
www.hatcocorp.com/

Hobart Corporation
www.hobartcorp.com/

Mid-America Restaurant Equipment
www.foodservice-equip.com/

North American Association of Food Equipment
   Manufacturers (NAFEM)
www.nafem.org/

■

OVERVIEW OF PRODUCTION AND DELIVERY SYSTEMS

The introduction of new foods and new equipment have made significant changes in our food production and service sections in today's foodservices. The result has been the shrinking down of kitchen size and expansion of service space.* This has saved on the cost of building space and space maintenance while increasing efficiency and allowing the use of fewer and less skilled workers in the production section. Kitchens now exist in which the production of a full menu is possible with only the use of steam equipment and

ovens; the Cynthia Bishop kitchen operated by the Texas Mental Health Department is an example. In some cases the change when value-added foods are used has been so dramatic that practically no production is needed. However, except for the expansion of dining area space, the changes in that area have been less dramatic, although new equipment has been able to make some minor changes.

The use of value-added foods, which require only limited treatment to be ready for service, has often resulted in the elimination of the butcher shop, the bakery, and to some extent the space required for the preparation of fruits and vegetables. New foods, such as food bases, have eliminated the need for the battery of steam-jacketed kettles needed for the production of stocks. The availability of many new frozen foods has increased the need for more frozen food space and eliminated the need for dry storage and refrigerated

*R. Ghiselli, B. A. Almanza, and S. Osaki, "Foodservice Design: Trends, Space Allocations, and Factors That Influence Kitchen Size," *Journal of Foodservice Systems*, **10**, 89–105, 1998.

products. New production methods such as the cook–chill method of producing tools in huge quantities and then holding them on inventory for later use has caused a division of kitchens into two areas: one for mass production, with a smaller area devoted to the day's menu requirements. Where such a system is used, more refrigerated space is needed.

Although space cost has been reduced and kitchens made more efficient, the change has not all been positive as far as cost is concerned. While the requirement for the amount of labor and its skill has been lessened, the added cost of value-added foods has to some degree eliminated this saving. Although less equipment either in units or size is required, the increased cost of adding these new features to present-day equipment has also reduced the saving.

Factors other than those relating directly to new foods or new equipment have also changed equipment needs. The Occupational Safety and Health Act (OSHA) and the Americans with Disabilities Act (ADA) have imposed space and equipment requirement changes in both production and service areas. (Hicks and Ward* have estimated that 15 percent of foodservice workers are physically or mentally handicapped. A large number of patrons are similarly handicapped and the needs of both must be met.) Work spaces and equipment must be made safe. Aisle space in dining areas must be widened. Nonsmoking regulations require special space allocation in dining areas. Sanitation codes require the installation of handwashing sinks and other features. All equipment and spaces must allow for easy, complete, and frequent cleaning. Toilet room regulations have made it necessary to install special equipment units and safety and assistance features. In some high-pollution areas special exhaust systems to remove smoke and odors have been required (Rowe†). Building codes and regulations have influenced equipment selection and equipment installation.

Equipment requirements depend on the basic menu. By writing a limited menu, one can eliminate the requirement for much equipment and also save on space. An elaborate menu requires much more equipment and space. Menu planners study recipes and note the equipment needed to produce the menu. Any special equipment needs are noted at this time and the equipment needed is added to the list of pieces needed. Menu items to be prepared from raw ingredients rather than value-added products increase equipment needs. Production capacities of equipment will depend on the quantity of food needed and preparation time. Equipment must be selected and arranged in a facility to produce maximum efficiency in the use of labor and to produce food of the proper quality. The ultimate objective is to establish a system that integrates equipment, people, and food to this end. A classic equipment purchase decision formula is presented below. Choosing between similar pieces of equipment might be done using Fig. 4-1.

Equipment can be very expensive, not only to originally purchase and install, but also to operate, and operators need to give detailed attention to evaluating equipment cost before purchase. A formula exists for deciding whether it will pay to purchase a piece of equipment:*

$$A + B/C + D + E + F - G = \geq 1$$

where

A = savings in labor over equipment life

B = savings in material over equipment life

C = equipment purchase plus installation costs

D = utility costs over equipment life

E = maintenance cost over equipment life

F = cost of equipment and installation cost at compound interest over equipment life

G = turn-in value of equipment at end of useful life

If the result is 1 or larger, the equipment will pay for itself in savings. Russell L. Bean, Marketing Manager

---

*J. Hicks and B. Ward, "Designing for ADA," *Foodservice Equipment and Supplies Specialist*, **47**(12), 38–44, Nov. 25, 1994.

†M. Rowe, "Kitchens Go Compact," *Lodging Hospitality*, **47**(8), 89–90, 1991.

*L. H. Kotschevar, B. A. Almanza, and M. Terrell, *Foodservice Planning: Layout, Design, and Equipment*, John Wiley & Sons, Inc., New York, 1999.

Type of item: _____   Manufacturer: _____

Model name and number: _____   Name of rater: _____

Instructions: Please indicate your evaluation of this item of equipment by circling the appropriate number for each question. Then total the subscores and final score as shown.

	Very Inferior	Somewhat Inferior	Average	Somewhat Superior	Very Superior
I. Operation of the machine					
A. Machine capacity	1	2	3	4	5
B. Ease of operation	1	2	3	4	5
C. Safety of operation	1	2	3	4	5
D. Quality of product	1	2	3	4	5

X = subscore for dimension I (add A–D) ☐

	Very Inferior	Somewhat Inferior	Average	Somewhat Superior	Very Superior
II. Machine cleaning and maintenance					
A. Amount of cleaning	1	2	3	4	5
B. Ease of cleaning	1	2	3	4	5
C. Frequency of maintenance	1	2	3	4	5
D. Ease of maintenance	1	2	3	4	5

Y = subscore for dimension II (add A–D) ☐

	Very Inferior	Somewhat Inferior	Average	Somewhat Superior	Very Superior
III. Machine reliability and service					
A. Durability and reliability	1	2	3	4	5
B. Energy efficiency	1	2	3	4	5
C. Manufacturer's warranty	1	2	3	4	5
D. Repair service	1	2	3	4	5

Z = subscore for dimension III (add A–D) ☐

Total score (add subscores X, Y, and Z) ☐

FIGURE 4-1 The foodservice equipment evaluation form, when properly completed, will assist operators in comparing similar pieces of equipment.

of Groen, has added a V or versatility factor, which takes into consideration increased food yields by reducing waste and saving on other equipment needs, such as reducing ventilation hood requirements or eliminating certain pieces of equipment. His formula then becomes

$$A + B/C + D + E + F - G - V \geq 1$$

■

TRENDS IN EQUIPMENT

Foodservice kitchen and serving equipment is constantly undergoing change, mostly for the better. The change has been gradual but has influenced such things as kitchen size, work patterns, product quality, production capacity, labor cost, and other savings.

For instance, more multiple-use equipment has appeared, such as tilting pans that can be used as a griddle, a steam-jacketed kettle, a braising kettle, a deep fryer, a thawing oven, and a bain marie, saving both space and cost while making it possible to increase the production capacity for many foods and reduce labor cost. The introduction of the combo-oven, which has the ability to generate its own steam and introduce it into the oven, has reduced the need for larger ovens which must have steam piped into it from an outside source. This oven can even eliminate the need for a thawing oven. It also often improves product quality. Higher inputs of energy into equipment and improved insulation have increased production capacity of cooking units and reduced energy costs.

The almost complete use of solid-state thermal controls in heating equipment today has improved equipment performance. The old thermal controls allowed a + or − variation of 10 to 15°F; the new ones now have only ±1 to 2°F variation. With such control and better insulation on ovens, it is now possible to put a rib roast into the oven, roast it to a desired temperature, then have the oven automatically change to a low holding temperature so that the roast can remain in the oven and cook no further and be at the desired state of doneness when served. We have seen new methods of heat development introduced; induction heating is a method in which part of the heating unit is in the utensil and the other part is in the stovetop. When these two are put together and the electricity is turned on, a magnetic lock locks the two together; heat is developed in both the locked stovetop and utensil elements, giving much more rapid heat input with less energy and far less heat loss than with traditional heating elements. We also have heating units that combine various methods of heating, such as infrared, quartz heating, steam injection, and microwave energy. Ovens are now built with better devices to spread the heat out more evenly. Convection heating moves heat around better and prevents "dead" spots from developing. Reflectors also move heat around, preventing dead spots. We have also made some strides in heat recapture. By putting heat pickup units into flues in which hot substances are being exhausted, it is possible to recapture enough heat to warm a large swimming pool. The heat given off by refrigeration equipment can also be captured and used to heat water. Although progress has been slow in this area, some is being made and in the future, as energy costs rise, will find far more use.

Equipment manufacturers have also introduced more automated features into their equipment, thereby increasing production capacity, eliminating heat loss, and reducing labor requirements. The filtering of deep-fryer fat, formerly a very dangerous and messy process, has been eliminated by automation. The use of the computer to control cooking times, temperatures, and other features has reduced labor needs considerably and given better control. Thus, upon computer signal, a computer-controlled deep fat fryer now lowers the filled basket into the hot fat, fries at a given temperature, and raises the basket at a set time. Eclair pastes can be put into an oven with a preset temperature of 450°F for 8 minutes and then the computer will change the temperature to 350°F, holding it for 30 minutes, and then turn the heat off, signaling the end of the cooking period. As this trend in automation continues, workers will become more button pushers than pushers and shovers. Although automation has made little change in basic production procedures, it does give better control and removes the possibility that a worker might forget to do a task at a prescribed time, make an error in setting a temperature, and so on. Better product quality results from such control.

Technology will continue to affect equipment trends, as in innovation in the use of light in cooking foods. For example, an oven that uses a combination of intense visible light to cook food from within and infrared light to brown the surface of the food cooks omelets in just over 1 minute, pizzas in $1\frac{1}{2}$ minutes, and steaks or chicken breasts in just over 3 minutes. The result is energy efficiency as well as increased productivity. We have also seen the introduction of better tools and small equipment. Knives and other manually used units now have better grips and are more easily maintained in top working order. Much manual equipment is now more cleanable. We also see much more "throw-away-after-one-use" items in the kitchen. Manufacturers give considerable attention to producing equipment that is more cleanable and sanitary.

Every facility exists as a complete system composed of a number of subsystems. The smallest subsystem is a work center, where one worker usually does one type of

TABLE 4-1   Systems Approach: Space Allocations for Various Menu Patterns

Subsystem	Conventional Menu		Modified Convenience Menu		Convenience Menu	
	ft^2	%	ft^2	%	ft^2	%
Receiving	120	3.00	120	3.00	120	3.00
Storage	1120	28.00	900	22.50	700	17.50
Preparation	1100	27.50	450	11.25	300	7.50
Cooking	470	11.75	300	7.50	250	6.25
Serving	800	20.00	800	20.00	800	20.00
Cleanup	390	9.75	350	8.75	350	8.75
Subtotal	4000	100.00	2920	73.00	2520	63.00
Unused space	—	—	1080	27.00	1480	37.00
(Variable total)	4000	100.00	4000	100.00	4000	100.00
Fixed space[a]	2000		2000		2000	
Total	6000		6000		6000	

[a] Fixed space is required for constant needs and does not vary in any operation. Janitorial storage areas, employee toilets, rest rooms, lounge area, and so on, would be located in this space.

work. Work centers are put together to produce sections, such as a bakery section, a cooking section, and so on, forming a larger subsystem. When sections are put together, we have the total system. If one subsystem malfunctions or has the wrong equipment, the entire system functions badly.

A typical foodservice has six subsystems—receiving, storage, preparation, cooking, serving, and cleanup—each with its own space and equipment, equipment arrangement, and space needs to perform adequately the subsystem's needs (see Table 4-1). Cooking may include several sections, such as bakery, meat cookery, garde-manger, and so on. The equipment usually put into each of these subsystems is discussed individually below.

■

RECEIVING EQUIPMENT

Before any food is received, it should be checked for quality and quantity (count or weight or both). It is then moved to storage or to the using area. (In the case of the former, it is included in inventory; if the latter, it is considered a direct charge to food cost for the day unless the kitchen does not use it that day and a kitchen inventory is kept.)

Scales

Scales must give accurate weights down to ounces or ounce fractions. A small scale may be used for the latter purpose, with a scale large enough to weigh the heaviest items for pounds and ounces, although some large scales can now do both jobs, handling weights of a 1000 or more pounds. Many scales also allow an invoice to be stamped showing the time and date of receipt along with the weight. It is best to see that large scales are level with their platform so that items can be rolled onto them without lifting. Factors to watch for in selecting a scale are durability, compactness, convenient design, accuracy, and a finish that promotes easy cleaning and safe use (see Fig. 4-2).

FIGURE 4-2 Receiving-area scale.

Moving Equipment

Dollies, trucks, and other mobile equipment may be needed to move goods. The type selected depends on the kinds and amount of items to be moved. In large operations, an electric or gasoline-driven truck may be used to move things; it may be equipped with an elevator to load or unload items from stacks. Pallet storage is popular in large operations. The shelving used for storage should be suited to the receiving equipment used.

Miscellaneous Receiving Equipment

Computers, desks, chairs, and file cabinets are also needed for some receiving subsystems. The specific items needed depend on the particular nature of each facility.

■ STORAGE AREAS

Adequate storage capacity—both dry and cold—is necessary to protect food, ensure proper sanitation, and maintain high food standards. The storage area provided should be designed to receive and hold the specific foods stored in it. Some foods are stored for longer periods of time than others. Fresh foods turn over faster than processed ones. Equipment used to move foods, shelving, and containers must be selected for the specific foods stored. Planning storage facilities prior to planning the menu usually leads to inadequate facilities.

Both dry and cold storage should be located near the receiving dock and the preparation and cooking subsystems. The flow of materials should resemble the pattern shown in Figure 4-3.

Refrigeration and Low-Temperature Equipment

Refrigeration and low-temperature equipment exist to minimize the deterioration of stored foods and to reduce the likelihood of contamination. This equipment keeps food cool or frozen to preserve flavor, color, texture, and nutritional elements. Chilled foods remain good much longer than foods held at room temperatures, and frozen foods keep much longer still. During the past several decades, foodservices have found it necessary to increase the amount of low-temperature storage provided and to reduce slightly the refrigerated areas. Refrigerators hold foods at temperatures of from

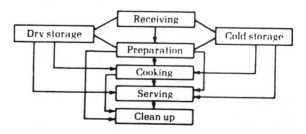

FIGURE 4-3 Flowchart of movement of materials from the receiving subsystem to the remaining subsystems of the foodservice operation.

about 34 to 40°F (1 to 5°C), whereas low-temperature units hold foods at temperatures of from about 10 to –10°F (–12 to –23°C). Units offering these temperatures come in a variety of types, including reach-in, roll-in, and walk-in models (see Fig. 4-4).

Four factors should be considered in comparing different refrigerated storage units: the ability of the unit to hold the relative humidity at 85 percent (to keep foods from drying out); the ability of the unit to control air distribution so that an even temperature is maintained throughout the unit, regardless of loading pattern; precise temperature control within a range necessary to protect the food and hold its quality; and compliance with the design requirements of the National Sanitation Foundation (NSF), including lack of seams, easily cleaned surfaces, and removable parts in interiors to help in cleaning.

Specialized cold-storage equipment may be required, such as quick-chill refrigerators that rapidly reduce the temperature of cooked foods to a safe temperature for storage and blast freezers that freeze foods quickly at –30 to –40°F (–34 to –40°C). Controlled thawing units offer sanitary and convenient overnight thawing of bulk frozen foods. Many operations today use ice makers. The ice may be block, small or large cube, or crushed. The capacity of these specialized units must be matched to the production needs of the operation.

Shelving and Containers

Shelving and storage containers are important elements in a good storage system. Properly sized adjustable shelving can increase storage space, and easy shelf removal is of primary importance in facilitating good sanitation. Considerations to bear in mind when selecting shelving or mobile storage equipment (such as carts, dollies, and roll-in racks) include suitability to need, strength, sanitation, safety, size, number needed, and cost. These factors are also applicable to the selection of storage containers; shape and capacity are additional factors to consider in choosing containers.

■

PREPARATION EQUIPMENT

Tasks such as vegetable and fruit cleaning and cutting, salad preparation, and meat cutting or trimming can be performed speedily and efficiently when the right equipment is properly used in the preparation area. Since the extent to which preparation or fabrication of food takes place on the premises depends on the basic menu pattern, most operations using conventional foods allocate considerably more equipment and space to preparation than do those using convenience foods. If conventional foods are used, efficient and effective food preparation must be achieved through proper equipment selection, location, and use.

Scales

Scales may be a necessary part of the preparation subsystem. Weighing food as it is removed from raw storage (especially in ingredient rooms) is most important. Good accounting records cannot be maintained unless this is done. Portion scales, flat-counter dial scales, and even large scales onto which items are rolled may be used. Accuracy is an important factor in selection as well as ease of use and cleaning. Scales

FIGURE 4-4　Cold-storage equipment: a roll-in unit.

a.   b.   c.

FIGURE 4-5 Food mixers: (*a*) counter model (5-qt capacity); (*b*) bench model (20-qt capacity); (*c*) floor model (60-qt capacity).

are also important in the cooking area for portioning and for ensuring that the proper amount of each ingredient is added to items being prepared.

Food Mixers

The upright food mixer is available in counter, bench, and floor models of different sizes and capacities (see Fig. 4-5). Mixer sizes vary from 5 to many quarts. Mashed potatoes, doughs, cake batters, whipped cream, icings, meringues, mayonnaise, and other products are prepared in them. Various agitators and attachments are available to speed the preparation of virtually every food item placed in a mixer (see Fig. 4-6). Horizontal mixers with interior rotating paddles are used for large-quantity production of such items as breads and cookies. Small units usually do not need them. Many mixers can handle various additional tasks, too. Attachments can be added to perform slicing, grating, grinding, or sausage extrusion.

Food Cutters

One type of food cutter, known as the Buffalo chopper, consists of a rotating bowl that moves food into the path of a spinning blade (see Fig. 4-7). As the food passes repeatedly through the blades, it is chopped into progressively smaller pieces. The longer the machine is allowed to run, the smaller the particles become. The Buffalo chopper's basic use is for preparing many items, such as chopped onions and cabbage, cheese for toppings, and meat trimmings for croquettes. It is easy to clean and contains no cracks or crevices that could harbor bacteria. Attachments are available for cutting, slicing, grating, shredding, and similar operations.

Another cutter, called the Qualheim cutter, chops, dices, makes strips, and does other work that saves much preparation time. The unit should be used only when there is a large quantity of work to do, since cleanup and reassembly times may be quite lengthy.

The vertical cutter and mixer (known as the VCM) chops, cuts, mixes, blends, emulsifies, purées,

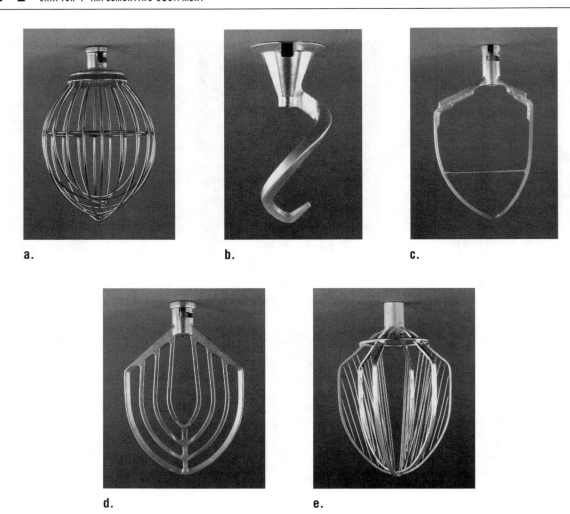

FIGURE 4-6 Commonly used agitators: (*a*) heavy-duty wire whip; (*b*) dough arm; (*c*) pastry knife; (*d*) flat beater; (*e*) wing whip.

and homogenizes foods in a matter of seconds. It works much as a home blender does; the chopping is performed by two knife blades that come up through the bottom of a deep bowl. A baffle extending from the bowl's lid is used for mixing and ensures that items are properly mixed and uniformly cut. The VCM prepares salads and cole slaws, emulsifies salad dressings, mixes batters and doughs, blends meats, chops cheese, and cuts bread crumbs. The operation may be completed in as short a time as it takes to switch the machine on and then off again. It has few

movable parts and is easy to clean. The machine is easily and quickly cleaned by adding water and a cleaning solution to the bowl, flicking the switch on momentarily, draining the wash solution, and rinsing the equipment with a hose as needed. The VCM tilts to facilitate emptying of the bowl. It is available in various sizes. A 25-qt model is sufficient for preparing up to 500 portions, and the 40-qt model is appropriate for preparing more portions than that. Large commissaries use 60-, 80-, or 130-qt sizes (see Fig. 4-8).

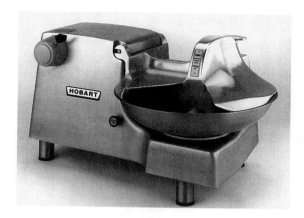

FIGURE 4-7 Food cutter.

FIGURE 4-8 Vertical cutter and mixer.

Similar to the VCM, but considerably smaller and more versatile, is the food processor. It consists of a bowl sitting on top of a small, powerful motor. The motor has the ability to drive any of a number of attachments at very high speed. With the blade attachment, objects can be puréed in a matter of seconds. Grating and slicing attachments are available as well. Batches to be processed can be as small as 1 cup and as large as 1 gallon. The food processor's small size makes it highly portable, and its ability to perform a variety of jobs has contributed to its popularity in many of today's operations.

and few removable parts, all of which can be disassembled easily for thorough cleaning (see Fig. 4-9).

Careful prescheduling of the slicer can greatly reduce required cleanup time. For example, prescheduling the slicer would mean cutting such items as celery, cabbage, carrots, and onions first, then progressing to slicing fruits and juicy vegetables, then going on to cheeses and cold meats, and finishing by

Food Slicer

The slicer is basically a circular knife on which items such as cheese, boneless meats, luncheon meats, vegetables, and breads can be sliced. A uniform, clean, straight slice of almost any reasonably firm product is possible with this piece of equipment. The item to be sliced glides back and forth on a carriage feeding into the knife. By adjusting the distance between the plate on which the product rests and the knife itself, the operator can adjust the thickness of the slice as desired. Slicers are available in different sizes, depending on the size of the knife blade. This machine should be obtained in the smallest size that can accommodate the largest item that must be sliced at the foodservice. Safety features should include guards around the blade

FIGURE 4-9 Food slicer.

slicing hot meats and other cooked items. This sort of scheduling allows minimizing cleaning of the machine between each food group; more efficient use of the slicer results, Of course, proper cleaning of the slicer between uses must always be enforced. If more than one person or production area wishes to use the slicer, the machine should be located centrally or placed on a portable cart for easy relocation.

Vegetable Peeler

The vegetable peeler is used when a quantity of vegetables or other hard items must be peeled. Hard root vegetables such as turnips, carrots, and potatoes are rapidly peeled by an abrasive, lightweight disk that spins around, removing the skins. Water flowing into the chamber removes waste as it accumulates. Peelers can handle from 12 to 70 lb in a single batch. A 15-lb machine peels 15 lb of potatoes in 1 to 3 minutes (see Fig. 4-10).

Meat-Processing Equipment

Meat-processing equipment includes meat saws, choppers, grinders, cubers, and tenderizes (see Fig. 4-11).

a.

b.

c.

FIGURE 4-10 Vegetable peeler.

FIGURE 4-11 Meat-processing equipment: (*a*) meat saw; (*b*) mixer/grinder; (*c*) tenderizer.

For operations that purchase meats already prepared, such equipment is not required. But some operations still cut meats from wholesale cuts or larger pieces of meat.

Miscellaneous Preparation Equipment

Other preparation equipment may be needed, as well—blenders, breading machines, can openers, coffee grinders, knife sharpeners, mobile carts, work tables (with locking casters, if mobile), and refrigeration equipment. A proper-sized waste disposal or waste collection system must be provided to ensure adequate sanitation and safety.

■
COOKING EQUIPMENT

The cooking and baking subsystem is the focal point of the food production system. The receiving, storage, preparation, and cleanup subsystems must adequately support the cooking and baking subsystem; therefore, equipment selected for these support subsystems must be designed with the needs of the cooking and baking subsystem in mind.

Ovens

Ovens in foodservices may be as large as several hundred square feet and able to cook food for thousands of portions at one time, or they may be as small as 2 ft² and able to cook only a few portions at a time. There are 6 different types of ovens: conventional, mechanical, convection, combo, microwave, and reconstituting.

Conventional Ovens

A typical conventional oven is heated by a lower heat source—and perhaps an overhead source as well—in an enclosed chamber. Transfer of heat occurs both by convection (of hot air moving in the chamber) and by conduction (when pans or other equipment in it

come into contact with hot surfaces). Variations on the conventional oven include the peel oven (for baking extra-crusty breads), the drawplate oven, and the deck oven (see Fig. 4-12). The deck oven is the most common of these. It consists of a stack of ovens built in decks or tiers, one on top of another. Small ovens situated under range tops may be the only ones used in small operations.

Mechanical Ovens

The mechanical oven evolved from the conventional one. It has a mechanically moving interior that shifts the foods around to different parts of the oven (see Fig. 4-13). One variety is the revolving tray or reel oven, which has trays that rotate in a circle around the oven interior. Another variety uses a moving or traveling tray, which passes through a long oven; different temperature ranges are possible within the long area traveled. The shelving on which foods are placed is power-driven and constantly moving. This arrangement offers several advantages: an even distribution of heat; lower costs (because of better heat distribution and more products baked per unit of heat used); and better baking results in terms of volume, color, and texture. Waiting time, temperature recovery time, and goods-shifting time are all said to be lower when these ovens are used—thus reducing costs. Many operations, however, do not have the volume to support these large ovens. The revolving-tray oven holds up to twelve standard-sized pans (18 in. × 26 in.) and the traveling tray ovens hold twenty-four such pans or more.

Another type of mechanical oven is the conveyor oven. A moving belt or track carries an uncooked item through a tunnel containing heating elements, and the item emerges from the oven as a finished product. Food can pass through as many as three different temperature zones—for preheating, cooking, and finishing. This oven is geared toward the high-volume operation that has a high peak demand, as exists in many pizza operations.

Convection Ovens

Many conventional ovens have stagnant heat areas where heat does not move. If heat is moved around by forced convection, cooking is more rapid, less heat is

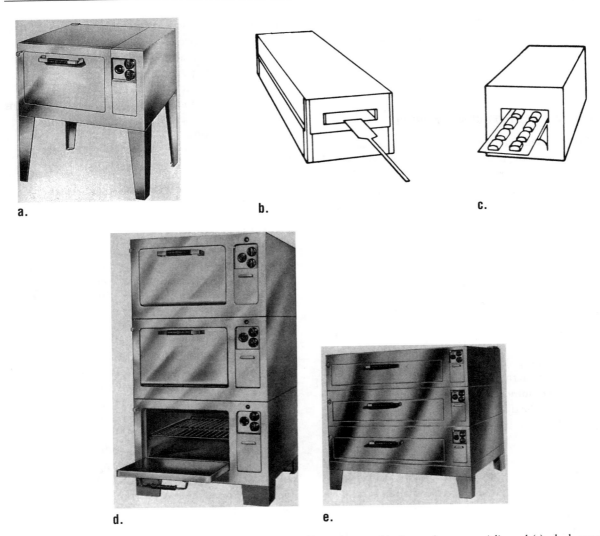

a. b. c.

d. e.

FIGURE 4-12 Conventional ovens; (*a*) all-purpose oven; (*b*) peel oven; (*c*) draw-plate oven; (*d*) and (*e*) deck ovens.

needed, and more food can be put into the oven. The convection oven—a conventional oven in which the heat is moved by a fan—allows these results to be achieved. Because of the movement of heat, the convection oven bakes or roasts at 50°F (10°C) lower, with a 25 to 30 percent reduction in cooking time. The amount of food that can be cooked per cubic foot of space is also increased.

Combo-ovens

The combo-oven gets its name because it cooks both with dry heat and with steam heat, making it possible to have it act as a dry heat oven, an oven that has enough steam to bake hard-crusted breads, or to act as a steamer. It also can act as a cook-and-hold unit. It produces a large quantity of food for the space it takes,

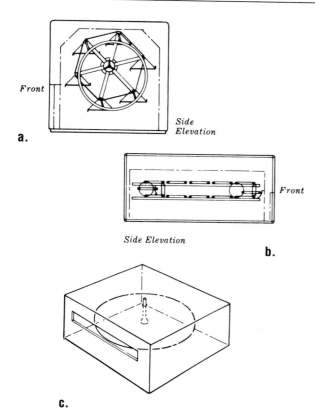

a.

b.

Front

Side Elevation

Side Elevation

Front

c.

FIGURE 4-13 Mechanical ovens: (*a*) revolving-tray oven; (*b*) traveling-tray oven; (*c*) rotary oven.

is energy efficient, and can reduce food shrinkage in baking because moisture can be introduced. Cooking times are also said to be somewhat shorter. Some facilities also use them as thawing ovens. Because they use solid-state heat control temperatures can be held to within narrow limits (±2°F for electricity, ±4°F for gas). Fans circulate heat, which gives them the same advantage that convection ovens have (see Fig. 4-14). They use either gas or electricity, as specified when ordering.

Microwave Ovens

Microwaves are a form of electromagnetic energy intermediate in frequency and wavelength between radio and infrared waves. When microwaves penetrate food, molecular activity or movement takes place

within the food, creating friction that heats the food internally. Metallic materials reflect microwaves and should not be used, as they can reflect microwaves back to the microwave tube and destroy it. Microwaves pass through glass and nonmetallic materials, so these materials are used to hold foods being cooked in a microwave oven.

Microwave cooking has not replaced conventional cooking in foodservice operations because they do not allow production of large volumes of food, and because they produce undesirable cooking reactions in some foods. The speed of cooking is often so fast that foods do not respond to heat as they do in other forms of cooking. When it is appropriately used, the unit can bring food from a refrigerated or frozen state to a servable hot state in a few minutes, thus allowing operations to prepare foods on order. The unit also has flexibility. With the proper adjustment of energy input, foods can be thawed, warmed, heated, cooked, or baked in a matter of a few seconds or minutes. It is especially useful in reconditioning or cooking small units, such as individually quick-frozen foods (foods in individual portions that are rapidly frozen at extremely low temperatures). When properly integrated into the

FIGURE 4-14 Convection combo steamer–oven. (Courtesy of Groen, A Dover Industries Company.)

food-production and food-service plan, it can do much to reduce costs and produce foods of high quality.

Reconstituting Ovens

Since many frozen foods are used today in foodservices, there is a major need for a unit that can quickly defrost them without damaging their quality. The reconstituting oven can bring −10°F (−23°C) items to a serving temperature in less than 30 minutes. Some units combine cycles of heat and refrigeration, which reduces drying on the edges while gently thawing the center. Other units use infrared waves and may reach temperatures as

high as 850°F (454°C) to bring foods from thawing to serving temperatures. Such energy is called *radiation* because the heat is transferred by infrared waves.

Ranges

The range, a combination of an oven and top-cooking units, was invented by the Chinese more than 5000 years ago. Modern ranges, heated either by gas or by electricity, are common in today's foodservices. The arrangement of top units may be open, closed, or some combination of the two (see Fig. 4-15). The menu of

FIGURE 4-15 Ranges and range-top combinations: (*a*) convection oven base with three rectangular hot plates on top; (*b*) rectangular and high-speed French hot plates on top; (*c*) griddle on top; (*d*) electrical coil units on top; (*e*) backshelf broiler or salamander attachable to the back of the range.

the operation dictates whether a heavyweight, medium-weight, or lightweight model is necessary. The selling price of the range should not be the deciding factor in determining which type of equipment to purchase. Many operations select light equipment because of its lower price, only to find that these units are not suited to the type of production needed. Instead it should be selected on the basis of its capacity, versatility, consistency of temperature, cleanability, serviceability, and dependability.

Griddles

Foodservices use either gas or electric griddles. They can come incorporated as part of a range, or they may stand alone on a platform. Food is cooked on the griddle surface in a small amount of fat, much as it would be cooked in a fry or sauté pan. Griddles are used to prepare large quantities of fried products quickly and with a minimum of labor. Splash guards contain grease on the griddle surface; they should be of appropriate height, and the grease troughs must be wide and deep enough to capture excess fat and debris. Griddles can be nicked or scratched, which causes food to stick on the cooking surface. Consequently, care should be taken when using metal utensils, and pots or pans should never be placed on the griddle.

Broilers

Broilers use radiated heat energy to cook. This is the same kind of energy with which the sun heats the earth. Short energy waves, closely associated with light, result from white heat.

Most broilers broil from heat that comes from above, but bottom or side broilers are also found. Heating from the top allows fat to drip down into collectors without flaming. The underheat broiler does not have this advantage, but it often yields a better charred product—one that also shows the markings from the broiler. Side broilers are good because the product can baste itself while broiling. Whatever broiler is used, the flames from it must not be allowed to burn fat. This could cause the formation of carcinogenic (cancer-producing) substances that could get onto the food. Broilers should be kept very clean since greasy ones can catch fire.

Steam-Cooking Equipment

Steam-cooking equipment increases efficiency in many kitchens by reducing cooking times and decreasing the amount of work required to produce certain items (see Fig. 4-16). Steam imparts heat rapidly and is economical to use. Because it does not burn food, steam

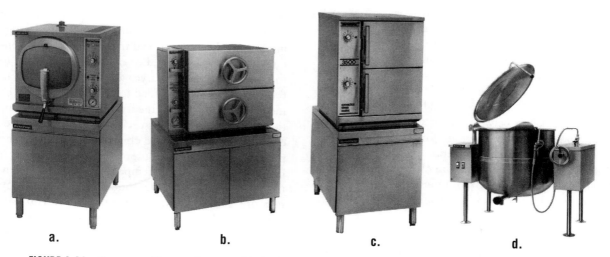

a.	b.	c.	d.

FIGURE 4-16 Steam cooking equipment: (*a*) high-compression compartment unit; (*b*) low-compression compartment unit; (*c*) convection steamer; (*d*) tilting steam-jacketed kettle.

cooking eliminates food scorching and the need for extensive pot-and-pan cleanup. Steam equipment may be self-contained, manufacturing its own steam, or it may receive steam from a central broiler.

Steam Cookers

Some units bring steam into a chamber where it comes directly into contact with the food. The steam vapor circulates in the chamber under pressure, in sufficient volume for good heat distribution. Low-compression compartmented units operating at 5 to 7 psi (pounds per square inch) are used for high-volume production. Some hold up to six 12 in. × 20 in. pans in each compartment. One-, two-, and three-compartment models are available. High-compression, compartmented units operating at 15 to 17 psi use smaller pans but cook much more rapidly. Small quantities can be cooked very quickly, resulting in a high-quality, freshly cooked product. One-compartment models are available as countertop units.

A convection steamer forces steam to circulate in a compartment under normal pressure. The circulating steam disrupts a vapor barrier that is created around cooking food, thereby decreasing cooking time. Since pressure does not build up, the door of the compartment can be opened at any time. Single-compartment capacity for a convection steamer is three 12 in. × 20 in. × 2.5 in. pans.

Steamers work well for many types of foods, with the exception of cakes and pastries. They can be of great use in defrosting frozen goods; covering the food prevents condensing steam from mixing with the product. After defrosting, the food can then undergo normal preparation. If possible, cooking different-flavored foods together should be avoided. Steam transfers flavors between foods, sometimes producing an undesirable combination of tastes. Perforated pans can be used to aid the circulation of steam, again shortening cooking times. The size of steamer to purchase depends on the maximum number of portions to be served and the portion size required per batch.

Steam-Jacketed Kettles

The steam-jacketed kettle works much as a double boiler does. Steam is generated and surrounds food contained in a separate compartment. But whereas the steam in the double boiler is not under pressure, the steam in a steam-jacketed kettle is, resulting in higher temperatures. Because the contact surface of the kettle is not at a high heat and because the heat is evenly distributed, food does not scorch easily. Nonetheless, browning meats and performing certain similar processes are possible in these kettles.

Capacities range from 8 quarts to 150 gallons. Steam kettles with capacities of 3000 or 4000 gallons are found in central commissaries. Rated capacities must be reduced by 15 to 30 percent to determine actual usable volume. Both stationary kettles, having a standard 1½-in. draw-off valve, and tilting kettles, designed to allow the transfer of foods to smaller vessels, are available. Cold-water cooling systems that inject cold water into the kettle jacket to cool hot items rapidly are optional. Some kettles are equipped with pumps so that liquid foods can be removed without having a worker do it.

The kettle should be thoroughly cleaned after each use. Bringing water to a boil in a dirty kettle can help loosen stuck-on food for easier cleaning. Stainless steel kettles are more common than aluminum ones; the latter type should not be scoured hard with abrasive materials.

Fryers

Deep-fat fryers are used to cook food in a bath of hot fat, producing a nicely browned, crisp outer coating with a nutty flavor and a completely cooked, moist interior. In the conventional fryer, fat is used to conduct heat from a gas-fired or electrical heat source to the food. The pressure fryer also cooks in deep fat, trapping moisture from food to generate steam, which increases pressure inside the fryer and reduces cooking times. This has proved especially successful for producing tender, moist fried chicken. Both regular and pressure fryers may have automatic, semiautomatic, or hand-operated features. Temperature control is essential, so a precise thermostat should be specified. Temperature recovery must be fast and accurate. Auxiliary safety controls—such as automatic turn off when the temperature of the fat exceeds 450°F (232°C)—may be needed.

The ratio of food to frying oil varies with the food item. Potatoes can usually be fried with a 1:6 ratio of potatoes to fat, but most other products are fried with a 1:8 ratio. Thus, a 15-lb fryer (holding 15 lb of fat) can fry 2½ lb of potatoes but only about 2 lb of other foods. In selecting the right fryer, the operation's basic menu pattern must be considered, along with the size of the food portion, the number of portions required per hour, and the amount of food the fryer can produce in an hour. On average, it is better to have two relatively small fry kettles than one large one, since only one small one need be used during off-peak periods; in this way, savings can be realized on heat and on reduced deterioration of frying oil.

In response to demand from high-volume operators, a conveyor fryer has become available. Food is placed on one end of a conveyor belt, dipped down into hot oil, cooked, brought back out, drained, and dumped into a holding receptacle. This makes possible a continuous flow of hot fried food during periods of peak demand.

Tilting Pans or Skillets

One of the most versatile pieces of equipment in the kitchen is the tilting pan or skillet. It can do the jobs of a range top, griddle, small steam kettle, stock pot, fry pan, and skillet. The tilting pan comes in various sizes. A large, heavy-duty pan has a depth of about 8 or 9 in. (see Fig. 4-17). It can tilt up to 90° on a horizontal axis. Stewing, simmering, frying, searing, braising, grilling, sautéing, boiling, defrosting, and roasting can be done in one. A great deal of heavy lifting and transferring of foods from one pan to another can be eliminated by this unit. Temperatures to use for best results are the following:

- *Simmering:* maximum of 200°F (93°C)
- *Sautéing:* 225–275°F (107–134°C)
- *Searing:* 300–350°F (149–177°C)
- *Frying:* 325–375°F (163–191°C)
- *Grilling:* 350–425°F (177–218°C)

Models may be mounted on legs, on the wall, on counters, on modular cabinet bases, or on casters. The tilting pan is easily cleaned with a mild detergent. Presoaking is recommended, if possible.

FIGURE 4-17 Heavy-duty tilting pan. (Courtesy of Groen, A Dover Industries Company.)

Cook–Chill Equipment

The foodservice system is one of the few businesses that still operates under the old guild system; it produces, sells, and services its products under one roof. The functions are not separated. This cannot be avoided. However, a partial breakthrough has been made with the introduction of value-added foods. Another partial method has been the introduction of the cook–chill system, where foods are mass produced in the individual unit, stored, and partial lots are later taken out for service. Schools, hospitals, correctional facilities, supermarkets, restaurants, hotels, and many other institutions now use this cook–chill method, maintaining consistent and satisfactory food quality while reducing food, energy, and labor costs.

The system takes kettle-cooked foods, seals them in tough, pliable casings at near-pasteurization temperatures, cools them rapidly to 40°F, and then holds them at just above freezing temperatures up to about 45 days. Or, foods such as poultry or meats can be sealed in tough plastic bags with seasonings and placed in a tank of circulating hot water bath to cook slowly over a period of hours. These bags are then rapidly cooled down and stored. Thus, when the cook–chill method is used, it is possible at one time to produce in much larger quantity a batch of food sufficient not only for one meal but for a number of meals. These foods can then be withdrawn from their refrigerated inventory in quantities sufficient to meet the immediate need, brought to the desired temperature, and served.

The system usually requires that kitchens be divided into two separated functioning units, one to mass produce the inventory and the other to withdraw a required day's or meal's supply, condition it, and serve it. Often, it is possible to take the ordinary kitchen and rearrange it quickly to produce the two

a.

c.

b.

d.

FIGURE 4-18 The CapKold Cook-To-Inventory process of cook–chill equipment: (*a*) inclined agitator kettle with pump on the bottom; (*b*) combination tumble chiller and cook tank; (*c*) mobile pump/fill station; (*d*) batch tumble chiller. (Courtesy of Groen, A Dover Industries Company.)

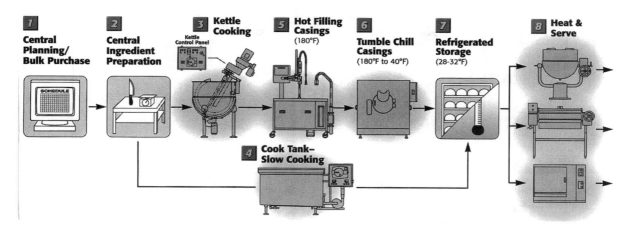

FIGURE 4-19 CapKold Cook-To-Inventory process.

units. Sometimes, larger equipment must be installed, and more refrigerated space is usually needed.

Equipment manufacturers have designed special equipment for this. Figure 4-18 shows an arrangement of such equipment manufactured by Groen, making up what they call their CapKold System. It has wide use. The upper line shows food being ordered and prepared for processing as normally done. It then is tank-, or kettle-cooked (see Fig. 4-18*a*). Upon the completion of cooking, the kettle uses a pump on the bottom to send the food to a portioner (see Fig. 4-18*b*) which puts the very hot food into plastic bags. The sealed bags then go to a chill tumbler (see Fig. 4-18*d*), which in an hour can bring a batch of bags down to 40°F. The batch then goes into refrigerated inventory, which is withdrawn in quantities as needed. Reconditioning can be done using steam kettles, steamers, microwave ovens, regular ovens, and so on. Figure 4-19 shows that food in sealed plastic bags can be put into a cook tank, where it slowly cooks in a circulating hot water bath until done. Chilling of these bags can be done in the chill tumbler and then handled from then on in the same manner as the kettle-cooked foods.

Small Tools for the Cooking Area

Knives, tools, utensils, pots, pans, and other small equipment are required in the cooking area as well as in other sections. These items must be selected carefully for durability, suitability for the job to be performed,

and cost. Good steel is essential in many cutting tools so that they hold a good edge and wear well (see Fig. 4-20). Well-designed handles on utensils and tools help relieve hand fatigue and increase efficiency (see Fig. 4-21). Working parts should be firmly fastened to grasps or handles. Weight must be considered in selecting pots and pans (see Fig. 4-22).

■

HOLDING AND SERVING EQUIPMENT

The serving subsystem handles the assembly of food, the holding of food (if necessary) for a short period of time, the arrangement of food on plates and dishes, and the presentation of food to the guest. Thus, the serving subsystem connects the heart of the house with the front of the house. Prior to service, food must be held at the proper temperatures to reduce deterioration and to allow the food to reach guests at its correct serving temperature. Success depends on having the right kind of serving equipment. Tableware, serviceware, napery, decorative items, and uniforms for service personnel are also matters that fall under the purview of this subsystem.

Holding Equipment for Hot Food

Hot foods should be held at temperatures of from 140 to 190°F (60 to 88°C). Each food has its own holding

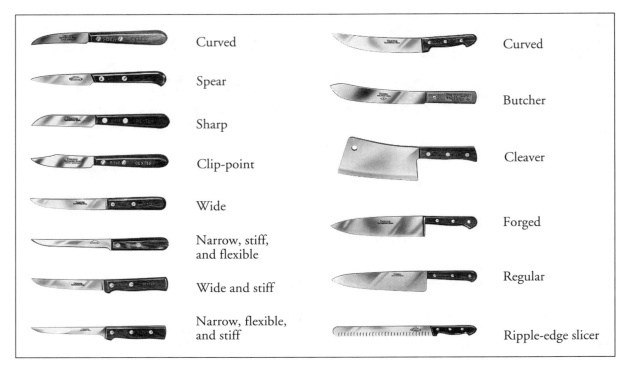

Curved

Spear

Sharp

Clip-point

Wide

Narrow, stiff,
and flexible

Wide and stiff

Narrow, flexible,
and stiff

Curved

Butcher

Cleaver

Forged

Regular

Ripple-edge slicer

FIGURE 4-20 Knives. (Courtesy of Russell Harrington Cutlery)

Broiler fork

Carver fork

Cook's fork

Cookie turner

Pie knife

Chef's steel

Sandwich spreader

Pizza cutter

Pan scraper

Dough cutter

Hamburger
or cake turner

FIGURE 4-21 Utensils and tools. (Courtesy of Lincoln Foodservice Products, Inc.)

Seasoning shakers Scoops Colanders Measures Cooking utensils—ladles, spoons, and skimmers

Graters and slicers China cap Tomato slicer Cheese slicer Cheese slicer

Onion petal slicer Can opener Apple cutter Lettuce slicer Potato slicer

FIGURE 4-21 Utensils and tools (*continued*)

temperature. Entrée and meat dishes usually are held at around 140°F (60°C), sauces and gravies at slightly higher temperatures, and thin soups, beverages, and some other thin liquids at 180 to 190°F (82 to 88°C). Health authorities require hot food to be held at temperatures above 140°F (60°C). Holding equipment for hot foods includes steam tables, pass-through holding units (located between preparation/cooking areas and the service area), undercounter warming drawers, bains marie, stationary or mobile heated cabinets, and in-frared warmers. Some holding cabinets are designed to display the cooked product while in use.

Holding Equipment for Cold Food

To reduce the growth of undesirable microorganisms, cold food should never be held at temperatures above 45°F (7°C). Cold foods should be served at around 40°F (4.5°C), and frozen foods should be served at around 24°F (−4.4°C). Undercounter refrigeration units, pass-through refrigeration systems, mobile refrigerators and freezer cabinets, and refrigerated display cases and storage drawers are among the many pieces that keep foods cold. Again, many units can be used to display and merchandise food, as well as to keep it at the proper temperature (see Fig. 4-23).

Sauce pans

Loaf pan

Sheet pans

Pie tins

Muffin tins

Covered roaster

Sauté pans

Fry pans

Double boilers

Egg pan

Round cake pans

Bain-marie pots

Stock pots

Stock pots with spigots

Roast pans

FIGURE 4-22 Pots and pans. (Courtesy of Lincoln Foodservice Products, Inc.)

FIGURE 4-23 Food-holding equipment.

Serving Equipment

Chafing dishes, mobile carts, and display cases are a few examples of holding equipment that may be used for service. Frequently, holding equipment decorated to please the customer's eye is used to merchandise the product. This is especially true in cafeteria and buffet lines, where guests make their own selections. Hot carts may be used to serve roasted meats and other hot foods. Cold carts may be used to display and sell hors d'oeuvres, salads, or desserts. Other equipment, such as a microwave oven, a reconstituting oven, or a steam cooker, is used for last-minute preparation or cooking.

Coffee Machines

Coffee is a popular beverage and should be prepared properly with the right equipment. Many kinds of units are used successfully.

Ice Machines

Ice machines are often installed in the serving subsystem's area to supply ice for such uses as drinking water, bar and beverage service, salads, desserts, and displays of cold foods. The ice produced can be in the form of cubes, cubelets, chips, or flakes. Machines are sized and designed as floor-mounted, wall-mounted, compact, or decorator models. The correct model for any particular operation depends on how the ice is used and how much is needed in any one 24-hour period. Among the more popular units are space-saving stacked units, continuous-flow units (machines under which portable bins can be wheeled), and portion-controlled ice maker/water dispenser units.

Beverage Dispensers

Beverage dispensers may dispense milk, punches and fruit-flavored drinks, carbonated beverages, juices, frozen carbonated soft drinks, and soft ice cream shakes. These units, fully automatic or semiautomatic, are easy to operate and require minimal labor input for a fast output of drinks. A wide range of models is available to meet almost any need. Units can be obtained for self-service or behind-the-counter service. All should be appealing and easily cleaned.

■

CLEANUP

Cleaning up is an ongoing process that begins before each meal and continues throughout preparation, cooking, and service; it is one of the most important subsystems in the total foodservice system, since it can be a determining factor in the success or failure of the operation. High labor costs, excessive breakage and loss of ware, undue maintenance, and poor employee morale can result if the cleanup section is not properly equipped and designed. Customers, too, can be unhappy with poorly maintained equipment.

Waste-Removal Equipment

Foodservice operations that prepare and serve large quantities of food generate a considerable amount of waste. Such waste should be easily and quickly removed as it accumulates. Waste and garbage cans often are used for this purpose. If so, they should be thoroughly cleaned each time they are emptied. The use of plastic liners can help make this a simpler process.

Disposers

A disposer can accept food wastes but not cans, bottles, or other hard materials. If only one disposer is used, it should be centrally located near areas where the greatest amount of waste is produced. Smaller units can be put into the individual kitchen sections to save steps, but ready access to disposers can encourage careless employees to waste materials. Disposers should be sized to meet the operation's needs. They should bypass rather than be hooked up to grease traps, since refuse going through the grease trap can clog it. Disposers are frequently used as part of the dishwashing system.

Compactors

A compactor—a box unit into which waste materials are fed—can accept hard materials as well as soft ones. It works by crushing or breaking. A hydraulic ram then compacts the crushed material into a solid mass. Containers holding the compacted materials can then be removed and thrown away. Volume of waste is reduced by up to 98 percent. The compactor usually is located on the loading dock, although portable models are available for movement from one location to another.

Pulpers

A pulper reduces waste by-products into a neat, semidry pulp that is odorless and free of vermin (see Fig. 4-24). Garbage volume is reduced by as much as

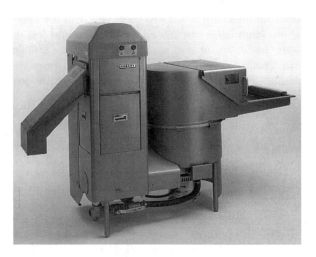

FIGURE 4-24 Pulper.

85 percent as the wastes are pulped underwater, converted into a slurry, and then rendered into a semidry pulp when the water is removed. Waste can be piped in its slurry state to a remote pickup point, eliminating the need for any physical handling. The quantity of waste determines the size of the original pulping unit required. There should be one unit at each of several points of waste generation. Pulpers will not accept metal, glass, crockery, or other hard items. A separate means should be provided for the disposal of such materials.

Dishwashing Equipment

The cleaning of serviceware—china, glass, and silver—is critical to any operation. Customers and health inspectors alike respond poorly to dirty dishes and utensils. To purchase a dishwasher of the proper size requires analysis of the number of pieces of ware to be processed each hour. Manufacturers overstate the capacity of washing machines by 25 to 33 percent. Five different kinds of washing machines are used in foodservices: counter and undercounter models, freestanding models, door models, rack conveyor models, and belt conveyor models (see Fig. 4-25).

Dishwashing machines come in one or more sections, usually referred to as *tanks*. A basic one-tank model has two cycles: a wash and a rinse. A two-tank model washes in the first tank and power-rinses in the second. A three-tank model uses one tank for a prewash, another for a wash, and the third for a rinse. Some larger models have a fourth cycle for the final rinse. A blower dryer is an optional unit on some larger models; it serves to speed drying, especially when dishware comes off a short belt at the end and must be removed before much time has been expended on drying.

In many models, the prewash and wash tanks are filled, but water flowing from the rinse section moves down into them, providing fresh water as old water is drained away. Most dishwashers have automatic temperature controls. Boosters are used to increase water temperature where and when needed. Prewash units usually operate at around 140 to 150°F (60 to 66°C); wash units, at 160°F (71°C); and the rinse, at 180°F (82°C) for 10 seconds or at 170°F (77°C) for

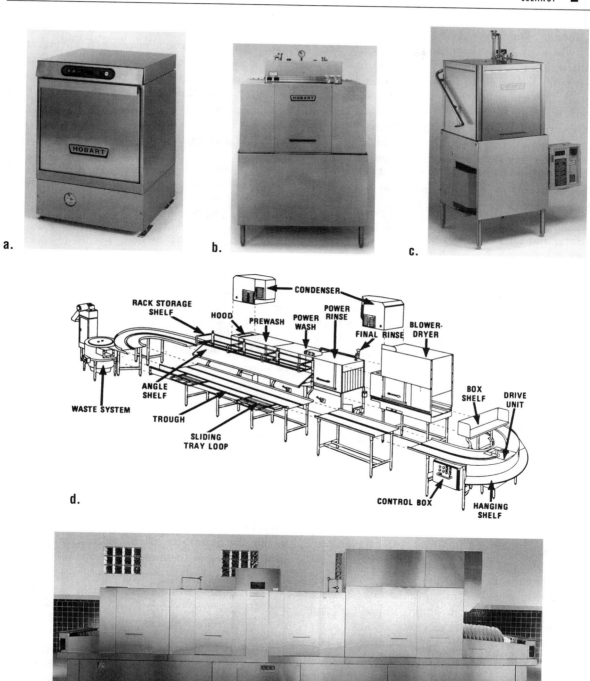

FIGURE 4-25 Dishwashing equipment: (*a*) undercounter model; (*b*) freestanding model; (*c*) door model; (*d*) rack conveyor model; (*e*) belt conveyor model.

30 seconds. Too hot a prewash or wash bakes on foods. In some of the newer low-temperature dishwashers, rinsing is done with chemical sanitizers at a water temperature of 140°F (60°C). Since this is the normal temperature of hot tap water, a booster heater (which is required by the 170 to 180°F machines) is unnecessary, allowing energy savings. Many machines have automatic detergent dispensers, which add detergent as needed, and automatic rinse detergent injectors. Rinse detergent is put into the final rinse water to give a better runoff of water for faster drying and reduced water spotting.

Counter and Undercounter Models

The smallest machines used are counter and undercounter dishwashers that are 24 and 30 in. wide. Manufacturers claim that the 24-in. model handles 300 plates or 540 glasses, plus silverware, per hour. The 30-in. model cleans an additional 75 plates or 60 glasses, plus additional silverware, per hour. These models are used in small operations in which only a few dishes must be processed.

Freestanding Models

Roll or hood machines are called *freestanding* models. Manufacturers claim that these models may clean as many as 570 plates or 1140 glasses per hour, depending on the model. These machines take up relatively little space.

Door Models

Dishwashers called *door* models can handle as many as 1325 plates or 2385 glasses per hour. They can be installed for either a straight-through or a corner operation. Corner units load in front and unload at the side (or vice versa). Straight-through units load and unload at opposite sides. This type of model usually satisfies the needs of foodservice operations that have approximately 100 seats.

Rack Conveyor Models

Dishwashers of the rack conveyor model have a belt or drive unit that moves racks filled with ware through the machine. One-, two-, and three-tank models supposedly can do up to 9000 plates or 16,000 glasses per hour. Machine lengths vary from 40 to 166 in. Blower dryers, vapor condensers, and vent connections are a few of the optional units that can be used with these larger machines.

Belt Conveyor Models

Flight-type or belt conveyor models have the greatest capacity of all. They are used in big installations where many pieces of ware must be processed each hour. A continuous peg-type belt on the machine eliminates the need for racks, allowing dishes, pots and pans, and other large units to be placed directly onto the belt. Silverware, cups, glasses, and other items are sent through in racks. Silverware is usually placed into canisters, eating end up. When the canister comes from the machine after cleaning, it is inverted into containers so that the handles are up.

According to manufacturers, typical capacities range from 5000 to 12,500 plates per hour in one-, two-, or three-tank machines. Models may be up to 60 ft long.

■

SUMMARY

The foods served, their quantity, and the type of service given determine equipment needs. If convenience foods are used, the equipment needed is different from that required for preparing foods from scratch. Equipment must be arranged and used properly and must be integrated with other equipment to that the various subsystems of the operation work efficiently. The average foodservice has six subsystems: receiving, storage, preparation, cooking, serving, and cleanup.

Raw materials, supplies, and equipment enter the operation through the receiving subsystem. Ordered food is checked here for quality, count, and weight. Adequate scales should be provided so that the weights of received items can be checked. In addition, transport equipment is needed to move foods to storage or using areas, and some office equipment is needed to perform various record-keeping tasks.

Storage areas should be planned to give adequate cold and dry storage. Cold storage should be placed near the receiving area and also near the areas that draw materials from it, thus achieving good workflow. Refrigerators, freezers, and other cold-storage equipment preserve food so that it does not deteriorate and lose its nutritional qualities. In the dry-storage area, appropriate shelving and containers are important in order to use storage space efficiently and safeguard the foods stored there. Shelving flexibility and container shape and size are important considerations in designing this area.

The preparation subsystem helps supply foods to the cooking, bakery, and salad units. Preliminary work is done here. Scales, mixers, cutters, slicers, and much other equipment should be selected to do the required jobs and to reduce labor. The vertical cutter and mixer (VCM) is helpful if volume is sufficient to justify its use. Vegetable peelers are used to peel hard root vegetables. Meat-processing equipment—saws, choppers, grinders, and tenderizers—may be needed if the facility does much of its own meat cutting.

The cooking subsystem covers food-cooking areas, the bakeshop, the pantry, and other areas where foods are prepared for service. The receiving, storage, and preparation subsystems, along with the service and cleanup subsystems, should be integrated with this subsystem. Cooking is the core of the entire system. Its equipment includes conventional, convection,

mechanical, microwave, combo, and reconstituting ovens; ranges; griddles; broilers; steam-cooking equipment (including steam cookers and steam-jacketed kettles); cook–chill equipment; fryers; tilting skillets; and knives, tools, utensils, pots, and pans.

The serving subsystem is the place where food is assembled, held for service, dished, and served to the customer. This subsystem joins the heart of the house to the front of the house and forms a vital link between what is done for the customer and what the customer experiences. Both hot-food-holding and cold-food-holding equipment is needed. Sometimes such equipment must also be used to display and to help merchandise foods; in this case, attention must be paid to the décor of the equipment. Other major pieces of serving equipment include coffee machines, ice machines, and beverage dispensers.

Cleanup consists of the washing of serviceware and pots and pans, the cleaning of the facility, and the removal of wastes. This is an important area from the standpoint of cost, maintenance of standards, and employee and guest satisfaction. Waste-removal equipment—garbage or waste containers, disposers, compactors, or pulpers—is used for waste removal. The dishwasher used may be a counter or undercounter model, a freestanding model, a door-type model, a rack conveyor model, or a belt conveyor model. The type of dishwasher selected should be based on the quantity of water that must be processed through the washer.

■

CHAPTER REVIEW QUESTIONS

1. Select five pieces of equipment and describe how they work and what factors are important in their selection. Give advantages or disadvantages of various types of this kind of equipment.

2. Compare the equipment found in an operation utilizing conventional foods as opposed to one using convenience foods. Describe the various subsystems in each one.

3. Plan a section (a subsystem) of a kitchen and place the proper equipment in it. Describe what items might be prepared or what might typically occur in this subsystem.

4. Take one piece of equipment and indicate important factors to know in its operation, maintenance, and cleaning.

■

CASE STUDIES

1. Using Figure 4-1 as your guide, evaluate a piece of equipment of your choice, noting the machine's operation, cleaning, maintenance, reliability, and service. Compare and contrast the results you obtain with those of others who have completed the form for similar pieces of equipment.

2. Jim Paige has been working for a major fast-food chain for the last ten years. while nurturing dreams of owning and operating a place of his own. Jim has decided that his nest egg is large enough to begin a serious investigation of his restaurant concept. He wants to open a full-service restaurant that caters to a market segment he is familiar with: the family. He wants to present a medium-priced menu featuring a fairly wide range of selections, while stressing "home-style" cooking. This includes a good seafood selection, steaks, sandwiches, chicken, and a salad bar. This menu will not present exotic dishes but will stick with traditional American preparations. The restaurant will be open for lunch and dinner. Jim knows that one of the first things he must do is get an idea of the cost of his project and its space requirements. Because most of Jim's experience has come in operations and not in restaurant development, his knowledge of equipment is limited to what he has dealt with in his fast-food operation. Jim has contacted a friend of his who teaches foodservice courses at the local college and has asked if the teacher happens to have a student who would be interested in helping him with his estimates. You are that student (there will be some pay involved!). What Jim is looking for is a rough idea of the equipment he will need for his operation and how it will be arranged in his kitchen (giving him an idea of the space he will need for the kitchen). He would like you to provide a list of necessary equipment, an explanation of your choices, and a rough layout of his kitchen-to-be, showing the placement of storage, work stations, and major pieces of equipment. What would your list, explanations, and sketch look like?

■

VOCABULARY

Subsystem

VCM

Reel oven

Mechanical oven

Horizontal mixer

Pulper

Broiler

Dollies

Peeler

Reconstituting oven

Rack conveyor

Tilting skillet

Buffalo chopper

Convection steamer

Radiation

Upright mixer

Compactor

Steam cooker

Convection oven

Belt conveyor

Cook–chill

Microwave oven

Tank

Counter dishwasher

Combo-oven

■ ANNOTATED REFERENCES

Battistone, S. R. 1991. *Spec Rite*. Cincinnati, Ohio: Food Service Information Library. (All kitchen equipment from ovens to fryers is reviewed in this book.)

Birchfield, J. C. 1988. *Design and Layout of Foodservice Facilities*. New York: John Wiley & Sons, Inc. (This text covers topics from facility layout to equipment implementation.)

Borsenik, F. D., and A. T. Stutts. 1992. *The Management of Maintenance and Engineering Systems in the Hospitality Industry*, 3rd ed. New York: John Wiley & Sons, Inc. (A chapter is dedicated to foodservice equipment maintenance.)

Ghiselli, R., A. Almanza, and S. Osaki. 1998. "Foodservice Design: Trends, Space Allocations, and Factors That Influence Kitchen Size," *Journal of Foodservice Systems*, **10**, 89–105.

Greaves, R. E. 1987. *The Commercial Food Equipment Repair and Maintenance Manual*. New York: Van Nostrand Reinhold Co., Inc. (All foodservice equipment is reviewed in this book, which is now out of print.)

Kotschevar, L. H., B. A. Almanza, and M. Terrell. 1999. *Foodservice Planning: Layout, Design, and Equipment*. New York: John Wiley & Sons, Inc. (A comprehensive text covering all aspects of the topic.)

Mill, R. C. 1998. *Restaurant Management: Customers, Operations, and Employees*. Upper Saddle River, N.J.: Prentice Hall, pp. 216–239. (Kitchen equipment is discussed in one chapter.)

Stevens, J., and L. Snowberger. 1997. *Food Equipment Digest*. New York: John Wiley & Sons, Inc. (Equipment contacts, including phone and fax numbers, are presented in this book.)

Further materials on equipment may be requested from manufacturers or their representatives. Hobart Manufacturing, for example, has extensive materials on implementing equipment according to the systems approach.

CHAPTER 5

Controls

LEARNING OBJECTIVES

By the end of this chapter, the reader will:

1. Understand who the buyer is and how he or she may make purchases for a foodservice facility in the marketplace.

2. Know how to select among purveyors and how to determine purchasing needs.

3. Recognize the value of using specifications in purchasing by setting the standard of what is needed by the foodservice operation.

4. Learn how to design and implement purchasing, receiving, storing, and issuing procedures.

5. Appreciate the role that the computer has come to play in the purchasing, receiving, storing, and issuing of food.

SELECTED WEB SITE REFERENCES

CLS Software
www.hospitalitynet.org

CMS Hospitality
www.cmshosp.com.au

Comtrex Systems Corp.
www.comtrex.com

Comus Restaurant Systems
www.comus.com

Eatec Corporation
www.eatec.com

Geac Computer Corporation Limited
www.geac.com

Instill Corporation
www.instill.com

Integrated Restaurant Software
www.rmstouch.com

Micros Systems, Inc.
www.micros.com

Once a menu is planned, a number of steps must be taken to bring it into reality. One of the first and most important steps is to purchase and receive the materials needed to produce the menu items. Skillful purchasing, coupled with good receiving and storage, can do much to maximize the results of a good menu. From purchasing the food from a purveyor who has a HACCP (hazard analysis critical control points) program in place, to implementing the same program (as discussed in Chapter 11) within the foodservice operation itself, management will help to ensure food safety until it is consumed by the patron.* In this regard, management must master six preliminary duties:

---

*In July 1996, President Clinton announced that the government's 90-year-old meat inspection program would be amended to include new methods for testing meat and poultry products for unsafe contamination. Instead of using just the old "hunt-and-poke" system, actual tests for bacterial or other contamination would be made. Other improvements to improve sanitation control at critical points were also announced. In 1997 the use of irradiation to destroy contaminating substances on meats and poultry was approved.

1. Know the market.
2. Determine the operation's purchasing needs, by department and category.
3. Establish and use specifications.
4. Design the purchasing procedure.
5. Design the receiving procedure.
6. Evaluate the purchasing task as a job description and fill the position.

■

KNOWING THE MARKET

Since markets vary considerably, a buyer must know the characteristics of the relevant local market. This requires some individual investigation of market suppliers of all types. To do a good job, the purchasing agent must also know a great deal about the products to be dealt with—for example, localities where the goods are grown or produced, the seasons of production, approximate costs of goods, conditions of supply and demand, laws and regulations governing the market, marketing agents and their services, processing stages, storage requirements, and grades. Since this means that the purchasing agent, or buyer, must maintain current information on changes in the market, it becomes important for this person to talk frequently with purveyors.

Today, many managers of operations may find their purchasing a nonproblem. The chain for which they operate only one of many units has set the purveyor from which to purchase. Price and other conditions are established. All the manager does is to set the amount desired, verify the delivery, and the head office takes care of the other requirements of purchase. In such cases, managers are limited to purchases to relatively small amounts and limited dollar amounts.

The Buyer

Buying demands of the buyer integrity, maturity, bargaining skills, and a moderate disposition. Buyers are subject to bribes and other inducements from dishonest purveyors who want to get their business. Attempts are sometimes made to force a buyer to purchase from specific sources as well. A rule often quoted is, "Treat your boss's money as if it were your own"; and another is, "Be friendly, be courteous, and socialize if it means improving your buying, but always stop at the point at which you would lose your right to remain a free agent." Many professional societies of buyers have ethical codes (to which members subscribe) requiring high ethical standards in the relationship between buyer and purveyor.

A buyer also must know a lot about the internal organization for which buying takes place. The buyer must know what specific items an operation needs and must be able to obtain them at a desirable price. The procedures for production and the ways in which items are used in the operation must be known, and the right type of item must be bought to suit it. For example, the appropriate item for a particular use may not be first quality but third. Ground beef from the lower grades has more flavor than that from the higher ones. C-grade tomatoes are preferred over A-grade tomatoes for soups and sauces. It is senseless to purchase perfect walnut halves when they are to be chopped and used in bakery goods; a broken walnut meat serves just as well. Knowing these things can help an operation reduce costs, improve quality, and enhance customer satisfaction. A buyer must also know storage requirements, space available, the financial ability of the operation to make certain purchases, storage times and temperatures, yield-testing procedures, and the way to establish a specification so that the right item is obtained.

Because of the importance of this position, special care should be taken in selecting the correct person for the job. The following is a sample job description for the position, including a synopsis of the responsibilities involved.

POSITION DESCRIPTION

Position: Purchasing Agent

Reports to: General Manager

Supervises: Purchasing, Receiving, Storing, and Issuing

Basic functions:

1. To keep management informed of market conditions, price trends, product availability, and methods of reducing costs.

2. To supervise, train, and motivate storeroom employees.

3. To implement corporate purchasing policies and procedures.

4. To maintain and upgrade the quality of items purchased.

5. To control purchase costs through competitive shopping for quality products.

6. To supervise all storeroom operations.

7. To maintain adequate records.

Responsibilities:

1. To carry out all administrative functions pertaining to payroll, scheduling, and recruiting. To maintain honest, reliable staff, and to promote personnel with potential.

2. To become a recognized "profit center" by recommending and implementing purchasing concepts that save money through contract purchasing, buying realistic quantities before price increases, and utilizing quantity purchase discounts and monthly liquor postoffs.

3. To maintain accurate written specifications on all products purchased, to understand these specifications, and to use them effectively in obtaining competitive price quotations and receiving products.

4. To have a written order and receiving schedule for all products. The appropriate forms should be used to prevent shortages and to minimize deliveries.

5. To maintain current market quotations on all products. Suppliers should be determined by the quality of their products, the dependability of their service, and the competitiveness of their pricing.

6. To set up a control system to ensure that all items being received are of designated quality, quantity, and price. Invoices and receiving records should be personally reviewed daily.

7. To supervise daily all issuing procedures, including ascertaining the following points: issuing hours are being enforced; requisitions are properly dated and signed, and products and quanti-

ties are clearly written; pricing (storeroom, direct, and meat tags) is accurate, and key controls are enforced.

8. To check storeroom procedures daily to ensure that stock is properly rotated, that items are properly priced, that sanitation standards are maintained, and that sufficient inventories are available for efficient operation.

9. To work closely with the chef with regard to items used, their amount, and when they will be required for food preparation. To constantly seek new or better products.

10. To work closely with the chef and the banquet department in recommending products that can be sold profitably, in discussing availability of seasoned items, and in estimating the time requirement necessary to obtain merchandise.

11. To visit local markets on a regularly scheduled basis, in order to be able to meet and select purveyors and to find and select the best-quality items available in the market.

Buying Methods

Buying methods depend both on the type of operation and on the buying market. Generally, buying methods can be grouped under two headings: formal buying methods, and informal buying methods. Formal methods are best and are most often used in large contracts for commodities purchased over time, while informal methods are more often used for casual purchasing when the amount involved is not large and when speed and simplicity are desirable. In formal buying, if a basic price has been established, prices do not vary much over the course of the contract. Prices may vary to a great extent with informal buying, however, as both prices and supply may fluctuate widely.

Informal Buying

Informal buying methods, also called *open-type buying*, are the ones most often used by small foodservice operators. This type of buying usually involves oral or computer negotiations. The buyer gets the price, based on the specifications of the operation, either over the phone, from the computer, or in person from a purveyor. Quality, price, and service are then weighed, and a decision is made to purchase or not. In using this system, buyers should obtain comparative prices from at least three to five different suppliers. While informal methods may vary from locale to locale, there are generally three types of methods used: the quotation-and-order-sheet method, the blank-check method, and the cost-plus method.

Quotation-and-Order-Sheet Method. The quotation-and-order-sheet method uses a tabulation of particular commodities wanted, in the quantities and quality needed. These specifications are usually listed in a column on the left-hand side of a sheet. Additional columns are available to record the prices quoted by various purveyors (see Fig. 5-1). Calls are made to the purveyors, and their prices are recorded in the column allowed for this. When as many prices as desired have been obtained, the prices are studied. A decision is then made to purchase something from a particular purveyor, and the price for that item is circled. Last, a secretary or clerk calls the purveyor to place the orders.

Blank-Check Method. Blank-check buying can occur when there is an extreme shortage of a commodity on the market or when some other condition exists that necessitates buying an item almost regardless of its cost. By this method, a buyer orders something without knowing its price, and then pays whatever price is necessary to obtain the item. Only reputable vendors should be given a blank-check order. Often, upon obtaining a price on an item, a vendor checks with

the buyers to see if the price is within an acceptable range.

Cost-Plus Method. The cost-plus method is often used in situations in which prices are not known or the market is unstable. Many vendors prefer this arrangement, since they do not have to add a safety factor to avoid the risk of losing money on commodities that fluctuate considerably in price. Instead, purveyors are free to buy at the most favorable price and then add on an amount necessary to cover costs and give a profit. The amount charged to the buyer over and above the initial cost of the item is usually based on a standard percentage. Thus, a vendor may purchase fruits and vegetables and charge the buyer the original cost plus 20 percent. If this is used, the seller's records should be open to the buyer.

Formal Buying

Formal methods, also known as *competitive buying*, involve giving vendors written specifications and quantity needs. Key stages of negotiations are done in writing; as the degree of formality increases, the proportion of negotiations reduced to writing increases, too. The invitation or request for bids may be made through newspaper advertising or by sending selected vendors a notice of buying needs.

Three formal methods are most common: the competitive-bid method, the negotiated method, and the futures-and-contract method. All are used to obtain the usual advantages of volume buying.

Competitive-Bid Method. In competitive bidding, sellers are invited to submit bids through some type of written communication (see Fig. 5-2). The vendors, in turn, send prices and other requested information on the commodities to the buyer. The cost-plus system is sometimes used in formal purchasing. Bids are opened at a specified time to determine awards. Only sellers that can meet the established purchase conditions of the buyer are considered in awarding bids. The invitation to bid usually contains a set of general conditions, such as terms of payment, discounts, method of delivery, billing requirements, and payment arrangements. In addition, the particular requirements for each item are set forth in a specification.

Produce Quotation and Order Sheet								
For use on _____ Delivery date _____								
					Price Quotes			
		Amount	Amount	Amount	Vendors			
	Specs	needed	on hand	to order	I	II	III	IV
Produce:								

FIGURE 5-1 Sample form for a produce quotation and order sheet.

INVITATION, BID, AND AWARD

Issued by
Manager
Happy Haven Restaurant

Address
1122 Supply Street
Happy Haven, Maryland

Date _____

Sealed bids in duplicate will be received at the above office until _____ ,
_____ for the items and in the quantities indicated for delivery on the dates indicated. Quantities indicated are approximate and may be reduced on instruction of the buyer. Increases up to 20 percent will be binding at the discretion of the buyer.

All items to be officially identified by the U.S. Department of Agriculture for class and quality. Costs of such service to be borne by vendor.

Items	Supplies	Quantity	Unit	Unit Price	Amount
1.	Chicken, fresh chilled fryer, 2-1/2–3 lb, ready-to-cook—U.S. Grade A To be delivered _____	500	lb		
2.	Chicken, fresh chilled fowl, 4-1/2–5-1/2 lb, ready-to-cook, U.S. Grade B To be delivered _____	100	lb		
3.	Turkey, frozen, Young Tom 20–22 lb. ready-to-cook, U.S. Grade A To be delivered _____	100	lb		
4.	Ducks, frozen roaster duckling, 5–5-1/2 lb, ready-to-cook, U.S. Grade A To be delivered _____	50	lb		

Vendor _____

FIGURE 5-2 Sample competitive bid form.

Negotiated Method. Negotiated buying is often used when vendors are hesitant to bid because of time restrictions, fluctuating market conditions, or high perishability of the product. In this case, negotiations may be conducted over the phone and confirmed later in writing. The negotiated method allows for quicker action because it is less rigidly structured than other formal methods. Several vendors may be approached to get a range of price, quality, availability, and other factors. This method is flexible yet still allows for competitive bidding.

Futures-and-Contract Method. Futures-and-contract buying is often used by large businesses that have sufficient capital and staff to contract for future delivery of commodities at an established bid price. The advantage of this system is that an adequate supply at an established price can be arranged in advance. As a result, problems with shortfalls and price fluctuations that otherwise could affect the availability of items required for the menu can be avoided. This system is sometimes used for short periods of time—perhaps only a week or a month—when commodities such as meats or fresh fruits or vegetables may be in limited supply. Items of more stable price, such as canned goods or even mature potatoes, onions, or cabbage, are often placed on contract for longer periods of time.

In some cases, the quantity desired under contract may vary, so the contract may specify that the amount purchased depends on the amount used during the relevant period. A price may be agreed on, along with quality and other factors, but the specific quantity to be used is not set (although a range is noted). Deliveries then depend on the inventory on hand. The maximum stock of an item is set by the buyer; then, when the stock diminishes to a certain point, the buyer reorders so that the stock, on delivery, reaches its maximum once more. In some cases, the purveyor drops by the operation regularly to bring the inventory of the operation up to a preestablished level. Either arrangement is often called *par stock supplying.* It is important in par stock supplying to maintain a safety stock—a quantity of the item sufficient to tide the operation over between the time of reorder and the time of delivery. Even maintaining a reserve stock is worthwhile in case delivery is late. This system is often adopted to handle daily-order items, such as

breads and dairy goods. All deliveries must be checked and all products rotated forward, leaving the newer stock to be used after the old stock is expended.

The Internet

Use of the Internet to survey the market for purchase needs, price and quality quotations, and other purchase functions is a new way for buyers to purchase. It is quick and easy and can lend itself to many of the informal and formal purchasing methods mentioned here. Thus, in using the quotation-and-order-sheet method, a buyer can quickly survey the offerings of a number of purveyors of a certain commodity. It is quicker and easier to use the Internet for this type of survey than to use the telephone, fax, or e-mail. It is very likely that a substantial amount of the needs of foodservices will be supplied by this procedure in the future.

Selecting Purveyors

Selecting a purveyor is an important step in purchasing. One of the first considerations is how able a particular purveyor is to meet the needs of the foodservice—not only in price, but in meeting delivery times, maintaining quality standards, and upholding reliability in other areas. Information about various purveyors can be obtained from other buyers. Visits can be made to their establishments, and opinions can be gained by inspection and by discussion with them. The buyer must attempt to ascertain how reliable and stable a purveyor would be under competition and under varying market conditions. Normally, vendors fall into one of five categories:

1. *Wholesalers* stock many kinds of products and may be able to provide good service, maintain good quality, and meet special needs.
2. *Direct sellers from manufacturers or packers* may be able to provide some savings by eliminating marketing agents and services.
3. *Local sellers*, such as *farmers and processors* can supply perishable goods such as fruits, vegetables, milk, bread, and bakery goods, possibly from local rural areas.

4. *Cooperative sellers* may be able to supply commodities produced by local farmers or others who have joined together to market their products.

5. *Retail food stores* may be able to sell at very low prices because they can purchase in much larger lots than a foodservice can.

Purveyors are also selected on the basis of experience. Gradually, those who do not satisfy the needs of the operation are weeded out. New vendors may be tried and retained. Some of the considerations leading to a decision to deal on a continuing basis with a specific vendor have been summarized as follows:

- How well does the purveyor anticipate needs? A purveyor is a buyer for customers and must look ahead and anticipate their requirements. The purveyor should watch market conditions and notify customers when market changes may occur.

- Can the purveyor offer adequate service in assembling the various items needed? Will the purveyor break quantities down to amounts the foodservice can use? Is the purveyor even willing to warehouse for preferred customers at times?

- Will the purveyor maintain adequate stocks? If supplies run low, can the purveyor obtain replenishments quickly?

- What credit terms does the purveyor offer? Are discounts available for prompt payment within a given time?

- Will the purveyor be able to deliver the goods in good condition and at times required?

- Will the purveyor use information about market conditions to protect customers or to make windfall profits?

- Does the purveyor offer a wide range of services and a wide range of goods? Savings can often be made by consolidating services and products; many purveyors now may be able to offer frozen commodities, fresh fruits and vegetables, processed foods, meats, and even some dairy or specialty products. Some purveyors even feature such items as prepeeled potatoes and packaged meats. A purveyor offering many things a foodservice needs can

simplify purchasing and reduce the number of contacts that must be made.

- Does the purveyor understand customers' operations? Can the purveyor recommend items that suit customers' needs, or are its services limited to merely taking orders?

The relative importance of these considerations varies from operation to operation. Buyers consider the following factors most important in evaluating potential suppliers:

- Supplier dependability
- Quality of delivery facilities
- A.M. delivery
- Absence of back orders
- Quantity discounts
- Cooperation in bid procedures
- Turnaround time
- Varieties of qualities carried
- Price ranges carried
- Varieties of products carried

In contrast, the least important factors are the following:

- Coupon refund offered by supplier
- Reciprocal buying
- Size of the supplier firm
- Free samples offered by supplier
- Weekend delivery

In some localities, the number of purveyors available to supply needs is very limited. Competitive bidding may not be possible, and the quality of service rendered may be poor. A difficult situation can develop, since there is no other practical source of supply, and since discontinuing buying would hurt the buyer. Sometimes, arrangements can be made for other purveyors not too far away to ship goods; when this is managed successfully, it may cause some nearby purveyors to shape up.

■

DETERMINING PURCHASING NEEDS

A menu dictates an operation's needs, and a buyer searches for a market that can supply those needs. After the right market is located, the various products available that meet the need are investigated. The proper product must be obtained to meet the production need and to yield the right kind of meal. Factors that affect purchasing requirements include the following:

- Type and image of the establishment
- Style of operation and system of service
- Occasion for which the item is needed
- Amount of storage available—dry, refrigerated, and frozen
- Financial resources and buying policies of the organization
- Availability, seasonality, price trends, and supply

In determining the right food item to meet a particular production need, the buyer should consider the skill of the employees, the processing method to be used, the suitability of the product for the required menu item, the storage life of the raw product, and other factors. The buyer weighs these factors and then searches the market to find a product that will best satisfy them. Usually, three general kinds of markets will be found to meet three principal foodservice needs: perishable, staple, and daily use or contract items.

Perishable Needs

The perishable market supplies fresh fruits, vegetables, meats, frozen foods, and dairy products. Prices vary, and so may supplies at times. It is wise to visit this market to inspect goods and ascertain market conditions. Phone calls may be helpful when an actual visit cannot take place. Informal methods of buying are used most often in this market. Perishables should be purchased to meet menu needs for a short period only.

Staple Needs

The staple food market can supply canned, bottled, and packaged products to be stored in dry storage without refrigeration. Formal or informal purchasing may be used. Because items are staples and can be stored, bid buying is frequent to take advantage of quantity prices. Staples packaged for foodservice use are generally in larger units than those packaged for the retail market. Because only 25 percent of this market goes to foodservices, however, packaging may favor the retail market. Sometimes purchases may have to be made of smaller, retail-size units.

Daily Use Needs

Daily use or contract items are delivered frequently on a par stock basis: stocks are kept up to a desired level, and supply is almost automatic. Milk, meats, vegetables, fruits, bread, coffee, and ice cream are among the items supplied in this manner. Deliveries may be made daily, several times a week, or somewhat less often. Most items are perishable, and the supply kept on hand should be designed to last only until the next delivery time. If an operation allows a delivery person to come in and replenish supplies automatically, management should verify the amount delivered. Also, the delivery person should put older stock first.

Quantity and Quality

Determining the quantity and quality of items to purchase is an important aspect of gauging the needs of an operation. The buyer should be informed by the chef or other members of the operation as to the quantity and quality of items needed. Management may establish quality guidelines and often may wish to supervise the ordering of quantities as well. After receiving this information, the buyer identifies the market and looks for the best possible item for the production need available at the best price. Delivery arrangements and other factors are usually handled by the buyer. Buying is not, therefore, the result of actions or decisions made by a single person, but is often the work of many. Buyers must act within the policies established by management in many matters relating to finance and delivery arrangements.

When considering the quantity needed, the buyer must keep certain factors in mind. First, the number of people to be served in a given period must be determined. Sales history data can provide this information. Portion sizes must also be known; these can be obtained from yield tests or standard portion lists (see Appendix E). Knowing the weights and sizes of the various products helps the buyer estimate quantities needed within fairly precise limits (see Table 5-1). Buyers need to know production results to be able to predict accurately how many portions a given amount of purchased goods will yield. They must also know the range of variability in yields. A purchase of veal may give one percentage of clear meat on one occasion and quite a different percentage on another. Cooking shrinkage may vary, too, causing predicted quantities to differ considerably from reality. However, buying a lot just to have plenty can lead to loss and problems in future use.

The desired quality of every product must be indicated to vendors so that the right product can be purchased. Grades, styles, appearance, composition, varieties, color, texture, size, relative absence of defects, and maturity should be specified. Quality standards should be established by management when the menu is first planned. Buying should then be a matter of following through to obtain products of the particular quality identified.

■

ESTABLISHING AND USING SPECIFICATIONS

Specifications describe the precise factors sought in an item in order for it to satisfy the production need. Specifications are usually given in writing and are filed in an easily accessible place. They an be given to buyers by computer delivery. They present a constant reference

TABLE 5-1 Weights and Sizes of Common Cans and Jars[a]

Can Size (Industry Term)	Average Net Weight or Fluid Measure per Can	Average Cups per Can	Cans per Case	Principal Products
No. 10	6 lb 3 oz (99 oz) to 7 lb 5 oz (117 oz)	12–13	6	Institution size—fruits, vegetables, and some other foods
No. 3 cyl	51 oz (3 lb 3 oz) or 46 fl oz (1 qt 14 fl oz)	5¾	12	Institution size—condensed soups, some vegetables, and meat and poultry products; economy family size—fruit and vegetable juices
No. 2½	27 oz (1 lb 11 oz) to 29 oz (1 lb 13 oz)	3½	24	Family size—fruits and some vegetables
No. 2 cyl	21 fl oz	3	24	Family size—juices and soups
No. 2	20 oz (1 lb 4 oz) or 18 fl oz (1 pt 2 fl oz) or	2½	24	Family size—juices, ready-to-serve soups, and some fruits
No. 303	16 oz (1 lb) to 17 oz (1 lb 1 oz)		24 or 36	Small cans—fruits and vegetables, some meat and poultry products, and ready-to-serve soups
No. 300	11 oz to 16 oz (1 lb)	1¾	24	Small cans—some fruits and meat products
No. 2 (vacuum)	12 oz	1½	24	Principally for vacuum-pack corn
No. 1 (picnic)	10½ to 12 oz	1¼	48	Small cans—condensed soups, some fruits, vegetables, meat, and fish
8 oz	8 oz	1	48 or 72	Small cans—ready-to-serve soups, fruits, and vegetables

[a] The net weight on can or jar labels differs among foods because of different densities of foods. For example, a No. 10 can contains 6 lb 3 oz sauerkraut or 7 lb 5 oz cranberry sauce. Meats, fish, and shellfish are known and sold by weight of contents of can.

point and are constantly being revised. If costs are given, their purpose is only to inform the buyer about what might have to be paid to get the desired quality. A maximum price allowable may be indicated.

Operations and management should define exactly what factors contribute to the overall adequacy of an ingredient to be used in a menu item. These factors can then be summarized to produce the specification. Size, grade, appearance factors, packaging, locality of production, and other elements may need to be defined. Trade and government specifications and standards relating to commodities should also be considered. Grade standards established by the federal government are the measures most often used, but at times trade or other standards may be used instead. Brands may be named to guarantee a certain quality. Good specifications simplify purchasing. The buyer and seller both know what is wanted when the specifications give all relevant facts. Better quality and lower prices follow when well-written specifications are used.

Standards of Quality

A quality standard indicates things such as maturity, flavor, and texture that contribute to eating satisfaction. A standard is a measure or benchmark used to evaluate quality, weight, value, or quantity. Standards have been established for many foods by the U.S. Department of Agriculture (USDA), the Department of Interior's Bureau of Fisheries, the Food and Drug Administration, and others.

Quality standards indicate the various characteristics of a specific grade of item. Three or more grades may be named, and in each the quality characteristics (e.g., maturity or texture) will differ to some extent. Not all items are graded on the basis of the same factors. Grades can be certified by federal graders; many meats, for instance, move on the market under federal grades. Eggs, butter, poultry, fresh and processed fruits, and vegetables are often graded. The purpose of quality standards is to make buying and selling more efficient by providing a means by which buyers and sellers can assess the quality of the items for sale without having to examine them. When the grade of an item is known, it is easier to agree on a selling price.

Different grades are used on the market. Federal grades consist of Manufacturing or Processing, Wholesale, and Consumer. Foodservices usually use the Wholesale grades but sometimes may move into the Consumer ones. Trade grades also exist on the market—for example, the butter scores of 93, 92, 90, and 89. Trade grade names are gradually being dropped; thus, the grades Fancy, Choice (Extra Standard), and Standard are being dropped for processed foods, and the grades A, B, and C are only used for first, second, and third quality. Butter trade scores are fast disappearing as the market focuses exclusively on AA, A, B, and C grades. Egg trade grades are seldom used. Only meat grades seem to be secure, although perhaps in the future these too will be subject to replacement by a different grade set, such as A, B, and C or No. 1, No. 2, and No. 3.

Voluntary and Mandatory Standards

Most grade standards are voluntary, which means that the item is not required to carry a grade. In fact, most processed fruits and vegetables, processed meats, and fresh fruits and vegetables do not carry a grade. Other items are graded, but only because the producer or packer wishes them to be—not because regulations require it. A shield is often used to indicate a federal grade stamp, and within the shield appears the grade of the item (see Fig. 5-3).

Because some items vary in quality more than others, a different number of grades may have to be used. Thus, some canned juice standards have only two grades, whereas most have eight. The number of grades depends on what it takes to move items on the market. Federal standards define factors that differentiate the quality levels in market products. Grade standards must be general enough to cover all kinds of products that may appear on the market. Thus, the grade standards for apples vary between apples grown in the state of Washington and those grown elsewhere, since the apples of the west take on more color. Peaches from Washington, California, Georgia, and Michigan differ significantly, and the standard must be broad enough to cover them all. Because a rather wide variation exists in the quality factors of citrus fruit, there are three standards: Florida, Texas, and California (including Arizona).

This USDA grade shield may be on the wrapper or on a wing tag on chickens, turkeys, ducks, and geese.

This is the USDA grade shield used on butter. It is printed on the carton and on the wrappers of ¼-lb sticks of butter.

Prime beef is the best and most expensive grade. Not many stores sell it. Most cuts graded Prime are very tender, juicy, and flavorful.

This U.S. grade shield may be used on canned, frozen, or dried fruits or vegetables. It is also used on a few related products, such as frozen concentrated juices, jams, and jellies.

Steaks and roasts of Choice grade are tender and juicy and have a good flavor.

Beef graded Select is not as juicy and flavorful as Prime or Choice, but it is fairly tender and has less fat than either of the higher grades.

The U.S. Extra Grade shield may be used on instant nonfat dry milk. It means that the milk has a sweet and pleasing flavor, and dissolves immediately when mixed with water.

Beef graded Standard has very little fat and a mild flavor. It lacks juiciness but is fairly tender because it comes from young animals.

This is the USDA grade shield for eggs. It may be found on the carton.

Beef graded Commercial comes from older cattle and is not very tender. It needs long, slow cooking with moist heat (pot-roasting or braising). If cooked properly, it has a good, rich flavor.

Although most fresh fruits and vegetables are sold at wholesale on the basis of U.S. grades, not many are marked with the grade in the grocery store.

FIGURE 5-3 USDA grades used as standards in buying foods.

A brand often is specified to identify quality. Thus, many restaurants insist on Heinz catsup because of the product's known quality and because many customers recognize this quality and approve it. A brand is not based on minimum standards for tenderness, color, freedom from defects, maturity, and so forth as the government grades are: a brand is only the quality the manufacturer says it is. For this reason customers must sample the product often to see if it supplies the quality they want. Brands are usually higher in price than unbranded items. Food buyers should accept only brands of reliable and established vendors.

There is a difference in meats and poultry between a quality grade, which the producer or packer voluntarily chooses to have appear, and the mark of inspection for wholesomeness, which often must appear under the terms of mandatory standards set by the USDA and other agencies.

Meat and poultry carry a circular stamp in which is written "Inspected and Passed." A number within the circle indicates where the inspection occurred. This means that the item is fit for human consumption. Some meats and poultry cannot be graded until they have been inspected and passed for wholesomeness. The USDA largely administers certification for wholesomeness, on authority given it under the (recently revised) Meat Act.

All meat and poultry (including processed products) that move across state lines must be inspected and passed. The products must come from healthy birds or animals that have been handled and processed under strictly HACCP conditions. The products cannot be adulterated and must be truthfully packaged and labeled.

Other Standards

Two other standards have been established by federal agencies. One, the "standard of identify," indicates exactly what a food is. If a food does not conform to the established standard, it cannot be called by the common name. Thus, if a mixture of meat and vegetables does not contain 25 percent beef, it cannot be called beef stew. It must be called vegetables with beef stew. Minimum (and sometimes maximum) quantities of various ingredients are specified in the standard, and the food must satisfy the stated requirements. The Food and Drug Administration, part of the Department of Health and Human Services, has jurisdiction over the establishment of the standards of identity.

The second standard established by federal agency relates to standards of fill and weight. The term *fill* refers to the quantity that must be in packaged goods, which varies according to the product and its form of packaging. Official weights have been established for different containers. Thus, a barrel used to package flour must contain 196 lb (two 98-lb sacks). A cranberry barrel is smaller and must hold 100 lb. Bushels of various items must weigh no less than a stated number of pounds, and other measures must also conform. Various agencies administer standards of fill, but most such standards come under the jurisdiction of the USDA. The Department of Commerce regulates standards of weight and size.

The Treasury Department has jurisdiction over the various standards established for alcoholic beverages. Originally this administrative department was charged only with collecting taxes and enforcing laws relating to alcoholic beverages, but from there it moved on to establish standards for such things as quality, packaging, and identity.

If all products do not meet a standard established by the government, the label must say so. For example, if a can of fruit does not meet the standard of fill, the label must inform the buyer of this. Similarly, a product that does not meet the standard weight must carry a notice of this fact on the label. Foods that do not meet the standard of quality must carry a statement such as "Below standard in quality." This does not mean the food is unwholesome, only that it fails to meet established quality standards. Although many foods need not carry a statement of grade, a food that does not achieve the minimum standard of quality established for it must indicate this failure on its label.

Specification Writing

As was noted earlier, specifications describe in writing and in standard terms the ingredients needed to produce the item on the menu. Only after the operation's needs and the market offerings are known can the specification for an item be written. The specifications should state exactly what is required to produce an item of the quality desired. Federal or trade standards of quality may be used, but these should only serve as models; usually, a foodservice has more specific needs than are named in these specifications. Thus, for example, the quality standard for fresh peaches may be too broad, raising the need for a narrower definition of quality. Since operations want items that meet special merchandising and quality needs, they write specifications that spell out these needs. For example, the Nut Tree Restaurant in California had a requirement for fresh pineapple that fell within a very narrow range of quality and, in fact, had to come from one specific area of Oahu in the Hawaiian Islands. The quality definition became so famous that it is now known as Nut Tree grade.

Writing the Specification

There are two major steps in writing a good specification: ascertaining the quality factors needed to give the right product and setting up the specification.

When the menu is planned, the amount and quality of the items on it are established. For instance, suppose that a particular chicken breast dish of fine quality is planned. To obtain proper-size chicken breasts for the dish, the operation may need a 4-lb fryer of grade A quality. Since only a few purveyors market a fresh-killed milk-fed bird that is fed solely on milk mash (which produces the very white, delicate meat needed for the item), the specification would also have to specify that only milk-mash-fed fryers are acceptable. The product also must be extremely fresh, so the specification should state that the birds must be fresh-killed. This means that delivery must be made within three days of slaughter, during which time the breasts must be held at 35°F (2°C). All details are covered in the specification so that the right product is obtained and no chances for error are introduced.

At times, a buyer or others concerned with defining the quality of the item needed may have to perform special tests to gain information. Four such tests can be used:

1. *Raw food tests.* These tests determine the best count, weight, or quality of the product (usually a fruit or vegetable).

2. *Processed food tests.* These tests check the yields and costs of different canned or processed foods. They consider quality, drained weight, count density, clearness of syrup, and other quality or physical factors.

3. *Butchering tests.* These cutting tests of meat, fish, and poultry determine portion costs after waste, trim, and by-products have been removed. The size of the portion and its appearance can also be judged. By preparing the item, the tester can judge its finished quality.

4. *Cooking tests.* These tests determine final portion costs as served. The product should be taken as purchased and given the required pre-preparation

treatment, then cooked, and then judged for portion size, appearance, quality, and any other factors (color, texture, form, and so on) that are considered important.

Once the characteristics required of the product are enumerated, the specification can be written. Usually food specifications include the following information:

- *Name of the item.* This should be the name used for the item in the standard of identity; in meats, this would include the name of the cut plus the institutional meat purchase specification number; the name used should indicate clearly what is wanted.

- *Quantity of the item needed.* This should be based on the factors previously discussed for deciding quantities.

- *Grade or brand wanted.* This should be stated in accordance with applicable standards; if a finer demarcation of quality is needed, appropriate details of this can be added.

- *Size of the item or number of items per package.* This may be expressed, for example, as a 12- to 14-lb ham or as twenty-four No. 2 cans per case.

- *Unit size on which the seller's price is to be based.* This may be a pound, a sack, a package, a case, or any other clearly defined unit.

- *Additional factors needed to get the exact item wanted.* These may refer to geographical area of production, count size, drained weight, sieve size, packing medium, syrup density, color, style, or other relevant characteristics.

Figure 5-4 gives a sample specification for canned asparagus.

In writing a specification, some people like to add descriptive details such as "tender, succulent, dark red, and free from blemishes, broken slices, or end cuts." All of these factors, however, are included in the standard for the grade; if the grade is stated, these characteristics must be present, since otherwise the product could not have obtained the grade named. Thus, it is

---

Asparagus, canned, all green stalks or spears

10 cases

U.S. Fancy or Grade A

24/2's

Quote price per case.

Style shall be stalks not more than $3^3/4$ inches in length; minimum drained weight shall be not less than 12 ounces; size of spears shall be extra large (mammoth). The product shall score 90 points or more. Pack shall be from current stocks. Inspector's certification of grade shall accompany invoice; all cases shall be marked with the certificate number.

FIGURE 5-4 Sample specification for canned asparagus.

superfluous to put such details in the specification—although some buyers like to add these just to remind vendors that they will be looking for these characteristics in the products they get. Specifications should be as brief and simple as possible, but they should not omit any essential factor needed to get the right item.

Using the Specification

The specification must be used properly to achieve the desired results. Having the specifications on file without using them accomplishes nothing. When specifications are used properly, the buyer, the receiving clerk, management, the storeroom personnel, and the purveyor all have copies and use them regularly in their daily jobs.

Copies of the specifications should be made available to all responsible personnel in written, printed, or computerized form. Copies filed alphabetically by food group can be kept in looseleaf notebooks, on 3×5 cards, or on computer disks. Today many of the best-run purchasing and receiving departments prefer to use computerized specifications.

Acceptance buying is used by some food services to purchase fresh, processed, or other foods. In particular, large buyers—state agencies, veterans' hospitals and federal agencies—tend to use this method.

It requires that the vendor receive the specification and send the specified item to a federal inspector, who also has a copy of the specification, for approval. If the item meets the specification, the inspector stamps the invoice and package with a shield; if it does not, the inspector rejects it, and the vendor must send a replacement that does meet the specification.

Specifications should be subject to constant revision and scrutiny. Information about products purchased should flow back to the buyer and make clear where changes are necessary. New products regularly appear on the market, and these should be investigated to see if they offer improved quality or lower cost. As these are tested and approved, they should be used in place of the other product, and the specification should be revised to indicate the new preference.

■

DESIGNING PURCHASING PROCEDURES

Proper buying requires a great deal of communication among buyer, receiving clerks, accounting personnel, and others. Management should take care to establish fluid lines of communication in this sphere to ensure that good products reach the consumer. It is usually the buyer's responsibility to establish purchasing process procedures as well as to monitor the procurement process.

The purchasing process involves a considerable amount of detail, including the exact item needed, the specifications for it, and the delivery requirements. The principal function, ordering, is most often accomplished through what is called a *purchase order*, which may be a handwritten or computer-generated document. On this form, the number, size, weight, and other pertinent information about the item are listed. Each form is numbered for reasons of security and consistency, and a proper signature is always required prior to purchase authorization. Generally, four copies of the form are produced: one for retention by the purchasing department; one for the accounting department; one for the purveyor; and one for the receiving department, to advise it of the order prior to delivery of the product. In some cases, a signed vendor's copy is returned in advance of delivery; at other times, the copy accompanies the delivery. Regular purchase orders cover a single delivery on a specified date, while open-delivery purchase orders arrange for the purchase of items over a period of time (see Fig. 5-5). Weekly and daily items are often purchased by open-delivery purchase orders.

A purchase record should also be maintained, listing each item ordered, its price, its purveyor, and any other information needed (see Fig. 5-6). This purchase price record is often kept on index cards, but increased use of microcomputers has allowed it to be kept on computer files as well. Various computer

PURCHASE ORDER

No. 1124

HAPPY COW DAIRY COMPANY
(NAME OF PURVEYOR)
123 Milk Can Lane
(ADDRESS OF PURVEYOR)
New Haven, Connecticut 06511
(PURVEYOR'S CITY AND STATE)

TO PURVEYOR: OUR PURCHASE ORDER NUMBER (ABOVE) MUST BE SHOWN ON ALL YOUR INVOICES AND PACKAGES.

DELIVERY DATE (S):
May 1-31, 2000

DELIVERY INSTRUCTIONS: Deliver fresh Mon. thru Fri. no later than 6 A.M. to storeroom receiving area.

TERMS: 1% 10-Net 30 days

FOR OFFICE USE ONLY:
CHARGE TO (ACCOUNT):   TOTAL TO DATE: BALANCE:

AUTHORIZED SIGNATURE

SERIAL MODEL, OR CIR SPEC. NUMBER	ITEM	QUANTITY ORDERED	QUANTITY RECEIVED	UNIT PRICE	EXTENSION
300.1	Milk Fresh Homo. 1 gal. Norris	2100 Gal	2250 Gal	$2.04	$4590.00

Date Received for Payment: 6/4/00
Prices, Extensions and Total: RC 6/4/00
Verified by: 2693
Date Paid: W. a. H.
Check Number:
Bursar's Initials:

SUB TOTAL: $4590.00
LESS DISCOUNT: $45.90
PURCHASE ORDER TOTAL: $4544.10

SHIPMENT RECEIVED AND VERIFIED BY:
5/31/00
CULINARY INSTITUTE REPRESENTATIVE   (DATE)

SHIPMENT RECEIVED AND VERIFIED BY:
PURCHASE REQUEST ORIGINATOR   (DATE)

Name and Address: Used for the name, address, city, state, and zip code of the purveyor.

Delivery Date(s): Filled in as required for regular or open-delivery purchase orders.

Delivery Instructions: The specific building, section, or address where delivery is desired.

For Office Use Only: For operations maintaining separate cost accounts.

Terms: Used to record applicable discount or other terms, as specified by the vendor.

Authorizing Signature: Signature of the person authorized to approve purchase orders.

Serial, Model, or Specification Number: On purchase orders for equipment, filled in with the serial, model, or catalog number; on purchase orders for food, filled in with the specification number.

Item: An accurate, complete description of the item desired; if a specification number is used, the description may be shortened, since the specification itself gives complete details.

Quantity Ordered: The quantity needed and the correct unit of sale (such as 50 lbs, 25 doz, 4 gal).

Quantity Received: The exact quantity received; on open-delivery purchase orders, all delivery slips for the period must be added, and the total quantity received must be recorded.

Unit Price: The dollar and cents figure per unit.

Extension: The number of units received multiplied by the unit price.

Subtotal: The figures in the Extension column summed.

Less Discount: The amount of discount (if any), entered in red ink or in parentheses.

Purchase Order Total: Remainder when discount is subtracted from Subtotal.

Shipment Received and Verified by (Culinary Institute Representatives): The signature, in ink, of the person actually receiving the merchandise.

Shipment Received and Verified by (Purchase Request Originator): Some establishments, when purchasing equipment and certain other items, require a purchase order request as authorization to prepare a purchase order. If a purchase order request is required, this section would be signed by the originator of the request. This would relieve the actual receiver from responsibility for having received the merchandise. This section would not be used on food purchase orders.

*Fixed space is required for constant needs and does not vary in any operation. Janitorial storage areas, employee toilets, restrooms, lounge area, and so on would be found in this space.

FIGURE 5-5 Sample purchase order, with line-by-line explanation of blanks.

PURCHASE PRICE RECORD

Date	1	2	3	4	5	6	7	8	9	10	11	12	13	14	15	16	17	18	19	20	21	22	23	24	25	26	27	28	29	30	31
Sep.																															
Oct.																															
Nov.																															
Dec.																															
Jan.																															
Feb.																															
Mar.																															
Apr.																															
May																															
Jun.																															
Jul.																															
Aug.																															

Specifications:

FIGURE 5-6 Sample form for a purchase-price record.

software packages are available that make recording purchasing information both fast and efficient.

RECEIVING

All too often, the benefits of an excellent purchasing plan are lost because of poor receiving. It does little good to perform careful calculations of quantities required or to specify the exact quality needed, and then order from a purveyor who should be able to supply your needs, if the receiving clerk takes any amount and any quality of goods that come to the door. Food cannot be sold profitably if it arrives in short weight or poor condition, or if it does not arrive at all.

Receiving Practices

Receiving practices may vary with different foodservice companies, but the following general principles governing the process are standard:

1. Check the invoice to see whether or not the products delivered agree with it.
2. Inspect the merchandise to determine whether or not it conforms with the purchase order and specifications.
3. Tag all meats with the date of receipt, the weight, and any other information needed to properly identify the delivery.
4. List all items received on the daily receiving report.
5. Accept the merchandise by signing the invoice and returning a copy to the deliverer. If a discrepancy exists, make note of it on the invoice and have the delivery person initial it.
6. Store or deliver the items received properly.

In large foodservices, receiving is a highly specialized, full-time job that should be supervised by the accounting department. In some foodservices, however, a full-time receiving clerk may not be needed, in which case the job may be combined with one in the storeroom. Proper authority and responsibility must be given to the person in charge of receiving, including jurisdiction over those who help receive and store delivered items.

At least seven practices can be identified as essential to good receiving:

1. *Be ready for the delivery.* The foodservice, the customer, and food quality (especially that of perishables) all benefit if receiving is efficient and moves with dispatch and order. Being ready helps to make this occur.

2. *Check the merchandise thoroughly.* Check all merchandise to determine whether it meets the criteria named in the purchase order, specification, or other receiving document. Open cases if they are discolored or damaged or appear to have been tampered with. Date canned goods before storing them, or date them as they are placed on shelves. Arrange goods so as to ensure first-in-first-out (FIFO) use. Many systems have been devised for checking produce quality, but careful examination is at the core of every one. Many receivers select a typical fruit or vegetable and cut it in half to check such characteristics as juice, absence of rust, and taste. If quality cannot be thoroughly checked because of insufficient time at the receiving dock or for other reasons, a thorough check for quality should follow soon thereafter. The supplier should agree ahead of time to this and should make good any deficiency.

3. *Weigh items separately.* When receiving bulk items, remove excess paper, cheesecloth, or ice. A scale weighs anything and everything placed on it. Excessive wrapping or ice adds superfluous weight. All iced items should be unpacked and checked for net weight.

4. *Weigh meat items separately, never together.* Meat is the most costly item received, and weighing each item separately can catch errors that could significantly increase food cost. Items should not be weighed together because 10 lb of missing tenderloins (for example) are not offset by 10 lb of ground beef.

5. *Tag meats on delivery.* This practice prevents disputes over weights with the supplier, the kitchen staff, or the butcher. Tagging meats also reduces the chance of spoilage or excessive weight loss,

since the tag can be dated for FIFO (see Fig. 5-7). It also simplifies calculations of food costs, since a good record of meat withdrawals can be obtained from the tags taken from meat as it comes from inventory.

6. *Check for quality.* Checking for quality is often neglected by the receiving clerk. It is important to see that the grade and quality of the merchandise agree with those shown on the invoice, purchase order, or other receiving document. Care is especially necessary when there are grades within grades.

7. *Store items promptly.* Check and store perishable items first. Exposure to high temperatures can lead to serious loss of quality or rapid spoilage. The quicker all merchandise is stored, the less chance it has to spoil, deteriorate, or get lost.

Because of large orders or the press of other duties, the receiving clerk does not always have time to make a complete check for quality; moreover, the clerk may not be capable of performing such a check. Meat, for example, is extremely difficult to judge and

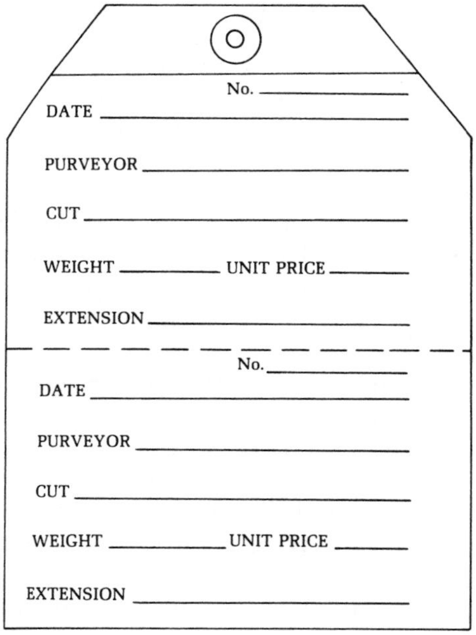

FIGURE 5-7 Sample meat tag.

requires checking by a person who is trained in meats. Persons capable of judging should do their part in checking. Quality checks can often be made on the basis of tests of random samples. For example, if ten cases of lettuce are received, the clerk could inspect several of the cases. It is wise to ensure that receiving clerks get training in quality determination so that they will be able to tell whether the quality of received items meets the operation's specifications or purchase order criteria.

Invoice Receiving

The most frequently used receiving method is called *invoice receiving*. In this method, the invoice accompanying the delivery is checked against the merchandise, along with the specifications, the purchase order, the quotation and order sheet, or other relevant document. Any discrepancies are noted and brought to the attention of the delivery-person, and action is taken immediately to correct the invoice. The quantity and quality of the products delivered should be checked against the purchase order and other receiving documents. This method is quick and economical; if correctly carried out, it is a fairly sure way of receiving goods properly.

Blind-Check Receiving

A blind receiving method involves giving the clerk a blank invoice or purchase order listing the incoming merchandise but omitting the quantities, quality, weights, and prices. The receiving clerk must insert data into the order on the basis of a check of the delivery. This approach forces the receiving clerk to make a serious check of the delivery and not merely to glance at it to affirm that the goods probably agree with the figures on the invoice. If no weight is on the invoice, the receiving clerk must weigh the items to get an accurate figure. A count is also required, as is a quality check. Another invoice with quantities, weights, quality, and prices is mailed to the accounting office. This invoice is checked against the one from the receiving clerk, and the figures in both are verified. Blind receiving is an accurate method of checking merchandise and verifying deliveries, but it takes longer and costs more, since it requires the clerk to prepare a careful and complete record of all incoming merchandise.

Partial-Blind Receiving

A hybrid of the invoice receiving and blind receiving methods is called *partial-blind receiving*. In this method, the receiving clerk is furnished itemized purchase orders, invoices, or other receiving documents with only the quantities omitted. When the merchandise is checked, the receiving clerk lists the quantity of each item in the space provided. This method is slower but more accurate than invoice receiving, and it is faster but less accurate than blind receiving. It is essential in blind receiving and partial-blind receiving either that the supplier's invoice does not accompany the merchandise or that the information on quantities be omitted or made invisible by a black area where such information would appear. Otherwise, there is no guarantee that the receiving clerk's numbers represent independent research.

Detailed purchase specifications should be available for all merchandise received. These should be duplicates of the ones used for purchasing. If tomatoes are ordered from a specification that reads: "Tomatoes, US No. 1, fresh, 5 × 6, pink, 20 lb minimum to the lug," the receiving clerk should be able to check grade, ripeness, and number in the case as well as weight. There are usually three layers of tomatoes in a lug, and if there are five tomatoes per row and six rows, there will be ninety tomatoes in the lug at an anticipated net weight of 20 lb. Good specifications permit good checking. It also is helpful to show minimum weights of packing containers in a chart or on some other form. The net weight of spinach and other greens should not be less than 18 lb, and the container weight should not a part of this. Florida oranges should have a net weight of 45 lb per $\frac{4}{5}$ bushel carton, plus the weight of the container. Having such information can help make receiving easier and quicker. Catching discrepancies is also favored by such a system. A case of whiskey that is missing one bottle, for example, can quickly be detected if it is weighed and the total contained weight is about 3 lb less than it should be.

Forms and Records

Recording incoming deliveries is as important as checking their quality and quantity. The form or style of doing this may vary. The three forms most commonly used for this purpose are the daily receiving report, the substitution invoice, and the request-for-credit memorandum.

Receiving Clerk's Daily Report

The receiving clerk's daily report lists the merchandise received in a form suitable for checking against the supplier's invoice (see Fig. 5-8). The daily report should list the following items:

- Date of delivery
- Invoice or purchase order number
- Supplier
- Number of units

- Quantity received
- Article
- Unit price (omitted in blind receiving)
- Total amount extension (omitted in blind receiving)
- Distribution of the delivery

Deliveries of food are divided into two groups: food direct and food stores. Many purchases are sent directly to the kitchen for use, which saves time by omitting the steps of storage and withdrawal. It also helps the foods involved retain their freshness. This is charged out on the receiving report as a food direct delivery, and it goes immediately into the cost of food

RECEIVING CLERK'S DAILY REPORT

INVOICE NO.	FROM WHOM PURCHASED	UNIT	AMT.	ARTICLE	UNIT PRICE	TOTAL AMT.	FOOD DIRECT	FOOD STORES
				TOTAL FOOD RECEIVED				

SIGNATURE

FIGURE 5-8 Sample form for a receiving clerk's daily report.

SUBSTITUTION INVOICE

FROM: Farmer Joe's Meats   DATE: 3/20/00

QUANTITY	UNIT	DESCRIPTION	AMOUNT	
25	lbs.	Ground Round @ $.84	21	00

RECEIVED BY: ___John Black___

FIGURE 5-9 Sample form for a substitution invoice.

for the day. Purchases sent to the storeroom or to meat or produce refrigerators are classed as food stores and become a part of inventory. If the receiving record is used as a consolidation of food direct and food stores for each day's deliveries, the consolidated amount must be equal to the total amount for the day's deliveries.

Substitution Invoice

A substitution invoice is primarily used when merchandise arrives without an invoice. It contains practically the same information as the receiving record (see Fig. 5-9). Indeed, some operations consider it a duplication of effort and as a result do not use it. When the supplier's invoice reaches the accounting office, the substitution invoice is compared with it as a basis for verifying and approving the delivery.

Request-for-Credit Memorandum

The request-for-credit memorandum, which is usually made in triplicate, lists discrepancies such as shortages in quantity or failure of the quality to conform to specification (see Fig. 5-10). The original is sent to the supplier, usually with the signed delivery invoice.

A copy is retained by the receiving clerk and another is sent to the accounting office. If a credit memorandum is submitted because of unsatisfactory merchandise, the following steps are usually taken by the receiving clerk:

1. Write up the request-for-credit memorandum, listing the reason why it is being made out.
2. Return the merchandise immediately.
3. Note on the invoice the reason for returning the merchandise.
4. Notify proper authorities so that menu changes can be made, if needed.

A good practice, when possible, is to have the delivery person sign the invoice for deficient merchandise. This can save explanations, telephone calls, and disputes.

Purchase Invoice Stamp

In some operations, a purchase invoice stamp is used on all incoming invoices (see Fig. 5-11). The receiving clerk stamps and dates the invoice, indicating receipt,

```
                    OPERATION'S NAME AND ADDRESS


      TO: _____        No. _____

      ADDRESS: _____

      CITY, STATE: _____    DATE: _____

         GENTLEMEN: Please send Credit Memorandum for the following:
```

INVOICE NO.	ITEM	QUAN.	UNIT OF SALE	UNIT PRICE	EXTENSION
					Total

```
      REASON:


                                      BY: _____


                                      TITLE: _____
```

FIGURE 5-10 Sample form for a request-for-credit memorandum.

```
Order No.      Rec'g No.
Date Rec'd _____          DATE REC'D_____
Quantity O.K. _____       QUANTITY O.K. _____
Quality & Price O.K. ____         PRICES O.K. _____
Ext. & Footing O.K. _____       EXTENSIONS O.K. _____
Entered _____            APPROVED BY_____
O.K. for Payment_____
```

FIGURE 5-11 Sample purchase invoice stamps.

and signs or initials the line "Quantity O.K.," showing that the correct quantity has been received. This invoice usually goes to purchasing for approval of prices and other factors, and it is then sent to accounting, where the invoice is compared with the receiving clerk's daily report. Then, if all checks pass, the invoice is approved for final payment.

Nonseller Delivery

It is becoming more and more common for sellers to deliver merchandise by carrier other than their own. This will especially increase with use of the Internet in which United Parcel Service (UPS), the U.S. Postal Service, or another carrier delivers the goods ordered. If possible, inspection should still be made according to desired procedures, but this may not be possible. In cases in which the buyer merely signs the delivery slip of the carrier and inspects later and then finds a discrepancy, the seller should be called and arrangements made for return of the goods or other disposition of the goods made as agreed to by the buyer and seller.

■
EVALUATING THE
PURCHASING TASK

Even after the food is served and the profit is made, the purchasing task is not over. How well was the purchasing done? Could something better have been used? What was the yield? Could quality be improved? Were values purchased that were not necessary? Were other values needed? Every purchasing system should include a good feedback arrangement so that the purchasing task can be improved. Just as recipes and specifications are constantly under revision, so should the entire purchasing procedure be. New products come onto the market, and new purveyors appear who may give better service, prices, or quality than their competitors. Many markets are highly dynamic, and planners must constantly be on the alert to note changes and modify the purchasing system used.

Operations often do value analysis—an analysis of what is needed against what is obtained. Value represents quality divided by price, or $V = Q \div P$. If the quality of a purchase is given a score of 4 and the price is \$2, the value in this case is 2 ($4 \div 2 = 2$). If a search of the market shows that a product of equal quality can be obtained for 50 cents less, the value of the latter product is 2.67 ($4 \div 1.5 = 2.67$).

Large factories set up value analysis departments. The personnel in these departments scrutinize what is purchased to see if a better product might be available, if some items are included that need not be, and if a better price might be obtained. The tendency of organizations to add such a department indicates that value analysis does help. It is not necessary, however, to set up a whole department. Buyers and others in the operation can practice value analysis themselves. By constantly seeking to improve the products purchased, they constantly upgrade purchasing performance.

■
PURCHASING AND
RECEIVING BY COMPUTER

Today more and more foodservice operations have become computerized, through the use of personal computers, laptops, or through outside services. In purchasing and receiving departments, great strides have been made in the last ten years toward reducing labor costs through computer automation. In most operations, lasers scan the universal product code (UPC) numbers of incoming boxed orders and automatically record the entries into the data bank of the computer for comparison with the purchase order and for inclusion in the inventory, or computerization by keyboard entry of the products received is done.

Perhaps the principal use of the computer in purchasing and receiving is in inventory control, verification of purchase orders, and par stock monitoring. Inventory control is achieved through connecting point-of-sale (cash register) data with storage files. Each time an item is purchased (or upon verification at the end of the shift, if so desired), a corresponding item is deleted from the inventory. Thus, the current inventory levels are monitored perpetually, and at any time the purchasing agent can ascertain the need for replenishment in any area by examining that area's inventory level.

Using the computer to process purchase and receiving orders has become widespread. In this system, each time a delivery is made, the delivery receipt is matched against the purchase order form on the computer for verification of size, weight, price, number of units, and so on. If the items do not match, the operator is alerted to the discrepancy. Once the delivery receipt is matched to a purchase order, the items are entered into the computer, and the inventory is immediately upgraded.

Using the computer for par stock ordering follows the same pattern, except that, as products drop below par levels in the inventory, a note is added on the printout or screen to alert the purchasing agent of the depleted stores, thus ensuring prompt reordering and maintenance of the needed par stock. Many programs have been devised that also provide for an automatic printout of needed items, by quantity, to create an immediate order form. Permanent records of the purchasing department, such as names of purveyors, purchase prices, and dates products were ordered and received, are also easily kept on file by means of the computer.

Another advantage of the computer in purchasing and receiving is its ability to perform rapid comparisons. By designing a program to store prices by purveyor, for example, a buyer can rapidly access information that can be used to identify the pricing trends of various purveyors. The same is true with respect to such information as timeliness of delivery, quality, and product omissions—all keys to making decisions on which purveyors to use.

------------------------------■------------------------------

STORING FOODS

Once food has been received, it must be properly handled to prevent waste and deterioration. While in many cases the responsibility for storing and issuing food is assigned to the same person who handles receiving and purchasing, in larger operations these may be separate jobs.

The task of properly storing and issuing food can pay for itself if it is conscientiously carried out by staff and management. While most manufacturing firms have recognized the importance of proper storage, many foodservices have not yet done so. Many storerooms in the industry are poorly planned and organized, poorly lit, and improperly secured—all of which contribute to lost time and money. In the worst cases, food products get stored in places where they are not found for months, by which time they are completely deteriorated. In many cases, the storerooms represent design afterthoughts and are put in inconvenient areas that would otherwise be surplus space. The main storage area often turns out to be the basement, where conduits, hot-water pipes, and other items offer both physical obstacles and high heat levels. Since the temperature in dry storage areas should not exceed 80°F (27°C) at any time, the placement of storage in relation to boilers, condensers, and heating units is an important issue. A storeroom should also be planned to promote smooth work flow, ease of maintenance, good security, and the preservation of value.

The correct storage temperature and storage humidity vary for different foods. If these are not kept at correct levels, an operator is inviting the outbreak of a foodborne disease. At the same time, a climate that is too dry in refrigerated or freezer storage can dehydrate the food and diminish its quality. Each of the three types of storage—dry, refrigerated, and frozen or low-temperature—has its own requirements for proper maintenance.

Dry Storage

A well-lighted, well-ventilated, dry, cool area is needed for good dry storage. It should be free of insects and rodents and should be clean, orderly, and well managed. The dry storeroom usually holds canned goods, flour, cereals, spices, sugar, and even some oils and shortenings (in sealed cans). The best location for it is near the receiving station and the kitchen, so that long hauls are not required and security supervision can be maintained.

Temperatures in the dry storage area should be between 50 and 70°F (10 and 21°C), with a minimum of 40°F (7°C) and a maximum of 80°F (27°C), 50°F being ideal, with a relative humidity of 50 to 60 percent. Proper temperatures preserve flavor and nutritive value and reduce waste and food spoilage.

Perishables such as potatoes, onions, fruits, and some vegetables need good ventilation and a temperature within the range from 40 to 55°F (4.5 to 13°C). They should be stored off the floor and away from the wall. Proper stacking and good arrangement on racks or skids can help facilitate movement of food items in and out. Bottom shelving levels should be at least 10 in. off the floor to ensure adequate ventilation and freedom from contamination by floor soil or water, as well as to allow for cleaning underneath. For adequate ventilation, all shelving should be at least 2 in. away from the wall, giving air room to circulate in the back. Such procedures offer the best chance for temperature control and for the elimination of odor transfers, moisture buildup, and rapid food deterioration.

Rodents and insects can cause heavy damage. Because these pests carry disease, all shipments should be inspected for their presence, and infested shipments should be refused. Vigilance and constant effort are necessary to keep the storeroom free of vermin.

Cartons should be stacked so as to prevent creating spaces that might serve as nesting places for mice or rats. Insecticides or rodenticides should be used only by qualified personnel, and poisonous materials should never be stored where food is stored or prepared. Poisons should be stored in a custodian's closet, in the furnace room, or in a locked cabinet; containers should be clearly marked and handled only when necessary. Poisonous materials include drain cleaners, lye, acids, insecticides, rat poisons, bug bombs, fly sprays, and roach powders.

Managing dry storage properly requires planning and time. The responsibility for such management should be given to one person who has complete control. No unauthorized personnel should be admitted to the dry storage area. Among the duties of the person in charge of dry storage are the following:

1. Keep the storeroom clean. At least once a week, the floor, tables, platforms, and shelves should be thoroughly cleaned. The floor should be swept and mopped according to schedule. A floor that slants toward drains facilitates mopping and drainage.

2. Inspect all shipments for damage, rodent or insect infestation, and spoilage, and take appropriate action if any is found. For example, the receiving station should be told to inspect items received more thoroughly. In addition, the purveyor might be warned, or the goods returned.

3. Place food in the storeroom as quickly as possible.

4. Date all packages.

5. Stack foods of a kind together to ensure good storeroom organization.

6. Place oldest stock in front and issue it first (FIFO). If the food items are stored in bulk, the bin containing them should be cleaned, and the new stock should be placed at the bottom of the bin. (This applies to drawers and bins in cooks' and bakers' tables, as well.)

7. Take frequent inventories.

8. Check frequently for evidence of damage, spoilage, broken or torn packages, and bulging or leaking cans. If any are found, they should be removed immediately, and the area should be cleaned thoroughly to prevent contamination of other foods.

Nonfood items that are stored in dry storage places include paper goods and reserve supplies of utensils and dishes. Cleaning supplies should have a separate storage place. Clean towels and uniforms should be neatly stored in metal cabinets, drawers, or other protected areas, away from dust and insects. Soiled linens and uniforms should be kept in hampers, laundry nets, or bags at appropriate points throughout the operation. They should be transported frequently to the laundry or to the place from which laundry trucks can take them away. Workers' personal belongings should be stored in spaces set aside especially for them, such as locker rooms, but not in the food storage area.

Refrigerated Storage

Sufficient and proper refrigeration is very important. Many sanitary codes require that the local health department approve all new refrigeration equipment prior to its use in a foodservice. The health department should be consulted before purchase is made of refrigerators or walk-ins to see if they meet required standards. The following perishable foods should be held under refrigeration until needed:

Fresh Foods	*Processed Foods*
Meat	Vegetable or fruit salads
Fish	Custards
Poultry	Gravies
Vegetables	Sauces such as hollandaise
Fruits (except tropical ones)	Sweet dessert sauces that can ferment
Eggs	Condiments that are perishable
Dairy foods	Cooked foods (especially meats)
	Foods prepared in advance
	Carryover foods

Walk-in refrigerators provide most of an operation's refrigerated storage because they typically have more permanent storage space than do reach-in units. Reach-in units are often used when refrigeration is needed in a local area near a work center. Six factors usually dictate refrigeration needs: space needed,

TABLE 5-2 Minimum Refrigeration Space Needed, Not Including Beverage Cooling or Frozen Foods

Number of Meals Served Daily	Recommended Capacity (ft³)
75–150	20
150–250	45
250–350	60
350–500	90

sanitation, air circulation, temperature, placement, and general practices and procedures.

Space Needed

The space needed for refrigerated storage depends on the number of meals served, the type of service, merchandising practices, the food itself, and delivery schedules. Table 5-2 lists recommended minimum refrigeration space requirements for an average-sized full-menu restaurant (not including beverage cooling or frozen foods).

Sanitation

Many foods will deteriorate, spoil, or become contaminated if they are not kept refrigerated. Clean refrigerated areas also help reduce the risk of food loss. Good cleaning habits include the following:

1. Wipe up any spilled food immediately.
2. Remove shelves, drip pans, meat hooks, ice trays, and containers, and wash them frequently with warm water containing a good detergent. Rinse and dry.
3. Defrost when any buildup occurs on coils or cooling surfaces. This improves efficiency and reduces operating costs.
4. Wash refrigerated area interiors often with warm water and a good detergent.
5. Wash rubber gaskets on doors (if present) with warm water and a good detergent or mild soap. This increases their life.
6. Wash the exterior of refrigerators with warm water and a good detergent.
7. Flush drainpipes with hot water and baking soda.
8. Retain a reliable service to clean, lubricate, and check mechanical refrigeration equipment periodically.
9. Check hinges, latches, gaskets, and other parts frequently.

Air Circulation

Air circulation is a must if the refrigerator unit is to work properly. Food items should not be overcrowded in the unit, and adequate space should be left between each product. Shelves should not be covered with paper in the reach-in unit, and no products should be stored against the wall or on the floor of a walk-in unit. Cold air should reach all items at all times.

Temperature

Temperatures in different parts of a refrigerated area may vary. The most perishable foods should be stored in the coldest areas, which are near the bottom (since cold air settles). Another cold area is located close to the coils. A thermometer should be located in the warmest area and checked frequently.

In some operations, separate refrigerators are used for various food groups. When this is the case, the ideal storage temperatures for each refrigerator are as indicated in Table 5-3. A refrigerator-temperature record

TABLE 5-3 Ideal Temperatures for Various Foods

Food Group	Temperature Range (°F)
Fruits	45–50
Vegetables, eggs, processed foods, pastries	40–45
Dairy products	38–40
Fresh meats	34–38
Fresh poultry, fish, seafood	32–36
Frozen foods	−10–0

Date	4/15/88		4/16		4/17		4/18		4/19		4/20	
Time	9	3	9	3	9	3	9	3	9	3	9	3
Freezer	0	−1	0	−1	+1	0	−1	0	0	+1	−1	−1
Fresh fish	33	34	34	36	33	35	33	32	35	33	32	34
Vegetables	43	41	43	42	41	41	40	40	42	43	42	42
Dairy	38	39	38	40	40	39	40	38	38	39	38	40

FIGURE 5-12 Sample refrigerator temperature record that has been completed twice daily for six days.

book may be maintained if needed (see Fig. 5-12). Some refrigeration equipment has automatic temperature recorders. Unnecessary door opening should be avoided, since this warms the refrigerator and increases power consumption. Walk-ins may be equipped with reach-in doors to reduce the amount of travel in and out. To save energy, some foodservices may find it desirable to schedule all entries for purposes of storing and withdrawing foods. Most foodservice operations have successfully attached door closures to refrigerator doors to ensure that they close upon exit.

Placement

The location of refrigeration units is extremely important. Since the amount of time spent fetching items for use in foodservice kitchens has been estimated to consume 50 percent of total labor hours, any design that brings the refrigeration units closer to the user is a great laborsaver.

The location of the compressors and condensers is also important. These units should have good ventilation to ensure the necessary airflow for cool operation, and they should be as close to the refrigeration units as possible. Many foodservice operations have chosen to place the compressors and condensers on the roof of the building—perhaps a good location except from the point of view of serviceability—or at another outside location. If this is done, winterization controls must be installed to keep the units from freezing in cold weather, and housings must be constructed to protect the equipment from rain and snow. A long run from the compressor to refrigerated areas consumes energy and should be avoided. Most health departments require drains for refrigeration units to be trapped, rather than be run direct into sewer lines, to prevent backup into the water supply. Freestanding refrigerators should be at least 6 in. from the wall for cleanability and air circulation, unless they are mounted on casters for moving. All units should be either sealed to the floor or mounted 6 to 10 in. off the floor for cleaning. Refrigerators should not be located near cooking equipment, and all hot pipes nearby should be well insulated.

General Practices and Procedures

The following practices and procedures should be among those set up for the use and care of refrigeration units:

1. Inspect deliveries before putting them into refrigeration units. Do not put spoiled, dirty, or vermin-infested products into the refrigerator.
2. Put food in as quickly as possible.
3. Shelves nearest the cooling unit or at the bottom are coolest. Use them for milk, meats, poultry, fish, and prepared dishes that contain these food items.
4. Use the refrigerator for perishable foods only. Jellies and tropical fruits such as bananas, pineapples, papayas, and avocados may not require refrigeration. The same is true for many root vegetables such as potatoes and onions.
5. Remove outside wrappings, which may contain soil or harmful bacteria. Packaged luncheon meats, butter, and cheese should be placed in airtight containers to prevent drying and discoloration.

6. Keep processed food and carry-overs covered to prevent them from drying out or transferring odors. This practice also prevents contamination from things dropping into them.

7. Be careful when placing warm food into the refrigerator. Food at the unit's center may retain enough heat to spoil. Warm food raises the interior temperature and may cause other foods to spoil. To cool food quickly, place the container in cold water or crushed ice, and stir frequently until it has cooled. Then cover the container, and place it under refrigeration. If a quantity of hot food must be refrigerated, make sure that it is stirred frequently while cooling. Such food should be poured into pans no deeper than 2 to 4 in.

8. Store foods that have strong odors in tight containers.

Frozen Storage

Frozen storage needs have been increasing over the past thirty years, most recently because of the large number of new products that are prepared and packaged in advance and require freezer storage.

Greater use of frozen foods has meant greater space allocated to the storage of frozen product in foodservice operations. Besides more space being dedicated to this product, equipment needs have increased from frozen cabinets to on-premises blast freezing units.

The increase in the use of frozen foods has come with a continual commitment by operators to improve portion control, labor costs, consistent quality, and scheduling of production. Limited menus have dictated the use of frozen foods, with operators having the frozen food storage space to buy advantageously and to provide needed variety.

With the use of greater convenience foods, operators have also taken advantage of preparing product on slower days, freezing it, and serving it on busier days. This levels the workload and helps to hold the line on labor costs.

The three-step process in freezing foods are (1) refrigeration to cook, (2) sharp or flash freezing, and (3) holding at low temperature. The faster food freezes, the better its quality, since sharp freezing keeps ice crystals small and undetectable.

With freezing occurring on-premises, operators must realize that placing newly frozen product with previously frozen product results in a deterioration in the foods held. Accordingly, there must be enough engineered freezing capacity, located separately and distinguished from storage capacity, to freeze food items in quantities necessary to meet menu requirements.

To distinguish between the new pieces of equipment available to operators, the correct terminology must be understood. The following is offered to clarify definitions.

Use and Temperature	Terminology
Equipment designed to store frozen food (operates at 0 to −10°F)	Frozen food storage Low-temperature reach-in Walk-in space
Equipment designed to do the actual freezing job (operates at below −20°F)	Food freezer Processing freezer Blast freezer Plate freezer Tunnel freezer

Frozen foods are successfully replacing a sizable portion of on-premises food production. As this trend increases, more frozen food cabinets are appearing. The cook, the baker, the salad maker, and counter workers are all using frozen products and need low-temperature storage in their work areas.

The following guidelines should be observed in work with frozen foods:

1. Keep frozen foods stored at a temperature of 0°F (−18°C) or less. The lower the temperature is and the less it varies, the better.

2. Cook thawed foods promptly. Ground meat or moist meat provides an ideal place for spoilage organisms to develop rapidly.

3. Never refreeze foods that have thawed. Nutritive value, flavor, and appearance suffer as a result of refreezing.

4. Insist that frozen items be frozen when delivered. Store only clean packages that have been inspected carefully. Products to be frozen should be packaged in a commercial freezer wrap to protect them while they are being stored.

5. Place old stock in front to be used first. Rotate stock and purchase new items frequently.

6. Bear in mind that frozen food will not keep indefinitely. Over time, it gradually loses nutrients, flavor, texture, and color (see Table 5-4).

Storage Controls

The use of correct storage controls ensures that food is appropriately handled before and during storage. One person should be given responsibility for seeing that good controls are instituted. This may be the receiving clerk, the storeroom clerk, or another person. Since good controls usually affect the storeroom most, the person in charge of it should see that the controls are implemented. Duties of the storeroom clerk include the following:

- Receiving deliveries, checking them into storage, and tagging or marking them with the date received and the price paid
- Maintaining orderly arrangement and cleanliness in storage areas
- Informing the supervisor when stocks are low or when they need attention

TABLE 5-4 Maximum Frozen Food Storage Periods at 0°F

Product	Storage Period
Sausage, ground meat, fish	1–3 months
Fresh pork (not ground)	3–6 months
Lamb, veal	6–9 months
Beef, poultry, eggs	6–12 months
Fruits, vegetables	One growing season to another

TABLE 5-5 Maximum Storage Times and Temperatures for Items in Dry or Refrigerated Storage

Food	Suggested Maximum Temperature (°F)	Recommended Maximum Storage Life	Remarks [a]
Beans, flour, rice	70	6 months	In original container or covered galvanized can
Candy (chocolate)	70	3 months	Wrapped or in original carton; may be frozen
Canned goods	70	12 months	In original containers
Cereals	70	6 months	In original packages
Cream-filled pastries, custards	36	Serve day prepared	Spoil readily; must be served the day prepared
Dairy products			
Butter	40	2 weeks	In waxed cartons
Cheese			
hard	40	6 months	Tightly wrapped
soft	40	7 days	In tightly covered container
Ice cream and ices	10	3 months	In original container, covered
Milk			
dried	70	3 months	In original package; if open, 38°F in tight can
evaporated	70	12 months	In cans; invert every 30 days
fluid	40	5 days	In original container, tightly covered
Eggs, whole	45	7 days	In original carton
dried	70	6 months	In original carton; if open, 45°F in tight can
whites	45	2 days	In tight container
yolks	45	2 days	In tight container; cover with water
Fish, fresh	36	5 days	Wrap loosely
shellfish	36	5 days	In covered container
Fruits			
Apples, pears, citrus	70	2 weeks	In original containers
Dried	70	3 months	In original containers
Peaches, plums, berries	45	7 days	Unwashed (wash before using)

- Assisting in food pricing and in the taking of inventory
- Ensuring that stocks are rotated and that foods are used promptly (see Table 5-5)

Stock Record Cards

Stock record cards are used to indicate the quantity of stock that exists on inventory (see Fig. 5-13). These cards are used for all items on hand and indicate how much has been received from purveyors, how much has been issued, and how much remains. The responsibility for maintaining stock record cards may be that of the storeroom, the accounting department, or some other department of the operation. Frequent spot checks to verify quantities on hand should be made.

Inventory Methods

Various inventory methods may be used. Overall, they serve to indicate the rate of stock usage, the amounts

NAME OF ITEM	TOMATO PASTE				UNIT	CAN	
LOCATION	MAIN STOREROOM				UNIT PRICE		
REMARKS	1 co = 6/10 106 S.G.			ITEM NUMBER	HIGH LIMIT	24	
					LOW LIMIT	6	
DATE	REC'D	ISSUED	BALANCE	DATE	REC'D	ISSUED	BALANCE
4/28/00	BF		9				
4/29		2	7				
4/30	INV		7				
5/3		2	5				
5/5	18	1	22				
5/7		2	20				
5/9		1	19				
5/11		2	17				
5/13		2	15				
5/15		1	14				
5/17		2	12				

FIGURE 5-13 Sample stock record card.

	TABLE 5-5 Continued			
Food	Suggested Maximum Temperature (°F)	Recommended Maximum Storage Life	Remarks [a]	
Gravies, sauces	36	2 days	In covered containers	
Leftovers	36	2 days	In covered containers	
Meat				
Cold cuts, sliced	38	5 days	Wrap in semi-moisture-proof waxed paper	
Cured bacon, sliced	38	1–2 weeks	May wrap tightly	
Dried beef	38	6 weeks	May wrap tightly	
Fresh meat cuts	38	5 days	Loosely wrapped	
Ground	38	2 days	Loosely wrapped	
Ham				
canned	38	6 weeks	In original container, unopened	
tender cured	38	1–2 weeks	May wrap tightly	
Liver and variety meats	38	2 days	Loosely wrapped	
Tongue, smoked	38	7 weeks	May wrap tightly	
Poultry	36	3 days	Wrap loosely	
Processed foods made with eggs, meat, milk, fish, or poultry	36	Serve day prepared	In covered container, spoil rapidly, so must be served day prepared	
Sugar and spices	70	3–6 months	In original package or covered galvanized can	
Vegetables				
leafy	45	5 days	Unwashed	
potatoes, onions, and root vegetables	70	7–30 days	Dry in ventilated container or bags	

[a]*Tightly wrapped* means that the wrapping material is placed as tightly as possible against the product; *loosely wrapped* means that the wrapping material does not have to touch the product.

of replacement units needed, the types and sizes of stock on hand, and the dollar value of stock on hand. Two inventory methods are used most commonly: the perpetual inventory and the physical inventory. Stock record cards are perpetual inventories if they are properly maintained, since they provide a perpetually accurate record of the amount of items on hand; as a result, they can help considerably in reducing losses in stock. In some operations, computers maintain the perpetual inventory. An actual count of what is on the shelves is called a physical inventory and is usually taken about once every month. A physical inventory is also taken periodically in operations that use a perpetual inventory, to verify the latter's figures.

A physical inventory is best taken by two people, one of whom should not be from the storeroom. If the outside person is from the accounting department or from another area, it is more difficult for the storeroom person to juggle figures during the count to cover or set up pilferage. The outside person counts the items on hand, and the other individual records the count on an inventory form (see Fig. 5-14). This procedure can also be reversed. Rechecks may be made on spot items to verify the figures obtained by the two people who conducted the full inventory.

Minor differences arising between a physical inventory and a perpetual one can be adjusted, but major differences should be investigated. The number of items on hand multiplied by the value of the item is called the *extended cost* or *extension* of the item. All extensions are totaled to give the value of the inventory. Different pricing formulas may be used. An average price may be used if prices have changed. Some operations use the actual price paid for goods, while others use the most recent price paid. Deciding what system of pricing to use is up to management.

Recently, physical inventory taking has been facilitated with the use of an electronic device that feeds information into a computer. The person taking the inventory punches into the device an item's code number and the amount on inventory. The device flashes this information into the computer, which processes the information.

FOOD SERVICE OPERATION

Physical Inventory ________ 19___ Page 1

Classification	Item	Unit	Quantity	Unit Price	Total Cost

TOTAL PAGE 1 ________

FIGURE 5-14 Sample physical inventory form.

ISSUING FOOD

Issuing is the controlled process of transferring foods from storage to a place where they can be processed. At the end of the day or recording period, the total value of goods issued should be known. This value plus the value of direct deliveries (foods that were delivered from receiving to processing for immediate preparation, without going to storage) gives the value of foods used during the period. Most operations give responsibility for food issuing to the accounting department; thus, the storeroom personnel who issue food are on the accounting department's staff. An accounting clerk may issue from inventories, and it is best not to combine such a job with the job of purchasing, receiving, or maintaining other accounts. If specific hours are established for issuing, people who work elsewhere may do the job.

Mechanics of Issuing

Issuing hours should be maintained according to a strict schedule. Personnel should know their daily needs in advance, since running back and forth to and from storage only wastes time and money. The issuing of foods should be a planned event that is carried out in an orderly fashion. This implies that, when an attendant is not on duty to issue food, the storage area is locked and only certain designated individuals have keys. The orderly issuance of all goods is enhanced by tagging merchandise appropriately and by using requisitions properly.

Tagging Merchandise Appropriately

Tagging merchandise appropriately simply means recording the date of receipt and the price paid on the goods themselves whenever goods arrive for storage. Taking time to tag items when they arrive for storage ensures that they will be stored in accordance with the rule of FIFO (first-in, first-out) and that they will be costed out correctly during inventory. All meats should already have been tagged with their weight and price by the receiving clerk when they first arrived at the operation. Such tagging emphasizes to personnel that the items in storage represent money and that this money is their responsibility.

Using Requisitions Properly

Using requisitions properly is similar to using checks properly at the local bank. Goods in storage represent money, so submitting a requisition form should be a mandatory step to take in order to obtain the goods, just as writing a check is a prerequisite for obtaining money from the bank. The requisition form contains space for the date, the name of the item issued, the weight and amount of the item, the price, and the signature of the person authorized to make the requisition

(see Fig. 5-15). Pricing the requisition is done by the individual responsible for issuing the food (someone from the accounting department). This task can be completed easily from the prices tagged on the products themselves in storage. Requisition forms are extremely important and should be handled carefully, just as checks are. One copy of the form returns to its maker, along with the issued merchandise; the other copy remains with the issuing clerk and is used to adjust stock record cards and to figure the daily food costs for the operation. Stock record cards, inventory procedures, and requisition forms constitute issuing controls as well as storage controls.

Issuing Controls

Good controls are necessary to prepare a meaningful daily food cost form (see Fig. 5-16). When this form is completed, a fairly accurate computation of food and supply use can be performed with a minimum expenditure of time and labor. Computing costs only once a month, for purposes of developing a profit-and-loss statement, often renders the information too out-of-date to be of practical use; as a result, the daily food cost computation becomes a very important tool. Any measure designed to correct an undesirable situation is too late to have much effect on the original problem after a month's time. Frequent checks are

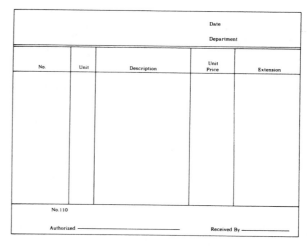

FIGURE 5-15 Sample food requisition form.

SUMMARY OF DAILY FOOD COST

DATE		(1) BEGINNING STOREROOM INVENTORY	(2) STOREROOM PURCHASES	(3) TOTAL	(4) STOREROOM ISSUES	(5) DIRECT PURCHASES	(6) GROSS COST	(7) LESS TRANSFERS	(8) NET COST	(9) SALES	(10) FOOD COST PER CENT	MONTH TO DATE NET COST	MONTH TO DATE SALES	MONTH TO DATE PER CENT
1/1		3500 00	310 00	3810 00	225 00	150 00	375 00	25 00	350 00	1050 00	33.3	350 00	1050 00	33.3
1/2		3585 00	400 00	3985 00	185 00	115 00	300 00	10 00	290 00	850 00	34.1	640 00	1900 00	33.7
1/3		3800 00												

*INFORMATION FOR COLUMN 2 AND COLUMN 5 COMES DIRECTLY FROM THE RECEIVING RECORD

FIGURE 5-16 Sample daily food cost form.

costly but may pay for themselves. Management must decide how often food cost reports and other cost reports should be made. Computers are often used to calculate food costs daily or even by the meal.

The calculation of daily food cost produces an estimate rather than the actual cost. Kitchen carryovers are not included, nor are items that make up daily supplies. If the bakeshop orders 1 lb of cinnamon and 1 gallon of vanilla extract, they are charged out as issues, but only a small part of each may actually be used. In the long run, however, the figures average out, and a fairly accurate cost computation is obtained. The omissions do not decrease values to any extent, for several reasons:

- Most operations have roughly the same amount of carryovers and daily-use supplies on hand from day to day. The variation is not great in either case.
- The running or to-date daily costs average out over an extended period.
- High cost figures result when poor forecasting occurs and there are high carryovers. If proper recovery is made or if food issues are lowered, the figures adjust.

COMPUTER USE IN STORING AND ISSUING FOOD

The computer has been used in large foodservice operations for several years to keep track of food storage and inventory information. With the advent of the personal computer, even the smallest foodservice operation can now successfully attain the same high degree of efficiency and economy as those larger operations.

Whereas in the past, products were tagged for storage by hand, today many foodservices utilize personal computers to print out labels at the same time that they create a data bank of information regarding the incoming order. These labels often include not only date and price (the principal requirements of tagging by hand) but the purveyor, the unit price, and other information valuable to the operation. The storage function has also been substantially aided by the development of recipe files on disk, which include the exact quantities needed for each menu item. Using these files, a chef or clerk can print out the needed requisition items for each meal—by recipe or in bulk—and the printout can serve as the requisition form. At the same time, more sophisticated programs can automatically delete these items from the storage inventory as they are electronically selected on the computer for use.

Because a principal attribute of the computer is its ability to perform repetitive tasks with a high degree of consistency and speed, such required functions as extending inventories (multiplying price times number of items) are easily accomplished by setting up a spreadsheet format on the personal computer. Most programs allow spreadsheet entries to be arranged in seconds alphabetically, numerically, by category, by date of purchase, or in any other format, simply by issuing a single command. Discrepancies can readily and rapidly be uncovered and traced by the computer as well.

■ SUMMARY

Controls for foodservice operations include the purchasing, receiving, storing, and issuing of foods. From purchasing the food from a purveyor who has a HACCP (hazard analysis critical control points) program in place, to implementing the same program within the foodservice operation itself, management will ensure food safety until it is consumed by the patron.

The purchasing and receiving of items are important functions that assure that the right amount of the right product is available at the right time. Management and staff must consider six areas in purchasing if satisfactory purchasing is to be achieved: (1) knowing the market, (2) determining purchase needs, (3) establishing and using specifications, (4) designing the purchasing procedures, (5) receiving, and (6) evaluating the purchasing task.

To buy well, the buyers must know the markets in which to purchase items, including supply, demand, pricing, laws and regulations, market channels, and the relative value of grades. Good buyers must possess many abilities and much knowledge and have contact with people who are knowledgeable about the market. Buyers should have integrity, maturity, bargaining skill, a knowledge of value, an even disposition, and a good sense of fair play. They must know the policies and procedures of management as well as how the operation itself functions. They must have an ethical as well as a good, down-to-earth, practical approach to problems.

Various buying methods will be used in purchasing. Informal methods, largely characterized by oral negotiations, often are used. The quotation-and-order sheet, the blank-check, and the cost-plus methods may be used. They facilitate purchasing and reduce time. Formal methods usually involve written details. Written specifications, invitation to bid, and other written purchasing techniques may be used. The formal methods most often used are the competitive bidding, the negotiated contract, and the futures and contract. Formal methods often permit lower costs because of volume buying. Either formal or informal buying methods may be used in Internet purchasing. Another important factor in purchasing is the selection of good purveyors who can provide the facility with what it needs at a fair price. Finding good purveyors often requires a search. The purveyor who serves the operation may vary from direct delivery—some operations even produce the items they use—to buying from a supermarket. There are advantages in each. Marketing agents provide services that must be evaluated in deciding on price. The service rendered and other factors that could or could not cost the operation must be evaluated. Large operations can purchase directly whereas smaller ones may have a hard time in finding a single supplier to meet production needs. Settling on a purveyor finally is determined by dependability, quality, variety, price, and economic stability of the purveyor's organization as well as factors such as sanitation, personality, and a knowledge of what the buyer needs.

Determining how much to purchase can be a challenge. The menu determines what is to be purchased. These foods can be categorized as perishable foods such as fresh fruits, vegetables, meats, and dairy products, staple goods, which are largely processed foods, although some staple foods, such as potatoes, cabbage, and some root vegetables could be included; and daily contract items such as coffee, bread, milk, ice cream, or others supplied on a daily or weekly basis. Various tests can be made to determine the quantity required, but these can vary. A constant check must be made if the quantity purchased is to meet production needs exactly.

One of the most important factors in getting a satisfactory performance from the purchasing function in an operation is to establish specifications and to use them properly. A knowledge of the standards established by the federal government and others is necessary to do this. Most standards are voluntary; a manufacturer or processor can use them if he wishes. Others are regulatory and must be used. If a product does not meet a standard, the label must state this regardless of whether the standard is voluntary or regulatory. Industry and brand standards often are used. These are usually less precise than the federal standards. The use of standards can simplify the writing of specifications and make them better understood. Simplicity, brevity, and completeness should be sought in writing a specification.

Specifications are the standards against which purchases are measured. To make specifications most effective, they need to be used correctly. The buyer, management, the receiving clerk, accounting department, and purveyor will all use the specification. For operations using acceptance purchasing, the specification is the basis for acceptance or rejection by the federal inspector.

The purchasing function is efficiently done when a good design of purchasing procedures is established. All personnel in the operation should help in setting this up, each contributing needs and viewpoints so all interests are protected and represented. The two most important forms used in purchasing are the purchase order and purchase record or purchase price record.

Receiving must see that what is ordered is actually what is received. If the quality, quantity, or other factors needed in an item are not checked to see that they

are there, the purchasing program fails. Various receiving methods are used to check incoming goods. Either invoice receiving, blind-check receiving, or partial-blind receiving can be used. When goods arrive at the receiving dock, they should be checked to see that they meet requirements and then should be delivered to the proper place. The receiving clerk must keep good, accurate records. The daily receiving report, the substitution invoice, and the request-for-credit memorandum are three forms frequently set up in the receiving department. A purchase invoice stamp may be used to indicate official receipt.

In the last ten years, great strides have been made toward reducing labor costs through computer automation. Perhaps the principal use of the computer in purchasing and receiving is in inventory control, verification of purchase orders, and par stock monitoring. Using the computer to process purchase and receiving orders has become widespread.

The storing and issuing of food are very important functions in the foodservice operation. Each provides, among other benefits, the essentials of control necessary for attaining a consistently profitable operation. But each must be performed step by step to attain the required end result. In dry storage, for example, cleanliness, ventilation, temperature control, and good organization are essential because contamination, deterioration, and waste can occur if the storage is improperly carried out.

Refrigerated or low-temperature storage is used to protect foods that are relatively perishable. Refrigeration slows down the growth of bacteria and microorganisms, while freezing suspends their growth. Good air circulation, proper temperatures, efficient location, and observance of general practices and procedures all contribute to the maximum protection of refrigerated foods.

If proper storage controls are instituted, lower costs will be achieved. This is especially true if only one person is responsible for both storage and control.

Inventory methods vary, but generally a combination of perpetual and physical inventory methods is used. A perpetual inventory provides, at a glance, the current status of any product. It is obtained by entering into the record books each addition to and deletion from inventory. The same result can be achieved through the use of computers at a much faster and more economical rate. A physical inventory is used to verify the perpetual record through an actual count performed by an inventory team, usually involving at least one person not from the storage department.

Food issuance procedures should give one responsible person control of assets on both a daily and a continuing basis. Control is usually assigned to the accounting department. Deliveries and issues need accurate tabulation to give the proper information concerning food cost and supply cost. Deliveries must be properly recorded, and a distinction must be made between direct purchases and storeroom purchases. Requisitions from the storeroom provide a measure of control and accuracy in the distribution to the various departments. Meat and some other high-cost items should be tagged immediately to help maintain proper control. The requisition forms used may vary, but they should give an accurate account of the quantity and value of goods issued. Stock record cards, inventory records, requisition forms, and other forms and procedures are all part of issuing and storage controls. From these records, the daily food cost is calculated. Precosting, immediate costing, or other methods for indicating costs before or immediately after they occur are better methods of management than working out those costs long after the fact. It is better to compute an estimated cost frequently than to calculate the actual cost precisely but infrequently.

Through the use of personal computers, many of the repetitious functions required in food storing and issuing can be carried out rapidly and without error. Such computers are used today in all types of foodservice operations, from the largest to the smallest.

■

CHAPTER REVIEW QUESTIONS

1. Study five nearby markets that sell food or other supplies used in foodservices. List them indicating what they sell. How do supply and demand affect these markets? Select five different foods or supply items and indicate the channels through which they pass before arriving at the foodservice.

2. What should a food specification contain? What is the use of a specification? Who uses it? What is acceptance buying?

3. Study the buying routine and procedures used in a foodservice. Who decides quantity and quality required? Who actually searches the market and makes the purchase? Set up in an organized manner the various forms and records used. Explain them in detail. Could a simpler procedure be used? Could any of these forms be simplified?

4. Discuss the three types of receiving methods for advantages and disadvantages. Suppose you were a consultant and found receiving procedures almost nonexistent, with cooks, wait staff, and others signing for goods delivered and no inspection being made. How would you approach management for correction and what would you suggest?

5. What are the six considerations to be viewed in judging refrigerated storage systems? Are there others? Judge a refrigeration system of your choice according to these six factors or any others that may be appropriate.

6. What is meant by perpetual and physical inventories? What control system must be used to make either work?

7. What does the daily food cost form indicate besides the cost of food used? Why is this form the overall information sheet for management? Is dependence on this report emphasized too much in view of increasing labor and other costs? Do you think the cost of labor and energy or other important cost factors should be included as well? In Figure 5-16, complete the form for items 1 to 10 if the following were true:

Storeroom purchases	$ 350
Storeroom issues	250
Direct purchases	175
Transfers	12
Food sales	1175

■

CASE STUDIES

1. Good Food Restaurants of Memphis has been in operation for nearly ten years. During that time their operations have expanded from the original location on Lewis Street to a total of five operations in the Memphis metropolitan market. Purchasing and receiving have been carried out, to date, by each individual restaurant.

   Until about six months ago, this system seemed to be working well. Around that time, however, a new manager who had been hired to operate the restaurant on Third Avenue stopped buying from the purveyor that had been supplying fruits and produce to the unit for several years. The new manager told the corporate office that the move had been made because another purveyor offered better service at competitive prices.

   Fresh Daily, the purveyor with the long-term relationship to Good Foods, had called the switch to the attention of the company office at the time. The corporate office told Fresh Daily that selection of purveyors was a decision it had always allowed the restaurant managers to make. The office suggested that Fresh Daily continue to bid for the business of the restaurant in question, however, and Fresh Daily continued to serve the other restaurants in the chain.

   Now, six months later, the corporate office has noticed that the food cost at the Third Avenue unit has risen slightly and is now almost 1 percent higher than that of the other restaurants. The menus at all units are the same, and traditionally the menu mix has been quite similar as well. Upon further inspection of the purchasing pattern over the last six months, the accounting department has established that most of the additional food cost could be attributed to higher expenditures in fruits and produce at the Third Avenue unit. The Third Avenue unit manager's response is that "service with the new purveyor is much better, and it's well worth the extra 1 percent in food cost." The accounting office suspects that there may be other reasons for the switch in

purveyors and the resulting increase in food costs, however—including the possibility of collusion and kickbacks to the manager of the Third Avenue unit. Since the purchasing has been left entirely to the restaurant manager's discretion in the past, without incident, the accounting office is at a loss to determine what steps to take to begin an investigation of the purchasing irregularity at the Third Avenue store.

   As the supervisor of the five restaurants in the Memphis area, you are responsible for determining why the food cost at the Third Avenue store is higher than that at the other units and for devising a system to prevent future problems. Give details of both the plan of investigation and the program for prevention.

2. Businessmen's Hotels, Inc., of Tulsa, Oklahoma, began operations in the oil boom years of the 1970s and rapidly grew to ten operations within the next few years. The hotels are all located in the oil-producing region of the Southwest, in cities such as Tulsa, Oklahoma City, Midland, Houston, and Denver. As oil prices went up, so did the occupancy rates at Businessmen's Hotels. Through early 1995, the company enjoyed occupancy rates companywide in the mid-80 percent range annually. Their foodservice operations, designed to cater to the oil-industry businessmen who stayed in the hotels, boomed as well.

   At the beginning of the summer of 1995, however, oil prices began to fall; by the autumn of 1995, oil prices had dropped to one-half of their previous level. By the spring of 1996, per barrel prices of oil had dropped to $10 to $12, and the industry as a whole was in a deep slump. Many oil companies laid off substantial numbers of employees and, as one way to reduce costs, eliminated most travel expenses. The Businessmen's Hotel chain's occupancy rates dropped to below 50 percent, as travel by people in their principal market was reduced. In some cases, room sales were as low as 30 percent of the total available space during peak weekday periods.

The foodservice operations suffered along with the dwindling room occupancies, since much of the former food and beverage business had been linked to businessmen staying in the hotels

During the boom years, the food and beverage operations had been thought of as an amenity to the hotels, since the real profits were coming from room sales. As a result, little attention had been paid to the profitability of the foodservice operations. Each hotel had designed its own system for foodservice purchasing, receiving, storing, and issuing; each operated independently from the other hotels in the chain. With the dwindling room occupancy rates, however, has come a need to redesign the foodservice systems to make them more nearly self-supporting.

The president of Businessmen's Hotels issued a memorandum on Monday that calls for the institution of strict storing and issuing functions, centralized in one location for cost savings (although he is not sure how this will help). As part of the program, the president also wants to use the hotel computer systems to detail the activities of the storing and issuing department more accurately and to eliminate unnecessary personnel from the payroll. Detail the steps you would take to begin to turn the foodservice storing and issuing department around. Explain the advantages and disadvantages of centralizing the departments, and respond to the president's request to computerize the department by outlining which functions are most readily convertible to computerization.

3. Chef William (Will) B. Fatt has decided to establish better controls in his kitchen. He has considered many options within the purchasing, receiving, storing, and issuing of foods and has decided to use the products offered from the Internet. In completing his research for him, you are asked to consider *www.chefdesk.com* as a source of information. What products does this site offer Chef Will? Would you recommend them for his use? Are there other links from this site that could be helpful for a foodservice operator? List two and describe why they might be helpful.

■

VOCABULARY

Primary market	Requisitions
Secondary market	FIFO
Retail	Perpetual inventory
Buyer	Direct deliveries
Grade	Vendor
Informal buying	Competitive bid
Blank check	Commodity
Specification	Wholesomeness
Invoice	Blind receiving
Purchasing invoice stamp	Formal buying
Negotiation	Par stock
Purchase order	HACCP
Cost plus	Direct seller
Ventilation	Purveyor

Staple food

Brand

Substitution invoice

Value analysis

Issuing

Low-temperature storage

Physical inventory

Dry storage

Quotation and order sheet

Shield-grade stamp

Futures-and-contract buying

Acceptance buying

Partial-blind receiving

Invoice receiving

Request-for-credit memorandum

Daily receiving report

Stock record card

Estimated versus actual food cost

Temperature record book

Irradiation

Internet

■

ANNOTATED REFERENCES

Coltman, M. M. 1989. *Cost Control for the Hospitality Industry*, 2nd ed. New York: John Wiley & Sons, Inc. (Purchasing and inventory controls, labor costs, computers, and more are discussed in this text.)

Dittmer, P. R., and G. G. Griffin. 1994. *Principles of Food, Beverage, and Labor Cost Controls*, 5th ed. New York: John Wiley & Sons, Inc. (All food and beverage controls are considered in this book.)

Labensky, S. R. 1998. *Applied Math for Food Service*. Upper Saddle River, N.J.: Prentice Hall, pp. 51–68. (Recipe conversions, costing, percentages, yield tests, and menu pricing are included in this text.)

Mill, R. C. 1998. *Restaurant Management: Customers, Operations, and Employees*. Upper Saddle River, N.J.: Prentice Hall, pp. 261–287. (Controlling costs is discussed in one chapter.)

Miller, J. E., and D. K. Hayes, 1997. *Basic Food and Beverage Cost Control*. New York: John Wiley & Sons, Inc. (From food and beverage costs to labor expenses, controls are reviewed in this book.)

Pavesic, D. V. 1998. *Restaurant Cost Control*. Upper Saddle River, N.J.: Prentice Hall. (Controls from cost ratios and financial analysis to yield cost and labor productivity analysis are considered.)

Reed, L. 1993. *SPECS: The Comprehensive Foodservice Purchasing and Specification Manual*, 2nd ed. New York: John Wiley & Sons, Inc. (Food purchasing is considered in total in this book.)

Stefanelli, J. M. 1997. *Purchasing: Selection and Procurement for the Hospitality Industry*, 4th ed. New York: John Wiley & Sons, Inc. (All aspects of food purchasing are covered in this text.)

CHAPTER 6

Service and Dining Etiquette

■

LEARNING OBJECTIVES

By the end of this chapter, the reader will:

1. Be able to distinguish between the various types of seated service.

2. Understand the types of self-service.

3. Learn the importance of manners and etiquette and be able to practice the same.

4. Value the use of computers in service and the impact they have had on service employees.

■

SELECTED WEB SITE REFERENCES

Carter-Hoffmann
www.carter-hoffman.com/

Chef Desk
www.chefdesk.com

Comtrex Systems Corporation
www.comtrex.com

Elo TouchSystems
www.elotouch.com

InfoGenesis
www.infogenesis.com

Restaurant Data Concepts
www.positouch.com

System Concepts Inc.
www.foodtrak.com

Touch Menus, Inc.
www.touchmenus.com

If all other functions in an operation are correctly done, but service—even service behind a cafeteria counter—is neglected, the establishment will fail. This is because service, more than any other single attribute of a food-service operation, is identified by customers as the reason for frequenting a foodservice. In a national poll in 1995, for instance, 85 percent of the participants identified service as the main reason they go to a certain foodservice. While in many cases customers blame poor service on individual employees, service is fundamentally the responsibility of the manager. The staff meets the customers one on one and delivers the service that customers remember, but management defines what form of service will be used throughout its system: how food is to be served, by whom, and in what manner. Commercial foodservices attain high standards not only with good food but with excellent service. Therefore, management must take the blame for failures in service as well as in food. Management is responsible for training personnel and for setting the standards of service. Even though these standards vary from one foodservice to another, certain principles must be observed if proper service is to be performed.

■

SEATED SERVICE

Seated service occurs when a guest remains seated throughout the meal. The amount of personal service given can range from the level maintained in fine-

dining establishments, where service personnel attend to every need, to that offered in fast-service outlets, where service, consumption, and payment are rapidly accomplished. The principle types of seated service used in foodservice establishments are American, English, French, Russian, banquet, counter, and tray. The type of service delivered depends to some extent on the menu, the desired atmosphere, the availability and skill of workers, and the desired market.

American Service

An inexpensive and fairly fast form of service is called *American*. It is easily learned by service personnel. The food is portioned in the kitchen and brought to the guests on plates or dishes, from which the guest eats. Side dishes such as bread and butter plates are used only when the main plate cannot accommodate the additional food. To add a bit more finesse to the service, personnel may include the bread and butter plates on the place setting and may serve rolls in a basket or plate covered with a napkin. The butter may be served separately, or it may already be present on the bread and butter plate. The coffee cup and saucer may be on the table from the outset, or they may be brought to the guest on request.

The service is simple. All solids are served from the guest's left with the server's left hand. Beverages are served from the guest's right with the right hand. When guests are a mixed group of men and women, service begins to the host's right with women (the oldest being served first), and then proceeds to elderly men, children, and finally men. When there are only men or only women, service begins at the host's right and moves counterclockwise around the table.

Placemats or tablecloths are used. Tablecloths should hang at least 12 in. over the table edge but not so far as to pose an obstruction to the guests. Silence cloths of felt, foam rubber, or other material are used under the tablecloth. Figure 6-1 shows a typical setting.

Operators may wish to vary the American setting by combining other types of service with it. Each operation must suit its own needs, and management must decide the service most appropriate for its customers. Simplifying the service makes it easier to introduce personnel who lack training in service.

English Service

English service is also known as *family, host,* or *holiday* service. It is an adaptation of the service found in English country homes. English service differs from other types of service in that the food is served on large platters or in bowls that are placed on the table for service throughout the meal (see Fig. 6-2). Of course, if they become empty, they may be refilled or removed to the kitchen. English service is not used often, but it may be found in clubs and in some other types of foodservices. Meats are carved by the host, while guests dish their own vegetables and side dishes.

FIGURE 6-2 English service may feature food served on large platters passed among guests.

The Canadian adaptation of English service does not include a waiter, so the host and hostess must be thoroughly adept and organized. When English service is informal, it is known as family service. Whatever the precise form of English service used, all items are served from the right and cleared from the left.

French Service

French service at one time was extremely elaborate. Because of the high cost of labor, however, the service has grown less popular. The equipment used is also expensive: a table-high rolling cart (guéridon) from which food is served; a chafing dish (réchaud) typically mounted on the guéridon and designed to prepare special dishes and keep food warm; and silver platters and trays used to bring food from the kitchen.

Two people wait on the tables, supervised by a captain who may oversee four tables. The principal waiter is called the *chef de rang*, and the assistant is known as the *commis de rang*. Both must be highly trained. The dress frequently includes white gloves. Duties of the *chef de rang* include the following:

FIGURE 6-1 American service setting.

1. Seat the guests, if a headwaiter or captain does not seat them.

2. Take guests' orders.

3. Serve all apéritifs if a cocktail waiter does not serve them.

4. Finish preparing the food at the table before the guests.

5. Present the check and collect the money.

Duties of the *commis de rang* include the following:

1. Receive orders from the *chef de rung*, take them to the kitchen, and order food from the kitchen.

2. Pick up food in the kitchen, bring it on a tray, and place it on the guéridon.

3. Serve food after the *chef de rang* prepares it.

4. Stand ready to help the *chef de rang* whenever necessary.

The *réchaud* is used to prepare food or keep it warm. If the *chef de rang* does not carve the meat, bone the chicken, or debone the fish, the captain may do it. All sauces or garnishes are prepared by the *chef de rang*. The wine steward, or *sommelier*, comes to the table to discuss the various wines to be served. This person takes the order and fetches the wine, after telling the *chef de rang* which glasses will be needed. The *commis de rang* secures the glasses and puts them in their proper places on the table.

The *chef de rang* uses a serving fork in one hand and a spoon in the other to transfer food from the platters or *réchaud*. The *commis* holds the guest's plate below the silver platter or *réchaud*, ready to receive food and catch spills. When there is only one waiter, the *chef de rang* must do all functions. The fork and spoon are then used in one hand to transfer the food to the plates. The rule in French service is usually that all food is served from the right, except butter. Bread plates, salad plates, and silver used on the left are placed from the guest's left. A left-handed waiter may have to serve completely from the left, which is difficult.

The setting for French service is a silence cloth, tablecloth, and top. The top is another tablecloth, easily removed after a meal and replaced for the next guests. A service plate about 10 in. in diameter is placed directly in front of the guest. An hors d'ouevre plate may be placed on this, and then the folded napkin. To the left of the service plate is the dinner fork. Above this is the butter plate and butter knife; in very formal French service, no bread is served, so the plate is not used. To the right of the service plate is the dinner knife, with the cutting edge toward the plate. A soup spoon may be next to the knife, or it may be brought with the soup, if ordered. Above the service plate, parallel to the table's edge is a dessert fork and spoon, but this can be omitted. Figure 6-3 illustrates a typical service setting. Knives and spoons are always put on the right, with the spoons to the right of the knives; forks go on the left. Guests use the silver from the outer edge in, as the courses proceed. Silverware also can be brought with the respective courses, such as the soup spoon with soup or the cocktail fork with the shrimp cocktail.

Wine glasses may be preset on the table or placed just before the wine is served. Sometimes, if more than three wines are served, three glasses may be arranged at the place setting and others added later after the used glasses are removed. Water is usually not served; if it is, as is true in the United States, the water glass is at the tip of the knife. Wine glasses are arranged next to this, and the progression is to move through the glasses from right to left for the various wines.

Coffee is served after dessert, and the coffee spoon may come with it. Fresh napkins may be given during the meal or immediately after finger bowls are used. Finger bowls—small containers holding liquid for cleaning the hands—may also be used during the meal,

FIGURE 6-3 French service setting.

with fresh napkins. It is not considered proper to have ashtrays on the table, since smoking during the meal in French service is disapproved. Salt and pepper shakers are not on the table either, because the seasoning is supposed to be correct as served. Dining should proceed in a leisurely manner.

Russian Service

Russian service differs from French in that all food is dished onto platters or serving dishes in the kitchen and brought to the table for service. Hotels use this service much more than any other for their higher-quality service. It is simple, yet elegant.

One waiter takes the orders, gives them to the kitchen, and does all serving. If a party of six is dining and there are three identical orders, these will be placed on one serving dish. A busperson may bring food to the table if there are a number of dishes, but this person does not take part in the service. The various trays are set on a serving stand. Empty plates for service are distributed in a clockwise fashion around the table by the waiter, at the guest's right, using the right hand. The service is reversed when all plates are down: food is put on the plates counterclockwise. The food dish is held in the left hand at the guest's left, and the waiter, using the right hand to hold a spoon and fork, places portions on the guest's plate. If it is soup, the soup is dished from a tureen into the soup bowl at the guest's place, using a soup ladle. Guests may direct the waiter on the size portions they want. Any unused foods go back to the kitchen.

Clearing is the same as for French service. Coffee is served from the right, following dessert, and a finger bowl and fresh napkin are presented to the guest at the close of the meal.

Russian service is best in an operation offering relatively few entrées, because problems arise in attempting to arrange different orders attractively on one tray. Another problem involves keeping food hot after leaving the kitchen and before service. Russian service is excellent for banquets, since all foods are the same.

The table setup for Russian service may be very much like French service, or it may be unique to the menu offered (see Fig. 6-4). A fairly large inventory of silver service is needed, as is true in French service. In

FIGURE 6-4 Typical Russian service setting.

both French and Russian service, there is a trend toward simpler table settings, with the necessary silver and glassware placed on the table only as needed for each course rather than the entire setting being used.

Banquet Service

Banquet service may be American, French, Russian, or a blend of these. Organization and speed are important, so foods are served at every table at the same time. The service should be based on this need, plus that of giving the guest the feeling of elegance and luxury.

American service is most commonly used, with a slight variation in that plates portioned in the kitchen are covered with banquet covers to keep the food warm and to allow stacking of many plates to a tray (see Fig. 6-5). If English service is used, the Canadian adaptation is most usual. Some caterers use family service for popular-priced banquets, bringing the food in on platters to be passed around the table. Russian service is often used for the nicest banquets. On such occasions, the waiter serves the entrée from a silver platter. Care must be taken to see that the platter is attractive and not messy, even down to the last portion served. The type of item being served is important. Good training can also help to make the service attractive. One problem is that much part-time help may be necessary. A short but informative training session can be given by the banquet manager before

FIGURE 6-5 Banquet covers
fitted on plates to allow stacking.

service begins. The banquet function sheet also can help inform personnel and ensure a smooth-running function (see Fig. 6-6).

Counter Service

Counter service is economical in labor cost, quick, and efficient. It appeals to guests who are in a hurry and will be found in fast-service operations—coffee shops, drugstore fountains, department stores, and others featuring convenience, low prices, and generally good food with fast, efficient service.

Guests are served at a counter. The menu is presented with a glass of water at the right hand side of the cover. The place setting known as the cover may or may not be present. Service personnel take the order as soon as the guest has selected the food, and this should be transferred at once to the preparation departments. If the cover is not set, it can be set while

the guest is deciding what to order. Speed and dispatch are desirable. Typically, a knife (with the cutting edge facing the space where the plate will be placed) and a spoon are set on the customer's right. A fork, with a napkin under or on top, is placed at the customer's left. The order is placed in the center of the cover. When service is over, the check is presented face down. Soiled dishes and tableware should be removed promptly. Dirty dishes are unappetizing to view while eating a meal, and other guests' appetites may be disturbed by them. They also can cause confusion, in that a guest sitting down while they are still there may be thought to have just finished eating and may not receive prompt attention.

Counter service is often coupled with table or booth service. Usually, special service personnel serve guests at these, while the counter service people serve only guests at the counter. The various counter shapes

NAME OF ORGANIZATION	DATE	TIME
TYPE OF FUNCTION	ROOM OR LOCATION	
NUMBER OF GUESTS	IN CHARGE	
RESERVATION MADE ON	CONFIRMED	
DEPOSIT RECEIVED	BALANCE	
HEAD TABLE SEATING	TABLES SEATING	
TABLE PLAN AND SET UP		
SEATING ARRANGEMENTS	SPACE FOR DANCING	
TYPE OF LINEN, CUTLERY, GLASSWARE		
FLOWERS SUPPLIED BY		
DECORATIONS (candles, favors, programs, place cards)		
MUSIC LECTERN	DAIS	MICROPHONE
CHECKROOMS CASHER	LOCATED	
MENU		
SERVERS REQUIRED AND TABLE POSITION		
TABLE 1 TABLE 2	TABLE 3	
TABLE 4 TABLE 5	TABLE 6	
PERSON IN CHARGE OF STAFF		
GRATUITIES		

FIGURE 6-6 Sample form for a banquet function sheet.

are important to speed and efficiency of service. The straight line counter is least efficient and takes the most labor. Rectangular or U-shaped counters are more efficient, since service personnel can work the front and both sides easily from one position.

Tray Service

Tray service is used in health facilities or in places where food must be delivered. A meals-on-wheels program furnishes such delivery service to elderly people. Some boarding houses and homes also use the service. Hotel room service is a variation of tray service. Food is dished onto plates, placed on trays, and then distributed to recipients, either from a central kitchen or from pantries located on each floor or wing (see Fig. 6-7). If the food comes from the

kitchen, the service is said to be *centralized*. If it comes from pantries, it is said to be *decentralized*.

Centralized Service

In the recent past, centralized service has been favored; but with the development of individual frozen food portions and many convenience foods that are easily portioned, decentralized service has taken on new importance. In centralized service, all trays are prepared in the serving section of the main kitchen. Here, under the supervision of the dietary department, proper controls, food temperatures, correct diet, and attractive food arrangement can be maintained. The trays may be distributed by conveyors to various floor levels, or they may be loaded onto carts and transported there. Either dietary personnel or nursing personnel distribute the trays to patients in

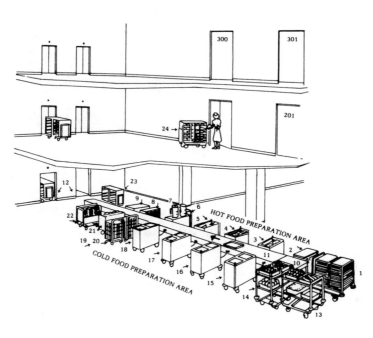

Dispensator

MOBILE EQUIPMENT AND
SYSTEM FOR HOSPITAL FOOD
SERVICE

1. Carrier and dispenser for trays
2. Mobile table for food requisitions
3. Hot food unit, dispenser for hot foods and heated dishes
4. Ditto
5. Ditto
6. Urn stand with urns for coffee and hot water
7. Dispenser for heated pots for coffee, tea, etc.
8. Dispenser for cups and saucers
9. Table for supervisor
10. Starting platform for trays
11. Conveyor belts for trays
12. Elevators
13. Carrier and dispenser for silverware, napkins, tray covers, setups
14. Dispenser for bread and butter
15. Dispenser for salads
16. Dispenser for pies or desserts
17. Dispenser for milk
18. Dispenser for ice cream
19. Dispenser for glasses
20. Dispenser for glasses
21. Dispenser for heated food covers
22. Dispenser for cooled food covers
23. Wagon for food trays being loaded
24. Wagon with food trays being unloaded

FIGURE 6-7 Tray service system.

health facilities. In some institutions, trays may be brought to a central dining room and given to patients or others as they come in to eat.

Decentralized Service

In decentralized service, food prepared in a central kitchen may be sent in bulk or in convenience food portions to service pantries, where it is placed onto dishes, trayed, and sent to those who must eat. Usually some type of mobile transport equipment is used to get the trays to where they are needed. In decentralized service, supervising all areas where food is being dished is difficult. Additional food handling is often required in decentralized service—once in the kitchen, and again in the pantries. Maintaining food quality may also be a problem. Some decentralized services require more labor-hours to do the same job than do centralized service units.

■

SELF-SERVICE

Self-service is called *informal service,* and it is becoming more popular. The guest often dishes up the food and either eats it in a place provided or takes it elsewhere to eat. Depending on where the food is eaten and the rules of the establishment, guests may or may not return dishes to a central area. If not, busing personnel do this. The different types of informal service are buffet, cafeteria, drive-in or take-out, and vending.

Buffet Service

In buffet service, food is put out for guests to select. Frequently, a long table is provided on which a quantity of food is displayed. Good merchandising effects are possible by presenting a mass of varied food types in a colorful display. Often, cold foods are first and hot foods last. Salads, sandwiches, seafood, soups, meats, vegetables, casserole dishes, and others are featured. The garde manger section of the kitchen often prepares many of the display pieces. Although buffet service is largely self-service, on some buffets a cook carves the meats or otherwise assists guests, particularly if service must be rapid. The price of the meal covers

the entire meal, although some operations may charge extra for beverages or dessert.

Showmanship is the key to a successful buffet. Themes such as holiday, a luau in Hawaii, Polynesian in the South Pacific, and Creole in New Orleans should be used. Centerpieces can be flowers, ice carvings, or decorated cakes (see Fig. 6-8). Lighting is important and can emphasize food. Spotlights on special foods can add much. Tables may be straight, L-shaped, U-shaped, or V-shaped. To speed service, some tables may be separate from others. For instance, the table where beverages are dispensed may be separate from the one where foods are served.

Many operations today are combining buffet with seated service. Guests are allowed to select cold foods (which are largely salad materials) from a buffet. They then order the rest of what they want from a

FIGURE 6-8 An ice carving added to a buffet is always elegant. Although this was done by a professional, ice molds are available for use and require less time, skill, and dollars, yet provide elegance to a buffet.

menu. Not all buffets are elaborate or stress showmanship. Many operations today use buffets to get a lot of people served in a short time. A salad bar is a good example. Others use buffets because they lack proper facilities to care for guests in another way. Thus, breakfast buffets may range from one in which guests can get only hot beverages, juice, and sweet bread products (continental breakfast) to one in which they can get scrambled eggs and other products as well. A hotel that caters to tours may find that one tour coming down to the regular dining room will swamp service and disappoint other guests. Therefore, the tour may be taken care of at one time in a separate dining room, using buffet service.

A *smorgasbord* is a buffet featuring Scandinavian foods, but a true one must have pickled herring, rye bread, and mysost or gjetost cheese. A hot entrée may be ordered from a menu. The entrée is served by service personnel after buffet foods are eaten. Russian buffets are used at receptions. These feature hors d'ouevre foods. Caviar in an ice bowl, rye bread, and butter are typical foods, along with many other tangy types. Often, if a buffet is used to serve the main foods, waiters later pass among seated guests with trays of food and beverages. This helps reduce congestion at the buffet.

The location of the buffet is important. It should be near the entrance and close to where guests will be seated. A central position allows service to be achieved from either side, although this complicates reservicing the buffet because of cross-traffic between servers and guests. It should not be located far back in a corner of the room.

Cafeteria Service

Cafeteria service may not be entirely self-service because the food is dispensed to guests, who carry it to a cashier or checker and then to a table where the food is eaten. Some cafeterias have service personnel carry the tray to the table. Often busing personnel remove the soiled dishes. Some cafeterias have a system in which people get a check from a checker and then pay as they go out. The most usual system is to pay at the end of the cafeteria line. Cafeteria service is popular because it reduces labor costs, thus allowing more

food to be given for the money. It also is speedy service, and guests can see what they are getting. Quality of food is an important factor in the success of a cafeteria. It must usually be moderate in price but should not in any way be low in quality.

Cafeteria service is most frequently used in massfeeding situations, such as at schools and military installations. Commercial cafeterias catering to the public are gaining in popularity, however, principally as a response to the demand for faster service. Today, many large corporations provide in-house cafeteria-style service for their employees, either on a contract basis with an outside foodservice operator or through their own foodservice department.

Normally, guests pick up a tray and their silver and napkin at the start of the cafeteria line. They then select the foods they wish. Usually, cold foods are first and hot foods last, but this can vary. A menu board plainly indicates the foods available and their price. Two basic cafeteria systems are used: the line, and the scramble.

Line System

The line system has people form a line and then move through, selecting foods. Service personnel serve and also replenish the counter. Emphasis is placed on standardized portions and speedy service. Colleges, in-plant foodservices, school lunchrooms, and even downtown eating places use such service. Counters may be L-shaped, straight, U-shaped, or even circular. Some line systems break up service into a speed line, where the foods offered can be quickly selected and served so that the line moves quickly, and a slower line where foods requiring slower service are offered. One of the problems with the line system is that a person who stands and ponders what to choose slows down the entire line. In some systems, bypassing is encouraged.

Scramble System

The scramble system has counters where specific kinds of foods can be obtained. They are not joined together in a line, so an individual goes where a certain desired food can be obtained, gets it, and then goes to another area. This system is sometimes called the *shopping center system*, since it resembles the organization of an enclosed shopping area. It is also called the *hollow-square system*. Once the guests have collected their food, they go to the exit of the square, where they pay the cashier.

Today modifications of this system are also referred to as *food fairs* or *food bars*. Combinations of food categories can be arranged together so that the customers spread out somewhat to alleviate line crowding. For instance, cold foods could be found at one station, salads at another, beverages at a third, and hot entrées at still another. This eliminates the need for customers to file past each selection whether or not they are interested in those foods.

Originally, it was feared that customers would find such a system confusing, but studies indicate that the customer quickly adapts to the spread-out stations offering different foods. Signage indicating the location of different products is the key to directing traffic and educating customers when using a scramble system.

An offshoot of the scramble system is the *food court*. First popularized in the World Trade Center in New York City, this system involves the establishment of several small satellite operations offering different food products. Each satellite prepares or finishes the products it offers on site. Emphasis is on menu variety provided by the myriad different choices. In many instances, these operations are backed by a central commissary or food preparation center for processing foods prior to their delivery to the satellites for finishing. This system may also appear under a variety of names. In Canada, for instance, it is referred to as a *village concept*, in reference to the cottage-industry production period in history, when different villages often produced separate foods. A good example of this type of system can be found in Boston's Faneuil Hall.

There are three definite advantages to the scramble system. First, there is no waiting line, so more people can be served in a given time than with the line system. Second, the system has greater flexibility. During low-volume times some sections can be closed; this would not work as well in a line system. A scramble system can also be enlarged more easily. Third, improved labor utilization occurs, particularly in the dispensing of hot and cold beverages at a central square and with foods guests can pick up themselves, such as prepackaged items.

Drive-in or Take-out Service

Drive-in or take-out service is largely self-service in that guests usually pick up their own food after ordering it. Variations include ordering and then driving to a pick-up window, where service personnel give out the food, and ordering in place and receiving carhop service. Guests may be expected to eat in their cars, or a small dining area may be provided. The menu is limited and specializes usually in one particular item such as chicken, hamburgers, tacos, or roast beef sandwiches. Because of the speedy service and low price range, this type of operation is popular.

Vending Service

Vending service utilizes machines to dispense food. Some vending systems dispense complete meals. Specific machines often dispense foods in a kind of shopping center arrangement. The fact that service from the machine is impersonal keeps many patrons from visiting such an operation. On the other hand, many industrial and institutional employees find them acceptable. Vending reduces costs. There are some disadvantages. The offerings must be limited, and some foods do not hold up well in the machines. Sanitation can be another problem. Some vending services dispense food that can be reheated in microwave, infrared, or quartz units.

Convenience Service

Convenience service has grown from a practically nonexistent segment to a healthy participant in foodservice in only a few years. It is characterized by the use of ready-prepared and ready-to-eat foods that the customer chooses and finishes preparing. The ready-prepared foods, such as sandwiches and popcorn, require cooking, which is usually done by a customer-operated microwave oven. The ready-to-eat foods, such as doughnuts, sandwiches, and salad items, can be heated in a microwave or eaten as they are. In either case, the system is essentially self-service. The growth in this industry has been tied to that of convenience food stores, which cater to neighborhood and passerby traffic. In some cases, the demand for food products in these stores has been so great that fast-food companies have placed miniunits in convenience food stores to capture the market.

In convenience purchasing, food products are either consumed on the premises or taken out. In many convenience stores, small seating areas have been set aside to accommodate the demand. The popularity of this system in convenience stores has led to the establishment of similar operations in supermarkets and office buildings. As more and more innovative methods of sustaining the shelflife of foods are perfected, this system will grow further.

SERVICE EMPLOYEES

The service employee is the person with whom the customer is most likely to have regular contact. Whether waiter or waitress, busser, maître d', *sommelier*, cafeteria server, or any other employee, this person is responsible for delivering not only the products but also the image of the establishment. Therefore, the value of the service employee cannot be overstated. This person is also the chief sales agent for the business. In a full-service restaurant, for example, each server may be responsible for as much as $150,000 to $300,000 per year in sales. A considerable amount of time and money may be required to train these employees properly. The National Restaurant Association estimated that the cost of training each server may amount to between $500 and $2500, depending on the type of service involved. Therefore, good foodservice establishments take great care in devising systems for the proper training and continuing motivation of service employees.

The service employee may be responsible for a variety of different tasks, depending on the type of foodservice. These may include delivering the food, taking orders, promoting products, removing dishes, and selling and opening wines or other beverages. In many cases, service employees are tipped by the customer for these services and derive a substantial portion of their wages from this source. In others, such as the cafeteria system, tips are often not involved, and the employees depend on the business for their entire remuneration.

In large foodservice operations, the responsibilities of hiring, training, and retraining service employees may be conducted by a human resources department. In smaller operations, the unit manager usually takes on these responsibilities. In either case, communication is the principal tool used. This communication may be primarily verbal, or it may involve pictures, photographs, charts, video presentations, interactive computer video, or a wide range of other methods.

COMPUTERS AND SERVICE

From hand-held ordering devices to computers in every aspect of service, the foodservice industry has benefited from improvements in technology. Although the trend toward automated systems began in the fast-food segments, table service restaurants have moved rapidly to capitalize on the advances.

Most facilities now have software systems that take advantage of preset keys for each menu item or a price-lookup (PLU) feature, or both. Under either system, menu items and prices are programmed into the computer and accessed by either an item key (preset) or a code number (PLU). The server taking the guest's order presses either the preset key or the proper code number and the item is automatically printed on the guest check or receipt. The price of the item is also recorded, and the bill is totaled. In many systems, the order is also printed or displayed in the kitchen. When the order is ready, the server is notified to pick up and proceed with service.

The improvement in speed is a result of two factors. First, the server does not have to enter the kitchen in order to place the order. This is done by the system, and many steps are saved thereby. Second, the server does not have to add up the check manually, since this is done automatically as the order is placed. The order is also recorded for future reference.

The improvement in quality of service is related to the development of computers. Because of the time saved, service personnel are able to spend more time with the customers. In addition, fewer errors are made in the totaling of a guest's check.

The consequences of these improvements in service are higher table turnover, greater guest satisfaction, and higher staff morale. Many operators claim to have achieved higher check averages and better tips for their employees as a result of the improved service. Clearly, the installation and full utilization of computers would benefit virtually every foodservice facility.

■

DINING ETIQUETTE

Over the past few years, "do your own thing" and "looking out for number one" have been the battle cry of the masses. What was traditionally considered as good manners was abandoned. While the intent of this text is not to teach manners, the goal of a person to treat and be treated with common courtesy should be paramount.

Each foodservice operator, then, must be aware of basic social graces. Those dining should have a sense of self-confidence in the varied social and employment situations in which they find themselves. It is important to understand a framework in which one can be comfortable and share respectful interaction with others. There is a need to be aware of multicultural customs and habits that are considered acceptable in regard to etiquette, as well. Accordingly, this section is very important for the student of hospitality management.

Invitations

Any invitations received deserve a prompt response. Most will say "Please RSVP," which is an international way of saying "Please let us know whether or not you will be attending our affair." If after accepting an invitation, you must decline it for some reason, you should phone your regrets as soon as possible and then follow that phone message with a note of regret.

Once an affair has been enjoyed, a thank-you note should be sent within one week after the event. It should always be a personal message rather than a printed card. Such notes are never out of place since they are a nice way to reciprocate thoughtfulness and acknowledge another's consideration. A phone call is another way to say thank you.

Thank-you notes are always obligatory upon the receipt of a present by mail, following a weekend visit, after receiving a gift when you are ill, for a wedding present, following a dinner by your boss, or when you have been the guest of honor at a dinner party. Letters of condolence sent to you are followed by a thank-you note as well.

The Gracious Host or Hostess

If you are hosting a party, all attention should be centered on your company. The male is the host and the female is the hostess. Everyone should be introduced to everyone else unless you are having a large party, in which case name tags may be appropriate. You should stay with your guests as much as possible and be alert to arguments, or uncomfortable situations among guests, to act quickly to rectify any problems.

If serving cocktails before dinner, the reception should not last longer than one hour, even if you have late guests. At least one nonalcoholic beverage should be available for those who do not drink liquor. You should never urge food or drinks on anyone, but supply the needs of your guests as best as possible.

If you are attending a function, invitations should be accepted promptly, as noted previously. You should arrive as close as possible to the time the event is to start, but never early. Be entertained and be entertaining, but don't be too helpful, as you are the guest. Don't overstay your welcome and extend your thanks warmly, but briefly, prior to leaving. Follow that thank you with another, whether by phone or written note.

Introductions

An introduction should bring people together, not alienate them. Use first and last names of each person, regardless of status, unless you are introducing a child to an adult. Traditionally, a man is presented to a woman and a younger person is presented to an older person.

If a name should be forgotten when meeting someone for a second time, simply mention that you have forgotten the person's name and apologize. This might best be stated: "I'm sorry. I have forgotten your name." In order not to forget someone's name, train yourself to listen carefully to a person's name the first time you are introduced, and you will not forget it.

Conversation

Listening is important in conversation as well as in remembering names. Above all else a good conversationalist is a good listener. Be curious about what others are saying and do not be self-involved. Concentrate on the conversation and look at the person with whom you are speaking. If you start thinking about what you want to say and how to respond in a certain situation, you will inhibit conversation by preventing spontaneity.

To be a good conversationalist, you must have a desire to please and be willing to share your sense of humor. Always bring up a topic when there is a lull in the conversation as others will be grateful for your contribution to what is being discussed.

Formal Dining

Depending on the type of service, the table is set with utensils to be used moving from out to in toward the plate (see Fig. 6-9). Thus, the soup spoon is located to the extreme right, while the salad fork is located to the extreme left. The water glass is placed above the dinner knife with wine glasses to its right.

The order of service at a dinner party is seating first by the hostess. Grace is said by the host followed by the serving of soup. The hostess then begins the meal with the lifting of her spoon. The carving and serving of meat takes place once the soup is finished and the vegetable is served followed by wine and bread or rolls.

FIGURE 6-9 Formal dining takes place at an elegantly set table such as the one shown here.

Once the main course is finished, the table is denuded, crumbed, and the salad is served. Dessert follows with coffee service and sugar and cream are passed. Cordials may be enjoyed with coffee or in a different room.

In the United States, a variation on the above is acceptable and involves the serving of the salad with bread prior to the meat being served. The European influence has the salad being served after the entrée. The choice is the decision of the host and hostess.

Ease at the Table

There are many tips to remember while dining, but the following should assist anyone attempting to improve his or her etiquette.

1. The time to begin eating is at the discretion of the hostess. When she picks up her fork to begin, everyone knows that it is time to begin. In larger parties the hostess should encourage guests to begin when the food is served so that it may be eaten while hot.
2. The hostess also should place her napkin first in her lap. Then guests should follow, with larger dinner napkins folded in two and smaller luncheon napkins laid open entirely.
3. The tipping of a soup or dessert plate is acceptable as long as it is tipped away from you.
4. Larger stemmed glasses are held with the thumb and the first two fingers at the base of the bowl. If glasses contain chilled wine, they are held by the stem. Small stemmed glasses are always held by the stem, while tumblers are held near the base. A brandy snifter is held in the palms of both hands. Little fingers should never be elevated in an affected manner while holding a glass.
5. As you take food from a buffet, you need not take anything you do not like since there is a wide choice of food. On the other hand, should a served plate contain food that you do not like, simply move it around on the plate and do not call your dislike of the food to the attention of anyone else, especially the hostess.
6. If there is gravy for the meat, pour it only on the meat. If there is rice, noodles, or dressing with the main dish item that is rich with sauce,

adding gravy to the starch is appropriate. Pouring catsup all over food is an insult to the chef and should be reserved for hamburgers and other nongourmet food.

7. As far as seasoning, it is an insult to the chef, professional or not, to shake salt and pepper indiscriminately over food one has not tasted first. Butter is for breads and not vegetables. The only vegetables that you are supposed to butter are baked or mashed potatoes.

8. Seating plans are established by the hostess and should never be changed, even for left-handed persons.

9. Bad food, gristle, pits, or other foreign objects in food should not be spat into a napkin. Roll the offending morsel onto a spoon and then onto the plate. Camouflage the unattractive piece with some remaining food on your plate.

10. If an accident should occur, remember that they happen to all of us. Try to be as inconspicuous as possible by lifting the bit of food with a convenient utensil. Whether or not to bring a stain to the attention of the hostess is a matter to decide at the time of the accident.

11. There are various degrees of reaching across the table that are acceptable. You should not stretch or rise from your seat to fetch something. If you have easy access to the salt and pepper and your neighbor on either side does not, it is polite to offer it to them before using it yourself. Pass the salt and pepper as a pair. When asking for something to be passed, ask the person closest to use it themselves and kindly pass it on to you.

12. Sit up straight throughout the meal and never put your elbows on the table while eating. They may rest on the table, though, between courses. The hand not holding a utensil with which to each or cut food should be placed in your lap while you eat.

■

SUMMARY

It is the responsibility of management to establish the type of service and the standard of excellence for the foodservice. The service should suit the operation and the type of food served. There are basically two types of service: seated and self-service. At times these may be mixed, but whichever type predominates determines the classification.

In seated service, customers are served by service personnel. There are several kinds. American service is a fast means of distributing food, as food is plated and portioned in the kitchen. The table setting also is simple. American service is inexpensive and easily learned by nonprofessional personnel. English service involves having the host and hostess participate in the plating of food at the table. Depending on the operation, service personnel may be used to assist. If they do not, the type of service may be called *Canadian*. French service is one of the most luxurious and expensive. It uses a great deal of expensive equipment and two waiters per station. The table setup is elegant but practical. In French service, food is prepared in front of guests.

Russian service differs from French in that portions of food are placed onto serving dishes in the kitchen, and waiters then serve the guests from these at the table. Russian service is next to French in elegance but it also is simple and works well for large groups.

Banquet service may use American, French, English, or Russian service, or blends of these. American service is often used because of its speed and simplicity. French service is used less frequently than Russian because of the cost and time. Russian service is used when the service must be elegant but must still meet the needs of banquet arrangements that reserve time for a speaker or for entertainment. Counter service occurs behind a counter and is much like American in the table setting. Speed is essential. Tray service may be centralized or decentralized. Centralized service has dishup occurring in the main kitchen, whereas decentralized service sends foods to smaller service units for final preparation and dishup.

Informal service and self-service are synonymous. It is used to simplify service requirements, give speed, and reduce costs. Buffet service is service from a table, where people help themselves. Some waiter service may be combined with this, such as the service of beverages and desserts by service personnel. Cafeteria service involves having customers pass in front of counters and select foods from them. Two types of cafeteria dispensing systems are used. One is the line system, in which customers stand in line and move along single-file, selecting foods. The other is the scramble system, in which required foods are dispensed from specific separate stations. The scramble system can be called the *shopping-center system* or the *hollow-square system.* Drive-in or take-out service is an informal type suited to the fast pace of modern living. Some service may be given by carhops or personnel who assist in removing trays. A vending service uses machines to dispense food. Although impersonal, it does answer the need of reducing service personnel costs and allowing food to be dispensed from remote locations away from production centers.

Convenience service has grown rapidly in the past few years. It is notable for its use of ready-prepared and ready-to-eat foods that the consumer

gives final production. The computer has made a significant difference in foodservices by providing the ability to speed service to the customer. Using this system, the server does not need to go to the kitchen to turn in orders, and fewer errors are made in accounting. The results of the introduction of computers to foodservices are greater customer satisfaction, faster customer turnover, and perhaps higher check averages.

Dining etiquette is important for any foodservice operator to understand. Although many rules are followed, this chapter introduced thoughts regarding invitations, the gracious host or hostess, introductions, conversation, formal dining, and ease at the table.

CHAPTER REVIEW QUESTIONS

1. Outline the essential features of American, French, Russian, English, and Canadian services. What are the advantages and disadvantages of each?

2. Demonstrate these five services before a group. Use dishes, trays, tables, or whatever is necessary to make the service as realistic as possible.

3. What are the advantages and disadvantages of centralized and decentralized foodservice? What are

the trends today? Compare line-type cafeteria operation with that of the scramble system. What are the advantages and disadvantages?

4. Dining etiquette is important for any foodservice operator to understand. Discuss the following six areas and note which rules were new to you: invitations, the gracious host/hostess, introductions, conversation, formal dining, and ease at the table.

CASE STUDIES

1. The Good Food Restaurant has been in operation since 1985 at the same location. From a small eatery utilizing only counter service, the operator has been expanded to include table service as well. At present, the operation has twenty-four counter stools and six 24 in. × 24 in. tables. The menu of Good Food has always highlighted the "home-cooked goodness" of the foods prepared by the owner's wife, Erma, and has been moderately priced. The restaurant has been open six days a

week (closed on Mondays) for breakfast and lunch, and it has closed daily at 3:00 P.M. Sam, the owner, has been the waiter for Good Food; when business was brisk, his daughter and son-|in-law pitched in to help. Good Food has provided the necessary income for Sam and Erma for twenty years.

Recently, however, the area where Good Food is located has undergone a transformation and is now the new "hot spot" part of town.

A number of new restaurants have entered the market, mostly offering fine dining, full service, and bar operations. Sam has decided that, rather than change with the market, he would rather sell out now and retire.

A group of young lawyers has purchased the building, equipment, and business from Sam and has hired you to manage the revised restaurant. As part of your job, they want you to devise the concept for the restaurant. The lawyers think the new restaurant should be more upscale than Good Food in order to meet the needs of the new market and its new owners. The new owners also would like to change the decor somewhat, but they are not interested in major renovation of the physical plant. Describe the types of operations possible for the new restaurant and how you would design the service system for the new concept.

2. As one interested in consulting, you have been asked to develop a list of Web sites involving etiquette. In completing your research, you have been asked to consider *www.yahoo.com* as a source of information. By using this search engine with the words *dining etiquette*, what Web sites do you find interesting? Why?

■

VOCABULARY

French service

Russian service

American service

English service

Canadian service

Banquet service

Counter service

Tray service

Réchaud

Guéridon

Chef de rang

Commis de rang

Tureen

RSVP

Finger bowl

Function sheet

Centralized service

Decentralized service

Straight-line cafeteria

Scramble system

Vending service

Convenience service

Human resources

■

ANNOTATED REFERENCES

Axler, B., and C. A. Litrides. 1990. *Food and Beverage Service*. New York: John Wiley & Sons, Inc. (From bus person to captain, all service positions are discussed in this text.)

Dahmer, S. J., and K. Kahl. 1996. *The Waiter and Waitress Training Manual*, 4th ed. New York: John Wiley & Sons, Inc. (All service aspects are presented in this book.)

Ecole Technique Hôtelière Isuji. 1991. *Professional Restaurant Service*. New York: John Wiley & Sons, Inc. (Table arrangements and service techniques are the basis for this text.)

Litrides, C. A., and B. Axler. 1994. *Restaurant Service: Beyond the Basics*. New York: John Wiley & Sons, Inc. (All aspects of service are considered in this book.)

Meyer, S., E. Schmid, and C. Spuhler. 1991. *Food and Beverage Service*. New York: John Wiley & Sons, Inc. (Service, cooking art, culinary terms, and more are presented in this text.)

Mill, R. C. 1998. *Restaurant Management: Customers, Operations, and Employees*. Upper Saddle River, N.J.: Prentice

Hall, pp. 128–155. (Developing high-quality service is presented in one chapter.)

Olsen, M., J. West, and E. Ching-Yick Ise. 1998. *Strategic Management in the Hospitality Industry*, 2nd ed. New York: John Wiley & Sons, Inc., pp. 259–280. (Managing service quality is the basis of one chapter.)

Human Resources

■

LEARNING OBJECTIVES

By the end of this chapter, the reader will:

1. Understand the human resources cycle, people management, and the issues of compensation and benefits.

2. Know how to assess the performance of foodservice personnel.

3. Learn how to write a job description and recruit, select, orient, and train a new employee.

4. Appreciate the legal environment and the many related issues, from lawsuits to sexual harassment.

■

SELECTED WEB SITE REFERENCES

American Society for Healthcare Food Service
   Administrators
www.ashfsa.org/

Foodservice Connection
www.foodservice.com

Hospitality Jobs Online
www.hotel-jobs.com

Hospitality Net
www.hospitalitynet.nl/

Job Search Websites
www.wku.edu/~hrtm/job-srch.htm

Monster Board Job Search
www.monster.com/

Tripod's Internship Center
www.tripod.com/explore/jobs_career/

Foodservice is a very people-oriented and people-intensive industry. Unlike manufacturing, where many repetitive tasks can be completed by machines, foodservice depends on the production of human beings for customer satisfaction and industry growth. Although it was once assumed in management circles that the proper way to increase productivity was to break each task down into the smallest possible repetitive tasks, today we realize that in many instances this system does not work (except for purposes of training). For whatever reasons, today's worker does not share the same attitudes and reactions held by workers in the past. Workers who lived and worked through the Depression regarded job security as their principal motivation for working. But today's baby boomers, who make up a large portion of the work force, may be more motivated by job satisfaction and opportunity than by security. Just providing a job to today's workers may not be enough; there must also be a reason for work.

Experts believe that job satisfaction will become an even more important issue in the future. While today's workforce has been inflated to accommodate the largest bulge in population ever known to this country, the workforce of the future will be different. Instead of there being too many workers, there will likely be too many jobs. In a labor-intensive industry such as foodservice, this may become especially true. In 1998–1999, a labor shortage combined with the already people-intensive criteria of the foodservice industry to create a real problem. To attract and retain lower-paid employees, wages were raised in 1997 from an average hourly rate of $6.05 to $6.37. Therefore, good foodservice managers today are learning not only the basics of personnel management, but also the finer points of what has come to be called *human resources development* in order to improve employee morale and hold workers.

■

HUMAN RESOURCES CYCLE

At one time American management assessed the skill of workers on the basis of their technical expertise. Yet this is not always true today, especially in the case of foodservice. Proficiency in foodservice depends on a combination of the ability to complete tasks and appropriate behavior. The most obvious example is the waiter or waitress who delivers service to a customer. But it is also true for heart-of-the-house personnel. For instance, a cook combines behavior with task in preparing a meal. While the task portion of the cook's job seems most obvious, each meal is made slightly differently; therefore, even with strict guidelines for preparation, each meal varies somewhat. This variance raises the behavioral aspect of service because, unless the differences in task are handled with dedication and sincerity, the meal will not be acceptable. Without a balance of the two requirements, a foodservice staff can become either technically perfect but hostile or very friendly but incompetent. The human resources cycle offers steps that, if properly followed, can help eliminate both potential problems. Both competence and proper behavior should be viewed as crucial to successful staff development.

The first step is selection. Without a properly implemented plan for selecting personnel as a foundation, all other steps will fail. The second step is performance, the establishment of criteria for determining whether or not a job is being done correctly. The third step is appraisal, the evaluation of performance standards, including the operation's rewards system and its method of development or training.

■ LEVELS OF PEOPLE MANAGEMENT

In almost every business, there are three levels of personnel management: corporate, general, and operational. Managers at the corporate level are interested in, among other things, the strategic planning (or anticipation) of human resources needs. At this level, the primary goal is to develop policies to determine where people fit over time. At the second level (usually handled in foodservices by general managers), the main objective is the proper allocation of human resources. Questions such as what kind of human resources are needed, where the needs are, and what arrangement will handle them most efficiently are addressed by this level of management. The manager at this level is interested in determining where each employee fits in, how well each can perform the duties required, how many duties each can handle, and so on. This is also the level at which skills development for personnel management takes place. In foodservice operations, however, all three levels can become the responsibility of the foodservice manager.

Strategic Human Resources Needs

Sometimes foodservices neglect to plan adequately for personnel management. We have all seen the results of failure at this level, such as a hastily trained server whose first day on the floor came too soon.

One of the best ways to anticipate the personnel needs of a business is to use a variation of the G.E. matrix, which was developed in the 1950s. In this system, a foodservice manager determines both the needs and the skills required in the operation by plotting them on a matrix.

Human Resources at the General Managerial Level

At the managerial level, the objectives are to find people who can fill the jobs, establish pay scales, and so on. The goal is to determine how to fit the employees' needs into those of the operation. While many managers feel that the key consideration for all employees is money, several other factors are at least as important to most employees. The reward systems involved in personnel management are determined at the managerial level, as are the requirements for and methods of promotion, incentives, recognition, and respect. Since in many cases an employee can choose from several job opportunities (in the worse case) or products only half-heartedly (in the best case), this level of human resources management is very important.

Objectives at the Operational Level

The human resources objectives of management at the operational level are often inadequately carried out, perhaps because management commonly regards the supervisory activities these objectives demand as time-consuming and distasteful. Yet their importance can be measured in direct on-the-job performance. If, for instance, performance appraisals of employees are conducted fairly and equitably, the staff will be more satisfied and will produce better. If only half-hearted attempts are made, however, or if the system is biased for any reason, the staff will respond with lower productivity. The operational level is also where training programs are instituted to develop more competent and productive employees. If the training is designed in such a way that management and employees both benefit from it, the program will achieve its aims.

NEEDS ANALYSIS

The first step in determining an appropriate human resources program for a foodservice is to perform a needs analysis. Here, the objective is to seek information about the operation and its personnel, in order to determine discrepancies between desired and actual performance levels.

The task goals of a foodservice staff are generally included in the mission statement of the organization. In a fast-food operation, for example, the company might establish as its principal objective the need to serve customers within 3 minutes of receiving their orders. All other activities then revolve around or depend on the initial goal of fast service. Production of food, type of service, training, and so on all reflect the priority of 3-minute service as the operation's mission. An objective evaluation of how often this primary goal is achieved determines in the most general terms the human resources needs.

EMPLOYEE RECRUITMENT AND SELECTION

Even though employees account for thousands of dollars in wages each year to the foodservices for which they work, many managers spend very little time on the procedures by which these employees are recruited or selected. Too often, the workers are hired only on the basis of last-minute newspaper advertisements. When this is done, the future of the foodservice depends on whoever answers the ads.

To recruit and select employees properly, the foodservice manager should establish both long- and short-term employee needs goals. In addition, a pool of available potential employees should be kept on file so that they can be reached at need. To provide for an orderly recruitment and selection process, the foodservice manager should design and have ready the proper type of recruiting material to be used for each job position. Many foodservices, for instance, design advertisements for schools, newspapers, magazines, bulletin boards, and in-house memos well in advance of their actual need. In addition to creating specific advertisements for recruiting, however, a foodservice manager must also maintain an adequate supply of applications, orientation booklets, policy statements, and procedural outlines for each job description, in order to be able to respond to demand as needed.

Too often, management assesses the people working in the organization only in terms of dollars and cents. This is wrong. The people in a foodservice can represent money in the bank, if properly managed, and a positive attitude toward staff will become even more important in the future.

The first step in developing a responsible attitude toward workers is for management to establish a commitment to the personnel in the organization. For many foodservice companies, this means thinking of employees in terms of the opportunity costs associated with them. An employee who is properly selected, oriented, trained, and managed with integrity will become an asset to the firm. If these steps are bypassed, on the other hand, or if shortcuts are taken, the opposite result is likely. One method of effectively displaying a commitment to employees is through a policy of internal recruitment and selection. Under this policy, prior to any external recruitment for job vacancies, a foodservice will post the jobs internally. By doing this and promoting deserving candidates, a manager can build the type of respect for integrity with the staff that is essential in a good operation. In addition, many foodservices have found that the current staff is a good source of recommendations for potential employees from the outside. Hiring people on the recommendation of other staff members can build a rapport in much the same manner as hiring internally.

For hiring from the outside, a foodservice should maintain records showing the degree of effectiveness of each type of advertisement used. This will help identify, over time, the most effective methods of recruitment, which can then be developed and reused. The outside sources of employees most often used by foodservices are newspaper, radio, and billboard advertisements, other foodservices (which may have a surplus of good applicants), schools, colleges, temporary help agencies, employment agencies, and former employees.

Employee selection can be effectively accomplished through various methods. In many cases, the methods used depend on the size of the staff available to accommodate the process. If the foodservice has a personnel department, for instance, the process can be lengthy; but if the manager in charge of operating the foodservice is also the person responsible for all of the hiring, quicker methods may be appropriate. Many foodservices have found that personal interviews combined with questionnaires are a good way to handle selection. Other foodservices use psychometric tests that analyze attitudes, values, personality, and beliefs to identify desirable personnel. Still others use skills tests or organizational records in selecting personnel. Today, many foodservices also use preemployment polygraph tests to determine the desirability of potential employees. At all times, the method used should be accompanied by reference checks of applicants for the position. In addition, a foodservice manager must scrupulously follow the guidelines established by the government for fair hiring practices.

■
ORIENTATION

Although orientation is often dismissed by foodservice management as being a time-consuming and insignificant portion of the human resources process, it can provide employees with a solid understanding of the job to be done. If the orientation clearly sets out the requirements and responsibilities of the job, the employee will know from the start what is expected. A good orientation should also include a complete tour of the facility, introductions to all personnel with whom the new staff member will be working, and a description of all company policies and procedures that are pertinent to the job.

Orientation should never be left entirely to another employee. While it may be necessary or even wise for an employee to lead the tour or handle the introductions, well-urn foodservices always involve management in the orientation process. In such foodservices, one manager is often designated as the training manager and bears managerial responsibility for orientation.

■
TRAINING

Training is one of the most effective methods of eliciting good performance on the job. As such, it should not be left up to other employees to perform. Training should be thought of as an educational process whose specific goal is to teach job competence.

The first step in designing an effective training program is to determine the critical incidents (most important aspects) of the job. The next step is to determine the exact criteria on which performance during training should be evaluated. This, of course, varies from one foodservice to another, as well as from job to job. In most cases, the criteria should address both behavioral and procedural objectives. The procedural objectives might include such points as how to perform a certain task properly and how much time to allow for the task. These objectives should be broken down into the smallest identifiable tasks, before being used for training. Each task should require no more than a single sentence to describe, and each should always begin with an action verb, such as "pick up the cup by the handle." This part of the training is often carried out in simulation before it is actually performed in situations involving customers or products of the foodservice; it should always be directed by a manager or supervisor in charge of training.

Several different methods of directing the employee during the first part of the training program have been successful. In some, all training is done at the work station; in others, training is conducted in classrooms or mock work environments.

After the stage of initial training is complete, most foodservices have found it advantageous to turn the new employee over to a current employee for apprenticeship training, which may last from a few hours to a few days. By observing the actions of a more experienced employee, the trainer can get a hands-on feel for the job before being called on to produce independently.

Training Tools

In addition to providing direct training by management and other personnel, foodservices use various audiovisual and classroom training tools. Depending on the position and the company, training may include classroom meetings in which the objectives of each job are described in detail. In addition, many foodservices today use videotapes to show how to perform tasks effectively. Others use computers, audio tapes, pictures, or presentations to describe task completion techniques. When designing a training program, management should investigate each method to determine how well it might work for its own foodservice.

Computers and Training

Computer-assisted learning is becoming more and more prevalent in foodservice operations. Merely surfing the Internet, management may be trained in new techniques of operation, learn about the competition, purchase needs, and discover new recipes for the markets being served. Beyond this, interactive computer software is available to teach and train in almost every conceivable subject area.

Once computer training has taken place, management should bring the trainees together in a group for discussion of concepts learned. Such group involvement requires active participation on each trainee's part. Social interaction takes place and communication skills are enhanced. Review of items learned requires quick thinking on the part of participants and instills ideas even further in one's mind. The result is practice in decision making as well as building member commitment to solutions worked out by the group.

Computers will continue to have an impact on all areas of foodservice operations. Managers are encouraged to familiarize themselves with both the software and hardware available on the market and to implement training packages where appropriate in management and leadership training as well as in sanitation and safety, nutrition, and other areas of professional development.

Behavioral Training

Foodservices sell both products and service. Therefore, training in foodservices is not limited to the issue of task completion. Because the industry depends on delivering service to customers, behavioral training is an essential part of the training process. This is especially important for the front-of-the-house personnel in a foodservice, but it is also important for the heart-of-the-house staff.

Behavioral training is not as easily accomplished as task training. And since most foodservice managers are themselves trained by a process of on-the-job training in each position, they may not be well-informed in behavioral training goals either. As a result, many foodservice managers find it difficult to determine the behavioral attributes needed for each foodservice job.

Behavioral training, like task training, depends on the introduction of clear objectives to each employee. Only after specific attributes are described in detail can the employee's possession and development of them be adequately measured. To identify these objectives, many foodservice managers must place themselves in the position of a customer or of another employee. By doing so, they can better recognize what is desirable behavior and, by extension, can determine what the objectives of the behavioral training process should be.

Part of successfully conducting behavioral training involves maintaining a concise record of progress and referring to it often. This allows the employee to get a sense of what kind of progress is being made. Taking note of specific progress also helps define areas that are in need of improvement. In this type of training, it is important for the manager to give immediate feedback on both progress and shortcomings. The immediacy of the appraisals tends to confirm the employee's objectives and enables the employee to adopt additional goals throughout the training process.

The final step in behavioral training for foodservice work is a complete evaluation of the employee's progress. Most managers have found that this is best accomplished one-on-one with the employee, in an open environment that invites discussion. This approach encourages self-analysis of training progress by the employee, including a determination of what additional goals need to be met. From this evaluation meeting, clearly stated objectives should be established to serve as the basis for future assessments of progress.

Records of these meetings should be made a part of the employee's permanent file. This "final" evaluation should not be the last that the employee receives, however; many foodservices have found that reevaluations every three to six months help spur the employee to achieve progressively higher goals.

Training the Trainer

Traditionally, training within foodservice operations has involved the "buddy system." That is, when a new hire is brought on, an older, more experienced employee trains the new. Although this may work in some situations, it is generally felt that this is not the best training method to use. Management's standards of excellence may not be told to the new employee and shortcuts or other inappropriate procedures may be taught to the new person.

Accordingly, management must find people within the operation who would like to train the new employees. If management cannot do this job each time, people who enjoy doing it must be found. These trainers should possess the following characteristics:

- Ability to communicate effectively
- Ability to motivate
- Enthusiasm
- Knowledge of job skills
- Knowledge of trainee abilities and skill levels
- Knowledge of learning principles
- Patience
- Understanding

The level of professionalism within the operation will be reflected by the trainer's ability to impart the operation's philosophy of hospitality to the new hires. With everyone training everyone, quality will slowly drop to the level of the least proficient employee. But with an effective train-the-trainer program, management is able to instill the standards it wants imparted to all employees and staff.

■
LABOR COSTS

Labor costs are among the most volatile expenditures that must be managed by a foodservice operator. This has become especially obvious in recent years in foodservice because of increased wage and salary rates, increased unionization, increased fringe benefits, and increased levels of mandated minimum wage. Therefore, learning how to manage labor costs properly represents a big step toward ensuring a profitable foodservice operation. In the final analysis, managers who effectively utilize their labor force are in many cases the same managers who operate the most successful foodservices. The basic tools for effectively managing labor costs are the schedule, staffing guides, and employee participation in job design.

Scheduling

The first tool a foodservice manager should master in order to begin to control labor costs is the schedule. The object of proper scheduling is to have the correct number of employees at work at any given time. Having too many employees on the clock results in a labor cost that is too high, while having too few leads to inadequate service of the clientele. To plan for the correct number of employees, the managers must make a schedule for each shift. Sometimes, weekly schedules are used; in other cases, monthly or even permanent schedules are created. Although arguments can be made in support of each method, most foodservice managers will find that the shorter the schedule is, the more flexibility the manager has in controlling labor costs.

For each job description (or employee category), a separate schedule should be prepared. To be most effective, the schedule should be easy to read and conspicuously posted. In addition, it should be posted early enough to allow for changes; in the case of weekly schedules, many foodservice managers post the schedule for the following week on Thursday—in plenty of time for all employees to see the schedule and arrange to make any necessary changes prior to the start of the schedule week.

Foodservices differ in their staffing needs, depending on the type of operation involved. In a foodservice where the level of customers can be predicted from day to day, the schedule may require little flexibility. On the other hand, in a commercial foodservice where the number of customers is less predictable, a labor force that can be adjusted to accommodate increases or decreases in customer volume is highly desirable. Many foodservice managers have found that the use of relief and part-time personnel provides the level of flexibility sought in such cases. Relief personnel are workers whose employment is intermittent, depending on the volume of business, and who are willing to be called on short notice when business or employee absences raise the need for extra personnel. Part-time personnel are those who work regularly, but only a few days per week or a few hours per day.

The use of part-time personnel has become especially widespread in fastfood operations. In some parts of the country, for instance, fastfood operations hire men and women to work only during the lunch hour, while their children are in school. In other cases, part-timers are employed after school or in the evening for short periods of time. While this tactic is certainly not limited to the fastfood segment of our industry—since part-timers also make up a large portion of the staff at many other types of foodservices—it is most prevalent there. A variation on part-time work that has evolved is called the *split shift*. Under this method, employees work the peak periods of lunch and dinner only, and are not employed during the hours in between. This approach is only possible, of course, when employees are willing to work such hours.

Staffing Guide

The first step in utilizing a schedule is to prepare a staffing guide for the foodservice. The purpose of this guide is to present an outline of basic personnel needs for a given period of time. Because the staffing guide varies from one foodservice to the next, it is wise for the manager to be in charge of developing this instrument. For the novice manager, however, various industry-standard staffing guides are available from a wide range of sources. For advice on predicting the number of employees that will be necessary for each job category, for instance, an inexperienced manager would be wise to consult the National Restaurant Association, the state restaurant association, or perhaps an accountant or other consultant.

To prepare a staffing guide, a foodservice manager must first determine the business's hours of operation and service. Once this is done, the number of employees required, by position, for each hour of operation must be determined. In staffing terms, these employees are referred to as *FTEs* (*full-time equivalents*). One FTE represents one full-time employee needed. Because the number of personnel required varies from day to day, a typical staffing guide is often constructed to portray the FTEs needed per week, thus taking into account both slow and busy periods. Once the total staff requirements are outlined, the foodservice manager can begin to use the scheduling method previously outlined.

Job Descriptions

A job description is a written account of the responsibilities involved in a job. In addition, it formally describes the tasks, duties, and performance requirements of each job. For a foodservice to operate at its highest level of effectiveness, emphasis must be placed both on creating these descriptions and on making certain that each employee knows the requirements of the job. Each job description should include the following items:

1. The title of the job and its classification in the organizational hierarchy.

2. A complete description of all tasks and duties associated with the job.

3. Performance requirements of the job.

4. Standards of measurement used in appraising performance on the job.

It is important that the job description reflect the real job; too often, foodservice job descriptions are outdated or ambiguous.

Job Design

The way in which a job is designed is often the single most important factor affecting how well the task is done. This can be especially true in the case of service operations. If the jobs in a foodservice are properly designed, the work is more likely to be completed successfully within an allotted time period. If, on the other hand, jobs are designed without consideration of how tasks interrelate, the opposite result can be expected. The first step in designing a job accurately and effectively is to analyze the task. Recently, many foodservice managers have discovered that one of the best ways to ensure high-quality job performance is to involve employees in designing the task.

Compensation and Benefits

How a foodservice manager compensates workers can obviously have a great bearing on the overall performance level they attain. Although to many managers compensation may mean only dollars and cents, this is not the only aspect of a good compensation policy. Such steps as establishing management integrity, encouraging staff participation in decision making, and offering fair evaluations of performance play a major role in the compensation package. Indeed, employees who feel unfairly treated by management often want more in the way of direct monetary compensations than do employees who experience job satisfaction.

The basic rules of compensation in the foodservice industry are set forth in federal and state statutes and sometimes in local ordinances. These regulations change periodically, and a good foodservice manager should be aware of those changes. Just to be aware of the existence and overall effect of regulations such as the Fair Labor Standards Act of 1938 and the Equal Pay Act of 1963 may not be enough, since many of the provisions that apply to foodservice employees differ in important respects from those that apply to other industries. Information concerning the most recent legal statutes governing pay can be obtained from federal and state wage-and-hour offices and from other government agencies.

The critical points to remember in establishing compensation packages for employees of a foodservice is that compensation is possible at three levels: wages, incentives, and benefits. Wages are dollars earned for completing the tasks assigned. In many cases, these are insufficient to generate the type of production required for foodservices, since so many foodservices have developed programs that offer incentives and benefits. Incentives—rewards for performance—often supply the motivational basis for employees to attain the standards required in foodservices. In some cases, bonuses and profit-sharing in return for excellent work are offered as incentives by foodservices.

In many cases, foodservices have also developed benefit programs to enhance the level of job satisfaction felt by employees. Benefit packages vary greatly from company to company and from position to position. In some foodservices, the principal benefits offered are those required by state and federal regulations, such as worker's compensation and unemployment compensation. In others, a long list of benefits can be found. Today, many foodservices are discovering that, to attract the type of employee who will stay with the company and be productive for a long time, more benefits must be offered. The most common benefit packages are insurance-related. Whether directed toward health insurance, life insurance, or retirement payments, benefits have proved to be a strong attraction in foodservices, especially with the cost of hospitalization rising as it is. Other benefits that may be offered by foodservices include health maintenance organizations (HMOs), fitness and conditioning programs, drug- and alcohol-dependency programs, pensions, low-cost short-term loans, tuition assistance programs, education assistance, flexitime positions, and part-time positions.

■

ASSESSING THE PERFORMANCE OF MANAGEMENT AND STAFF

As part of its regular human resources plan, every foodservice should have a method for evaluating the performance of management. In addition to supplying the company with valuable information about the progress of personnel, these evaluations also provide a responsible basis for establishing criteria for raises and promotions. Various types of performance evaluations have been successfully used in foodservices. Among the most prevalent types are skills audits, performance appraisals, and assessment centers. Once it has established standards for the level of job performance expected, management should measure performance, compare it to the standards, and then make an evaluation.

Skills Audits

Skills audits are a qualitative method often used to assess the abilities of management. First, a workshop of managers and supervisors is organized, and participants are divided into peer groups. Then, aided by facilitation, these groups brainstorm the performance standards, skills, priorities, behaviors, and task resolutions for each job description. By working in groups, people in management are challenged to define what their roles should be in the foodservice.

Performance Appraisals

The performance appraisal is perhaps the most widely used method of evaluating both workers and management. Performance appraisals are based on progress toward a clear set of goals, which can be established for three principal reasons: compensation, promotion, and career development. The goals should be established by management in consultation with the staff—not by management alone. In the case of a foodservice, the goals should be both procedural and behavioral.

After the goals to be accomplished have been set, specific methods of measuring progress toward the goals should be established. Like the goals themselves, the measurement scale should be determined jointly by management and staff, and the measurement itself can be performed by either of the two. Various performance appraisals have been successfully conducted, including management appraisals of staff, staff self-appraisals, staff evaluations of management, group evaluations, and management self-evaluations. Often it is beneficial to combine two or more of these methods

in order to eliminate any bias that may otherwise skew the appraisal. One method often used in foodservice is for the staff members to evaluate themselves individually at the same time that management is evaluating them. An open discussion of both the consistencies and the inconsistencies that exist between the two evaluations has often fostered future employee growth and development, as well as better communication between employees and management. After the goals and measurement methods are established, regular evaluations are performed to determine the progress of each person involved.

Several negative factors can influence the performance appraisal system, and a manager should be aware of the threat by each. Among the most likely errors in an evaluation is to note the most recent act as typical, when in reality it may be isolated. Favoritism can also play a distorting role in performance appraisals unless closely monitored, as can the "halo effect," in which the evaluator recognizes only performance levels in a favorite portion of the job. To eliminate these pitfalls, many managers utilize a system of spot-checking appraisals at random to determine their reliability. While the system is not foolproof, it has proved valuable to many foodservice operators.

Assessment Centers

Assessment centers were originally designed by the military during World War II for training agents. Since then, various businesses, including many foodservice organizations, have adopted such centers as an effective means of evaluating personnel. To date, the assessment center format has been used primarily to test the ability of management to make decisions, yet the process can also be used to evaluate employees as potential managerial candidates.

In using the assessment center system, small groups of six to eight managers (or prospective managers) are formed, and each group is asked to perform in a variety of situations that either simulate on-the-job management or call for role playing by the participants. While participating, the members of each group are evaluated by an observer on the basis of various attributes displayed.

One of the greatest strengths of the assessment center is the opportunity it affords to evaluate the interpersonal skills of each participant. Since the test situations call for the development of group decisions, observers are able to determine how management personnel will interact with other members of management or with the staff.

■

LEGAL ENVIRONMENT

All liability in a foodservice operation is traceable to risks, most of which can be prevented. Taking steps to minimize these risks will help to prevent costly law suits and maximize profitability. According to the law, liability is an enforceable responsibility that one has for another. Liability is not just there, then, it is created. Foodservice operators create liability by inappropriate preparation and service of food, unsafe conditions in the foodservice operation, or poor judgment in the service of alcohol. Since liability is created, it can be prevented.

In this section areas are discussed in which the operator can prevent liability through proper management and consideration for the staff and employees hired as well as the markets served. The legal environment in which all operators find themselves today is a difficult one worthy of careful attention and understanding.

Fraud

Foodservice operations have often been the victims of crime. Since most criminal laws relating to operations stem from individual state legislation, the discussion here will have to be general. Management should check with local counsel to ensure the correct understanding of laws.

Foodservices have been victimized by "skips," credit card fraud, bad checks, and other crimes. Although the electronic age has all but eliminated many of these crimes, management must understand that fraud is a criminal offense and must be reported. For example, using stolen, forged, or altered credit cards to obtain services from an operation can be punishable

with fines up to $10,000 and imprisonment for up to ten years.

Staff and employees must be trained to understand what to do when fraud is suspected within an operation. The criminal statutes provide strong weapons in the operation's fight against persons seeking to defraud. Enforcement of these penalties acts as a deterrent against such criminal activity.

Damage Claim

In working with insurance companies, law enforcement officials, and counsel, foodservice operators may be entitled to damage claims in a variety of situations. Patrons may also be looking to be paid damages for situations in which they find themselves.

Again, the legal environment today is very specific as to who gets what and how much. For example, management that wrongfully serves intoxicating beverages is in serious trouble and will pay dearly. There are also cases of slander, choking victims, discrimination, and facility negligence. All of these can cost an operation its very existence.

Foodservice management must be aware of its responsibilities and take care to execute those responsibilities. With employees, patrons, government, as well as the community, all having and making demands on the foodservice operator, he or she will have to be aware of this environment if long-term success is to be achieved.

Lawsuits

Modern society seems to be encouraging everyone to file a lawsuit. Each person is told that he or she has certain rights and that has led many to courts where battles are fought to prove those rights. The law, in general, requires foodservice operators to exercise reasonable care to protect patrons from personal injuries and harm to property caused by poorly maintained premises or lack of safety precautions.

All public accommodations must adhere to local studies regarding building codes, fire regulations, and safety. Management is not required to take extreme or expensive measures to guard against every accident, but they are expected to exercise reasonable care in the supervision of employees and management of premises to help ensure patron safety.

For example, if a manager notices that a floor mat in the entrance to the operation needs to be tacked down but does nothing about it, he or she may be liable if someone trips over it. The same is true for the mopping up of spilt water. Unless a sign is posted noting the wet area, management will be responsible if a person slips and falls.

Lawsuits can be extremely time consuming and expensive. Management has a duty to provide safe premises, a duty to safeguard property of patrons, and a duty to police conduct of patrons, employees, and third parties. When such duties are not carried out, management is asking for a lawsuit that will be much more difficult to handle than the original duty was.

Legal Liabilities

As stated above, liability is the responsibility that one person has to another that is enforceable in court. The foodservice operator's liability consists of duties and obligations owed to patrons for safe premises, a safe environment, and safe food.

There are three approaches a patron can take when having suffered from eating or drinking unfit food or beverages: negligence theory, breach of warranty theory, and strict liability theory.

Negligence theory requires that the injured patron be able to prove that the negligence on the part of the foodservice operator resulted in unfit food being served which resulted in illness.

Breach of warranty theory is most commonly used and states that the patron only has to prove that the operator served an item that caused illness. In other words, the patron does not have to prove that the operator was negligent, only that the operator breached the contract of sale by serving unfit food.

Strict liability theory does not require that negligence or breach of warranty be established, but simply that the unfit food caused the person to become ill. Legal liabilities for foodservice operators can be complex and difficult to understand. The responsibility to learn these liabilities is important, though, and counsel should be consulted if local and state regulations are not clearly understood.

Sexual Harassment

The issue of sexual harassment in the workplace has become a major concern in every industry. Foodservice operators must develop effective policies to deal with this serious problem that could result in liability. Guidelines have been written to specify the steps that employers should take to prevent sexual harassment. They are:

- Raise the subject of harassment with all employees.
- Express strong disapproval and develop appropriate penalties.
- Inform employees of their rights to complain and how to do so.
- Develop methods to make everyone fully aware of the problem and how to prevent it.

Foodservice operators can be held responsible for any employees involved in sexual taunts, lewd or provocative comments and gestures, and sexually offensive touching by co-workers and patrons.

The legal environment is strict and real in today's foodservice industry, and management, staff, and employees all need to be aware of the consequences for not understanding it.

Gender Equity

Most everyone has heard the phrase: "Our company is an affirmative action, equal opportunity employer and does not discriminate against persons on the basis of race, religion, national origin, sexual orientation, gender, marital status, age, or disability." Such a statement made when hiring suggests that the foodservice operation is interested in hiring the best possible person for the job.

There is much discussion today that men make more money than women for the same job done. There are positions that may have been traditionally held by females in the kitchen or elsewhere. Gender equity recognizes that the job to be accomplished can be done successfully by either male or female. Pay equality is also an important consideration. Only by treating everyone equally can the statement made above be realized as the company's policy. Foodservice operators are encouraged to follow suit.

■

SUMMARY

Foodservices are people-oriented businesses that require skill in knowledge of various methods of human resource development. The goals and objectives of the baby boomer workforce differ in some ways from those of earlier workforces; therefore, new personnel practices are required to fulfill the needs of foodservices.

The human resources cycle involves selection, performance, and appraisal. Each step is critical to the development of a good foodservice staff, and each must be considered on both procedural and behavioral levels.

The three levels of personnel management are strategic (corporate) management, general management, and operational management. Each has different objectives. Strategic management focuses on planning for the future; general management is involved with filling the job, designing compensation programs, and so on; and operational management is primarily concerned with training and with evaluating on-the-job performance.

Before beginning a personnel management program, a foodservice should develop a needs analysis to determine what types of personnel and skills are needed. During this process, the most important objectives of the organization should also be set down.

Even though employee recruitment and selection strongly influence how the job is completed, recruitment is often treated as a last-minute task. A good recruitment program should include establishing the integrity of the company in the community and among employees, adopting a policy of internal hiring, and seeking potential employees by the most effective means available. Selection can be based on interviews, questionnaires, psychometric testing, skills analysis, skills tests, or various other methods. Many foodservices combine several selection methods in an effort to obtain the best employees.

Orientation of new employees to the foodservice is a critical task that should be carried out by management. Included in the orientation are a tour of the facility, introduction to the other employees, and a discussion of all relevant policies and procedures of the company.

Training is one of the most important aspects of developing a good foodservice organization. It should be performed by management rather than by staff. The training supervisor should establish specific goals to accomplish in the course of training, and these goals should have both procedural and behavioral components. Numerous training tools are available to foodservice managers, including videotapes, audiotapes, computers, pictures, classrooms, and guest lectures. Training the trainer is an important concept in the modern foodservice operation.

Labor costs are very volatile in foodservices, and among the basic management tools that must be mastered to deal with these costs are scheduling, staffing guides, and employee participation in job design. The schedule provides an operational budget for labor costs over a period of time. A staffing guide indicates how to plan for scheduling. Job descriptions are written outlines of job responsibilities that are used as the basis for evaluating job performance and for identifying training needs.

Compensation packages consist not only of wages, but also of incentives and benefits of various kinds. Today, many foodservices are finding that they must develop more and better compensation packages to attract desirable personnel to their companies.

Skills audits, performance appraisals, and assessment centers are all methods of evaluating performance. A performance appraisal, the most common of these, identifies the specific goals that a job was designed to achieve and evaluates the progress of the employee in attaining those goals. Management and personnel should both be involved in the analysis process.

The legal environment in which all operators find themselves today is a difficult one worthy of careful attention and understanding. Foodservices have been victimized by "skips," credit card fraud, bad checks, and other crimes. Although the electronic age has all but eliminated many of these crimes, management must understand that fraud is a criminal offense and must be reported. In working with insurance companies, law enforcement officials, and counsel, foodservice operators may be entitled to damage claims in a variety of situations. Patrons may also be looking to be paid damages for situations in which they find themselves. The law, in general, requires foodservice operators to exercise reasonable care to protect patrons from personal injuries and harm to property caused by poorly maintained premises or lack of safety precautions. The foodservice operator's liability, then, consists of duties and obligations owed to patrons for safe premises, a safe environment, and safe food. Foodservice operators must develop effective policies to deal with sexual harassment, and gender equity recognizes that the job to be accomplished can be done successfully by either male or female.

■ CHAPTER REVIEW QUESTIONS

1. What is the definition of liability as it relates to the foodservice operator? What are the three theories regarding liability? Which is most commonly used?

2. Sexual harassment is a continuing issue in society. What does it involve, and how should a foodservice operator deal with it on a day-to-day basis?

3. How should the foodservice operator effectively manage labor costs? What are the basic tools used? Discuss each and provide examples where appropriate.

■ CASE STUDIES

1. As head of a recruiting firm, you have been contacted to do research on labor, its availability, and to report to a foodservice operation all you can find about turnover of foodservice workers. You are not certain where to look, but you are told to begin at *www.reidsystems.com* on the Internet. What can be said about employee turnover in the foodservice industry? Report on one other labor-related issue of your choice from the research completed.

2. Dianne Ross has taken over a foodservice operation that is known throughout the company as a problem. The operation has among the highest rates of absenteeism and turnover of any in the organization. In addition, the unit's sales and productivity have been declining of late.

   On her first day on the job, Dianne immediately notices that the employees are reluctant to talk to her. In fact, they avoid Dianne. After investigation, she finds out that this is due to the management tactics of her predecessor, who often singled out employees and publicly berated them for their poor performance on the job. Moreover, the former manager virtually was unapproachable with any problems concerning either the operation or personnel.

   Since Dianne has graduated from a hospitality school in which she was required to take a few courses in personnel management, she knows the answer: She must begin to develop a sense of faith and fair play with the staff. Since Dianne is so busy with the day-to-day operations (which were left in a shambles), she has asked you to help her design a program to turn the personnel policies of the operation around. Dianne has asked you to make an outline describing the key elements in a human resources program for her foodservice. Because she will have to introduce this program to junior management in the organization, she has also asked that you give some details on the importance of each step. How would you begin to develop such a program for Dianne?

■

VOCABULARY

Human resources	Gender equity
Human resources cycle	Recruitment
Orientation	Internal recruitment
Performance appraisals	Skills tests
Job design	Critical incidents
Skills audits	Simulation
Assessment centers	Behavioral training
Job descriptions	Benefits
Employee selection	Equal Pay Act of 1963
Needs analysis	Fair Labor Standards Act of 1938
Compensation	Profit-sharing
Negligence	Benefit packages
Labor schedule	Sexual harassment
Staffing guide	Liability
Labor cost	

■

ANNOTATED REFERENCES

Berger, F., and D. Ferguson. 1990. *Creative Techniques for Hospitality Managers*. New York: John Wiley & Sons, Inc. (Creative thinking in the hospitality industry is discussed in this book.)

Drummond, K. E. 1992. *Retaining Your Foodservice Employees*. New York: John Wiley & Sons, Inc. (Motivating employees to do their best is the basis for this text.)

Go, F. M., M. L. Monachello, and T. Baum. 1996. *Human Resource Management in the Hospitality Industry*. New York: John Wiley & Sons, Inc. (From communication to decision making, this book covers motivation, training, conflict resolution, and other topics.)

Hinkin, T. R. 1995. *Cases in Hospitality Management: A Critical Incident Approach*. New York: John Wiley & Sons, Inc., pp. 127–200. (Cases on diversity, sexual harassment, staffing, and training are a part of this text.)

Lundberg, D. E., and J. R. Walker. 1993. *The Restaurant from Concept to Operation*, 2nd ed. New York: John Wiley & Sons, Inc., pp. 163–206. (Staffing and training matters are considered in this book.)

Marvin, B. 1994. *From Turnover to Teamwork*. New York: John Wiley & Sons, Inc. (Employee motivation and retention are the basis for this text.)

Mill, R. C. 1998. *Restaurant Management: Customers, Operations, and Employees*. Upper Saddle River, N.J.: Prentice Hall, pp. 288–369. (Human resources are discussed in chapters on employee selection, training and development, and motivating the employee.)

Newman, D. R. 1998. *Human Resource Management*. Upper Saddle River, N.J.: Prentice Hall. (Sections on understanding yourself, human resource management, customers, plus others, are featured in this book.)

Tanke, M. L. 1990. *Human Resources Management for the Hospitality Industry*. Albany, N.Y.: Delmar Publishers, Inc. (Human resources are considered from A to Z in this book.)

CHAPTER 8

Product and Profits

■

LEARNING OBJECTIVES

By the end of this chapter, the reader will:

1. Be able to control quality within a foodservice operation.

2. Understand issues of security within a foodservice operation.

3. Learn of the principal aspects of accounting, including the profit-and-loss statement, balance sheet, break-even analysis, and budgeting.

4. Be able to distinguish between labor, food, and beverage costs and control the same.

■

SELECTED WEB SITE REFERENCES

Apple Computer, Inc.
www.apple.com

Compaq Computer Corp.
www.compaq.com

Dell Computer Corp.
www.dell.com

Digital Equipment Corp.
www.digital.com

Gateway 2000 Inc.
www.gw2k.com

Hospitality Industry Technology
    Exposition and Conference
www.hitecshow.org

Hospitality Solutions International
www.hsi-pos.com/

HOST Group
www.hostgroup.com

InfoGenesis
www.infogenesis.com

Intel Corp.
www.intel.com

International Association of Hospitality
    Accountants (IAHA)
www.iaha.org

International Business Machines, Inc.
www.us.pc.ibm.com

NEC Corp.
www.nec.com

Sony Electronics, Inc.
www.sony. com

Squirrel Companies, Inc.
www.squirrelsystems.com

■

CONTROL OF FOOD COST

The best-planned menu in the world will do an operation little good if it is not produced as intended. Failure to follow through on an excellent menu is a common fault in foodservices today. The food must come into the dining room with the proper quality, while yielding the right profit.

Management must see that proper controls are instituted to obtain production efficiency, to minimize waste, and to reduce costs. Standardized recipes, tests for yield, quality controls, precosting methods for menu items, production schedules, security procedures, and the right selling price are all necessary for proper managing of products.

Standardized Recipes

A standard is a measure used to evaluate a quantity, quality, or volume. We obtain a value of something and then compare this value to the standard to see whether it is equal to, exceeds, or falls below the value of the standard. These results tell us that the item measured is, respectively, equal to, better than, or worse than the standard.

A standardized recipe is a formula established by management that yields a known quantity of product of a known quality. In quality and quantity, it meets standards that have been established by management as desirable for the foodservice. The calculation of food cost and selling price cannot be made with any accuracy unless recipes are standardized to use only specific ingredients in known amounts, resulting in a known yield of portions. Using a standardized recipe reduces the risk of poor food handling and preparation failure. Such a recipe should establish standards of presentation and service, as well as of quality and quantity. A well-standardized recipe also can help train employees in good work methods and in basic food production principles. It also is a safeguard against the dominance of a chef or cook who keeps the recipe used a secret and leaves management to guess at costs and yields. Using the standardized recipe allows others to fill in or the chef and produce successful foods.

Standardizing any recipe begins with analyzing it thoroughly. Certain standard proportions in food production must be observed. For example, the proper ratio of sugar to flour in a butter cake is 1.25:1. A recipe in which the amounts of sugar and flour vary too much from this ratio probably will not produce a good item. Checking a recipe against these standard proportions can thus give a good clue as to whether it is properly balanced or needs revision.

After analysis, the recipe should be tested. For this step, all ingredients must be carefully weighed or measured, production time should be carefully noted, and the condition of the item as it goes through production should be described. Taking notes during testing allows the testers to work back later to see exactly what happened. Mixing procedures, preparation and cooking times, temperatures, major utensils used, and the method of serving should be precisely recorded, as well as yield and portion size (see Fig. 8-1). Dividing portion size into total yield gives the number of portions obtained. Nutritional facts may be stated on the standardized recipe. If possible, a picture of the finished product should be taken and perhaps also a picture of a portion, so employees will know how both

---

Preparation time: ¾ hour   Cooking time: 2½ hours
Major utensils used: covered 12″ × 20″ × 4″ pan, scale, steam cooker, oven
Proper method of serving: serve on dinner plate
Nutritional facts: 470 calories per portion; protein 33 g; fat 23 g; carbohydrate 33 g; calcium 87 mg; iron 6.1 mg;
   Vitamin A 3700 I.U.; thiamine 0.3 mg; riboflavin 0.46 mg; niacin 8.96 mg; ascorbic acid 63 mg

Yield Count: 24 servings   Portion Measurement: 1 cup   Total Yield: 5½ qt

Ingredients	Weights	Measures	Counts	Method
Beef cubes, cut in 1- to 1½-in. pieces	7½ lb			Combine flour, salt, pepper, and paprika. Dredge beef cubes in seasoned flour and save remaining flour.
Flour		2 cups		
Salt		3 Tbsp		
Pepper		½ tsp		
Paprika		2 tsp		Heat cooking fat in a 12″ × 20″ × 4″ pan. Add floured beef cubes and brown at 375°F (moderate oven) for approximately 1 hour. Stir once during browning; add remaining flour just before stirring.
Cooking fat		¼ cup		
Whole cloves			12	Place 1 clove each in the end of 12 of the onions.
Onions, small whole			24	
Water, hot		2 qt		Add water to browed meat mixture. Stir thoroughly. Add onions, potatoes, carrots, and bay leaf.
Potatoes, small	4 lb EP			
Carrots, cut lengthwise, then into thirds		16	16	
				Cover tightly and cook in oven 1½ hours or until meat and vegetables are done.
Bay leaf			1	
Brussels sprouts	2½ lb			Cook Brussels sprouts, drain, and fold into stew.

FIGURE 8-1   Standardized recipe card for beef and brussels sprout stew.

are supposed to look. The whole recipe is then critiqued, changes are made if necessary, and the item may be retested.

If the people performing the tests know food production, they usually need not do more than three tests, and often one or two tests are sufficient. To perfect some recipes, however, many tests are needed. For example, more than thirty tests were needed to develop the 100-portion Navy recipe for chiffon cake. The recipe worked well in small quantities, but increasing the ingredients to give 100 portions in the same ratios yielded an unsuccessful product. Consequently, a completely different recipe, varying in ingredient ratios from the small-quantity one, had to be developed by trial and error. It often is desirable to submit a recipe for evaluation to people other than those doing the testing—even customers.

A recipe that is properly standardized can be modified to give more or fewer servings. The method for modifying the quantity of ingredients to increase or decrease portions appears in the conversion charts in Appendix E. No recipe is ever finished; each time it is used, it should be reevaluated. Whenever a recipe is improved, food standards rise. Recipes can also become

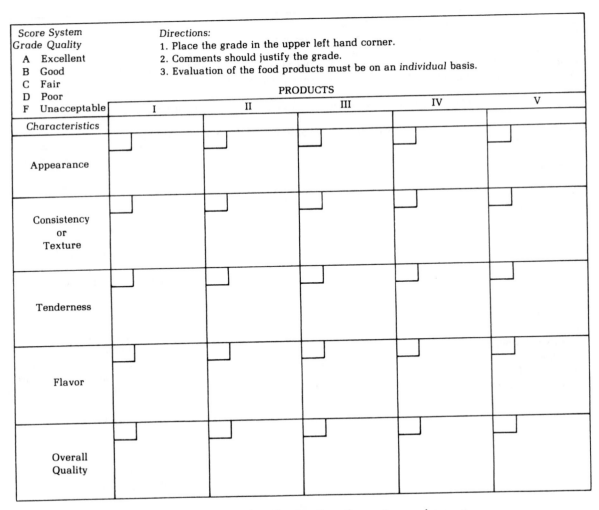

FIGURE 8-2 Sample evaluation form for scoring products.

outdated as new materials appear on the market. For example, many recipes still in use today call for the addition of soda to a liquid before adding it to the product. This step was specified because many years ago soda was more coarsely ground than it is now, and dissolving before adding was necessary to get it properly distributed in the product. The soda of today is fine enough to be sifted in with the flour. Yet people still go on dissolving the soda, just as others did fifty years ago, not realizing that the practice is no longer necessary.

In scoring a product, analysts must evaluate the factors of flavor, texture, color, tenderness, appearance, form, and temperature. Score sheets are usually used (see Fig. 8-2). Some testing facilities have tasting rooms; judges are carefully selected for their abilities to taste, see, and evaluate. The committee that accepts the final recipe also accepts the standard of presentation it establishes. Testing is the backbone of all standardized recipes. When the entire menu has been translated into standardized recipes, all recipes should assume the same form and contain the same information (see Fig. 8-3).

a. Ingredients	Measures	Method
Mayonnaise	3 cups	Preheat broiler. Combine mayonnaise, Parmesan cheese, and Worcestershire sauce.
Parmesan cheese, grated	1 cup	Blend well and place ½ tsp of mixture on small crackers. Brown lightly
Worcestershire sauce	2 Tbsp	under broiler.

b. Action	Measures	Ingredients
Preheat broiler. Combine and blend well	3 cups 1 cup 2 Tbsp	mayonnaise, grated Parmesan cheese, and Worcestershire sauce.
Place Brown	½ tsp	mixture on small crackers. lightly under broiler.

c. Ingredients	Measures	Method
Mayonnaise	3 cups	1. Preheat broiler.
Grated Parmesan cheese	1 cup	2. Combine mayonnaise, Parmesan cheese, and Worcestershire sauce. 3. Blend mixture well.
Worcestershire sauce	2 Tbsp	4. Place ½ tsp of mixture on small crackers. 5. Brown lightly under broiler.

d. While the broiler is preheating, combine 3 cups mayonnaise, 1 cup grated Parmesan cheese, and 2 Tbsp Worcestershire sauce. Blend well, and place ½ tsp of the mixture on small crackers. Brown lightly under the broiler.

FIGURE 8-3 Forms of recipe presentation: (*a*) descriptive form, giving ingredients followed by modification, then measures, and then method; (*b*) action form, giving narrative action with listed measures and ingredients; (*c*) standard form, giving modified ingredients first, measures second, and method third; (*d*) narrative form, giving ingredients and methods together.

Yield and Quality Tests

If costs are not known, potential profit cannot be known. To identify costs, it may be necessary to run yield tests so that the cost per portion can be determined. Yields recorded should reflect those obtained under actual operating conditions.

A yield test is often done to determine the number of usable products obtained from a recipe, as compared with the total amount of materials purchased to make the product. If the item cost is not at or below a desired maximum, it must be removed from the menu. Information gained from yield tests may include the extended cost, the portion cost, the unit cost, and the cost factor. These are defined as follows:

- *Extended cost:* total cost of the material tested

- *Portion cost:* cost per serving; total yield divided by portion size gives the number of portions, and the extended cost divided by the number of portions gives the portion cost

- *Unit cost:* cost per unit as purchased, usually stated on the purchase invoice (for example, 1 gal of ice cream)

- *Cost factor:* ratio of portion cost (the computed cost per serving after the yield test is done) to unit cost (the original cost per unit as produced or purchased), obtained by dividing the portion cost by the unit cost; the cost factor allows analysts to

determine the new cost per portion when the item's price increases or decreases, without having to redo yield tests

Figure 8-4 shows a sample yield test card recording the results of a yield test involving chocolate ice cream. A 5-gal container gives approximately 145 portions if a No. 16 scoop is used. It is wise in making yield tests to check to see if the actual yield matches the computed yield. Once the extended cost has been obtained, analysts can calculate the portion cost by dividing the extended cost by the number of portions. In the present example, $5.45 ÷ 145 portions = $0.038. It also is sometimes desirable to obtain a cost factor, which is equal to the portion cost divided by the unit cost. In Figure 8-4, this is shown as $0.038 ÷ $1.09 = 0.035. If the price of ice cream increased to $2.18 per gallon (unit cost), management could immediately ascertain the new cost per portion by multiplying the new unit cost by the cost factor of 0.035. In the example illustrated in Figure 8-4, the new portion cost would be $0.0763.

The important formulas to remember in yield testing are the following two:

$$\text{cost per portion} = \text{extended cost} \div \text{portion yield}$$
$$\text{cost factor} = \text{cost per portion} \div \text{unit cost}$$

A selling price can be calculated from this information. If the portion cost is divided by the desired

---

(handwritten annotation:) total cost ÷ # servings = amt/: that unit cost = yield (cost per portion)

Product: Chocolate Ice Cream			
	Quantity 5 gal	*Unit Cost* $1.09	*Extended Cost* $5.45
Size of Portion	No. 16 scoop		
Average Portion Yield	145		
Cost per Portion	$\dfrac{\$5.45}{145} = \$0.038$	$\dfrac{\text{Extended Cost}}{\text{Portion Yield}}$	
Cost Factor	$\dfrac{\$0.038}{\$1.09} = \$0.035$	$\dfrac{\text{Cost per Portion}}{\text{Unit Cost}}$	

(handwritten annotations: "ratio", "cost", "reference to be compared if item price ↓ ↑ in future", "gal", "1 gal")

FIGURE 8-4 Sample yield test card.

percentage of food cost, the selling price is obtained. In the example illustrated in Figure 8-4, the selling price at a food cost percentage of 25 percent would be 15 cents ($0.038 ÷ 25% = $0.15 or 15 cents).

Sometimes a selling price factor is obtained, and the selling price is computed from this. The selling price factor is calculated by dividing 100 percent by the desired food cost percentage. When this factor is multiplied by the portion cost, the result is the selling price. Thus, if a food cost percentage of 25 percent were desired and the portion cost were $0.038, the calculation would be (100% ÷ 25% = 4; 4 × $0.038 = 15.2 cents).

Besides yield tests, patron acceptance tests should continually be made on food items. There are a number of ways to ascertain customer acceptance of menu items. One of the simplest is to keep track of the number of portions that are sold, by checking sales slips. The totals can be recorded daily; then the total monthly sales of each item can be compiled from the daily records. Such records are often kept on a filed copy of the menu. The daily sales record sheet should also give the price of items, the combination of items offered, and such general information as customer count, serving time, runout time, amount of carryovers, weather, and special events that might have affected sales (see Fig. 8-5). If an item is not selling, it may lack customer appeal, it may be of poor quality, or its price may be too high. Keeping the daily sales record gives management information on items that sell well and on those that do not.

A good sales record also helps management to forecast amounts that will probably be sold and to estimate the number of personnel and the amounts of materials and supplies that will be required to meet production needs. Running out of items too early may

Coffee Shop

Day and Date: Tuesday, April 4, 2000 — *Weather:* Raining

Customer Count: 160 — *Events in Town:* Parade

Meal: Dinner, 5–8 PM

Menu Item	*Price*	*Sales (portions)*	*Run-out Time*	*Carry-overs (portions)*
Filet of Sole	XXX	35	7 PM	---
Chef Salads	XX	45	---	---
(etc.)				
Banana Cream Pie	X	30	---	12

Comments: Promote desserts on Tuesdays to match sales during rest of week.

Coffee Shop No. 000624

DATE	NO. GUESTS	TABLE NO.	WAITER
4/4/00	3	14	HJ

NO.	ITEM	FOOD	LIQUOR
2	Filet of sole	XXX	
1	Chef Salad	XX	
2	Banana Cream Pie	X	

Food
Tax
Liquor
Total

FIGURE 8-5 Sample sales record sheet and sales slip.

indicate poor forecasting, while a large carryover may signify overforecasting, lack of patronage appeal, an excessively high price, or poor quality. The projection of sales on any one day should be based on the sales of the item on that day of the week in the past, modified by the other factors maintained on the sales record that management believes may be influential. For example, on Saturday night many steaks might be sold, whereas on Sunday fried chicken might be the big seller. Suppose that a big convention center is located across the street from the foodservice and that a big boxing event is to occur there on Saturday night and a gospel revival meeting (which many elderly people are expected to attend) on Sunday. These two events alone would cause management to stock up for bigger steak sales on Saturday night and for extra chicken sales on Sunday.

Management also can perform other tests to determine correct portion sizes, the most pleasing color and flavor combinations of food, and good food arrangement. Good testing gives information that can lead to better management decisions, without guesswork.

Precosting Menus

Precosting is a procedure used to determine costs and establish production quotas before actual food production begins. It is important to know this information before food is purchased so that any necessary changes can be made. Some useful data for precosting can be obtained from accurate yield tests. The cost of a portion multiplied by the number of portions sold gives the dollar food cost for all portions sold. The same number of portions sold multiplied by the selling price gives the dollars in sales that the item brings in. The food cost divided by the total sales then gives the food cost percentage of this item. Often precosting is done to obtain just such information. The following example illustrates the procedure for obtaining a food cost percentage:

- 40 portions of soup at a cost of $0.15/portion = $6.00 food cost
- 40 portions of soup sold at $0.60 each = $24.00 in sales
- $6.00 ÷ 24.00 = 0.25 or 25% food cost percentage

Most foodservices know their costs and establish a desirable food cost percentage so that the expected costs will be obtained and a predictable profit will result. Again, knowing the food cost percentage of an item enables management to determine whether the item's price meets the standard of the desired food cost percentage. If it does not meet this standard, management must determine which price is better for the operation. No menu ever consists of items that all have the same food cost percentage. Some are higher than the desired standard, and others are lower. Management expects by a proper sales mix to achieve the desired percentage. If this does not happen, management is in trouble. This explains why achieving a correct sales mix is important in producing a menu.

Here is an example of how sales mix and various food costs work out. Item 1 has a food cost of $0.25 and sells for $1.00, giving a food cost percentage of 25 percent. Ten units are expected to be sold, for a sales yield of $10 and a total food cost of $2.50. Item 2 costs $1.50 and sells for $4.50, giving a food cost percentage of 33.3 percent. Twenty units are expected to be sold, for a $90 sales yield and a total food cost of $30. The total sales are now $100, and the total food cost is $32.50, for an overall food cost percentage of 32.5 percent. The rule, then, is to find the food cost of an item and to multiply this by the number expected to be sold, and then to multiply the selling price by the number expected to be sold. The results of these computations give both the food cost and the sales for each item. Next, all food costs and all sales should be totaled. By dividing total food costs by total sales, the analyst can calculate the expected overall food cost percentage.

In a typical operation, total food costs for a day are forecasted, and so are sales. Management estimates the expected overall food cost percentage. Thus, if sales were forecast (estimated) to be $1000 and food costs to be $430, the estimated overall food cost would be 43 percent ($430 ÷ $1000 = 0.43) (see Table 8-1). If this forecast of food cost is unacceptably high, management must either replan the menu to achieve a lower overall food cost percentage or alter the menu to achieve a different sales mix. To accomplish the latter purpose, more higher-gross-profit items must be sold to increase sales dollars and fewer lower-gross-profit items must be sold to decrease food costs.

TABLE 8-1   Precosting the Menu, Item by Item

Menu Item	Portions to Be Sold	Food Cost per Portion	Extended Cost	Sales per Portion	Extended Sales	Food Cost Percentage
1	10	$0.25	$ 2.50	$1.00	$10.00	25
2	20	1.50	30.00	4.50	90.00	33
•						
•						
•						
Total			430.00		1000	43

Precosting of daily menus and *du jour* menus should occur before the individual menu is written, but cycle menus must be precosted before the entire cycle begins—a considerably more difficult job. Menus used continuously should be precosted every thirty days, or whenever food costs change considerably. Precosting employees' menus is also necessary, regardless of whether the food comes from the regular menu or is prepared especially for the employees. Computers today can simplify the job of precosting.

A failure to sell all items that were forecast to be sold can raise havoc with precosting calculations. Suppose that in the previous example, only ten (instead of the predicted twenty) $4.50 items were sold. Now the food cost remains at $32.50 but sales are only $55. Table 8-2 compares the two situations: The expected food cost in situation A is 32.5 percent, but in situation B (omitting consideration of the value of ten carryovers

TABLE 8-2   Comparison of Food Cost and Sales Totals

Menu Item	Situation A		Situation B	
	Food Cost	Sales	Food Cost	Sales
1	$ 2.50	$ 10.00	$ 2.50	$10.00
2	30.00	90.00	30.00	45.00
Total	$32.50	$100.00	$32.50	$55.00

of the $4.50 item) it is about 59 percent. In cases in which sales are not achieved, management can find itself in trouble with few remedial options. The bottom line is that forecasting must be accurate. The only benefit to be salvaged from early detection of a radically off-the-mark forecast is that management is advised of the problem much sooner than if it were to wait until the end of the month to see what happened.

Production Sheet

The production sheet is a final control needed before the menu begins to govern kitchen production. It tells the purchasing department and the kitchen what food items to purchase and prepare for any particular day. Because overpurchasing and overproducing are the chief causes of high food costs, the production sheet is an extremely important control device. By promoting proper purchasing and production, the production sheet reduces waste and, thus, reduces the food cost percentage.

The production sheet must name the item to be prepared, the portion size, the number of portions, the mass of portions (by weight or measure), the time and place of service, and the recipe number for each item. Employees' meals, banquets, luncheons, and in fact every meal served must have a separate production sheet. Only through such organizing can purchases be planned ahead and materials made available when needed. Overloading refrigerators, using excessive labor, producing too much food, and wasting food through poor planning are all eliminated when the production sheet is used properly (see Fig. 8-6).

Selling Price

The determination of selling price is based on information gained from a yield test. Such a determination is necessary in menu planning for sales made to customers. It is not necessary if customers' meals are included with lodging for a single price, as in the American plan or in hospitals. Nonetheless, management in such institutions must have much the same information to arrive at an overall selling price as is needed to calculate selling prices on menus. Selling prices vary with the quantity of food served, the type of service, and other factors. The selling price must cover costs in these operations, as well as giving the needed profit. If it does not, the operation does not last long. Different methods are used by foodservices to determine selling price. Among the most common of these are the multiplier method, the pricing factor method, and the planned profit method.

Multiplier Method

The multiplier method is a handy rule-of-thumb technique, but it is of dubious value because it bases selling price on food cost alone. Based on what the food costs, the menu sets the price at three or four times this, or any of various other multiples of the cost. There are two approaches to using the multiplier method: actually multiply food cost by a factor derived from dividing the desired food cost percentage into 100 percent; or divide the food cost by the desired food cost percentage. For example, if a food cost percentage of 40 percent were sought, the analyst could compute the desired food cost factor (100% ÷ 40% = 2.5), and multiply this by the food cost of an item to get its selling price. Thus, if the food cost is $1.20, the calculated cost is $1.20 × 2.5 = $3.00. Using the same example, the second approach to the multiplier method gives $1.20 ÷ 0.40 = $3.00 as the selling price.

The problem with this method is that not all food items involve the same amount of labor or other costs for their production, although the assumption in using food cost as the sole criterion is that they do. Thus, an operation's management might not know that it is losing money on certain items that have much higher than normal nonfood costs associated with them. Even if proper cost calculations were made, management would still not know. Too many operations depend on the multiplier method. This may lead the management astray, as well as price items unfairly to customers.

Pricing Factor Method

The pricing factor method offers a solution to some of the problems posed by the multiplier method. For this

Name of Operation_____		Time of Service_____		Date of Service_____	
Menu Item	*Size of Portion*	*Number of Portions*		*Quantity of Portions*	*Recipe Number*
Beef and Brussels stew	8-oz ladle	24		5½ qt	XX
_____					
_____					
_____					
_____					
_____					
_____					

FIGURE 8-6 Sample production sheet.

method, each food item is placed into one of two categories: foods needing little or no preparation (Nonprep); and foods needing a significant amount of preparation (Prep). Thus, milk would be classed as Nonprep, and beef stew would be classed as Prep. The raw food cost is recorded and summed for both Nonprep foods and Prep foods (see Table 8-3). In the situation represented in Table 8-3), the operation has budgeted $200 for all food—30 percent (or $60) for Nonprep, and 70 percent (or $140) for Prep food. Total raw food cost accounts for 40 percent of sales.

Next, labor cost is placed into Nonprep and Prep categories. The labor cost of pantry workers, fry cooks, bakers, and others doing actual preparation is charged to Prep only. This is logical, since these workers are paid only to prepare foods. The costs of employees such as busboys, waiters and waitresses, dishwashers, and others responsible for all food (Prep and Nonprep) served in the operation are charged in proper proportions to both Nonprep and Prep categories. This, too, is logical, since these workers are paid to see that both Nonprep and Prep foods are served. If total labor costs account for 30 percent of sales (or $150 in this case), and if the operator notes that 40 percent of the total labor expense is found to be incurred in preparing foods, then $60 (40 percent of $150) should be charged to Prep only. The rest of the expense ($150 − $60 = $90) should be allocated to both Nonprep and Prep, in the same proportions used

for the total raw food cost—30 percent to Nonprep foods, and 70 percent to Prep foods. This is logical, since each food category should reflect Nonprep labor expenses proportionate to the cost of the food allocated to that category. Thus, the remaining labor expense is noted as $27 (30 percent of $90) in Nonprep, and $63 (70 percent of $90) in Prep.

Other expenses (accounting for 20 percent of sales in this example) are $100 and are allocated to the two food categories according to the 30:70 raw food cost ratio. Other expenses include $30 (30 percent of the $100) under Nonprep and $70 (70 percent of the $100) under Prep. Finally, in this example, profit is distributed, as 10 percent of sales, according to the same 30:70 ratio. On this basis, the Nonprep category receives $15.00, and the Prep category gets $35.00.

Different steps must be taken to arrive at selling prices for the Prep and Nonprep foods. For any Nonprep food, the operator would divide the total sales for this food group (which is $132) by the total cost of food not requiring preparation ($60, in this case). The $132 is the sum of raw food cost, labor cost, other expenses, and profit. The pricing factor thus becomes 2.2 ($132 ÷ $60 = 2.2). In a similar manner, the pricing factor for food requiring extensive preparation can be determined by dividing $368 by $140. This gives a pricing factor of 2.6 ($368 ÷ 140 = 2.6).

If this foodservice operator wished to change prices on an existing menu, he or she could simply multiply

	Nonprep (Foods)	Prep (Foods)	Total	Percentage of Sales
TABLE 8-3 Pricing Factor Method				
Raw food cost	$ 60	$140	$200	40
Labor			150	30
Prep employees		60		
Nonprep employees	27	63		
Other expenses	30	70	100	20
Profit	15	35	50	10
Total	$132	$368	$500	100
Pricing factor	2.2	2.6		

the food cost of the prepared portion by the appropriate factor. For example, if a half-pint of milk in a disposable carton were purchased for 4.5 cents, the cost could be multiplied by the factor of 2.2 and a selling price of 10 cents could be established. On the other hand, if a chicken pot pie, requiring a great deal of preparation labor, cost $.50, the sale price of $1.30 would be determined by multiplying $.50 by 2.6.

Once these factors have been determined for any single operation, they may be used to update menu prices, until a significant change occurs in either the ratio of one food group to the other or in the proportion of direct preparation expense to the total labor expense.

The immediate results of using this simplified, yet much more effective, method of pricing menu items are:

1. A predetermined food cost percentage (in this example 40 percent) can still be achieved even though two or more price factors are used.

2. Because of the use of two or more price factors (2.2 and 2.6), menu prices will be uniquely different from other foodservice operations, enabling the operator to compete very effectively for his share of the total market.

3. The allocation of food into categories to which can be added direct labor expenses enables the user to immediately identify menu items that are being sold for a loss or that are not carrying their fair share of expense.

4. The elimination of loss items and the inherent fairness of prices charged to the customer invariably increase sales volume and profits of the operation.

Planned Profit Method

The planned profit method of the Texas Restaurant Association is a third method of setting a selling price. The basics of this system involve creating a well-planned budget that gives a preview of expected financial operations over a given period in terms of operating (overhead) costs, actual raw food costs, labor costs, and planned profit. Operating costs include expenses such as advertising, telephone and utilities, laundry, insurance, rent, taxes, employee meals and uniforms, office

supplies, linens, and china. Actual raw food costs are determined from standardized recipes, with the cost of all extra trimmings included. Labor cost includes wages and salaries paid to employees. Finally, profit is figured as an expense. Profit must be included if it is desired.

After establishing the budget, the operator determines the menu prices needed to meet the expenses by calculating—per volume dollar of business—the operating cost percentage (operating costs ÷ total sales), the labor cost percentage (labor cost ÷ total sales), and the planned profit percentage (planned profit ÷ total sales). These three percentages are then summed and subtracted from 100 percent to give the desired raw food cost percentage per dollar volume of business. By adding together operating costs, labor costs, planned profit, and desired raw food cost, the operator obtains a figure for total sales. By dividing the actual raw food cost (figured from the standardized recipe of a menu item) by the desired raw food cost percentage, the operator can compute the selling price of the menu item. For example, suppose that the following percentages have been computed:

Operating cost percentage
(operating cost ÷ total sales) = 37.4%

Labor cost percentage
(labor cost ÷ total sales) = 17.0%

Planned profit percentage
(planned profit ÷ total sales) = 15.0%
69.4%

The sum of these percentages is 69.4 percent. The desired raw food cost percentage is therefore 30.6 percent (100% − 69.4% = 30.6%).

To arrive at the selling price of an item, the ingredients' costs are summed. Suppose that for item C the following ingredient costs are identified:

Ingredient I	$0.12
Ingredient II	0.05
Ingredient III	0.09
Ingredient IV	0.11
Ingredient V	0.01
	$0.38

The total ingredient cost of item C is therefore $0.38. This is divided by the desired raw food cost percentage (which is 30.6 percent) to arrive at an indicated selling price of $1.24 ($0.38 ÷ 30.6% = 1.24).

If this method of calculating selling price is used, the operator is sure of making 15 percent profit—as long as the costs have been accurately calculated and do not change. Because in reality expenses continually change, however, the operator must review selling prices carefully and frequently in order to make this system work.

Three rules may be helpful in setting menu prices:

1. Maintain a realistic price spread between the highest- and lowest-priced items. Receiving a smaller profit percentage on a high-priced item may still result in a higher check average and more income, although the raw food cost percentage is relatively high. A saying in the foodservice business is, "You don't operate a business to make percentages, but to make dollars."

2. An item sold at or below cost to attract customers is a loss leader. Such an approach is good for a drugstore where other items may be purchased. It is not good for a restaurant or other profit-oriented foodservice, where nothing else is purchased to give a profit.

3. Control costs and review menu prices, but do not neglect sales volume. A 10 percent increase in sales will add more to profit than is often expected. This is because fixed costs remain unchanged, and little more labor is needed to handle the extra volume; relatively few additional expenses are incurred, too. Consequently, profit can rise as a proportion of sales. Fair menu pricing, with good merchandising techniques, builds volume.

Menu Engineering

A fourth method of pricing takes into account the popularity of individual items and their respective contribution to profits. This method was popularized by Don Smith and Michael Kasavana. First, overall cost of the product is calculated. This would include the cost of the item from the standard recipe, plus the cost of any garnishes and any other food included in the price of the meal, such as bread and butter or a salad. This figure represents the potential cost if no food were wasted, incorrectly proportioned, or stolen. It is subtracted from the price being charged on the menu, providing a contribution margin. The contribution margin represents funds that pay labor and overhead costs and contribute to profit. In some cases, a food item may have a high cost on a percentage basis, but its contribution margin will still be higher than that of other items. The menu items are then classified by whether they have high or low contribution margins.

Next the popularity of the item is judged. This is done in terms of the menu mix, the proportion of items sold in each category. An item with a high contribution margin and high sales has more of an impact on profit than does an item with a low contribution margin and high sales. The menu mix average is determined by dividing the number of sales of each item by the total number of items sold. Any item that has a menu mix percentage of below 70 percent of the average is deemed to have low popularity. Any menu mix percentage above that level is considered to indicate high popularity.

At this point, each item is judged to have either a high or low contribution margin and either a high or low popularity. On the basis of these two classifications, items can be separated into four different groups:

- High contribution margin–high popularity "Stars"
- Low contribution margin–high popularity "Plowhorses"
- High contribution margin–low popularity "Puzzles"
- Low contribution margin–low popularity "Dogs"

Specific menu items can then be evaluated with respect to their performance on the menu and their ability to contribute to the overall objectives of the operation. Stars are the real performers on a menu, contributing significantly to the bottom line. Plowhorses sell well but do not contribute a lot. These items may be attracting customers because they are seen as good values; increasing their price should only be done very carefully. They can be turned into Stars, but only if people value them enough to pay a higher price. The Puzzles are not selling, but if they did they would contribute substantially to profits. These are items that may merit some experimentation in price, placement on the menu, or sales effort from the staff. The Dogs are not selling well and do not contribute either. Serious thought should be given to replacing them on the menu with something more promising. Ideally, all items would be either Stars or Plowhorses, and food cost and profitability would be excellent. Because of ever-changing customer tastes and preferences, however, menu engineering must remain an ongoing process. It is a process that can provide insight into customers' buying patterns as well as into controlling production costs.

■
QUALITY CONTROL

Quality control in a foodservice facility relies on the establishment of standards. To be meaningful, these standards must apply to all aspects of an operation—from the moment the menu is planned to the moment the finished product is presented to the customer. The standards should apply to all edible products and their contact points. An effective quality-control program results in better overall control of the product and increased customer satisfaction.

Perception of quality is affected by a consumer's background and lifestyle as well as by the surroundings in which the meal takes place. In addition to psychological influences, tangible attributes of the product such as taste, texture, color, appearance, and nutritional value all contribute to the perception of quality. Quality guidelines have been defined to some extent by the Food and Drug Administration and the U.S. Department of Agriculture, which have established standards (for determining whether a product is safe for consumption) and grades (for defining the composition of the product).

Comparisons of various products provide a useful technique for determining whether a product meets the standards of a particular establishment. It is not unusual for purveyors to offer samples for exactly this purpose. Once the product is in the foodservice establishment, it is up to the management and employees to maintain the quality of the product. Management has the final responsibility of establishing standards of quality and in seeing that they are achieved.

Receiving and Storage

The product may go through any number of processes once it is inside the doors of an establishment. Inspection during receiving is the first step in any quality-control program. Purchasing specifications establish standards for the quality of items to be used. At receiving, the product must be inspected to ensure that it matches what was ordered in number, weight, and quality. The packaging should be examined for breaks or leaks. Frozen good should be checked for signs of thawing or freezer burn.

During storage, a number of factors can influence the quality of the product. The shelf-lives of foods vary, and this must be taken into consideration in arranging and maintaining storage. Fresh foods tend to have very short shelf-lives—in some cases as brief as 24 hours—so ordering exactly the amount needed is essential. Frozen foods have longer shelf-lives, but they deteriorate over time as well. Maintaining proper storage temperatures for frozen or refrigerated foods is imperative. Carefully watching temperatures during thawing is important also. For the most part, products should be allowed to thaw under refrigeration, unless they can begin cooking while still frozen. Dry goods and groceries tend to last a fairly long time; however, poor storage conditions (such as exposure to moisture or high temperatures) can have a negative impact on the quality of dry foods and can shorten their storage life. In many cases, proper observance of the first in, first out (FIFO) inventory system helps keep inventories fresh.

Production and Holding

Once the product has entered the preparation phase of production, various elements work together to ensure its continued high quality. Here again the standardized recipe plays an important role by outlining the correct procedures, timing, and use of equipment for proper presentation of an item. To ensure consistency of quality standards, the recipe should be followed scrupulously. Sanitation standards establish a basis for maintaining quality during preparation as well. Proper handling of the product and rigorous observance of sanitary procedures are necessary. In this respect, the equipment, too, must be kept sanitary and in good working condition. Tools and utensils must be properly cleaned to avoid contamination of food during preparation; equipment should be subject to a regular maintenance and cleaning schedule and should be repaired immediately in the event of breakdown. A cook who does not possess the correct tools in proper condition cannot be expected to produce a high-quality product.

Once cooking has begun, control of temperatures and timing becomes important. Ovens need to be checked on a regular basis for the accuracy of their thermostats. Even if the oven is gauged properly, loss of heat from a bad seal or from cold spots can have a negative effect. In any event, cooks should be encouraged to carry and use thermometers to check internal cooking temperatures and holding temperatures. Temperatures of deep-fat fryers need to be monitored closely. High temperatures, along with exposure to air, contribute to the deterioration of fat, resulting in soggy and discolored deep-fried products. The fat should be filtered after each shift and covered when not in use. The use of timers is extremely helpful. In a busy kitchen, it is not unusual to lose track of how long something has been cooking. The timer serves as a reminder that helps avoid costly overcooking or burning of the food.

Observing proper temperatures during holding is necessary not only for sanitary reasons but to maintain quality. Foods, whether they are to be served hot or cold, must be held at the proper temperature. It is not unusual for lettuce to wilt or for sauces to separate if their temperatures are not controlled. Products have a holding life that must not be exceeded. Some foods hold much better than others, without diminishing in quality. This can be a function of the preparation technique used or of the actual contents of an item. In many cases, recipes can be modified to facilitate holding.

■

SECURITY

Unfortunately, theft is a fact of life in the foodservice industry. Ignoring the possibility of its existence can wreak havoc on food costs and prove disastrous for an operation. The best approach is to foster prevention by removing temptation and opportunity. Many managers may be aware that theft exists in an operation without being aware of its extent. Considering the number of failures in the industry, controlling theft should be a major goal of any organization. Any effective effort to control it will translate directly into profits.

There are multitudinous opportunities for theft in purchasing and receiving, if proper controls are not maintained. The thief can be an employee or a purveyor; in some cases, they may be working together. One major danger arises when the purchasing agent is also in charge of receiving. Common types of thievery include kickbacks, fictitious companies, credit memo games, invoice "errors," weight shortages, and quality substitutions. Separation of functions can help avoid many of these problems. Many of the systems and procedures described in Chapter 5 were designed to eliminate opportunities for theft.

Storage areas can be prime targets for dishonest activity. In many operations, they are located close to a door, and long periods of time may elapse during which other people are not in a position to observe activity there. Storage areas are the bank vault of a foodservice establishment and should be secured accordingly. Limited access and keeping storage areas under lock and key can help considerably. Assigning keys to only those who need them and changing locks when they leave the operation are good preventive measures. Imagining the position of a thief and pondering opportunities can bring lax areas to the manager's attention. Maintaining perpetual inventories can help alert management to and pinpoint theft problems in storage.

Food in process can be particularly difficult to safeguard. Management should have a pretty good idea of raw product requirements for daily preparation. Overproduction should be discouraged. Eating policies should be made clear and enforced for all personnel. Theft is theft, whether it is being done by management or staff; double standards only set a bad example. Food should be given to servers only with the proper documentation. Supervisors must be vigilant against cooks' providing food for their or others' personal consumption. Following standardized recipes, adhering to production schedules, and observing correct menu ordering procedures will go a long way toward preventing theft.

COMPUTERIZATION

Computers are beginning to take a larger role in controlling and managing the product. Purchasing databases provide information on market costs and item availability. It is now entirely possible for purchasing agents to access information directly from the computer of the purveyor to allow for direct ordering. This information will aid the foodservice operator in forecasting and should cut down on the transit time of all products from the grower to the consumer.

Once the product is received, quantity and price information can be fed into the foodservice's system and placed in an ingredient file. This makes it considerably easier to maintain a perpetual inventory (a time-consuming process when done by hand). The food production supervisor thus has immediate access to current information on in-house item availability, without having to take a walk through the storage facilities. This information is also available for precosting menu items and determining food costs. Recipes are pulled from a database containing standardized recipes, and then the recipes are exploded to the proper quantities. The product cost of a menu item is then determined individually, as well as on a gross production basis, through integration of data from the ingredient and recipe files. And all of this is done on the basis of yesterday's or today's prices, not those from six months ago. Armed with this information, the supervisor can modify recipes, change a menu plan, adjust production schedules, or change purchasing decisions in an informed and timely manner. Computerization of these functions will improve overall control over the product, contribute to product quality, and further limit opportunities for theft. Computers can also be used to control times foods are cooked and baked and to set temperatures.

PRINCIPAL ASPECTS OF ACCOUNTING

In the simplest terms, profits are the excess of income over expenditures during a given period of time. If income has exceeded expenditures over that period, a profit is realized. If expenditures are not properly managed, however, a net loss may result, regardless of the amount of income realized. In the foodservice industry, which is notable for the failure within the first few years of operation of three-fourths of the new businesses that enter the field, learning how to manage income and expenditures can make the difference between success and failure. For the foodservice manager who wishes not only to be among the surviving entrants in the field, but also to be among the successful ones, management of profits is mandatory.

The foodservice manager must begin the task of managing profits by becoming proficient with some basic accounting tools, including profit-and-loss (or income) statements, balance sheets, break-even analysis, and budgeting. In addition, the foodservice manager must learn how to calculate and control labor, food, and beverage costs.

For ease of communication and comparison, some basic assumptions of accounting are generally accepted. The most important of these for the foodservice manager to learn relate to the business as an entity, the fundamental accounting equation, the theory of the going concern, the use of cost and money as bases for measurement, the concept of consistency, and the concept of conservatism. The following descriptions offer working definitions of these terms:

- *Business entity:* concept of separation of the business from the person who provides the assets. For a foodservice manager, this means that purchases or sales made on behalf of the business should always be kept separate from those of a personal nature. No business funds should be used for personal purchases and no personal funds should be used for business, unless the transaction is listed as a loan or other capital source to the firm.

- *Fundamental accounting equation:* has three components: assets, liabilities, and capital. Assets are the resources of the business; liabilities are the debts; and capital is the amount invested in the firm. All transactions recorded in accounting can be stated in terms of how these three components are affected. Any increase in a business's assets, for instance, must be balanced by a corresponding decrease in another asset account or by an offsetting increase in liabilities or capital.

- *Going concern:* concept based on the assumption that a business will continue to operate for an indefinite period of time. This assumption provides the basis for recording assets at acquisition cost, rather than at market value, because the purchase is assumed to be intended for continued operations.

- *Money and cost:* closely related concepts. All items are recorded at cost, based on the assumption that items recorded at their cost value are less subjectively valued than items recorded at a relative value such as a market appraisal value. All money entries are made in terms of today's dollars, rather than in terms of some reference-year-value of the dollar. Although we know that the dollar has less buying power now than it had in 1978 or 1997, we still record the value of a dollar today as $1, just as we did in those years.

- *Consistency:* theoretical basis for using the same methods for all accounting transactions. For instance, if straight-line depreciation has been used in the past, it should also be used in the future; any change in accounting methods should be properly noted at the time it is made, and a sound business should refrain from introducing repeated changes.

- *Conservatism:* concept based on the theory that the values of transactions should not be overstated. Instead, moderation and restraint should be used in estimates and appraisals of all accounting transactions.

Profit-and-Loss Statement and Its Preparation

The profit-and-loss statement is a record of the activity of a business covering an entire specific period of time. In some nonprofit organizations, profit-and-loss statements are called *statements of revenues and expenses*. The profit-and-loss statement reflects each transaction—whether a cost or a sale—that took place during that period of time. For a foodservice manager who is experienced with the profit-and-loss statement, the record can provide a window through which to see the past, in order to prepare for the future.

A profit-and-loss statement is prepared by entering each item of income or expenditure in an appropriately labeled category. The major categories used in foodservice operations include sales (or income), cost of sales, controllable and capital expenses, and fixed costs. To make the profit-and-loss statement more meaningful, many foodservices break these major categories down into subcategories such as food income, beverage income, cigarette income, laundry costs, insurance costs, and payroll costs.

The first step in preparing a profit-and-loss statement is to enter the sales or income figures for the period, followed by the cost of goods for the period. To allow computation of the cost of goods for the period, two inventories must be taken. The opening (or beginning) inventory is taken at the start of the period (which often covers one month), and the second (or ending) inventory is taken at the end of the period. In each case, the items are first counted and then multiplied (or extended) by the price paid for them. When all items have been extended, a total dollar value for the inventory is complete.

To complete the computation of the cost of goods sold during a period, the accountant must add all purchases to the beginning inventory, and then deduct the ending inventory (which has been computed in the same manner as the beginning inventory). The result is the cost of goods consumed during the specified time period. This total should then be listed in the appropriate columns on the profit-and-loss statement as the cost of goods sold (or consumed).

The following simple equation illustrates this calculation:

$$\begin{array}{l} \text{Beginning inventory (+\$)} \\ + \ \text{Purchases (+\$)} \\ - \ \text{Ending inventory (--\$)} \\ \hline \text{Cost of goods sold} \end{array}$$

Deducting the cost of goods sold from income leaves the gross profit.

The next step in preparing a profit-and-loss statement is to list all controllable expenses that are not products, including operating wages, payroll taxes, utilities, supplies, and laundry. These items are often referred to as *controllable* expenses because the amount of each is to some extent under the control the food-service manager. After controllable expenses are deducted from gross profit, all fixed costs are then deducted from this total. In the profit-and-loss statement, fixed costs comprise items that are present at a set cost each month, regardless of the volume of business done. Deducting fixed costs from income after controllable expenses provides a figure called *operating profit*. In some foodservices, additional items are then deducted; in others, these items are treated as either controllable expenses or income (as appropriate). After all of these steps have been completed, the bottom-line figure is called the *net profit* for the period.

Balance Sheet

Unlike the profit-and-loss statement, the balance sheet is a statement of assets, liabilities, and owners' equity at a particular point in time. The balance sheet is often completed for a foodservice on the last day of the month, quarter, or year (depending on the accounting cycle used). As its name implies, the balance sheet must balance; therefore, the total amount of assets must equal the combined total amount of liabilities and owners' equity. The assets are listed on the left-hand side of the sheet, and the liabilities and equity are listed on the right-hand side. If the two sides of a balance sheet are not equal to one another, there is an error somewhere in the calculation.

The assets on a balance sheet are divided into four categories: current assets, fixed (or long-term) assets, other assets, and funded reserves. The current assets include all cash and all assets that can be converted readily into cash within a short period of time, generally one year. Current assets thus include cash on hand, cash in savings and demand deposit bank accounts, and marketable investments, but not receivables or inventories. The fixed assets include capital improvements, furniture, fixtures, equipment, land, and building structures. Items listed as other assets on a balance sheet include such items as prepaid expenses, escrow accounts, and prepaid fees. Funded reserves listed on a balance sheet include depreciations and other amortized items.

The liabilities listed on the right-hand side of the balance sheet include both short- and long-term obligations to pay. Current liabilities include such items as accounts payable, accrued expenses, and the current portion of any fixed liabilities (such as mortgages). Long-term liabilities are debts such as the amount of total mortgage owed, capital leases, and deferred income taxes.

The owners' equity portion of the balance sheet reflects the difference between total assets and total liabilities. If the assets outweigh the liabilities, the equity is a positive figure. The reverse is true if the liabilities are larger. In many new businesses, the liabilities may be larger for some period of time; but in ongoing businesses, the assets should always outweigh the liabilities. Included in the equity portion of the balance sheet are such items as capital stock, capital surplus, reserves for replacement, and income over the previous time period.

Break-Even Analysis and Its Calculation

Good foodservice managers work out strategic operating plans for their businesses. For most foodservice managers, the plan includes a determination of the operation's break-even point. The term *break-even analysis* refers to a determination of the point at which the operation will break even (with no profit, but also no loss). The break-even is not difficult to determine, if the correct steps are followed.

The first step in determining a break-even point is to decide in the case of each cost whether it is either fixed or variable. By definition, fixed costs are costs that remain constant from month to month. Variable costs are costs that change every month in response to the amount of business done during the period. For instance, a mortgage payment is a fixed cost, whereas the amount of money used to purchase food supplies varies with the volume of business. In some cases, a cost category may contain both fixed and variable components, and these must be separated before a break-even point is computed. For instance before a calculation of the break-even point is attempted, labor costs that would exist whether or not any customers come in should be separated from those that depend on the volume of business.

There are two methods of calculating the break-even point for a foodservice. The first is algebraic and can be expressed as the following formula:

$$\text{break-even point} =$$

$$\text{fixed costs} \div \left(1 - \frac{\text{variable costs}}{\text{sales}} \right)$$

As an example, consider the ABC restaurant, which had the following relevant costs and sales last period:

fixed costs = $25,000

variable costs = $60,000

sales = $100,000

If these figures are inserted into the formula for break-even analysis, the result reads as follows:

$$\text{break-even point} = \$25,000 \div \left(1 - \frac{\$60,000}{\$100,000} \right)$$

$$= \$25,000 \div (1 - 0.60)$$

$$= \$25,000 \div 0.40$$

$$= \$62,500$$

The second method of calculating the break-even point is by means of a graph. This involves first establishing costs on one axis and sales on the other. The costs should be graded in realistic form from a point below the lowest potential costs to a point above the highest possible costs. Once these have been determined, the next step is to enter the fixed costs. For this example, the figures used in the preceding formula will be used again. Therefore, the fixed costs are $25,000, so a straight line is drawn across the graph at that point. Then sales are recorded by drawing a diagonal line from the origin of the graph (at 0) to the actual sales figure (here, $100,000). The final step in this method of determining the foodservice break-even point is to draw a diagonal line from the point at which fixed costs intersect the cost axis (here, $25,000) until it crosses the sales line. This intersection point represents the break-even point for the foodservice. As this line is extended forward toward $100,000 in sales, the gap between costs and sales widens, reflecting the progressively greater profit that can be made as sales increase.

Budgeting

Budgeting is one of the principal procedures used to manage profits in a foodservice. The primary purpose of budgeting is to establish the goals of a business in written form. Budgets also provide one of the most often-used frames of reference for analyzing the performance of a foodservice.

The following rules are observed by most foodservice managers in establishing their budgets:

1. Involve everyone who is responsible for achieving the results of the budget in the process of establishing goals through the budget.

2. Base the budget on appraisals of potential performance, not on unachievable goals.

3. Develop a budget that effectively assigns responsibility to people who possess a considerable degree of influence over the budget's success rather than to people who have little bearing on the outcome.

4. Plan to make the budget valid over a range of levels, rather than appropriate for achieving a single goal.

5. Review and update the budget periodically, and revise it when necessary.

6. Break the budget down into periods of coverage that are short enough to allow timely appraisals of progress toward the goals.

7. Include information systems—for accurate and timely input on how the foodservice is living up to the budget—as part of the system.

Many foodservice managers have found that the process of establishing the budget can be as valuable as the budget itself. This is true because, by analyzing alternative courses of action in the process of establishing a budget, foodservice managers develop a keener awareness of the needs of their operation. At the same time, establishing a budget forces the foodservice manager to examine what is necessary to achieve the profit levels being forecast in the budgets and establishes a basis for future comparisons. If used properly, the budgeting process can also act as a self-evaluating tool for a foodservice organization. In fact, many foodservices have found that an optimal method of using budgets is to create them monthly as part of the process of closing the operation's books. In this way, the performance of the past month can act as a guide in determining how to budget the next month; at the same time, it can provide the occasion for the self-evaluation and goal-setting needed to spur management to accomplish higher goals.

■

WAGES, TAXATION, AND GOVERNMENT REGULATION

Since so many decisions made in day-to-day operations are contingent upon federal, state, and local mandates, the foodservice operator is in partnership with the government in running the foodservice. These regulations greatly affect profits and should be a concern of every manager.

Accordingly, as minimum wage bills are introduced, taxation reform is made, and the government, in general, determines policy on such items as workplace smoking, business meal deductibility, Internal Revenue Service (IRS) tip audits, tourism promotion, payroll taxes, family and medical leave, immigration laws, OSHA reform, health insurance, blood alcohol concentration (BAC), food safety, legal reform, and a host of other policy mandates, the foodservice operator is inundated with information regarding the serving of food to the dining public.

The best recommendation for any manager is to become involved with trade associations at the local, state, and national levels. Trade associations keep the operator abreast of the latest trends, laws, and potential acts of intervention. Accordingly, the foodservice operator is better prepared to make logical and correct decisions in the ever-changing atmosphere in which business is conducted in modern society.

■

SUMMARY

Such factors as reducing waste, increasing productivity, and using food efficiently should be considered in planning a menu. The standardized recipe, which gives a known quantity of food of a known quality, can be used to maintain standards and control production. Using such recipes also gives an operation greater flexibility. Other important factors influencing what may be put on a menu are yield, quantity, and quality tests. These indicate the yield of the items from quantities purchased and give an accurate cost estimate. The quality of items can also be checked to see if it meets the operation's standards.

Menus should be precosted. This can come both from costing standardized recipes into portions and from yield tests. Knowing the cost of an item allows an analyst to calculate a selling price and then the food cost percentage, the percentage of sales that the food represents. By precosting and knowing ahead the cost of items, management can remove or add food items in order to increase or decrease the food cost percentage until it reaches a desired level. The production sheet is a control needed before the menu becomes a reality; it instructs the purchasing department and the kitchen to purchase and prepare specific food items for a particular day.

Another factor to consider in managing the product is the determination of the selling price of menu items. Among the different methods used to determine selling price are the multiplier method, the pricing factor method, the planned profit method, and the menu engineering method of Kasavana and Smith.

Quality control depends on the establishment of and adherence to standards. These standards need to be followed from the moment the product arrives at the foodservice until the moment it is placed in front of the customer. In purchasing and receiving, arriving products must be inspected, and then proper storage must be provided. In production and holding, standardized recipes, sanitation guidelines, equipment maintenance, and temperature controls all help maintain quality.

Security has become an area of major concern as it becomes more difficult for foodservices to make a profit. Prevention—through the removal of opportunity and temptation—is the best approach. Theft is most likely to occur during purchasing, receiving, or storage. A comprehensive security program can virtually eliminate the problem of theft.

Computers are being used more frequently to control and manage the product. Through the use of databases, the food item can be tracked from purchasing to receiving to production to service. Computer advances will make it much easier to integrate many of the other important procedures into the operation.

Profits are the excess of sales over costs. In order to make a profit, an operation must carefully manage both sales and costs.

Understanding several principles of accounting is mandatory for knowing how to manage profits. Among these principles are the business entity concept, the fundamental accounting equation, the theory of business as a going concern, money and cost at constant values, consistency in accounting methods, and conservatism.

The profit-and-loss statement is a record of business activity for a *specific period of time;* the balance sheet is a record of business at a particular point *in time.* Break-even analysis is important for determining the amount of sales a company must make to reach a point of no profit and no loss—the break-even point. This can be determined in two ways: through an equation, or through a graph.

Budgeting is any method of planning for sales and expenses. A budget constitutes the written goals of a foodservice.

The foodservice operator is in business with the government since so many decisions made in day-to-day operations are contingent upon federal, state, and local mandates. These regulations greatly affect profits and should be a concern of every manager.

■

CHAPTER REVIEW QUESTIONS

1. Eric Burton has just retired from his job as an airline pilot. He has spent a number of years in a service industry and would like to pursue a second career as a restaurant owner. He finds the idea of playing host to all his old friends very appealing. He admits that he doesn't have a detailed knowledge of the restaurant business, so he has hired you to be his manager. He will still be your boss, and he intends to play a major role in making decisions, although he will trust your judgment.

   Eric realizes that proper management of food production is crucial to the success of his operation. Consequently, he is very interested in the process of costing and setting menu prices. Eric would like you to give him a brief explanation of the methods you are familiar with, identify which one would be your choice, and tell why. Draw up an outline of your explanation and your advice.

2. Standardize a recipe that needs it. Include nutrition information on it. Set the recipe up to give five, ten, twenty-five, and fifty portions. Include temperatures, planning instructions, yield, portion size, file number, and other required information.

3. Give the advantages and disadvantages of the three methods discussed for setting selling prices.

■

CASE STUDIES

1. Lydia Lunchcounter has just taken over a branch restaurant with a problem. While some years ago the operation was one of the most consistent performers in the entire company, recently the profits have dipped to the point where the restaurant is now losing money. Lydia, known throughout the chain as a turnaround specialist, has been sent in to make the operation profitable again.

   After observing and analyzing the operation for a week, Lydia is ready to begin the turnaround. She has separated the task into several different categories, including upgrading and retraining of service personnel, revitalization of the décor and physical plant, and profit management. As a new member of Lydia's management team, you have been given the responsibility of overseeing the management of profits. It is an opportunity to shine and to establish yourself with Lydia for the future.

Outline the steps you would take to determine how the foodservice is currently performing and how you would begin to establish workable goals for the future. Because the last management team seemed to operate pretty much from the seat of their pants, your agenda will include establishing budgets, a break-even point, and a profit-and-loss statement based on the current month. The list that follows contains the pertinent figures you will need. You should not only complete the analysis called for in each step, but also provide a short description of how the figures were arrived at, which members of the staff you consulted for information, and how you plan to monitor progress. Include a break-even graph in your work.

Sales; $80,000 (food, $60,000; beverage, $20,000)
Fixed costs: $26,000
Variable costs: $43,000
Prior month inventory total: $35,000
Inventory this month: $41,000
Purchases during month: $23,000 (food, $19,000; beverage, $4,000)
Expenses
    Operating wages: $12,000
    Payroll taxes: $3,000
    Insurance: $800

Utilities: $1,400
Laundry: $400
Repairs: $600
Travel: $300
Rent: $1,800
Depreciation: $2,000

2. Current issues in foodservice management are constantly before operators. In the past you have done research on wages, taxation, and government intervention. You have commented on workplace smoking, business meal deductibility, Internal Revenue Service (IRS) tip audits, tourism promotion, payroll taxes, family and medical leave, immigration laws, OSHA reform, health insurance, blood alcohol concentration (BAC), food safety, legal reform, and a host of other policy mandates that face the modern foodservice operator.

As an expert in the field of current governmental affairs as they relate to the foodservice industry, then, you are asked by the local hospitality trade association to discuss the number one challenge facing the industry today as it involves governmental intervention. You decide to first look on the Internet at *www.restaurant.org* to identify the many areas being considered under Governmental Affairs. Using the list generated from your research, you then speak to an industry leader in your community to assist you in choosing the number one challenge. Present your findings, quoting the industry leader where appropriate.

■

VOCABULARY

Profit
Yield test
Standardized recipe
Extended cost
Unit cost
Multiplier method
Pricing factor method
Menu mix

Security
Profit-and-loss statement
Balance sheet
Break-even analysis
Budget
Food cost
Beverage cost
Business entity

Fundamental accounting equation

Going concern

Money and cost concepts in accounting

Raw food cost

Food cost percentage

Precosting

Portion cost

Cost factor

Planned profit method

Contribution margin

Quality control

Database

Concept of consistency in accounting

Conservatism

Inventory extension

Beginning inventory

Ending inventory

Cost of goods

Fixed costs

Variable costs

Equity

Break-even point

Government regulation

■

ANNOTATED REFERENCES

Coltman, M. M. 1991. *Financial Control for Your Foodservice Operation*. New York: John Wiley & Sons, Inc. (Cost controls in foodservice operations are the basis for this text.)

Coltman, M. M. 1994. *Hospitality Management Accounting*, 5th ed. New York: John Wiley & Sons, Inc. (From understanding financial statements to ratio analysis, it is discussed in this book.)

Cote, R. 1997. *Understanding Hospitality Accounting*, 4th ed. East Lansing, Mich.: The Educational Institute of the American Hotel & Motel Association. (From theory and practice to the use of computers—it is all covered in this book.)

Dittmer, P. R., and G. G. Griffin. 1994. *Principles of Food, Beverage, and Labor Cost Controls*, 5th ed. New York: John Wiley & Sons, Inc. (An authoritative text on the subject matter.)

Messersmith, A. M., and J. L. Miller. 1992. *Forecasting in Foodservice*. New York: John Wiley & Sons, Inc. (All aspects of forecasting are given consideration in this book.)

Mill, R. C. 1998. *Restaurant Management: Customers, Operations, and Employees*. Upper Saddle River, N.J.: Prentice-Hall. (Chapters related to controlling costs are included here.)

CHAPTER 9

Property and Promotion

LEARNING OBJECTIVES

By the end of this chapter, the reader will:

1. Know the issues of foodservice design, including construction and environmental considerations.

2. Learn how to maintain the foodservice facility considering codes, the external environment, security, and other requirements.

3. Be able to distinguish between marketing and merchandising within a foodservice operation.

4. Understand the merchandise of food from both the "front of the house" and "heart of the house" considerations.

5. Recognize the impact the computer has had in foodservice marketing and merchandising.

6. Be able to identify the new industry growth segments.

SELECTED WEB SITE REFERENCES

CaterMate Event Management Software
www.catermate.com

CaterWare Inc.
www.caterware.com

Hospitality Sales & Marketing Association
    International
www.hsmai.org

National Restaurant Register's Menu-Online
www.onlinemenus.com

Newmarket Software Systems, Inc.
www.newsoft.com

World Wide Waiter
www.waiter.com

PROPERTY COSTS

The costs associated with physical space occupied by a foodservice establishment continue to be significant. Yet most foodservice operators remain largely uninformed about this topic. While most foodservice operators can readily recall the ups and downs of the energy industry in the past, and the effects on customer counts of the long gas lines of 1974 or the tumbling oil prices of 1985–1986, few associate those turbulent changes with dramatic cost fluctuations in their own businesses. Fewer still have any idea of what to do to control those costs. Yet the foodservice manager can and must learn the basics of energy management and

conservation, of building management, and of maintenance management in order to ensure a profitable operation in the future.

In most foodservice operations, the manager is the chief engineer and energy specialist. While this is not true, of course, in large properties such as hotels, resorts, military installations, and some institutional operations or mixed-use facilities, most foodservice managers do not have the luxury of organizational structures that managers in those operations enjoy. In addition, the foodservice manager in most operations is also responsible for building and grounds maintenance. In many establishments, this alone can be an awesome task. In many older buildings that provide an attractive ambiance for restaurants, for instance, managers must be aware of and able to solve problems ranging from leaks in the plumbing, roof, or walls to whether or not the structural loads associated with equipment additions can be maintained.

At one time the costs associated with plant/property engineering and maintenance were viewed as fixed, since they varied little from year to year. Now the costs associated with these duties can range from 2 or 3 percent of total sales to 10 percent or more. In addition, these annual expenses may represent up to 30 percent of the original investment for the operation; in fast-food operations, this average is about 24 percent. In other words, the costs associated with engineering and maintenance could equal the original costs of establishing the business every five years or so. Utility costs average another 2 to 4 percent of sales per year.

■

FOODSERVICE DESIGN

The type of foodservice establishment is a major factor in planning the design of the facility. Each market demands a different physical layout emphasizing the operation's particular attributes of type and style. A commercial fine-dining foodservice design, for instance, would differ significantly from a university cafeteria design. Each type also calls for specific variations in the amount of space devoted to various functions in the design. This means that, while the heuristic (or rule-of-thumb) assumption that 25 percent of the total foodservice space should be devoted to the kitchen may be valid in some instances, it is not always so. Most variations in kitchen design, however, can be roughly categorized into two principal types: the straight-line flow and the functional flow.

Flow of Work in the Kitchen

Establishing a manageable flow of work through the kitchen helps speed production and reduce labor. Since much of the actual labor time in kitchens is spent fetching or retrieving items for use, an efficient design is important. Each facility requires a different specific pattern based on the anticipated production level of each menu item, the types of equipment used,

and the expertise of the staff. Although the details of each pattern will differ, eight basic rules to follow have been identified in establishing flow in work centers, sections, and the entire layout:

1. Functions should proceed in proper sequence directly, with a minimum of crisscrossing and backtracking.

2. Smooth, rapid production and service should be sought, with minimum expenditure of worker time and energy.

3. Delay and storage of materials in processing and serving should be eliminated as much as possible.

4. Workers and materials should travel minimum distances.

5. Materials and tools should receive minimum handling, and equipment should receive minimum worker attention.

6. Maximum utilization of space and equipment should be achieved.

7. Quality control must be sought at all critical points.

8. Minimum cost of production should be sought.

Straight-Line Flow for Kitchen Design

The straight-line flow for kitchen design is most efficient when adopted by an operation producing a restricted number of similar products; it is often referred to as the *assembly-line flow method*. The word *straight-line* refers more to the type of production than to the actual design layout (which may be in any of various shapes). In terms of production, however, products proceed in a line from one area of the kitchen to their final stop in another area. Interruptions are short-term, and there is no stop-and-stare. This system can be found in fast-food operations, in military and institutional foodservice settings, and in large cafeteria operations.

Functional Flow for Kitchen Design

Also called the *process* method of design, functional flow system is used most often when many and various products are made to order. The functional design system assumes that numerous unrelated products are being prepared simultaneously in different areas of the work center, with concurrent peaks and valleys of workloads in each area. This method is most commonly found in commercial kitchens that produce from menus.

Kitchen Design Space Allocation

The exact amount of space needed to complete any specific function in a foodservice varies, depending on volume, type of service, number of workers, type of supplies, storage requirements (especially in remote areas that receive limited deliveries), and so on. Storage for most operations can be allocated on the basis of the total kitchen space multiplied by 3 to 5 ft^2, and storage areas an be broken down as shown in Table 9-1.

Receiving, Storing, and Issuing Requirements of the Plant Property

In the amount of space they devote to the functions of receiving, issuing, and storing, foodservices differ considerably. In each case, however, the location of storage space should be close as possible to receiving and production. In larger operations, the receiving area has a receiving dock large enough to accommodate the trucks that are used to deliver products to the foodservice. In most locations this means having a dock that is 40 to 44 in. above the driveway. The receiving

area should be sufficiently large to accommodate the number of products likely to be received in a day. In most climates, accommodations should be made at the dock to protect deliveries from inclement weather conditions. It is also important to install a door at the receiving dock that is sufficiently large to allow a hand truck or four-wheel dolly to pass through it easily. In most cases, this means a door at least 3½ ft wide, although 4-ft-wide doors or double doors with no center post are suggested.

The receiving, storing, and issuing areas should be large enough to contain not only the products but also the equipment used in those departments. For many foodservices, the receiving area should contain bulk weighing scales, food-testing equipment (such as fat analyzers), and hanging scales. In larger foodservices, a separate office is provided for receiving and contains a variety of modern equipment used in the process. For example, computers may be used to scan incoming deliveries and add the products immediately to inventory totals. The personnel working in this department must be well-versed in the grades and types of products available in the foodservice industry today and must maintain specific standards established by the operation.

Foodservice Offices

The types of offices used in the industry are almost as numerous as the types of foodservices themselves. In many small foodservices, there may be no formal office at all; while in larger operations, offices may be provided for each department. There may, for instance, be receiving department offices, personnel offices, service management offices, and maintenance offices. The amount of space allocated for office use depends on the style and type of operation, as well as on the available space.

As computers are used more and more in foodservices, the amount of space used for offices has changed. Today, even small convenience foodservices use computers for various aspects of the operation and thus require additional office space. In most operations, cash and receipts are stored in the foodservice office, and this function determines the size of space needed. A small, one-person office may only require a 6 ft × 8 ft space.

TABLE 9-1 Storage Areas as a Percentage of Total Storage

Type of Storage	Percentage of Total Storage
Dry storage	30
Refrigerated storage	25
Frozen storage	10
Beverage storage	15
Refrigerated beverage storage	5
Nonfood (china, linen, paper, silver, etc.) storage	15

In many foodservices, the location of the office is determined by the function it performs. The office should provide ample working space for the personnel involved, allow access to the workplace, and satisfy any special requirements of the foodservice (such as including space for conferences or file storage). In foodservices in which only one office is used, the location should be such that the office is central to as many aspects of the operation as possible. If the public is meant to frequent the office, it should not be located in the kitchen.

The equipment needs of a foodservice office vary with the other facets of the business. In some cases, only a wall shelf, a calculator, and a small desk with drawers are needed. In others, the office must accommodate several filing cabinets, computers, printers, copiers, and so on. With the growing use of personal computers in foodservices, each operation should carefully consider adopting this resource in order to remain competitive. Certainly, the management located in the office should be supplied with all the equipment needed for a smooth and efficient operation.

Kitchen Equipment Space Allocations

The size and configuration of the kitchen depends on the style of service, the equipment space requirements, and the size of the operation. Choosing commercial kitchen equipment and designing the configuration of work areas around the equipment can require a great deal of expertise and experience. Often the design is worked out by a contract architect, a consultant, or by a designer on the staff of an equipment supplier. In some smaller foodservices, however, the owner or manager may be responsible for designing the kitchen equipment configurations.

Commercial kitchen equipment comes in various sizes. A range/oven combination, for instance, can be purchased in widths of anywhere from 24 in. to several feet, depending on the number of burners and/or flat grills associated with the item. The same is true (although with a lesser degree of variation) of equipment depths. The height—at least in terms of the working surface—remains fairly standard. Most equipment can be set either on legs or on casters, depending on the foodservice's needs. Many operators prefer that equipment be placed on casters, to be movable, and these ensure ease of cleaning.

The varieties of equipment required in an operation depend on the type of cooking to be performed. In foodservices that specialize in making sautéed foods to order, for instance, there is a greater-than-average need for open burners for skillets and sauté pans. In institutional foodservice operations, in which large quantities of foods are prepared together, a minimum number of open burners may be required, thanks to the reliance on such bulk production equipment as tilting braising pans. An inexperienced operator should consult with other foodservice operators and with experts in the field (either consultants or reputable equipment suppliers) prior to making decisions about equipment purchases.

Dishwashing

Several plant/property considerations must be included in space allocation planning for dishwashing equipment. The most important of these considerations is the type of system to be employed, the location of the system, and the issue of sound pollution.

While dishwashing was at one time performed mostly by hand, most foodservice operations today use automated dishwashing equipment. Since one of the principal requirements of a dishwashing system is that it be able to handle the load during peak hours, the load for that period should be used as the basis for comparing the adequacy of various dishwashers to meet the foodservice's needs. The operator must also decide whether an electrical or gas-fired unit would be best for the operation, and whether the dishwasher should be conveyorized (primarily of value in larger operations) or hand-propelled. The number of stainless-steel dish tables to have available on either side of the unit (dirty dishes on one side, clean dishes on the opposite) must also be determined after an investigation of the number of dishes handled during peak load periods. To conserve energy, an operator must consider the various new low-temperature dishwashers on the market today that utilize both temperature and chemicals for cleaning.

The location of the dishwasher is important to the flow of the entire kitchen. Servers or clearers who remove dishes from tables should be able to drop dishes off at the dishwasher station without entering the main work area of the kitchen. In this way, traffic in the kitchen can be limited. Because the function of servicing the cook and service areas with clean dishes is essential, the dishwasher should be situated so as to reduce the amount of time spent carrying dishes to these areas. Locating the dishwashing system properly will also reduce breakage because of the improved flow of both dirty and clean dishes. Because the dishwashing area is noisy, the sound should be contained within the specific washing area. In many foodservices, separate dish rooms are designed as part of the kitchen operation, to prevent noise pollution of either the main kitchen or the dining areas.

Waste Disposal

Waste disposal is an essential function of every foodservice operation. Careful consideration should be given to the types of products served, the size of the operation, the location of the waste system, and the types of waste (food, paper, plastic, or cans) before any single system is implemented. In addition, the state and municipality in which the foodservice is located will likely have codes or laws regulating how waste may be handled. Among the possible disposal methods for foodservices are compaction, incineration, pulping, use of in-sink-erators (trough, sink, or cone), or use of garbage cans. Any of these systems (other than simply placing garbage in cans or dumpsters for pickup) will reduce the odor and bacterial contamination of waste.

In many instances, hybrid waste disposal systems are used. For example, large institutional, hospital, resort, and military foodservices often use dumpsters for nonfood products and pulping for food waste. In some states, pulping—a system that creates a liquefied waste by grinding foods with a recirculating water system—is now required by health departments. Other locales restrict the amount of pulping because of overloaded sewage disposal facilities. A foodservice operator who is unfamiliar with either the local codes for waste disposal or the availability of different systems can seek information from architects, consultants, or foodservice equipment companies. In many areas of the United States, a foodservice operator can find consultants who specialize in waste disposal.

Dining Room Space Requirements

Effective dining room design must take various needs into consideration. The amount of seating space required, traffic corridors, décor provisions, customer facilities, and service stations all affect the total amount of space needed. The two most important topics to be considered during allocation of dining room space are the comfort of the diner and the efficiency of the service personnel.

Seating Space Considerations

Space for dining room seating is generally calculated on the basis of square feet per diner. The size of dining room tables can vary greatly. It is important to remember when choosing tables that both the size and the shape affect customer comfort and efficient dining room flow. Tables that are 24 to 30 in. square are often used in bar/lounge seating, but those sizes are generally too small for comfortable dining. In a dining room, 36 in. × 30 in., 36 in. × 36 in., 36 in. × 40 in., and 36 in. × 42 in. tables are commonly used. Square or rectangular tables, because of their straight edges, are preferable in cases in which it may be necessary to move tables together to accommodate larger groups. Aisle space should be calculated on the basis of the types of equipment to be used in the dining area—taking into consideration, for instance, the specific dimensions of such items as rolling service carts. Adequate aisle space must be maintained, of course, whether or not dining room service carts are to be used. Most designers recommend setting aside a minimum of 24 in. between chairs to ensure adequate aisle space. That distance is, in fact, required by code in many states and municipalities for safety reasons.

Recent government rules specify special aisle and seating space to accommodate patrons with disabilities.

Seating Arrangements

The seating arrangement to adopt for a foodservice facility depends on the type of operation and the desired capacity. In some facilities—fine dining rooms, for example—freestanding tables that seat four customers (called *four-tops*) are the norm. This arrangement may not be the best for other types of establishments, however. Wall-oriented booth seating, for example, accommodates more guests than do freestanding tables in a rectangular dining room. In cafeterias, institutional foodservices, and military installations, long tables accommodating eight to twelve (or more) guests may be the most common type used. In coffee shops and some other types of foodservices, counter seating may be the best choice, providing the style of comfort desired by the customer and allowing optimum use of facility space. The average size of the party of anticipated customers should also be considered, as this varies greatly in different types of foodservice operations. Each type of seating arrangement requires different space allocations for table height, size, width, and so on; the operator is responsible for matching all the ingredients correctly to create the efficiency and ambiance desired.

■

ENVIRONMENTAL CONSIDERATIONS

Making customers physically comfortable during their visit to the foodservice operation may be as important to the operation's success as offering high-quality food and courteous service. While in the past foodservices were able to accomplish this goal with a minimum of expenditure, today such items as heating and air conditioning have a major impact on the bottom line. How each of these areas is handled in a foodservice depends on the type of operation, the climate, the locale, the building's construction, and other factors. In new foodservice operation construction, a great deal of thought is given to such features as heat loss or gain, automated or computer-controlled systems, energy-efficient equipment, and alternative sources of power. In most existing foodservices, however, major renovations to incorporate such attributes are not feasible, so other measures must be adopted to provide the highest possible comfort for the customer.

Structural Changes That Reduce Energy Costs

Many changes can be made in existing foodservices to reduce energy cost and increase comfort. For example, adequate insulation of walls, windows, ceilings, and doors can markedly improve the efficiency of existing

HVAC (heating, ventilating, and air conditioning) systems. Double or revolving doors at entrances are another relatively minor alteration that can improve the internal environment. Adding doors to the kitchen to lessen airflow into the kitchen's exhaust system can greatly reduce energy costs.

The type and placement of air ducts in both the kitchen and dining areas should be given due consideration. It is important, for instance, that tables not be placed directly beneath fresh air ducts, since customers do not want a constant stream of hot or cold air rushing over them during dining. In the kitchen, air ducts should be placed in work areas in a layout designed to maximize comfort for the workers. In the kitchen, the air ducts must not be placed near the kitchen's outside exhaust system (designed to remove smoke from cooking areas); otherwise, a substantial waste of energy could occur from the exhausting of newly conditioned air. Similarly, if the thermostat is located in a particularly warm spot (in the sun, for example), the air conditioning system will overwork itself trying to compensate for the temperature at that spot rather than for the temperature in the room as a whole.

Using a heat recirculation system (to capture some of the heat produced by equipment for use elsewhere), solar energy, or conventional means of energy conservation can lower the size of the monthly energy bill. One of the first steps a foodservice operator should take is to inspect and correct the thickness of insulation in the floors, ceiling, and walls to reduce heating or cooling loss. If the foodservice operator is not well versed in environmental control, specialists should be consulted for a thorough evaluation of all available options.

Kitchen Exhaust System

In most instances in which cooking is involved, foodservices are required to install and maintain kitchen exhaust systems to remove smoke, grease, and odors. These systems vary in size from small units with $\frac{1}{4}$-hp motors to very large units with motors that have a capacity of several horsepower. In most states, foodservices are required to install exhaust systems that have no rough edges. In addition, new units must have welded seams, to satisfy established building codes. Operators should be aware of the specific requirements governing exhaust systems in their state or community. This information can usually be learned from a municipal building inspector or from the local health department.

An operator either can purchase a prefabricated exhaust system with motor and roof- or wall-mounted exhaust or can have a custom unit built for the foodservice. This decision should only be made after a thorough investigation of the variabilities of the operation, the costs of different exhaust units, and the opinions of specialists in the field. The type of metal to use in constructing the unit depends on the local and state codes governing the system. In some states, galvanized metal is allowed; in others, the unit must be made of stainless steel. Most foodservice equipment companies can provide the necessary answers concerning exhaust systems, as can independent experts.

The exhaust system should be mounted in such a way as to cover (usually with an additional 3 to 6 in. on either end) the cooking equipment completely. In many foodservices, units are secured to beams in the roof and suspended by cables or steel bands over the cooking line, where they penetrate the ceiling. In addition, they penetrate the roof, to exhaust the air and smoke outward. The area around the roof penetration (where the exhaust throws the air) can be a likely place for leaks to develop. Proper flashing must be mounted at this point, either by the exhaust firm or by a separate roofing contractor.

The size of fan to use in the exhaust system can be determined by evaluating the kitchen cooking equipment involved. This determination may involve the participation of an architect or engineer, especially in larger foodservices. Many smaller foodservices allow the equipment company to determine the airflow needs of the exhaust. In either case, the unit's construction and location can substantially affect the heating and air conditioning costs of the operation. If the exhaust is located near the dining room, for instance, or operates in a manner that causes it to draw air from the dining room, that room's conditioned environment can be disturbed, leading to energy waste.

Some of these problems can be overcome by designing a kitchen exhaust system that balances air intake with air outflow. In all exhaust systems, the air coming into the system (usually through hollow panels in the system's walls) should be balanced with the air outflow. In some locales, this incoming air must be tempered through installation of a heated makeup air (incoming air) system; in others, air directly from the outside can be used.

■
MAINTENANCE OF PROPERTY

In order for a foodservice operation to function properly, the plant/property must be kept clean and in proper working order. This is best accomplished through a regular maintenance program. While many foodservices provide maintenance only after a problem develops, this approach to corrective maintenance is both expensive and nonproductive. Instead, operators should practice a system of preventive maintenance that stresses proper regular care and servicing of all aspects of the foodservice property.

Equipment Maintenance

To institute a program of preventive maintenance, a foodservice manager must first make a list of all equipment to be included in the program. This list should include not only food-preparation items in the kitchen, but also all heating, air conditioning, plumbing, lighting, and other equipment. On the basis of this list, the foodservice manager should compile maintenance and repair data pertinent to the equipment listed. These data may include steps to be taken during maintenance, specific items to be checked or replaced, and descriptions of how service is to be performed (e.g., "check grease level or exchange filters"). From the list of equipment and specifications, the foodservice manager can develop a checklist for regular use. This checklist should specify dates for inspections and make provision for signature of the party who provides service, whether a member of the staff or a representative of an outside agency.

The equipment maintenance checklist should be entered or otherwise developed on the basis of a calendar of events. Many foodservice operators actually use a calendar for this purpose, writing directly on it the items to be serviced on the dates marked; others use separate dated files or even computerized reminders. On the checklist, the exact maintenance to be performed at different intervals must be specified, and each item should be discussed comprehensively in terms of which services can be performed as regular maintenance, which require additional service, and which involve replacement of parts. The form used to record service should also provide a space for listing the condition of the equipment when serviced, leaving ample room for comments that can be followed up on during the next servicing. Space should also be provided for listing service purchases and the costs associated with those purchases. This form should be kept as part of the permanent record file of the operation, as it provides a good record of the condition of each piece of equipment used in the foodservice.

Other Maintenance of Property

While the proper care of equipment may be the most easily recognizable need in a foodservice maintenance program, it is not the only one. The building (both inside and out), the parking lot, and the grounds should all be covered by regular maintenance programs as well. Indeed, since these are more likely to be visible to the public than the equipment is, maintenance of them may influence the profitability of a foodservice more quickly than would a program for equipment maintenance.

A foodservice manager should establish the same type of program for interior plant/property maintenance as is established for equipment. This means that prevention, rather than correction, is the goal. A calendar of maintenance items should be established that includes both daily and less frequent items. Like the equipment list, this list should specify the exact duties to be performed during maintenance. Items on the list should range from duties performed by the regular foodservice staff (such as after-shift cleanups) to work done by outside agencies (e.g., carpet cleaning or window washing.) The list should also cover maintenance of all surfaces of the plant/property, including ceiling tiles, walls, floors, doors, glass, and any decor

items used in the foodservice. In many foodservices, separate lists are maintained for each area of the facility. In such a case, one list may be kept for kitchen maintenance other than equipment (such as walk-in cleanups and wall washing), another for the dining room, and a third for public facilities.

The list for areas outside the building should include walls, sidewalks, the roof, ducts, windows, doors, and awnings, among the other items. Some tasks must be performed to maintain appearances (such as window washing, painting, and otherwise beautifying the property); others must be done to ensure customer safety (such as oiling door hinges and inspecting the roof). In the parking lot, a regular schedule of light-bulb changing, sweeping, and litter removal should be established. To a large extent, the maintenance of property grounds depends on the type of property. In some instances, lawns must be mowed and weeded; in others, trash dropped by passers-by must be picked up.

■

EXTERNAL ENVIRONMENT

Creating a pleasant external appearance can encourage customers to frequent the establishment. While many aspects of the external environment must be left to specialists (such as architects or designers with expertise in building construction appearance), others are within the province of the foodservice manager. One example is the creation of a proper environment around the trash-holding area of the foodservice. There is probably no quicker way to discourage customers from visiting than by improperly protecting the trash area. This portion of the facility should be kept away from the customers' view—either by location or by fencing or other means—and should be kept clean and neat, with no small pieces of trash left littering the area and calling attention to it.

Parking and Parking Lot Lighting

The proper planning of parking lots is important to any foodservice operation. In order to design this portion of the property effectively, a foodservice operator must be able to appraise the external appearance of the establishment from the point of view of a customer. Curb appeal is especially important to foodservice operators in the commercial segment. For example, curb cuts (entrances to the parking lot) should be designed to offer ample turning space and minimal stacking of traffic in the street at the entrance. The parking lot must be adequate for the size of the establishment. In many locales, the amount of parking a foodservice establishment must provide is dictated by code, usually in terms of a formula based on the seating capacity of the operation. The external environment of the plant/property depends not only on regular maintenance of the grounds and building exterior, but also on the efficient construction of outside facilities.

Other considerations that are important in planning a parking lot for a foodservice are walkways, lighting, landscaping, accessibility to the building, and visibility of the trash removal and receiving areas. The most important aspect of walkway construction is that the walkway should clearly direct the customers to the main entrance of the foodservice establishment. The parking lot (and hence the walkways) should be located as close to the front door as possible. Steps (and ramps for the handicapped) should be carefully planned.

The arrangement of lighting in the parking lot is significant for both aesthetic and security reasons. For the safety of the customers, no portion of the lot should be improperly lighted. Several different types of adequate lighting are available today, including mercury, metal halide, and sodium. Each variety has advantages and disadvantages that the foodservice operator must weigh in the context of other variables in the operation. Lighting should be connected to timers or electronic activators so that it switches on when needed, to reduce energy consumption.

Many foodservices have found that well-landscaped parking areas and grounds can add considerably to the overall ambiance of the facility and thereby increase business. In the parking lot itself, the type of surface employed should be adjusted to the climate, season, and type of operation. A number have found using a curb-side parking system is appreciated by many patrons, especially upscale operations.

Signage and External Environment Merchandising

In many foodservice facilities, the external environment is shaped as an advertising aid to promote the restaurant. Theme-oriented commercial foodservices, for instance, might include in the design of the external building or grounds a continuation of the motif presented inside. In some cases, foodservices incorporate a lot of glass windows—in part to provide light and ambiance for the customers inside, but also to draw the attention of potential guests outside. In other operations, the outside of the building is illuminated with indirect floodlighting to call attention to the facility. The entrances to a foodservice represent another area of the external environment that requires substantial attention. The entrances should be designed to allow freedom of entry and egress as well as give a positive appearance to the property.

The external environment can be complemented by signage. In some instances, signage is required because of poor visibility of the plant/property itself; in others, the signage simply supports the impression made to the public by the building and grounds. Since the signage is such a visible advertisement for the foodservice, care should be taken in its design. Foodservice operators must also take into consideration the various city, county, and state signage codes in force in their area. These codes were created as a means of preserving the tone and integrity of the community, and they dictate in many instances the type, size, color, and construction materials for signage that can be used. Lighting of these signs can be accomplished in a variety of manners. Signs may be illuminated from the inside (with neon or internal light bulbs) or from the outside (with spotlights directed onto the signs). How illumination is achieved should be determined by how well each prospective method would promote the overall ambiance desired.

■

CODES AND OTHER REQUIREMENTS

Each locale has codes or other ordinances that apply to various aspects of foodservice operations. While these codes may seem restrictive and burdensome to many foodservice operators, they actually provide sound bases for management of the property. General codes affecting foodservices include building, construction, fire, and safety codes. Building codes, for instance, affect the types and locations of businesses that are permitted. Construction codes govern the materials that may be used. Fire codes (especially important from a customer safety standpoint) regulate the building size, number of entrances, types of materials, and manner in which foodservices are constructed.

In addition to these general codes, certain industry-specific codes dictate many of the plant/property requirements of the business. Probably the most important of these to a foodservice are health codes.

The maintenance practices discussed heretofore have two important goals: establishment of a clean and efficient operation and adherence to health codes. In most locales, health codes dictate the types of storage required for many foods and the temperatures of storage required. How often refrigeration units should be cleaned, for instance, may be based on health maintenance codes. For many foodservice operators, these codes offer a good basis or reason for instituting maintenance policies. Code provisions relating to employee hair length and restraints, for example, give the foodservice manager sound reasons for instituting hair hygiene rules for the staff. Codes can help the manager establish operating hours, quality of products, and customer adherence to sanitation as well. Requirements dictating that foodservices must use inspected products, for instance, provide safety for customers from various food-borne diseases. In each locale, foodservices are governed by codes that affect daily business. A good foodservice manager must be aware of and adhere to each relevant code provision. Generally, an outline of these requirements is available through the city, county, or town offices in which the facility is located.

■

SECURITY

A foodservice operation often constitutes a major investment. Therefore, the security system employed to protect the operation should be chosen with great care. And since operations vary greatly in their security

needs, a system designed specifically for the foodservice should be implemented. For some foodservices, the operational needs might include internal security of a valuable china, art, or silver collection that must be counted and secured at the end of each shift. For other foodservices, the principal needs might only extend to security of the cash and the equipment. In all types, however, there is an overriding need for security of the customer.

External Security

The first line of external security for a foodservice and its customers is the outdoor lighting system. A foodservice that is well-illuminated is usually significantly protected against crime and vandalism. In addition, well-lighted external property reduces the possibility of accidents that can injure staff and customers.

The building's doors and windows provide the most likely points of attack for vandalism and burglary. In designing a foodservice with effective security built into these two areas, such measures as using metal instead of wooden frames should be adopted, and alarms should be mounted on the doors and windows. The locks on doors themselves can provide different degrees of security. For instance, a dead-bolt or frozen dead-bolt lock provides greater security than do many other types. The type of door is also a significant variable. A hard-core wooden door, for instance, provides more security than a hollow-core wooden door; and a metal door provides more security than a hard-core one.

Alarm systems have been used effectively to deter crimes during times when the foodservice operation is closed. Many different types are available, including microphoned premises, contact systems, and light beams. In choosing a particular alarm system, the foodservice operator must consider how the system is interfaced with the local police department or sheriff's office. A sensible alarm system should contact the authorities—or the alarm company headquarters—prior to notifying the foodservice manager.

Internal Security

Most foodservice operations employ internal security systems to protect valuables. In many operations, for instance, a safe for storing cash and other valuables at night is located in the manager's office. Some properties have gone a step further and effectively installed safes that require two keys to open, can only be opened during certain times of the day, or can only be opened by armored car services.

While cash receipts have the most obvious need for security, other aspects of a foodservice operation require a great deal of security as well. The inventory of a foodservice, for instance, is generally a very valuable asset of the operation, and without a well-designed plan to secure it, it can be lost at great potential cost. Locks on walk-in refrigerators, food storage rooms, reach-ins, liquor storage areas, and supply storage areas should form an integral part of a foodservice's security system. In addition, regular maintenance and spot inventories of each of these areas should provide a double-check for security. Even maintaining the proper level of inventory can help prevent loss, since keeping too much on hand tends to create the impression that a small quantity will not be missed. The foodservice operator should strive to diminish the likelihood of internal theft by employees by eliminating opportunities for it.

Fire Security

Protection of both the customers and the plant/property from fire is another area of great concern to the foodservice operator. All foodservices should be equipped with fire alarm systems to detect fires as early as possible, both in the public areas and in the kitchen. In addition, since the kitchen is the most likely area in a foodservice for a fire to start in, the operator should install a fire-extinguishing system there. In many locales, such a system is now required by law.

Various fire-extinguishing systems can be installed, depending on the needs of the foodservice. In some cases, a fire-extinguishing system is mounted in the kitchen exhaust system, over the cooking equipment. This system can detect higher-than-normal heat levels and automatically release an extinguishing agent such as CO_2 (carbon dioxide) or halon gas. A CO_2 unit extinguishes fire by discharging a foam or powder containing CO_2 over the area to smother the flames, while a halon unit draws away and consumes the oxygen in

the area, thereby eliminating the fire. In an efficiently designed exhaust-hood system, the filters in the system help prevent the possible spread of fire as well as remove grease and dirt from the air being exhausted. To operate efficiently, these filters should be regularly cleaned to reduce grease buildup, and the entire duct system should be included in the maintenance program of the property, so that the interior is thoroughly cleansed periodically of potentially hazardous buildup.

For the safety of both employees and customers, a foodservice facility should provide ample fire exits. In most locales, the number, location, and size of these exits are dictated by codes or laws that base their requirements on the size of the building and the type of construction used in the property. These exits should always be kept clear of obstructions such as tables and chairs and should be clearly designated by appropriate signage. For illumination during electrical outages, battery-operated emergency lighting systems are usually located near exits.

PROPERTY CONSTRUCTION

Constructing a foodservice facility of any size can be a major undertaking. To carry out this process properly, operators who lack significant experience in construction should solicit the aid of a licensed architect at least, and probably also a designer, a construction supervisor, and several other specialists. The number of specialists to engage depends on the complexity and size of the project. A large foodservice construction project might include—in addition to the architect—an interior decorator, a sign contractor, an interior designer, a foodservice equipment consultant, a site surveyor, a construction manager, a structural engineer, a mechanical engineer, an electrical engineer, a fire protection consultant, a parking consultant, a landscape architect, a lighting consultant, and various suppliers of products. Moreover, any project will call for the involvement of a number of subcontractors, working under the general contractor or construction manager, to build different facets of the facility. In smaller foodservices, these various consultants are often eliminated in favor of a triumvirate of the architect, the designer, and the construction manager or contractor. The number and degree of involvement of outside experts in each foodservice construction project will depend on the operation.

Construction Planning

The planning of a foodservice operation can (and probably should) take several months to complete. During this time, the architect and the owner, with perhaps a designer and a construction specialist acting either as an outside general contractor or as an in-house manager, will meet to develop the property. The architect is generally the project team leader, and the ultimate responsibility for the development rests with this specialist. It is the architect who must take the ideas developed in group meetings and put them on paper to give the project visual definition. While many of the subcontractors may draw up additional plans and submit these to the architect for review or revision, and while the designer may develop plans for specific aspects of the property, the architect oversees the operation in general. For this, the architect receives a fee ranging from 1 to 5 percent of the total project cost. The percentage tends to be higher on smaller projects, as the total costs are less.

During the process of creating the project on paper, the development team goes through several identifiable phases. The first phase is called *schematic design*. During this stage the architect, designer, and owner develop the general scheme of the foodservice. The drawings are generally single-line and limited in definition, and they concentrate on perspectives rather than specifics. It is important that a feasibility study, a menu, and other operating information be available to guide the planning. The second phase (called *design development*) is characterized by detailed drawings that indicate specific aspects of the mechanical, plumbing, equipment, structural, electrical, and fire protection systems. Building materials are also selected during this phase.

After the design documents (drawings produced during the design development stage) are approved, the construction documents (or construction drawings) are made. These detailed drawings identify all aspects of the facility exactly as it is to be built. They must be extremely detailed in most cases, since the workers who build the facility will depend on these drawings for their understanding of what is to be built. Computer programs exist which can do all the details of drawing any plan.

Site Analysis

Many foodservices fail because of the site chosen for them to occupy. The old axiom of "location, location, location" regarding real estate fully applies to foodservice facilities. This means that, unless the location is right, the facility will probably fail, regardless of all other conditions. To identify a desirable foodservice site, an operator should take into consideration the visibility of the site from the vantage point of potential customers. For a commercial facility, this usually means visibility from the streets or highways that pass the site; for an institutional operator providing foodservices to a college student clientele, it may mean visibility to the greatest volume of walking traffic; and so on. The operator should also consider the accessibility of the site. Far too often, a good site is located and built on, without providing adequate access for the proposed users. This may mean, in the case of a commercial facility, that potential customers passing by on the street find it difficult to drive to the plant/property. To a military foodservice operation, inaccessibility may be a matter of the distance that the majority of troops must walk to get to the facility, and the obstacles they may encounter along the way. To an industrial plant foodservice, inadequate access may involve difficulties workers experience with getting up or down flights of stairs, elevator congestion, and so on.

Careful site analysis should include a visual plan of the site prior to development of the project. Temporary drawings should be made that convey the general impression of the planned facility on the site, to foster discussion and analysis of various issues associated with the venture. Consideration of the natural dimensions of the property, water, sewer, and gas connections, and various other characteristics of the site is essential to this process. In addition, the steepness of slope (slope can be determined by dividing the rise, or elevation, by the run, or distance), wind exposure, and type of ground surface (swamp, rock, clay, sand, or other) should be considered when planning a food-service. Many operators who are inexperienced with foodservice construction and planning have to engage a site analysis firm to test these various aspects of the proposed location. This should be done prior to land purchase, if possible.

Construction Process

The length of the construction process for a food-service facility depends on the type of facility, the materials used, climatic conditions, and other factors. In some parts of the country, for instance, the short outside building season considerably prolongs construction time. The entire process should take from eighteen months to three years, depending on the project's size and complexity. Of that total, the planning process averages from six months (for smaller foodservices) to two years (for large operations).

In most construction projects, a construction schedule is necessary to ensure that adequate progress is made. This schedule should outline the specific times—in days, weeks, or months, depending on the size of the project—by which various elements will be constructed. A good construction schedule will ensure, for instance, that someone will not be trying to lay a floor at the same time that the ceiling contractor is trying to install the ceiling. The schedule should also develop overall guidelines to enable each of the separate subcontractors to plan their work schedules on different jobs.

Construction Materials

The materials used in foodservice construction vary widely. Among the criteria used to determine the selection of materials are needs, climate, availability, cost, and energy efficiency. During the planning process, decisions about which types of materials to employ should be carefully considered and evaluated by the architect, the owner, the designer, the construction manager and other interested parties. In some cases, the city or state has imposed restrictions on what types of materials may be used and on how each may be employed.

The types of materials used can have a great bearing on the cost of the foodservice construction project. This is true not only of the exterior or structural elements of the project but also of the details of interior construction. For example, choosing a floor surface to use in the kitchen could involve selecting from among several different options, each at a different cost. The kitchen floor could consist of poured concrete with sealant, quarry tile, vinyl asbestos tile, a poured acrylic floor cover, or some other type. Each option presents a different cost factor to consider. Because of the high number of variables involved, many architects and contractors have developed computer programs to analyze them. The resulting rapid and efficient analysis of such options is one good reason for a foodservice operator to consult specialists when planning a facility.

During the planning process, a foodservice operator must consider the wear and tear on the materials used in construction. It may be cheaper overall, in some cases, to use more expensive materials from the outset than to replace less costly materials regularly as they wear out. Because of the many decisions that must be made in planning the materials to use in constructing or renovating a foodservice operation, the careful planner will seek and value professional assistance.

■

NEW INDUSTRY GROWTH SEGMENTS

The foodservice industry is ever-changing and ever-growing. From new markets offering food for immediate consumption, such as convenience and grocery stores through delivery and takeout, to booming market segments, such as the cruise and gaming industries, catering businesses, prisons, and business and industry contracts, foodservice appears to have unlimited potential for the future. These segments are each discussed in the following paragraphs.

Convenience and Grocery Stores

Americans are spending less time cooking every year and are continually looking for more convenient ways to feed their families. Convenience and grocery store concepts of providing food for hungry mouths are sprouting up everywhere.

The most recent venture has been the local gas station, which serves fresh panini, three-cheese pesto, and double espressos along with its usual selection of octanes. For Americans ravenous in appetite but starved of time, these new venues provide quick, good-quality cuisine convenient for the person who has an empty refrigerator at home.

In 1996, Americans consumed an average of 4.1 commercially prepared meals per week, up from 3.8 in 1991, according to the National Restaurant Association. Although most people think that traditional restaurants served these meals, the new trend is known as home-meal replacement and involves operations offering a highly movable feast that can become dinner elsewhere, as people in the United States want food fast but restaurant-quality fresh. Many grocery supermarkets now provide a space with tables and chairs where people can take purchases and consume them on the premises.

Home-meal replacement is the critical battleground between supermarkets, convenient stores, and restaurants for the consumer food dollar. Again, in 1996, of the $691 billion that Americans forked over for food, 46 percent was for dishes bought outside the home, with more than half of that going to takeout.

Delivery and Take-out

As competition grows within the foodservice industry, restaurateurs are developing delivery systems that can dispatch their finest dishes to one's home for consumption. Originating from the pizza delivery concept, these new models for success are involving entirely separate businesses taking patrons' orders, picking up the food at any of a number of restaurants, and delivering the cuisine to the address desired.

The kitchen increasingly seems to be a place to pursue cooking as a hobby, not a daily grind. In sit-down foodservice operations, more than half of the meals ordered are being taken home to be eaten. Restaurants are becoming the supermarkets of prepared foods. In the same sense, grocery stores are becoming more like take-out restaurants.

Add to all of this the Internet and the ability to surf menus from home and order for immediate delivery. New gourmet stores are being opened that are upmarket grocery outlets offering 400 plus items prepared by 35 or more on-site chefs and bakers daily. Such concepts are taken a step further by the company. Noting that maids now come into homes to clean them, why not send chefs in to cook for them? Where will this end? Research suggests that opportunities are unlimited in the foodservice industry for those wishing to work hard and smart.

Cruise Industry

With cruise lines building new ships and expanding into new markets, this industry segment offers exciting opportunities for the adventurous foodservice operator. While the cruise industry was traditionally off limits to Americans, since ships were registered only in foreign lands, today's modern fleet includes operations managed by Disney and a new, larger ship named World City. The latter will actually have three separate hotel complexes on board when launched upon completion. These magnificent ships offer complete foodservices just as a resort would. In fact, the buffets on board alone tempt many to travel on sea today since the food is so good. The cruise industry is one worth looking at for anyone considering a career in the hospitality industry.

Gaming Industry

Another exciting career is to be found in the gaming industry. From Biblical accounts of "the casting of lots" to Native Americans gambling among themselves in a manner adopted by their people hundreds of years before the white man's law had any effect, gaming has been here a very long time.

The outcome of gaming in the United States can be seen in the casinos, river boats, resorts, and magnificent foodservice operations built to support these pleasure palaces. Again, the buffets are lavish but food is served to meet everyone's need.

Foodservice operations in the gaming setting should attract the guest. Prime rib sold for the price of a regular sandwich brings the hungry gambler into an operation. High customer counts usually generate increased gaming revenue.

Guest convenience is another reason for the importance of food and beverage operations within the gambling setting. Management wants the patron to spend as much time as possible in gaming activities. Thus, the dining experience must be user-friendly so that the gambler does not wish to go to another gaming facility to dine.

The gaming industry is glamorous and offers such facilities as a buffet, room service, snack bar, coffee shop, gourmet room, specialty restaurant, catering facility, showroom, main bar, portable bar, lounge, and casino service bar. Operators must be willing to work hard, but the rewards are great in the gaming industry.

Catering

Another growth segment that is attracting sharp operators is catering. Besides being a professional executive, the caterer must be a salesperson willing to work the operation as well as to market it. There are convention and sales functions to cater. Weddings offer tremendous opportunity. Kosher catering has grown in recent years. Beverage sales is another important aspect of the catering opportunity.

The caterer must think success to be successful. From the initial contact with a client to the execution of a fully planned and prepared event, the caterer must be organized and good with people. Good-quality management is what catering is all about and experience over time will create a fine catering operation for an industrious person.

Prisons

This growth segment is much like any other noncommercial foodservice operation, such as hospitals, assisted-care living facilities, and nursing homes. In other words, the audience will be with you every day, and creativity and imagination must be evident in the menu to treat the foodservice patron correctly.

The challenge is also in nutrition and correct preparation of the cuisine. Since the patron will be eating solely in the foodservice operation within the facility, it is important that a balanced menu as well as one that is nutritiously sound be served. The hours to be worked are regular compared to other foodservice operations, but the challenges to offering creative and new menus every day are more evident in noncommercial foodservice operations such as prisons.

Business and Industry

The business and industry segment was known as contract foodservice in the past. It involves establishing foodservice relationships within business and industry to serve meals at corporate headquarters, within buildings, and at meetings of businesspeople. The segment has grown extensively and can be highly profitable.

The foodservice operator must be able to meet the needs of highly important and time-efficient customers. When the meal is served, it must be done quickly and with high quality. There is little room for error in this segment as the patrons are accustomed to the best and expect the same from their foodservice provider.

■

FOODSERVICE MARKETING

To promote a foodservice professionally, the operator must carefully plan the enterprise's goals and how to meet them. This is especially true of foodservices in which both production and service aspects take place under one roof. In such a case, the foodservice operator is selling products *and* services—not just one, as a manufacturing concern might. The products are the foods and beverages offered on the menu. The services are the less tangible attributes of the operation: ambiance, lighting, sounds, smells, friendliness of the staff, manner of delivery, and so on. Like those of any other service operation, these attributes are perishable (cannot be stored for use when desired), are simultaneously produced and consumed, and vary to the extent that each service person may deliver them in a different manner.

It is up to the foodservice operator to package the products and services in a manner that is most pleasing to customers. The first step here is to determine who the customers are and what they want. The process of answering these questions and devising a program to meet the needs identified is called *marketing*. Although this book deals primarily with internal promotional concepts—and while marketing is mostly external—it is important to cover a few of the basic concepts of marketing as well.

Marketing is not the same as advertising. In fact, an advertising program is only one aspect of a good marketing plan. Marketing involves building in customers' minds the business image that management would like them to perceive, while advertising is simply one method of conveying that image to customers. Other marketing methods include the correct packaging of products and services to meet demand and the proper presentation of products and services.

Determining the likely customers of a foodservice—or the *market*, as it is often called—can be accomplished in several different ways. The most common is through a market research survey, which collects demographic information concerning the operation's locale. From this information, the foodservice operator can determine such critical factors as the educational levels of potential clientele, food taste preferences, age, income, number of persons per household, and average amount of money spent dining out. On the basis of this information, a marketing plan can be built. If the survey indicates, for example, that most potential patrons prefer steak dinners and full service, the operator would be wise to establish this type of operation. This rule about responding to market interest applies not only to commercial foodservices but to all other types of foodservices as well. It is just as important for a military foodservice to cater to the preferences of its clientele as it is for a commercial operation to do so.

In addition to the basic market research survey, more involved methods can be used to determine the market for a foodservice. One popular method used for foodservices is a focused group study, in which meetings are conducted with the potential patrons to identify specific likes and dislikes about an existing or planned foodservice. This can be especially beneficial for determining the attitudes and spending habits of members of the prospective market. Lifestyle analysis is another method of market research that has gained in popularity. It is based on the assumption that, by studying the lifestyle of the market—amounts of time spent on various leisure activities, preferences for lifestyle, and so on—an operator can better determine how to tailor the operation to fit those with that lifestyle.

Yet simply studying the market in which a foodservice will be located does not supply enough information to plan the foodservice properly. The operator must also plan to position the business in the market. Since the market in which a foodservice is located probably offers a variety of dining options to choose from, the foodservice operator should pinpoint the segment of the market for which it will compete. The public is likely to associate the operation with a few salient features such as price, type of foods sold, and ambiance; therefore, the operator should determine how to present these features before entering the market. This is at the heart of positioning a business. For instance, if several steak houses are already in the market offering moderately priced meals, the operator might consider positioning the operation to fill a segment where the competition is less heavy, such as by opening a gourmet burger business. Although this type of fine-tuning may seem most appropriate to commercial foodservice operations, it is equally useful for all other types of foodservices.

Once the market and the positioning of the business in the customers' minds are established, the operator can begin to promote the business, both externally (through a marketing plan) and internally (through merchandising).

■

MERCHANDISING FOOD

If an operation is to succeed in pleasing customers and building sales, merchandising must occur. This is true not only for commercial operations that must attract customers, but also for institutions that serve a captive audience. Many operations fail to realize the importance of good merchandising and, as a result, have poor merchandising systems. The operation that does a little extra in merchandising often beats out the competition, even when the quality of food and service may be slightly inferior.

The areas of chief importance to merchandising are the training of employees, the planning of special promotions, the featuring of menu items, the creation of atmosphere, the development of good public relations, and the use of promotional material.

Training Employees

The term *merchandising* may be narrowly defined as the selling of merchandise, but in a broader sense it is the presentation of factors that entice people to buy. Many operations could sell more food if they used better merchandising techniques. Employees can be invaluable in getting people to buy. They usually have the first contact with the customer and consequently have the best opportunity to sell. Suggestive selling has been found to be especially effective in foodservices. The key to suggestive selling is to do the thinking for the customer by recommending à la carte items as the menu is presented.

The job of service personnel is a lot more than order taking.* Personnel should be trained to make suggestions to customers highlighting the high-gross-profit items on the menu or the items that should be moved. Salesmanship, not "pushmanship," is desired. Good salesmanship can help please a customer more than is generally realized. A smart service person often can sense what might please a customer, or a few well-directed questions or comments may give enough information to know what the customer would be most pleased with.

Because service is the aspect of a foodservice that is most often recognized by customers as the determining factor in whether or not to return, care should be taken in the selection, training, and retraining of service personnel. It is certainly a good idea for a foodservice operator to devise a strategic plan for merchandising products and services, but unless that plan is carefully carried out by the service personnel it is doomed to failure. A proper merchandising training program, therefore, must take into account not only the products and services offered, but also the behavior of the personnel delivering those items to the public.

Planning Special Promotions

Business in a foodservice may fluctuate, and low-volume periods may cause a major loss of profit. Building business at these times can take up some of the slack and help meet overhead and labor costs that exist whether or not there is business. Three methods are used to build foodservice business with special promotions; personalizing service to include special events in customers' lives; developing special promotions around holidays; or capitalizing on local events that lend themselves to special promotions.

Special Day Celebrations

All customers have specially important days in their lives—birthdays, anniversaries, graduation days, or visits from friends or relatives. Some operations offer patrons cards to fill out identifying such special days. These cards are then filed and used to remind the customers to come in to celebrate; some type of arrangement in keeping with the occasion is offered to make the proposal more attractive. This is all done through the mail, with a tickler file reminding the operator to notify the guest of the impending special event. Even if the guest does not come in, the reminder is good public relations.

Holiday Celebrations

Holiday promotions can be quite successful. Mother's Day, New Year's, Thanksgiving, and other holidays are special days on which many people want to eat out. Special menus can be printed and sent to customers or shown around the operation; they also can be put into newspapers or otherwise advertised. Suitable tags on service personnel, table tents, and verbal reminders are also effective. The event can be emphasized in advance by special decorations.

Local Happening Celebrations

Local events offer many opportunities for special promotions. A football game, a convention, or a

---

*Suggestive selling can also occur behind a cafeteria counter, in a takeout operation, or even in a hospital room. It need not be confined to a waiter or waitress in a commercial operation.

community project may provide the occasion. A "winning-team special" can be used to feature the first event, and a "convention caper dinner" can be used to feature the second. Even offering special snacks after visits to theaters or nearby museums can help bring people in.

Other Special Promotions

Special events are not needed for special promotions. For instance, as the baby-boom generation has grown older, a small boomlet of children of these baby boomers has developed. This provides a perfect market for a foodservice operator to promote. Because the baby boomers themselves have grown up dining out, they are used to going out to eat; but because they now have children, their dining habits have changed. Now the boomers are looking for foodservices that cater to the special needs of parents and young children. A special promotion aimed at children—such as one offering balloons, toys, or coloring books—can be very effective in capturing this market.

A foodservice operator should plan both routine and special events to promote the business internally through merchandising. Routine (or everyday) activities such as music or other entertainment have often been successfully used, as have the use of banquet or meeting-room facilities. Myriad possibilities for promotions can be discovered when planning a merchandising program. For information on activities that others have successfully used, an operator can contact such organizations as the National Restaurant Association or the American Hotel/Motel Association.

Featuring Special Menu Items

Operators at all levels of the foodservice industry feature food items in their merchandising programs. One secret of merchandising food is to develop several things that are especially distinctive and make the operation stand out. These may include certain foods, special service, or special uniforms. Cocktails served in a distinctive fashion or with a special flair can be enough to compel attention and bring business back. Special food combinations offered at a good price during low peak times can help. Women shoppers

may be enticed to come in for a special combination of small pieces of cake and tea. Other foods that can be emphasized on the menu are hors d'oeuvres, salads, and desserts.

Hors d'Oeuvres

Hors d'oeuvres can attract customers and raise check averages (see Fig. 9-1). Instead of simply leaving a patron to eat a hard roll while waiting for the soup to be served, the operation may present a lazy Susan jam-packed with hors d'oeuvres for sale. The popularity of hors d'oeuvres is especially high if a cocktail is offered with it. The price can be right for one or both. Even though the profit may not be as great as with some other items, the buildup of the check average can help greatly.

Salads

Salads have become a major product for many foodservices. This is especially true because of changes in American dining and health habits that emphasize more selective and healthful dining. For many diners today, the salad has become not just an addition to the meal, but the meal itself. In order to capture this market, foodservices have developed merchandising programs that emphasize salads. In fine dining, this may mean placing salads on a cart and wheeling it around the dining room in colorful display for guests to see,

FIGURE 9-1 Attractive hors d'oeuvres are a profitable addition to any foodservice operation.

or carrying salads on a tray through the dining area. At other times, the salads are prepared at tableside with flair, to capture the attention of diners.

In other types of foodservices, salads are promoted through the use of salad bars that offer a wide range of products from which customers can make their own special salads (see Fig. 9-2). The degree of acceptance of such promotional tools can be seen in the regularity with which salad bars are offered today. Salad bars can now be found in fastfood establishments, in institutional settings, and even on jumbo airliners. Whether the salad is served at the table or through a bar, garnishes can give it color and texture. Pickles, anchovies, parsley, and many other items are offered for this reason.

Desserts

Desserts can be high-profit items, but many people today avoid them, thinking of the calories. It is possible, however, to plan desserts that contain few calories but still satisfy the almost universal desire for something sweet, delicate, and pleasing to eat at the end of a meal. A group of different desserts can be placed on a cart and rolled to guests' tables for them to see and select from. The trays on which desserts are placed should be carried so that other patrons see them. A few special desserts can become so well known that customers come to the foodservice especially for them.

FIGURE 9-2 Salad bars promote customer satisfaction, as the diner is allowed to select the food desired. (Courtesy of Chef Jeremy Corson.)

Creating Atmosphere

Service and atmosphere are important aids in selling a foodservice. Atmosphere is important in giving guests total satisfaction in their dining. Both external and internal atmospheres establish in patrons' minds an opinion about the operation that can strongly influence whether they will be pleased by the experience of eating there or not.

External

The external atmosphere of an operation is affected by the neighborhood in which the operation is located, as well as by the appearance of the building and grounds. In some foodservices, the interior approach to the dining area is treated as part of the external atmosphere. In a nursing home or institution, the interior approach might include the entrance to the building, the halls, the elevators, and other areas through which patrons must pass to get to the dining area.

Cleanliness and neatness are two very important factors in creating a good external atmosphere. There should be no debris or clutter in the surrounding yard. Drive-ins should be particularly careful to see that paper and other debris are quickly picked up. Waste receptacles where paper goods and other materials can be deposited should be located conveniently and in plain view. These receptacles should be neat and have a good cover. Parking lots should be kept orderly and should be plainly marked as to entrances, exits, and areas for parking. The building should be nicely landscaped and properly maintained. Colors should be inviting. The sign should be informative but also should carry the logo or emblem by which the operation is identified. Signs of franchise operations may remind guests of pleasant visits to other operations in that chain. If an operation is located in a poor neighborhood through which people must pass to get to it, it is extremely important for management to make the place stand out and give the external area a neat and inviting appearance. A good image is extremely important, and the contrast to close surroundings can help create such an image.

Internal

The internal atmosphere of a foodservice is created in the area in which patrons dine or obtain their food.

Because customers' eating pleasure is affected by their immediate surroundings, the operator who creates a pleasant, cheerful, and imaginative internal atmosphere will do much to satisfy customers.

The first step in creating an attractive internal atmosphere is to be sure that the customer is comfortable. The proper temperature should be maintained, with no cold drafts or hot air striking people. Sunlight should not create glare. Lighting should enable people to see well but should not be too bright. Contrasts of light in the area can give a pleasing effect. Restrooms should be readily available and immaculate.

The entire area should be clean, neat, and well-maintained. Furnishings should be uniform in style and should contribute to the overall décor. Tablecloths, if used, should hang neatly and evenly from the tables. If tablecloths are not used, place mats should be neatly set. Table settings should sparkle, and the entire place should convey a feeling of good organization and cleanliness.

A second step in achieving good internal atmosphere is to create a desirable mood for the operation. Some places want to appear busy and yet efficient. Bright colors, a simple décor, and plenty of light can help do this. Cheery colors and lots of light should characterize operations geared toward families. Soft colors and dim lights help create a suitable mood for luxury operations. Low background music of good quality can be used to enhance a mood. Atmosphere can be further developed by the décor. Colors should follow through from walls to furnishings, draperies, and even tabletops.

Creating the right mood can do much to help win customer approval. Eating pleasure is greatly affected by how a customer feels. School lunchrooms, nursing homes, hospitals, and other institutions can raise the morale of patrons by providing a bright, cheery atmosphere. Other aspects—such as employee's uniforms, hairstyles, and attitudes—contribute to an overall mood effect.

Developing Good Public Relations

Public relations is defined as the total effort of an operator to win and keep the goodwill of customers and the community. By maintaining goodwill, the operator encourages people to view the operation favorably.

A community wants to feel that a foodservice is a part of it and is working toward its general welfare. By doing what it can to promote the good of the community, a foodservice can build tremendous goodwill and can take a position as an important member of the community family. Management should do what it can to promote community affairs and should join local groups such as the Kiwanis, Elks, or Rotary. Belonging to church groups and taking part in PTA and other activities also mark the operator as belonging. Supporting charitable drives by utilizing posters in the operation or even putting clip-ons on menus is good public relations. If the operator is in a tourist area, the business may distribute maps and other materials that are provided by the local chamber of commerce. Getting involved with the local welcome wagon agency, which offers free gifts from participating businesses to new community residents, is also a good idea. Foodservices from time to time have good-quality food that they cannot use. Giving this to charitable organizations builds goodwill. Helping with food donations at Thanksgiving and Christmas also helps. Some operations sponsor Little League baseball teams or adult bowling teams.

Having good community relations is furthered by promotional news stories in the press and other media. If an operation can work up some good material of human and community interest, the media will use it. Matters of potential media interest include the following:

- Opening a new restaurant
- Complete remodeling or expansion
- Announcing a new appointment at the executive level, the retirement of an old employee, or honors awarded to one
- Inviting a celebrity to visit, with an acceptance received
- Holding a special affair or meeting at the operation
- Offering new services, such as catering or a take-out service

A news story should cover who, what, where, when, and why. The important facts should come first, followed by less important details. In this way, a story that is too long can be shortened without destroying the basic story. By keeping in touch with television stations, radio stations, and other communications media, the operation increases its chances of getting a news story about it aired.

Every guest who enters the operation provides it with an opportunity for promotion. One of the first aspects of promotion is to be sure that the guest enjoys the experience of eating there. Word-of-mouth advertising has been said to be the most effective form of promotion a foodservice can use. Suggestions can be solicited from guests on how better to serve them. Offering to help with social affairs—such as preparing special cakes, dishes, or other items they might need, or sending one's chef or head baker to give a demonstration at a woman's club meeting—also helps let guests know that their business is appreciated. Pleasant, efficient service and a cordial thank you and goodbye leave guests with a final impression that the operation appreciated their visit.

Using Promotional Materials

Publicity can play an important role in merchandising, and advertising can play a part in publicity. For some time McDonald's would not advertise, depending on word-of-mouth, and building customer volume by offering a good product at a reasonable price. With growing competition, however, McDonald's turned to advertising and found it an effective aid. Newspaper and magazine articles, direct mailings, and radio and television spots are effective with some markets. A recent study, however, showed that less than 1 percent of gross income is spent by the industry on advertising and that a substantial part of this did not achieve the results it should have because it was directed to the wrong market. Advertising must be carefully thought out and directed toward specific market to achieve its objectives. Many things also can be done within an operation to advertise, such as posting information materials on backbars, in windows, or around the operation, using table tents and menu clip-ons, or even arranging for personnel whom customers see to wear badges.

Backbar, Window, and Other Signs

Informative materials on backbars, in windows, or around the operation can be quite effective in announcing things management wants customers to know.

Coming promotional events such as special dinners on Mother's or Father's Day or a special bakery promotion can be broadcast by such signs. Showing a special dish of good gross profit in a colored photograph can improve sales for that item. In some operations, mobiles or small decorative signs hung by dark (almost invisible) thread can be used to advertise. Operators must be careful, however, not to clutter their dining area by using signs. A place's appearance is cheapened by having too many signs, too gaudy signs, or signs that are inappropriate to the décor of the operation.

Table Tents and Menu Clips

Table tents and menu clip-ons can be effective merchandisers, doing much what the signs previously discussed would do. Special luncheon or dinner features, the presence of a room kept open for late supper service, a catering service, or a cocktail lounge feature can be announced through table tents. A menu clip-on is a little printed or illustrated card that is affixed to the menu. It should not cover up other portions of the menu or clutter the menu's appearance.

Waiter and Waitress Badges

Waiter and waitress badges can be an effective means of attracting attention. They can carry the same types of messages as signs or clip-ons might. In one operation, a badge bearing the letters *AFFFT* was worn by waitresses. Whenever a curious customer asked what the letters meant, the answer was "Anyone for french fries today?" As with all promotional materials, management must be creative in devising badges and similar items.

Computers and Foodservice Merchandising

Today, many foodservices use in-house computers that allow them to undertake many merchandising efforts that they formerly could not attempt. One example is that of the menu itself. Since the menu, if pleasantly created and designed, can act as a promotional tool for the foodservice, care should be taken in its design. Many foodservices have discovered that this can be accomplished by using a computer in the store. The commercial art packages available for computer-assisted design, for instance, can give the operator the ability to create menus immediately, according to need. These same programs can create table tents or other artwork for internal marketing.

Using various desktop publishing software on computers, operators can also generate an in-house newsletter or newspaper to promote special events or foods. This same software can also be used to promote a newsletter for the staff, and this is now being done in many larger foodservices to raise employee morale.

Computers can be used to merchandise a foodservice in many other ways as well. One such method might be the computerization of regular customers' preferences for table, food items, beverage, and so on. By using a computer in this way, a foodservice could print out relevant information, based on daily reservations, for the servers to use. One operation that does this has found that it not only enables the operation to anticipate its needs better, but also gives the customer the feeling of receiving personalized attention.

In foodservices, the computer has been employed as a merchandising tool to assist in the ordering process. By speeding up ordering in such foodservices as cafeterias and employee dining facilities, the computer ensures that the customer is accorded better service. This may involve either staff- or customer-activated computer ordering. Interactive video (the combination of computers with videodisk players) has also recently been utilized to enhance customer satisfaction through merchandising. In such a case, a customer touches a screen to activate certain commands from the computer-controlled videodisk. The screen might offer information on food products, local events around town, or other information that could serve to merchandise the operation.

■
SUMMARY

Energy costs can represent a substantial portion of overall foodservice expenditures. The manager must be well versed in property energy and maintenance costs as well as in procedures that can be used to reduce these.

The design of a foodservice facility can largely determine its viability. Both labor and food costs can be affected by the foodservice facility's design. The design should follow a functional pattern, especially in the kitchen area. The type and purpose of the operation will point toward making this design either straight-line or functional flow, depending on the type of operation.

The space devoted to specific aspects of the foodservice is important to the future success of the foodservice. A heuristic approach suggests how foodservice space should be divided in the kitchen and the public areas to achieve maximum efficiency and comfort.

Among the more important environmental decisions facing a foodservice operation are the choice of the type of HVAC system, the location of air ducts, and the manner in which kitchen fumes are exhausted.

Proper maintenance of a foodservice facility must be done on a preventive rather than corrective basis. This approach allows maintenance personnel to identify and correct possible problems with the plant/property before the operation suffers any significant inconvenience. Developing a preventive maintenance checklist is the first step in establishing such a system.

The external environment of a foodservice—including such elements as the outside appearance of the building, the garbage-holding areas, the parking lot, lighting, signage, and security—can significantly affect the viability of the operation.

Property construction may vary a great deal, depending on the foodservice's type and function. Architects, designers, contractors, and other specialists should be employed in the analysis of such decisions. The site and materials to be used in construction help determine the costs and needs of the project as well. During construction, a schedule should be set up that clearly identifies when different parts of the construction must be performed.

The foodservice industry is ever-changing and ever-growing. From new markets offering food for immediate consumption, such as convenience and grocery stores through delivery and takeout, to booming market segments, such as the cruise and gaming industries, catering businesses, prisons, and business and industry contracts, foodservice appears to have unlimited potential for the future.

Both marketing and merchandising are forms of promotion for a foodservice. Marketing usually deals with external activities of the foodservice, while merchandising more often concentrates on internal promotion. Development of a marketing program is based on accurate descriptions of the market itself (which can be created from a market survey) and of customers' likes and dislikes. Once this information is compiled, a foodservice operator should plan the operation's position in the market so as to take advantage of the market's likes and dislikes in the best possible manner.

Merchandising should use as many factors as possible to entice people to purchase food. The following six activities are especially important in merchandising: training employees, planning special promotions, featuring special menu items, creating atmosphere, developing good public relations, and using promotional materials.

Personnel who come in contact with customers must be given special training if they are to merchandise well for the operation. Suggestive selling should occur to increase check averages. Customers are often far more pleased with their food and service if service personnel take the time to find out their needs and see that they are met.

Planning special promotions—such as giving special attention to guests' birthdays, anniversaries, family parties, and other special occasions, featuring special foods and meals for holidays, or building around athletic or other community events—can do much to merchandise.

Special menu items can be promoted by display, by promotional efforts of personnel, or by advertising. Three foods that lend themselves to such promotions are hors d'oeuvres, salads, and desserts, although other foods can also be used for this purpose. An operation can differentiate itself from others and can draw customers by having just a few special foods on the menu.

The proper atmosphere must be created by developing a good external appearance. Neatness, organization that facilitates entry, and special décor can be used to develop a good external atmosphere. Colors, signs, landscaping, good maintenance, and a host of other factors can also contribute to this. An attractive internal atmosphere is created by combing good décor, good colors, good lighting, neatness, and organization. Establishing a logo and following it through can help to set up an image that customers recognize and associate with the operation.

Good public relations can be developed by giving special attention to guests (such as through direct mailings), by helping guests in their special events, and by offering other personalized attentions. Community relations should also be developed to elicit the attention and goodwill of people in the community toward the operation.

Promotional materials inside an operation can take the form of signs, mobiles, clip-ons, table tents, waiter or waitress badges, and other items. They must be used with care and taste, however, to avoid giving the operation a cluttered or overly commercial appearance.

The use of computers in foodservices has greatly enhanced the ability of operations to cater to their customers' needs. This is done through creation of menus, table tents, and other merchandising displays, through newsletters, or through computer-assisted ordering systems.

■

CHAPTER REVIEW QUESTIONS

1. Select a kitchen of your choice within a foodservice operation and note how space is allocated to the functions of receiving, storing, and issuing as compared with space available for equipment. Look at the space allocated in the dining room. What recommendations would you make to use the space more efficiently? Use the tables for space allocation as they are presented in this book in your answer.

2. Discuss merchandising and the factors that can be used to promote business. Differentiate among the six factors discussed in this chapter.

3. How does an operation's atmosphere merchandise food? What is meant by external and internal factors in atmosphere? Find pictures of outside and inside areas of operations and indicate good and bad features as far as external and internal factors in atmosphere are concerned.

■

CASE STUDIES

1. Smokey's Restaurant is located in an old historic mansion near the downtown market in Charlotte, North Carolina. Smokey's specializes in sautéed items, most of them cooked personally by the owner, Smokey Robertson; consequently, the restaurant uses a lot of open gas burners in the kitchen. The restaurant has been open about fifteen years, and Smokey and his family have always made a good living from the business. Over the past four or five years, however, Smokey's profits have dropped by about 30 percent annually, despite the fact that sales have held their own.

   The drop in profits is due to the increased costs of energy and maintenance in the old building, which was built in 1846. Originally, Smokey compensated for the increased costs by raising prices. After three price increases, however, Smokey feels that he can no longer pass the costs on to his customers without doing serious harm to his business. Already, while his sales have remained about the same, his customer counts have dropped following the last two price increases; Smokey knows that if he increases prices again, he risks alienating his customer base further. Instead, he must begin to address the problem of high energy costs directly.

   The building is made principally of brick and stone. The roof was newly installed only three years ago. The parking lot represents a continual drain on funds because the gravel used for it tends to wash away with heavy rains, so that Smokey has to regravel the lot at least once (and sometimes twice) per year. Smokey has avoided paving the lot because of the cost involved, rationalizing that a few dollars spent on gravel now beats the major cost of paving. The floors in the building are old oak plank and have formed a major part of the décor since Smokey opened for business. Floors in the kitchen are the same. For part of the year, Smokey can use the outdoor patio attached to the rear dining room for additional seating, and during those months his volume increases dramatically. He would like to figure out a way to

use it yearround without destroying the ambiance of the setting.

   The restaurant seats 100 during cold weather and 145 in warm months because of the patio. It is a full-service operation with a small bar. There are lots of windows in the facility, as you might imagine in an old house. Service is offered on both the first and second floors

   Smokey feels that he does not know how to go about renovating for energy savings. Other than putting more insulation in the walls and maybe the roof (things he has read about in the newspaper), he simply does not know what to do. His banker (a regular customer) has promised to lend Smokey the money required to complete the upgrade if sound reasons for changes can be offered to the bank's loan committee

   Outline a program of maintenance and renovation that would help reduce and control Smokey's energy expenses. Include an equipment maintenance program, based on what you assume Smokey would be using in the kitchen given the menu items mentioned. In describing the program, justify the expense of both loan money and labor hours.

2. The Best Food in Town restaurant opened about six months ago. After an original flurry of activity, sales in the restaurant have consistently declined to a point where it is no longer feasible to operate. Because the owners are certain that their products are as good as or better than any others in town, they are confused as to how and why this happened

   The restaurant is located in a converted service station at a busy intersection on the edge of town. The location seems good, especially because of the high level of visibility inherent in the location. The menu is primarily that of a steak house and closely resembles that of another restaurant in the area that has been in existence for sixteen years. In fact, Billy and Eric Idel, the two brothers who own "Best Food," both used to work at the other restaurant as cooks, so they are certain that their food is at least comparable. But while the other restaurant continues to flourish, Best Food is going downhill rapidly.

In an effort to stem the downward slide, the brothers have increased the size of the steaks they offer by 1 to 2 oz each. Although this was a risky proposition because of the damaging effect on food costs, the brothers thought it a wise move in order to attract more business. The increase has had little effect, however, as few customers even seem to know about it.

To build their business, Billy and Eric must devise a way of distinguishing themselves from their competition and of promoting their restaurant as something different. They have asked you to devise a merchandising plan to promote their restaurant. In order to do so, you must first answer the following questions for the owners. What are the first steps to take toward the goal of creating a desirable image in the customers' minds? How should the brothers merchandise their products to emphasize the increased portions offered? Suggest specific promotional programs for the restaurant.

3. As the new engineer on staff in a very large food-service operation, you are asked to research preventive maintenance for the HVAC systems. You are referred to *www.afgo.com* and are to report what their services are. What could the people at this Internet site offer you?

4. As a leader in foodservice, you have recently surfed the Internet and found *www.tubeworks.com* and are very excited. What products would you recommend foodservices use and why?

5. Many new ways of merchandising foodservice operations have surfaced on the Internet. One such way is seen at *www.entreeonline.com*, which features many different restaurants. You are asked to develop a similar Web site but are asked to make improvements. What would those improvements be, and why would you recommend them?

■

VOCABULARY

Heuristic
Straight-line design
Functional flow design
Four-tops
Waste compaction
Waste incineration
Waste pulping
Curb appeal
HVAC
Kitchen exhaust
Preventive maintenance
External environment
Curb cuts
Marketing
Merchandising
Hors d'oeuvres
Table tents
Focused groups

Suggestive selling
Codes
Dead-bolt lock
Internal security system
External security system
Halon gas
CO_2
Schematic design phase
Design development phase
Construction documents
Site analysis
Slope
Construction schedule
Computer-assisted artwork
Menu clip-ons
Positioning
Market research
Home-meal replacement

■

ANNOTATED REFERENCES

Feltenstein, T. 1992. *Foodservice Marketing for the 90s*. New York: John Wiley & Sons, Inc. (All aspects of merchandising and marketing are covered in this book.)

Kaplan, M. 1997. *Theme Restaurants*. Glen Cove, N.Y.: PBC International, Inc. (Merchandising food and creating themes is the basis for this text.)

Katz, J. B. 1997. *Restaurant Planning, Design, and Construction*. New York: John Wiley & Sons, Inc. (From concept to construction, this book details the development of a restaurant from beginning to end.)

Kazarian, E. A. 1989. *Foodservice Facilities Planning*. New York: John Wiley & Sons, Inc. (Facilities planning and management are the topics discussed in this text.)

Kotler, P., J. Bowen, and J. Makens. 1999. *Marketing for Hospitality and Tourism*, 2nd ed. Upper Saddle River, N.J.: Prentice Hall. (The subject of marketing for the entire industry is overviewed, with market areas for restaurants discussed on pp. 392–393 and restaurant sites on pp. 478–481.)

Kotschevar, L. H., B. Almanza, and M. Terrell. 1999. *Foodservice Planning: Design, Layout and Equipment*, 3rd ed. New York: John Wiley & Sons, Inc. (Planning and designing a foodservice operation and equipment implementation are presented in this book.)

Lefever, M. M. 1989. *Restaurant Realty: A Manager's Guide*. New York: John Wiley & Sons, Inc. (From planning and opening a restaurant to daily operations, many aspects of restaurant marketing and management are presented in this text.)

Mill, R. C. 1998. *Restaurant Management: Customers, Operations, and Employees*. Upper Saddle River, N.J.: Prentice Hall, pp. 48–100. (Developing a marketing plan is presented in one chapter, promoting the operation in another.)

Newland, L. E. 1997. *Hotel Protection Management*. Spokane, Wash.: TNZ Publishers, Inc. (Guest protection and reasonable care are considered in this book.)

Pegler, M. M. 1989. *Successful Food Merchandising and Display*. New York: Retail Reporting Corporation, pp. 130–189. (Chapters on food display for fast foods and food courts are considered.)

Reich, A. Z. 1997. *Marketing Management for the Hospitality Industry*. New York: John Wiley & Sons, Inc. (Marketing strategies are considered in this book.)

PART II

Quantity Food Production and Management

Cooking Principles, Methods, and Trends

◼

LEARNING OBJECTIVES

By the end of this chapter, the reader will:

1. Understand cooking principles as they relate to heat transfer, time and temperature, and energy.

2. Be able to distinguish between the various methods of cooking both with and without liquid.

3. Learn about the various heat sources for cooking and innovations that are taking place in cooking.

4. Recognize the trends in cuisine around the world.

■

SELECTED WEB SITE REFERENCES

Cater Ease
www.caterease.com

CuisineNet Cafe
www.cuisinenet.com/café/index.html

Electronic Gourmet Guide
www.foodwine.com

Food Network: Cyber Kitchen
www.foodtv.com

Foodservice and Hospitality
www.foodservice.ca

Internet Food Channel
www.foodchannel.com

Cooking is basically the application of heat to food to make it more appetizing, digestible, palatable, and safe to eat. Heat has been applied to food for centuries, but only recently have new methods and equipment been developed to aid cooks in producing quality food in a short period of time.

To prepare food properly, a person must have knowledge of basic cooking principles and methods. Food reacts or changes when heat is applied to it. Thus, in cooking, a roast not only changes color but changes texture and flavor as well. A rare steak is best when it is still fairly red inside but no longer has the soft, jelly-like consistency of raw meat. A cake is the product of a delicate balance of the gelatinization of starch, the coagulation of protein, and the action of a leavening agent. A soup depends on the solubility and flavor development of the ingredients used. The ability to produce a wide variety of fine food takes years of training and learning.

■

COOKING PRINCIPLES

Cooking is a science, as are chemistry, microbiology, physics, and geology. Of these, cooking is most closely related to chemistry and physics. The properties of many ingredients used in cooking resemble those of chemical substances, since they react under certain conditions. For example, baking powder and baking soda react to give off carbon dioxide gas. Mayonnaise is an emulsion made by using egg to bring tiny globules of oil into a state of suspension. Heat causes many reactions to occur and is the focus of the discussion in this chapter.

Heat

Heat is measured by a unit called the *calorie*, defined as the amount of heat required to raise the temperature of 1 gram of water by 1 Celsius degree. It is a one-thousandth part of the Calorie [or kilocalorie (kcal)], which is the amount of heat required to raise the temperature of 1 kg of water by 1 Celsius degree. The present system in the United States uses the Btu (British thermal unit) to measure heat. This is the amount of heat needed to raise the temperature of 1 lb of water by 1 Fahrenheit degree. About 4 Btu equals 1 calorie, since 1 Btu equals 0.252 calorie; 1 Calorie or kcal equals 3968 Btu.

The Celsius method of measuring heat uses a scale based on the range between the temperature at which water freezes ($0°C$) and the temperature at which water boils at one atmosphere of pressure ($100°C$). The distance between freezing and boiling is then divided into 100 equal units—the Celsius degree. In contrast to this, the Fahrenheit scale identifies $32°F$ as the freezing point of water and $212°F$ as the boiling point of water. Thus there are 180 Fahrenheit degrees between boiling and freezing ($212 - 32 = 180$). This means that a Fahrenheit degree is 100/180 of a Celsius degree, and that a Celsius degree is 180/100 of a Fahrenheit degree. These work out to the fractions $\frac{5}{9}$ (0.55) and $\frac{9}{5}$ (1.80), respectively.

To convert temperature readings from Fahrenheit to Celsius, subtract 32 from the Fahrenheit reading and multiply the remainder by $\frac{5}{9}$ or 0.55. For example, 50°F equals 10°C (50 − 32 = 18 × $\frac{5}{9}$ = 10). To convert temperature readings from Celsius to Fahrenheit, multiply the Celsius value by $\frac{9}{5}$ or 1.80 and add 32. For example, 40°C equals 104°F ($\frac{9}{5}$ × 40 = 72 + 32 = 104).

When we speak of Btu or calories, we are talking about a quantity of heat. When we say that it takes 30,000 Btu (about 7500 calories) for an electric range to bring a stockpot containing 10 gallons of soup from 50°F (10°C) to boiling (212°F), we are indicating the quantity of heat needed to do this job. Only about 13,500 Btu actually performs the heating; the rest of the heat is lost from the range top. A kilowatt of electricity produces 3412 Btu. Thus, it takes nearly 10 kilowatts or 34,120 Btu of energy to do the job.

Comparing Fuel Heat Values

A gas range is about as efficient as an electric range, if the usable heat of the gas range is considered. The gas range must produce more actual heat than the electric range to do the same job, however, because it must exhaust some heat with its combustion gases. Electricity is said to be 100 percent efficient, while gas is said to have an efficiency of from 75 to 80 percent.

Competitive pricing makes the cost of electricity and gas about the same in many localities. A kilowatt-hour of electricity often sells for about 4.5 times more than a cubic foot of gas, if prices are competitive. For example, if electricity sells for $0.03 per kilowatt-hour, natural gas must sell for around $0.007 per cubic foot to be competitive ($0.03 ÷ 4.5 = $0.00694).

Suppose that a kitchen needed 20,000,000 Btu in one period. If electricity sold for $0.03 per kilowatt-hour and gas for $6.94 per therm (1 therm equals 100 cubic feet of gas or 1,052,000 Btu), what is the cost of electricity and of natural gas to furnish 20,000,000 Btu? For electricity the cost is

$$\frac{20,000,000 \text{ Btu}}{100\% \text{ efficiency} \times 3412 \text{ Btu/kilowatt-hour} \times \$0.03} = \$175.84$$

and for gas the cost is

$$\frac{20,000,000 \text{ Btu}}{75\% \text{ efficiency} \times 1,052,000 \text{ Btu/therm} \times \$6.94} = \$175.07$$

In terms of usable heat, steam equipment has a slightly higher efficiency than gas-fueled or electrical equipment. As an energy source, steam has an efficiency of 100 percent, compared to electricity at 100 percent and natural gas at 75 to 80 percent. Steam is sold on the basis of 1000-lb lots; thus, 1000 lb of steam must sell for nearly the same price as a therm of gas to be competitive with it, and it must cost approximately 290 times more than a kilowatt-hour of electricity.

Sometimes steam is rated on the basis of horsepower. For instance, a kitchen's demand might be said to be 5 horsepower per hour. A horsepower of steam equals 34,000 Btu, so the requirement in this example would be 5 × 34,000 = 170,000 Btu.

As we move into the future and energy becomes more scarce and costly, we will have to become more knowledgeable about the energy efficiency of various kinds of equipment—as well as about the efficiency of the various kinds of fuel available to us—so that we can compare their costs and determine which provides the maximum yield for the energy and money expended.

Heat Transfer

Heat moves by conduction, by convection, or by radiation. In cooking, all three means are used, although the first is perhaps the most commonly used. Heat flows from relatively warm portions or areas to relatively cool ones. When two substances reach the same temperature, there is no flow of heat between them.

Conduction. Conduction is the movement of heat through material by passing from one molecule of matter to the next, spreading eventually through the whole substance. An electrical element buried in an iron griddle top becomes hot; and since the element is touching the iron, the heat flows to it and then spreads throughout the griddle. If a match is lit and held at the end of a metal rod, the heat will flow along the metal until it is felt at the other end of the rod. In cooking, we transfer heat through a burner into the pot, from there into the water or fat, and from there into the food. Thus, several layers of conducting media are involved. Copper and aluminum are very good conductors of heat, and iron is not bad. Because stainless steel is a relatively poor conductor, stainless steel frying pans and other equipment made of this metal usually have copper-clad bottoms; the copper readily picks up the heat, and then transfers it evenly through the pan. In other cases, a top and bottom layer of stainless steel is used to make the utensil, and a thin layer of iron, copper, or aluminum is placed between them to spread the heat taken in by the bottom layer of stainless steel, spreading the heat evenly into the top layer that touches the food.

Convection. Convection is the movement of heat in liquid or gas. Hot air expands, making it lighter than cold air and causing it to rise. Cold air descends, where it in turn becomes heated. The result in an enclosed space is a circular motion of air picking up heat, carrying it upward, cooling, and then sinking back down to get more heat. In the convection oven, heat is moved by a fan that distributes or convects the hot air throughout the oven. A convection oven works more efficiently than a regular oven, which depends on the natural convection of hot air. Natural convection does not spread the heat out as well. Convection can also occur with hot liquids such as water or fat. A refrigerator works because a liquid (the refrigerant) picks up heat from inside the refrigerated chamber and moves it away.

Radiation. Radiation is the transfer of heat by wave energy. Radiated heat travels about 186,000 miles per second, so it takes the sun's heat about 8.3 minutes to reach the earth from the moment it is radiated. Radiated heat is tiny particles of energy that flow from a hot body to a cooler one. One person who is warmer than another in a room will radiate heat to the other person.

Broilers and toasters cook largely by radiation. To get the highest amount of radiation from the heating element, the level of white heat should be reached. An object glowing with red heat gives off some radiation in the infrared wave-energy range, but an object glowing with white heat gives off considerably more radiant energy. This is why some broilers contain ceramic particles that rapidly reach white heat when heated up—a temperature about 6000°F (3316°C).

A dark object absorbs more heat than a lighter one does. This is why aluminum pans containing frozen foods are colored black on the outside. Because a shiny object reflects radiated heat away, heat reflectors are used in ovens to bounce radiant energy back at the food being cooked. By the same token, a shiny, bright aluminum pan will not bake a pie as well as a duller pan will. If we want to wrap something in a way that helps hold the heat in, we put the shiny aluminum side of the wrapper inside, so that it bounces the heat back. Conversely, if we want to keep heat from getting in, we put the shiny aluminum side of the wrapper on the outside.

Time and Temperature

Time and temperature must be controlled to get the desired cooking results. If a food is cooked at a lower temperature, the cooking time must be extended. Raising the temperature may shorten the cooking time. Through the manipulation of the relationship between temperature and time, food can be cooked to the proper doneness. Too high a temperature cooks a cake on the outside before the leavening gas has fully developed inside. The cake eventually cracks on the top because the gases expanding inside, after the top has become firm, break through all at once. A steak can be overcooked on the outside and undercooked on the inside because of too much heat. On the other hand, many foods come out soft and pasty when insufficient heat is used to cook them.

Large pieces of meat are best cooked for longer periods of time at lower temperatures. This results in better flavor, more juiciness, and a better texture. Smaller pieces of meat should be cooked at slightly higher temperatures, because the longer cooking times associated with lower temperatures can lead to dried-out meat. Tough cuts of meat and vegetables are best cooked for longer periods of time, in order to tenderize or soften them. Extended, low-temperature cooking dissolves the connective tissues in meat, while high heat can make meat less tender by causing excessive shrinkage and a toughening of some of the protein.

Overcooking can destroy nutrients, as can too much heat. Steam destroys nutrients faster than cooking in water because it is higher in temperature; however, since steam does not leach away or dissolve as many of the nutrients as cooking in water does, the overall nutrient loss is usually less. Temperatures must be adjusted to the particular food being cooked (see Fig. 10-1). A potato baked at 350°F (177°C) is soggy and has a poorer flavor than one baked at 425°F (218°C). Some people prefer a fairly high heat for broiling a steak, since this improves the outside flavor as a result of caramelization of some of the substances.

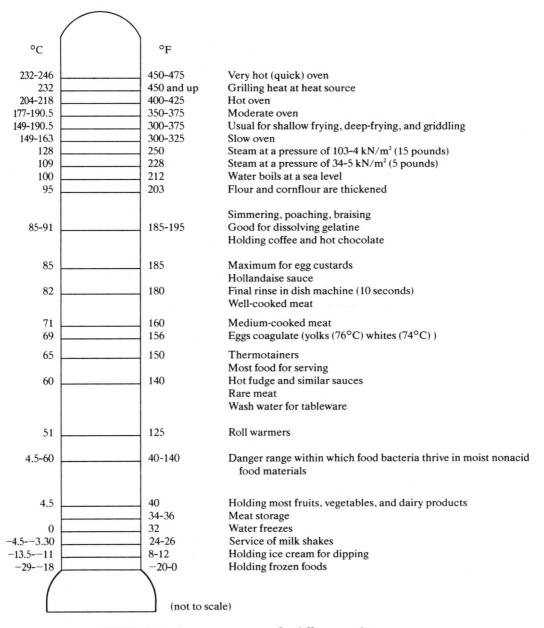

°C	°F	
232-246	450-475	Very hot (quick) oven
232	450 and up	Grilling heat at heat source
204-218	400-425	Hot oven
177-190.5	350-375	Moderate oven
149-190.5	300-375	Usual for shallow frying, deep-frying, and griddling
149-163	300-325	Slow oven
128	250	Steam at a pressure of 103–4 kN/m² (15 pounds)
109	228	Steam at a pressure of 34–5 kN/m² (5 pounds)
100	212	Water boils at a sea level
95	203	Flour and cornflour are thickened
		Simmering, poaching, braising
85-91	185-195	Good for dissolving gelatine
		Holding coffee and hot chocolate
85	185	Maximum for egg custards
		Hollandaise sauce
82	180	Final rinse in dish machine (10 seconds)
		Well-cooked meat
71	160	Medium-cooked meat
69	156	Eggs coagulate (yolks (76°C) whites (74°C))
65	150	Thermotainers
		Most food for serving
60	140	Hot fudge and similar sauces
		Rare meat
		Wash water for tableware
51	125	Roll warmers
4.5-60	40-140	Danger range within which food bacteria thrive in moist nonacid food materials
4.5	40	Holding most fruits, vegetables, and dairy products
	34-36	Meat storage
0	32	Water freezes
−4.5--3.30	24-26	Service of milk shakes
−13.5--11	8-12	Holding ice cream for dipping
−29--18	−20-0	Holding frozen foods

(not to scale)

FIGURE 10-1 Oven temperatures for different cooking purposes.

Energy

Somewhere in the future, we are not going to have the supply glut of energy we have today. Coal and nuclear fission seem to be the best present answers to our future fuel needs. We have enough coal to last for over 500 years and enough uranium to last for about 200 years. Thorium, plutonium, and other heavy molecular substances can also be used to produce energy. Hydrogen fusion (two atoms of hydrogen to make one of helium) may be our best future supplier of nuclear energy. Solar energy, the manufacture of gas and other combustible products from organic materials, the extraction of oil from shale or tar sands, the use of geothermal energies, wind energy, tidal energy, and energy from the heat of the seas are all possibilities. All will undoubtedly contribute to our energy needs in the future. While we will probably not run out of energy, the energies we use in the future will differ from those we used in the past. Energy is also going to be much more expensive.

We have been an energy-wasteful nation because we have had so much of it, and because energy was so cheap. Often, it was less expensive to use energy than to add insulation or build better buildings. We could over-specify needs in lighting, in heating, and in cooling because the cost of doing so was not great and the additional benefits were desirable. Now, this must stop. In addition to finding new sources of energy to supply our ever-increasing needs, we have to start saving energy and obtaining maximum yields from energy use. We can no longer afford to waste energy. We will have to learn to make do with less, because the supply will be limited.

Programs to save energy have been established by the National Restaurant Association, the American Hotel & Motel Association, and others. Many food-services, hotels, and motels set up energy-saving programs. It is not unusual to achieve a 25 percent savings without too much effort. Both employer and employee must bear some responsibility for reducing energy use. The program for practicing this joint responsibility will have to be carefully planned and implemented, if energy is to be saved.

Responsibility of Management

Management must originate, plan, and implement energy-saving programs. To do so, it must have an overall grasp of what is involved and what is possible. No program will be successful if it does not pay off within a reasonable amount of time, and management should focus first on conservation steps that give the biggest payoff.

Most of the energy used in a foodservice is expended on heating and air-conditioning the facility, cooking food, and heating water. Much less energy is used for lighting, refrigeration, and other purposes. Producing savings in areas where large amounts of energy are used is therefore going to pay off more than producing proportionate savings in areas where small amounts of energy are used. Nonetheless, the small areas of use should not be omitted from attack. For example, reducing the number of lights or changing their wattage is a simple energy-saving step; and in any program of attack, the ease with which a program can be implemented should be a significant factor to consider.

Management should take time to study the problem and develop an overview of the whole establishment's use of energy. In some cases, an energy reduction in one area would simply require an energy increase of the same size in another. The advice of the utility companies serving the operation should be sought, since they are knowledgeable about how energy balance can be used to achieve savings. Management should also consult with the engineering division (if there is one) and should allow engineering to help plan and assess any energy-saving programs.

Ideas on how to save energy should be invited. Perhaps installing a solar heater would reduce the cost of heating water enough to allow long-term savings; installing curtains to stop airflow through doorways of walk-in refrigerators or freezers might likewise be desirable. A computer might be used to shut down some equipment during peak energy-use periods, in order to reduce the demand rate. Many operations pay not only a basic charge for the amount of electricity they use, but also a charge based on the maximum demand made for electricity at any one time during a period. Thus, reducing this maximum demand can often save money. It may be feasible, without creating any hardship for personnel, to make the building slightly less warm in winter and slightly less cool in the summer, thus saving energy. For every Fahrenheit degree in temperature saved in building heating or cooling, about a 3 percent saving in energy occurs. Cold-water detergents coupled with a sanitizing rinse in iodoform may provide an energy-efficient alternative to washing dishes in hot water. There are many possibilities that management should seek out and evaluate. When an idea seems to warrant implementation, management should see that a plan for accomplishing this is established and carried out.

A good program of preventive maintenance and repair can save energy. Employees of the engineering department should set up a schedule to see that such regular procedures as frequent defrosting of refrigeration coils, removal of lint and soil from around refrigeration condensers, testing limit switches for proper working order, ensuring the presence of a bluish white flame on gas burners, and checking to see that good gaskets are in use on refrigerators and steam equipment so that a complete seal is created (a paper put into a closed refrigerator or steamer door should be held firmly and should not be able to be pulled out) are maintained. Lights lose efficiency very quickly when they gather dust and soil. Consequently, lights and fixtures should be cleaned periodically to save energy. A light that is old uses a lot of energy but does not produce much light. This lowered efficiency can be ascertained when a light globe shows an inside darkening. It is better to throw such a light away and install a new one than to wait for it to burn out. Some operations now have a program in which lights are all changed at a specific time, regardless of whether they are still burning. The saving in labor of going around one time to change all lights in a facility—rather than spasmodically changing them one at a time—and the saving in energy previously discussed make such a program pay off.

It is management's responsibility to set up a program to instruct employees on the need, the ways, and the means of saving energy. No program will be successful if employees are merely told that they should save energy. They need instruction, guidance, and supervision.

After management has surveyed the situation and decided on a course of action, it sets the plan up and instructs employees to implement it. But management's job is not yet done. Employees need counsel and attention, and management must check to see that the plan is being followed.

It is wise to set up a chart that shows how many Btu of energy have been used during a month and what the cost of this energy is. This information can be put on a graph and placed prominently where employees can see it. This helps keep employees interested in the conservation program and lets them see how well they are doing.

Responsibility of the Employee

The focal point of any energy-saving program must be the employees, since it is primarily they who use the energy. Much energy is lost because employees turn equipment on too far in advance of use. The heat-up times of equipment should be ascertained, and employees should be instructed not to turn equipment on until only a period equal to the heat-up time remains before the equipment will be needed. Employees also often fail to plan production properly. As a result, they frequently use equipment when only a partial load is inside it. Scheduling the use of a grill, oven, or other unit can save energy by holding back production until a full batch is processed. One study found that the use of electricity in a bakeshop could be reduced by 27 percent simply by replanning the use of ovens, baker's stoves, and other heating equipment. Too often, energy-consuming equipment stands idle, waiting for items from some other part of production to be readied for it.

Employees often turn equipment on with the thermostat set extra high, thinking that this will make the unit heat faster. It will not. Then the employee goes on working and often forgets to turn down the heat, letting the equipment overheat and forcing workers to turn it off so it can cool down to the proper temperature. This is both wasteful of energy and hard on the equipment. Also, if an operation has its electric bill figured on a demand basis, employees should not turn all electrical equipment on at once. Turning on one or two units that are needed first is the correct procedure. When these heat up, several other units can be turned on next, and so on until all needed equipment is ready. This avoids setting a high peak demand at the start and saves money as well as energy.

Energy can be wasted by careless employees in a number of ways. An employee should know that when water boils, it is as hot as it will ever get. Keeping the heat up and letting the boiling continue does not cook the food any faster: it only wastes energy. Over three times as much energy is needed to keep something on full boil as to keep it on simmer, and yet the cooking time is about the same for both. When an item boils, the excess heat goes through the product and is lost. No extra heat goes into the food. Then, extra energy must be used to operate fans to move this lost heat so that it does not become a problem for the employee.

Employees should be taught how to maintain equipment and do simple repairs on it. It should not be necessary for an engineer to fix a gas burner to ensure a proper flame; it is a simple enough task, requiring only proper adjustment. Many times equipment wastes energy because it is dirty or because it is improperly handled and maintained. Operational and maintenance manuals should be available for employee use. Supervisors should see that they are used and should instruct employees on the proper handling of equipment.

Employees should be taught to organize ahead for withdrawals of food from walk-in refrigerator and freezer units and even for withdrawals from reach-ins. Open doors can let a lot of cool air escape. Similarly, opening an oven door leads to a reduction in oven temperature of about 30 Fahrenheit degrees for every 5 seconds the door remains open.

More heat is needed to raise the temperature of water 1 Fahrenheit degree than to raise the temperature of almost any other substance an equal amount; yet employees run it freely, thinking little of the energy going down the drain. Much energy is saved if water temperatures in the boiler are reduced; 160°F (71°C) is much too hot to put one's hand into; 140°F (60°C) water serves just as well for many kitchen uses.

Employees should be aware of how costly it is to operate equipment. The energy needed to run a high-power fan can cost over $3 per day; furnishing the hot water for a pot-and-pan sink costs about $11 per day; energy for ranges and ovens can run as high as $9 per unit per day. Compiling a list of costs and publishing these may bring about increased awareness of the cost of energy. Perhaps then a dishwasher will not be left running when nothing is going through it, and other units may be shut down when they are not needed.

Employees should feel that they have an obligation to see that maximum yield is obtained for the energy used, and they should realize that energy-saving steps can occur only through their actions. Saving energy benefits not only the operation but also the country and, ultimately, the employees themselves.

Responsibility of the Manufacturer

In the future, manufacturers undoubtedly will pay a lot more attention to the energy efficiency of their equipment than they have in the past. A small gas pilot left on for 24 hours, 365 days a year, can cost as much as $70 each year in some high-energy-cost areas. Automatic-lighting pilots are needed. Temperature sensors should be installed so that, when foods reach a specific temperature, the heat shuts off or shuts down to a continuing temperature. Better conducting materials and better insulation are also needed. More efficient burners than are produced by drilling holes into pipe are possible. Some refrigerators have automatic defrosters that rely on electric heat to remove the ice on coils, while at the same time heat that could be used from the condensers is allowed to dissipate into the area.

More attention should be paid to ensuring that energy in equipment reaches the items to be cooked rather than go out freely into space. Heat reflectors or other heat-directing equipment could help. Better thermostats are needed, too. A solid-state thermostat is currently available, but few manufacturers put it on equipment, since it is more expensive than conventional thermostats. Manufacturers should alert buyers to the fact that, by paying slightly more for equipment, they can obtain a substantially better pay-off in energy saving. In the past, the low cost of energy did not make it practical for businesses to invest more in energy-saving devices and controls; now, however, with the higher cost of energy, buyers are more willing to pay the extra initial cost if they are aware of the benefits.

■
COOKING METHODS

A good chef or cook knows that *mise en place* is one of the most important things to learn in the profession. The term means "put in place," and it refers to organizing the job beforehand, getting all the materials needed, and then keeping things in order, cleaned up, and moving toward the final goal of production. Items needing no cooking—such as chopped onions, carrots, and grated cheese—should be ready and easily accessible. All materials to be used should be on hand, as should all necessary tools and utensils. Tools should be sharp and in good condition.

Another important thing to know is which cooking method is best to use for a particular menu item. Many times recipes indicate which method to use, but professional cooks and bakers have learned that specific products are best produced through certain special cooking methods. These methods can be separated into two general categories: cooking with liquid, and cooking without liquid.

Cooking with Liquid

Liquids used in cooking may be water, milk, tomato juice, wine, broth, or stock. Cooking in liquid is called *moist-heat cookery*. Meats contain a fibrous substance called *collagen* that holds them together and makes them tough. Because collagen can be dissolved by moist heat, many tough cuts, such as beef chuck, are cooked in liquid. The connective tissue in vegetables is called *cellulose*, and this too can be softened by moist heat. Most other foods can also be cooked in liquid. Even tender items such as fish may be cooked in this way to give them special moistness or to develop their flavor. Some vegetables are baked—but only those containing enough moisture to produce the effect of a moist heat in baking.

The most common methods of cooking in liquid are boiling, blanching, braising, poaching, simmering, steaming, and stewing. Some of these involve only slight differences in cooking methods.

Boiling

Boiling involves cooking substances by submerging them in water that is maintained at a sea-level temperature of 212°F (100°C) (see Table 10-1). A disadvantage of this method is that cooking in water leaches out valuable nutrients, which may be lost if the cooking water is discarded. Such nutrients can be saved, however, by using the cooking liquid for a stock, soup, or sauce. Usually foods are boiled for a

TABLE 10-1 Boiling Temperatures at Various Altitudes

Altitude Above Sea Level (ft)	Boiling Point of Water	
	°F	°C
0	212.0	100.0
500	211.1	99.6
1,000	210.2	99.0
2,000	208.4	98.0
3,000	206.6	97.0
4,000	204.8	96.0
5,000	203.0	95.0
6,000	201.2	94.0
7,000	199.4	93.0
7,500	198.5	92.5
8,000	197.6	92.0
9,000	195.8	91.0
10,000	194.0	90.0
12,000	190.4	88.0
15,000	185.0	85.0

minimum time only. Vegetables, for instance, may not be completely cooked when removed from the boiling water; instead, they are sent to the steam table, where they finish cooking while being held. Many items are cooked in part by being dropped into boiling water, which shortens their overall cooking time. Spaghetti, other pastas, and cereals are cooked by being dropped into boiling water and stirred so that they avoid sticking together. Hard boiling is never recommended because it wastes fuel and can toughen some items and break up others (such as soft vegetables). Meats are seldom boiled, but many are simmered instead. Eggs, poultry, and fish are not boiled, either, because the high boiling temperature harms the proteins they contain.

Blanching

Blanching involves subjecting an item to boiling water for only a short time. It may be called *scalding*, although scalding usually refers to dipping a food into water or allowing a hot water bath to strike the food for a very short time. Scalded milk is milk brought just to boiling and then taken off the flame to reduce its temperature rapidly. Kidneys and chicory may be blanched to remove some of their strong flavor, and liver and sweetbreads may be blanched to remove the external membranes. Some fruits and vegetables, such as peaches and tomatoes, may be blanched or scalded to remove the outer skins. Because blanching can stop enzymatic action, some fruits and vegetables that are to be frozen are blanched first.

Blanching also commonly refers to the nearly complete cooking of vegetables in deep oil, after which the cooking of the blanched vegetables is completed in a second step just before service. Parboiling is another form of blanching, but in this case the item remains in the water for a longer period of time. Again the purpose is to cook the item partially before giving it other treatment. Thus, squash may be parboiled before it is baked. Parboiling may also be done to firm a food before it is handled further. Even though blanching, scalding, and parboiling do not usually produce completely cooked food, they are special methods of liquid cooking that are classed as separate cooking techniques.

Braising

Braising involves cooking food in a small quantity of liquid. Fowl (old chicken), meat, vegetables, and other foods that must be cooked for a long time are often braised.

There are two methods of braising: *brun* (brown), or *blanc* (blond). *Brun* involves first searing the meat to give it color. It is erroneously thought that searing seals in a meat's juices; studies show that this is not true. Searing is usually done in a small quantity of fat, and the meat is turned frequently until it is well browned on all sides. Some foods may be dredged in flour before being seared. A blond product is braised in the same way as a browned one is, except that it is not seared, that is, it is not placed in a small quantity of hot fat to gain color first. After searing, if it is done at all, the product is put into a covered pan, known as a brazier, and liquid (water, stock, thin sauce, or a combination of these) is added to the level of one-third of the height of a roast or two-thirds of the

height of smaller food products. Often, a mirepoix (50 percent chopped onion, 25 percent chopped carrots, and 25 percent chopped celery) is added, along with seasonings. The mixture is then brought to a simmer, on top of the range, after which the brazier is tightly covered and placed in a 300 to 325°F (149 to 163°C) oven until the product is well done or tender. The leftover liquid may be used to make a sauce or pan gravy to serve with the item.

Pot-roasting, swissing, fricasséeing, and jugging are all forms of braising. Pot-roasting is usually a *brun* product, and a smaller quantity of liquid may be used for it than for some other braised items. Swissing is very similar to pot-roasting but is done to smaller meats. A fricassée is often a braised product, although sometimes it refers to a stewed product. Jugging is a form of cooking that takes place in a covered pot. It is often used with game meats, especially hare.

Poaching

Poaching is cooking in liquid at a temperature below the boiling point of water. It amounts to a low simmer. The food may or may not be covered with liquid, although the liquid should be at least halfway up the food. Poaching is usually done in a shallow pan in stock, court bouillon (for fish), milk, or some other liquid. Food may be basted with the liquid. Fish, eggs, and certain other foods that require gentle handling and low-temperature cooking are often poached. Poaching poultry increases its flavor, moistness, tenderness, and yield over roasting, particularly in larger birds. This method takes only half the time of roasting and requires less cooking space. Turkeys may be poach-roasted with excellent results. For this, turkey cut into parts is placed skin-side-down in pans, partially covered with water or stock, and poached in an oven.

Simmering

Simmering involves immersing food in a liquid and cooking it at a temperature of between 185 and 205°F (85 and 96°C). Often, the liquid is brought to a boil, the food is added, and the product is then held at the simmering temperature until cooked. At other times, the liquid and the product are brought slowly up from room temperature to simmering and then held there.

This is the procedure used in making a clarified consommé or bouillon. Meat placed in cool water and brought from the cold stage to a simmering temperature gives slightly more flavor to stock than does meat added to already hot water and cooked. Simmering is usually described by a chef as "let the liquid smile, not laugh out loud." "Laughing out loud" is thought of as hard boiling. In simmering tiny bubbles appear from time to time and only some slow motion is apparent. Meat, poultry, and some other items are more tender, juicier, and more flavorful when simmered. The cooking liquid may be used to add flavor. Since braising is also cooking below a boil, it can be considered a simmering method.

Steaming

Steaming involves cooking foods in boiling-hot water vapor. This method favors the retention of nutrients. In free-vent steaming the food is placed over or in live steam and allowed to cook. Many vegetables, potatoes, and other foods cook well this way. Free-vent steaming can also be done in a waterless cooker, a very heavy kettle fitted with a tight lid. No water is added to the cooker, but the food generates enough of its own liquid to produce steam. Carrots, cabbage, and some other high-moisture vegetables cook well by this method.

Special equipment may be used for steaming vegetables. For example, steam can be introduced under pressure, to raise the temperature and cook the food more quickly (see Table 10-2). High-pressure steamers of this type cook with a pressure of around 15 psi; lower-pressure steamers cook at around 6 psi; in the former type, some vegetables cook in a matter of seconds. Food cooked in the lower-pressure units reach doneness a bit more slowly, but still much faster than when boiled.

Cooking can also be accomplished with forced steam convection, which feeds steam in around food with great turbulence, increasing the speed of heat transfer to the food without requiring the heat to exceed 212°F (100°C). The theory underlying this method is that the force of the steam convection breaks a naturally formed barrier that exists between the steam and the food in a more static steamer.

Steaming may reduce shrinkage, as well as preserve form, flavor, and nutrients. There is no burning

TABLE 10-2 Steam Temperatures at Different Pressures		
Gauge Pressure (psi)	Temperature	
	°F	°C
0	212.0	100.0
2	218.3	103.5
4	224.6	107.0
6	229.8	109.9
8	235.0	112.8
10	239.4	115.2
15	249.6	120.9

or boiling over, which reduces the amount of cleaning required. The same-sized pans can be used for steaming as are used during holding in the steam table; thus the number of pots that have to be washed is reduced. Almost any kind of food can be steamed, but foods that pack and some doughs and batters do not respond as well to steaming as to other methods. Steam can also be used to reheat foods quickly, and it is a good way to thaw foods prior to giving them other treatment.

Stewing

Stewing involves simmering small cuts of meat, disjointed fowl, or cut-up meat in liquid. It is distinguished from braising in that braising typically involves whole or sliced food, whereas stews are made up of chopped or cubed food. Another difference between stewing and braising is that, in stewing, the item being cooked is completely covered with water, stock, broth, or other liquid, whereas in braising only the amount of liquid needed for service (2 oz per person, typically) is used. Stewing usually involves thickening the liquid with starch and serving it, together with its cubed, cut, or chopped contents, as a one-dish course. Braised foods, on the other hand, are served with garnishes and vegetables.

Stewing is done in a heavy pot to ensure even cooking, and it usually takes place on top of the range or in a steam-jacketed kettle. Depending on the color desired in the stew, the food to be cooked may or may not be browned first. Thus, stews are either *brun* or *blanc*. Browning results in a darker-colored stew such as a brown ragoût. The light-colored Irish lamb stew, on the other hand, requires no browning of ingredients. Whether or not the cooking vessel is covered is another variable that affects the end product. Greater reduction (through evaporation) of the liquid takes place in a stew that is allowed to vent freely, and the result is a thicker, heavier stew.

Meats, poultry, seafoods, fruits, vegetables, and various combinations of these are stewed. Stews may be seasoned in different ways, and usually an operation develops its own stew of distinctive flavor. Many stews are low in cost and yet highly popular with customers.

Cooking Without Liquid

Methods of dry-heat cooking (cooking without moisture) are usually employed for tender meats, fish, and poultry, for some less-tender products of these, and for moist food products. Meats high in connective tissue do not cook well in dry heat. Low temperatures are usually used, although broilers may generate temperatures of up to 6000°F (3316°C). Overcooking or excessive heat can dry products out and toughen them; lower-temperature cooking allows some of the connective tissues to soften before the item is done. Tough products that must be cooked a long time in dry heat come out tough, dry, and unpalatable.

Major methods of cooking without liquid include baking, barbecuing, broiling, grilling, ovenizing, roasting, and frying (both deep-frying and sautéing). Frying and sautéing use no liquid other than hot fat, which may be viewed as presenting a special case of dry-heat cooking because of its particular traits as a cooking medium.

Baking

Baking involves cooking in dry heat in an oven. Vegetables, meats, doughs, batters, fish, and poultry are often baked. Baking temperatures normally range from 250 to 475°F (121 to 246°C), although higher temperatures can be used (see Table 10-3).

TABLE 10-3 Oven Temperatures

Description	Temperature		
	°F	°C	Gas No.
Very slow	250	121	½
Slow	300	149	1
Medium slow	325	163	2½
Medium (moderate)	350	177	4
Medium hot (moderate)	375	191	5½
Hot (moderately hot)	400	204	7
Hot	425	218	8½
Very hot	450–475	232–246	10–11½

Some hard-crusted breads and other products must be baked with steam. This keeps the crust soft during part of the baking so that a heavier, crisper crust is formed later when the steam is turned off.

Some ovens are heated in stacks, with the heat rising from the bottom level to the top stack. The temperature difference may be 5 to 10 Fahrenheit degrees (2.7 to 5.6 Celsius degrees) between the bottom level and the top level of the stack. Foods requiring relatively high baking temperatures are baked in the lower stacks, and foods requiring relatively low baking temperatures are baked in the upper ones. The heat in baking is transferred mostly by natural convection, except in convection ovens. Some radiation can be obtained, as well, by using metal reflectors. A casserole is said to be baked, but it is actually cooked by a form of braising in the oven. Baking and roasting may be the same thing.

Barbecuing

The term *barbecuing* is often applied to a method of broiling foods while basting them with a special sauce. Real barbecuing, however, involves roasting foods in a covered pit. Some barbecue ovens duplicate this, introducing a smoke that gives special flavor to the product. The sauce added is usually quite tangy. In oven- and pit-barbecuing, cooking temperatures are kept quite low; but in barbecuing over a grill or in a broiler, the temperatures are higher. Barbecued food is normally very popular.

Broiling

Broiling involves cooking by radiant heat. Food is placed on a rack above, below, or between the heat source(s), which may use gas, electricity, or charcoal. The proper temperature to maintain is governed by the fattiness, tenderness, and size of the food. Broilers should be preheated but should not be left on constantly unless used constantly. Food is usually dipped into oil or brushed with oil to prevent it from sticking and to reduce evaporative loss. Depending on the thickness of the food and the intensity of the heat, food is placed 3 to 6 in. from the heat source. Frozen foods and very thick pieces are cooked even farther from the heat source, so the heat will not be too great. Flames should be controlled. Charring (burning of fat from broiling) is suspected by some authorities of causing cancer.

Meats, poultry, seafood, fish, many vegetables, and some fruits are prepared by broiling. A 2-in.-thick steak requires about 10 minutes of broiling on each side. Some cooks prefer to use a higher temperature at the end to develop a rich, brown color. Broiling takes skill; creativity and expertise are required to produce a good broiled product. Pan-broiling is a method of cooking food on the hot, dry surface of a pan in which conduction, not radiation, is the process of heat transfer. Any fat present during pan-broiling is discarded before it can accumulate.

Grilling

In its broadest sense, the term *grilling* refers to cooking on a fat-coated surface or griddle at fairly high temperatures of 325 to 400°F (163 to 204°C). The French consider grilling to be synonymous with broiling, and we hear the word used in this sense also. The term *mixed grill* is often used for broiled meats. Grilling and frying or sautéing are very similar. For grilling, food is put on a lightly greased griddle and fried. Additional fat may be added. Griddle-broiling involves cooking foods on a hot, dry surface, as in pan-broiling. The fat is removed as it forms around the product; similarly, in pan broiling, the fat is poured from the pan as it forms.

In both grilling and griddle-broiling, the product is cooked or browned on one side and then turned and cooked or browned on the other. The item should not be punctured, covered, or cooked in any liquid. Eggs are grilled at 325°F (163°C); French toast, at 350°F (177°C); and pancakes, potatoes, fish, and meat, at 400°F (204°C).

Ovenizing

Ovenizing involves placing food on greased pans, frequently dribbling fat over it, and baking it in an oven. Fairly high temperatures are used—325 to 425°F (163 to 218°C). Some smaller foods may be cooked at even higher temperatures. The finished food looks much like fried or sautéed food. Ovenizing is used in quantity cookery to produce large amounts of food with relatively little labor. Breaded foods need more fat added than nonbreaded ones do. In either case, however, the total amount of fat used is less in ovenizing than in sautéing (frying) or deep-frying.

Roasting

The term *roasting* originally meant cooking on a spit over an open fire, a practice that would be called *broiling* or *barbecuing* today. Today, it means cooking food with dry, indirect heat in an oven. The term *baking* is frequently used to mean roasting, but it may also be used differently: we bake a ham, but we roast a sirloin; bread and other bakery products are baked, but nuts are roasted. For roasting, meat is placed, fat side up, on a rack or trivet (perforated underliner) in an uncovered pan that contains no liquid. As the meat cooks, the fat bastes it. Because the meat is raised above the bottom of the pan by the rack or trivet, it avoids braising in its own juices and fat. The meat should be fitted to the pan. Leaving too much bottom surface exposed allows the drippings to char and burn.

Using a temperature as low as 300°F (149°C) gives the highest yield and the best flavor. Some chefs start roasts at 450°F (232°C) to sear the meat and give it a nicely browned outer surface; they then drop the heat to the desired roasting temperature. Smaller pieces of meat may be roasted at higher temperatures to limit their drying out.

Meat is considered well done when it has an internal temperature of around 170°F (77°C) or more, medium done at around 160°F (71°C), and rare at 140°F (60°C). Aged meat may have to be cooked to lower internal temperatures for rare and medium, since the color changes in such meat occurs at a lower temperature than in unripened meat. A large roast should be removed from the oven when its internal temperature is about 15 to 25 Fahrenheit degrees (8.3 to 13.9 Celsius degrees) away from the desired degree of doneness, since a temperature rise in the meat's interior occurs after removal from the oven. Pork, rabbit, and some other meats should always be cooked to well done to ensure that trichinae are destroyed.

Frying

Frying involves cooking foods in a fat or an oil. It adds a rather nutty, pleasing flavor to the food, as well as crisp texture and good appearance. Over- or underfrying is easily possible. A fry cook possessing considerable skill makes a delicious and wholesome product; a poor fry cook produces greasy, dry, hard, tasteless products. The term *frying* may indicate either deep-frying or sautéing. In this text and in the trade, however, the word *fry* most often means sauté; if deep-frying is meant, the term *deep-frying* is usually used.

Deep-Frying. Deep-frying involves cooking foods by immersing them in hot fat. Done properly, this method produces tasty, attractive products. Fat is kept from penetrating the food either by a coating of flour, batter, or dough that forms an impenetrable crust when the item begins to fry, or by the use of sufficiently high fat temperatures to cause the product itself to create a barrier of steam that prevents grease penetration. Problems arise if the temperature of the fat drops below the level required to keep the vapor barrier around the food: the result is a soggy, greasy product. To deep-fry successfully, a cook should know something about fats and oils, cooking temperatures, and the proper techniques to obtain high-quality products.

A cooking fat and a cooking oil are very similar chemically. They differ in that a cooking fat has a higher percentage of saturated fatty acids, while a cooking oil contains a higher proportion of polyunsaturated and monounsaturated fatty acids. In terms of cooking there are differences as well. Products deep-fried in cooking fats have a slightly different flavor—one that is preferred by several fast-food companies for cooking french fries—than products deep-fried in cooking oil.

An oil can be made to approximate a cooking fat through the process of hydrogenation, whereby the unsaturated portion of the oil accepts additional hydrogen atoms and, therefore, increases the proportion of saturated fat in the oil. The higher the proportion of saturated fat in a cooking oil or fat, the more solid it will appear. For instance, at room temperature butter is relatively solid and opaque, a cooking fat is soft and cloudy, and an oil is liquid and translucent.

An oil is made into a solid fat by being heated under pressure, having ground nickel added to it, and then having hydrogen gas bubbled through it. An atom of nickel picks up a hydrogen atom and fixes it at an unsaturated point. It then leaves, picks up another hydrogen atom, and fixes it at another unsaturated point. Thus, the nickel does not join the fat permanently but acts as a carrier to help the reaction along. We call this type of chemical agent a *catalyst* and we call this process *hydrogenation*.

Oils and fats are called *saturated* (if all places are filled on the carbon chain), *monounsaturated* (if one place is not filled), or *polyunsaturated* (if more than one place is unfilled). Hard fats from meat, coconut oil, and cacao (chocolate) oil are usually saturated. Some vegetable oils, lard, poultry fat, and a few others may contain combinations of saturated and unsaturated fats and oils in varying ratios. Fish and some vegetable oils are quite unsaturated. A monounsaturated or polyunsaturated oil does not perform as well in frying as does a saturated one.

Other factors also are important for frying. A good frying oil should not have a smoking point (the temperature at which decomposition occurs and fumes are given off) below 425°F (218°C). Satisfying this condition requires starting with a good oil or fat that contains a large proportion of saturated particles. Stabilizers are also added. One of the best frying oils on the market contains a silicone compound that helps prevent the oil from breaking down and thus keeps it in good condition for frying. Most frying fats and oils contain antioxidants to reduce the chances of oxidative rancidity (or chemical deterioration) of the fat.

To produce a fat or oil, three fatty acids must react with one glycerol molecule to form a triglyceride. Basically, fats and oils are mixtures of triglycerides. Any fatty acid that is not linked to a glycerol molecule in a fat or oil is referred to as a *free* fatty acid. Removing free fatty acids from a fat gives it a higher smoking point and makes it a more durable cooking medium. Butter, lard, fish oils, and other fats contain many free fatty acids, which is why they smoke at relatively low temperatures and do not make good frying products. If pork, fish, and some other products are fried in a good frying fat, their fat with its free fatty acids gradually mingles with the good oil and lowers its quality.

When a fat or oil breaks down, its fatty acid chains break away from the glycerol molecules and become free fatty acids. When a fat breaks down extensively and smokes, acrolein—a white vapor that causes the nose, eyes, and throat to smart—is formed. A fat that is badly broken down assumes a dark color, smokes at a fairly low temperature, and gives food a sharp, acidic flavor. If breakdown is allowed to continue further, the fat begins to foam, and pale yellow bubbles crawl up and over the fryer.

Fats may not always break down; instead, they may join together to form waxes or resins. Resinous products may form on griddles or in deep fryers for this reason. Fats or oils that join together are called polymers. The formation of polymers is considered harmful to fat. Using fats properly and following recommended techniques helps prevent fat breakdown and the formation of polymers. Deteriorated fat should be discarded, since good foods cannot be produced in it.

Allowing foreign matter such as charred particles or crumbs to accumulate in a frying fat speeds the fat's deterioration, so the rule is to filter fat frequently. Fry kettles and equipment should be kept very clean, as well. Fat in a deep, narrow fryer will not break down as quickly as fat in a shallow, broad one with a much greater exposed area. Because metals such as copper, brass, and iron speed deterioration, fryer surfaces are usually made of nickel, chrome, or stainless steel. Salts such as common table salt and the curing salts used in cured meats also speed deterioration. Therefore, salt must be kept away from the fryer, and cured meats should not be fried in the deep fryer. Fat should be stored in a cool, dry, dark location in airtight containers. New oil or fat should never be put into uncleaned containers. Rancidity or other undesirable reactions are likely to occur if these frying products do not receive good care.

At the time a fryer is being loaded with fat, the thermostat should be set at 250°F (121°C) or lower.

Fats will melt first around the heating elements and, if the temperature is too high, will scorch there. Once a fat breaks down, it quickly causes other good fat to go through the same process. Even though only a small amount is scorched, the breakdown of the entire mass of fat will be more rapid. A 15-lb fry kettle holds 15 lb of frying oil or fat, and this is the correct quantity to use for frying in that particular kettle. The ratio of fat to food is usually 6 to 10 lb of fat to 1 lb of food. French-fried potatoes and a few other foods can be fried at the lower ratios, while some others must be fried at the higher ones. A good fry kettle holding 15 lb of fat can fry about 2 to 2.5 lb of french-fried potatoes at one time, but only about 1.5 lb of chicken.

If about 15 to 20 percent of the fat or oil in the fryer is used up each day and new fat is added—along with good filtering, proper frying techniques, and correct temperatures—no fat need ever be discarded. The new added fat is sufficient to keep the whole mass of fat fresh and in good condition. This daily addition of fat without discarding the old is called *turnover*. Obtaining the right turnover and using proper frying methods will produce good fried products and save money.

Foods to be fried must be prepared differently according to the food and the final product desired. Some require little final preparation. For example, potatoes for french-frying are often soaked in cold water to prevent them from darkening. They are then drained well and fried, with no further treatment. Other foods may have to be breaded, dipped in batter, or wrapped in dough before being fried. High-moisture foods need a heavier, tighter covering than do drier products. Firm foods need less covering than do soft foods. Food to be covered should be properly cleaned and dried first.

For breading, five trays or containers need to be set in a row: the first contains the product to be breaded; the second contains flour; the third contains a moistening mixture of egg wash (eggs plus liquid, usually milk), plain egg, or plain liquid; the fourth contains bread crumbs, cracker crumbs, corn meal, flour, or a prepared mix or some mixture of these; and the fifth (which may be a rack) holds the breaded product. Uniformity of crumb size in the fourth container is necessary. The finer the crumbs, the finer the coating will be. Breaded products should be allowed to dry for

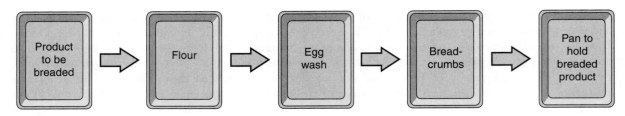

FIGURE 10-2 Breading procedure and setup.

at least 15 minutes in the final pan or rack to give a coating that adheres best and puts as little moisture as possible into the fat. The fat bubbles less when drying has taken place (see Fig. 10-2).

Batters need to be thick to adhere properly to food and give a good cover on the end product. Mixes are often used that only require the addition of water. Batters can also be made from scratch. Dough wraps must be done tightly to seal the product in and to prevent excessive fat penetration. The deep-fried apple turnover and the Chinese egg roll are examples of dough-wrapped products.

The temperature to use for deep-frying is dictated by the quantity of food to be fried in each batch, the temperature of the food, and the kind of food. The lowest frying temperatures are around 325°F (163°C), and the highest are around 425°F (218°C)—the latter being unusual but desirable for making puffed (souffléed) potatoes. A large quantity of cold food added all at once to the fryer drops the temperature, so original temperatures may have to be higher to prevent the food from cooling the fat so much that the food becomes soaked with grease and is poorly cooked. Large pieces of food are usually cooked at lower temperatures to prevent them from being overfried on the outside by the time the inside is done. Some thin items may be fried at high temperatures because they would dry out with extended cooking in fat. Food fried at the right temperature is golden brown, crisp on the outside, and low in retained fat. Foods fried at too high a temperature develop a dark brown, hard crust and a burned taste, and often have an uncooked center. Foods fried at too low a temperature become fat-soaked, unattractive, and unpalatable.

Fats subjected to heat should not remain unattended, since they might burst into flames. The flashpoint of a fat or oil is around 600°F (316°C). The melting point of most cooking fats is about 120°F (49°C). The smoking points for suet, butter, lard, and cottonseed oil are 235°F (113°C), 325°F (163°C), 345°F (174°C), and 420°F (216°C), respectively.

Deep-fry baskets should be loaded away from the fryer to prevent moisture or food particles from dropping into the fat. The baskets should not be overloaded. The lowering of the food into the fat must be done slowly so that the fat can adjust. If done too quickly, the fat may flow out of the kettle. After it is fully immersed, the basket is given a slight shake to loosen food that might be sticking together and to adjust it into the available space. The food may be lifted out during cooking and given another shake to distribute portions to cook all areas. When the food turns a golden brown, the basket is lifted. At this point, the food may be given a slight shake and the basket left to drain momentarily above the fat. Food should never be held in the basket over the fryer for prolonged periods of time, as this will lead to excessive fat absorption by the food and a greasy taste. Batter-dipped items should be put into the fryer after the basket has been lowered, to prevent the food from cooking onto the basket. The finished product should be held in a dry, warm place or under an infrared lamp. Products should only be salted immediately before service. Salting ahead reduces the crispness of the product. The following summary of proper deep-frying procedures was put together by Market Forge.

Acceptable Fried Foods Result from Proper Frying Procedures

1. For good color and flavor, use only good- to excellent-quality fat.

2. Room temperature foods fry more uniformly than chilled or frozen foods.

3. Keep thermostat temperatures below 400°F for good products and minimum fat breakdown.

4. Thick foods require lower frying temperatures and longer cooking times.

5. Breaded products usually provide a crisper, longer lasting crust than do batter-treated products.

6. Reasonable workloads aid in producing high-quality foods and prolonging the life of the fat.

7. Fry pieces of similar size at the same time, to ensure evenly cooked pieces of uniform color.

8. Lower battered foods into fat slowly, to prevent splattering.

9. Turn doughnuts and fritters just once.

10. Fill frying baskets to only one-half to two-thirds of capacity—no more.

11. Always cover fat that is not in use.

12. Refrigerate fat that is to be held for more than a few hours. This helps lengthen the life of the fat.

13. Do not mix foods in a fryer, as there will be a flavor transfer.

14. Foods of high sugar content absorb more fat.

15. Foods high in moisture absorb more fat.

16. The greater the surface area of food exposed, the greater the amount of fat absorbed.

Proper Operation of the Fry Kettle Prolongs the Life of the Fat

1. Never melt fat at a temperature higher than 250°F. Otherwise, scorching will occur.

2. To revitalize fat, add 15 percent new fat every frying day.

3. The food-to-fat ratio can be a minimum of 1 lb food to 10 lb fat and a maximum of 1 lb food to 6 lb fat.

4. Reduce the kettle temperature to 250°F, if the kettle is to stand idle for ½ hour or more.

5. Never allow elements or burners to be exposed while melting fat.

A Clean Fry Kettle Is Essential to Avoid Excessive Fat Breakdown

1. Strain and clean fat once every 8 hours or once a day.

2. Thoroughly clean kettle, elements, and burners once a week.

3. Never salt food over fat.

4. Always skim off floating particles between batches.

Miscellaneous Fryer and Fat Information

1. Always operate a fry kettle under an exhaust fan.

2. Shortenings that are liquid at room temperature are more digestible than ones that are solid at room temperature.

3. Hydrogenated fats and oils are treated with hydrogen and therefore keep longer. They are solid at room temperature.

4. Vegetable shortenings can be used for frying or baking, and also have a higher smoke point than fats.

5. Fat that is in the process of decomposing does not maintain necessary frying temperatures.

6. Fat should be checked daily for foreign flavors.

7. Fryers should recover temperature within the first third of total cooking time.

8. Thermostat calibration should be checked at least once every six months.

9. When checking the temperature of fat in a fryer, place the fat thermometer's bulb approximately 1 in. below the fat's surface for a proper reading.

10. A fat that has a low smoke point will decompose quickly at frying temperatures.

Pressure Frying. A form of cooking that combines steam-pressure cooking and frying is called *pressure frying*. In this method, food is immersed in hot fat, a lid is tightly screwed down on the kettle, and the steam released during cooking builds up a pressure that helps cook the food and make it more tender. At a certain point, the pressure of moisture trying to leave the food equals the pressure buildup inside the kettle, resulting in an equilibrium of pressures and the retention of more moisture in the food. Since the moisture cannot leave the food, 75 percent of the item's natural juices remain inside it, as compared to 50 percent in conventional deep-frying. Pressure frying can also reduce fuel costs.

Sautéing. Sautéing is cooking in shallow fat at a temperature of from 335°F (168°C) to 425°F (218°C). The fat is usually about ⅛ inch deep in the pan. Sautéing is a quick cooking method; indeed, the word *sauté* comes from the French infinitive *sauter*, which means "to jump." A *sauteuse* or sauté pan is straight-sided to keep fat from getting out too easily. The pan can be covered so that the food can also be braised in it. Foods that are sautéed and then braised are often only seared briefly in fairly hot fat before braising begins. A pan with slanting sides called a *poêle*, or frying pan, is used when the item is to be flipped (as potatoes or an omelet may be) or turned with a spatula.

Many sautéed foods leave a crusty deposit on the bottom of the pan. This is deglazed by adding wine, stock, or other liquid and reheating the pan while removing the encrusted materials with a fork. They then go into solution with the liquid and make what is technically called a *fond* but is more often termed a *sauce* or *au jus.* An already-prepared sauce or thickening agent may be added at this stage, or the fond may be naturally reduced. Reducing the fond involves lessening its volume by boiling it. The reduced sauce is then served with the sautéed food. Deglazing is also done after roasting to save and use drippings that might otherwise be washed down the drain. Many foods are sautéed, including fish, meat, potatoes, vegetables, and poultry.

Pan-frying is another word meaning the same thing as sautéing. It usually refers only to food cooked completely by the process—that is, with no subsequent braising.

Stir-Frying. This is a variation of sautéing in which a wok is used rather than a sauté pan. The wok has curved sides and a rounded bottom in order to diffuse heat and make tossing and stirring of food product easier. Because items to be stir-fried are cut into small pieces, which acts as a means of tenderizing the food, the food does not need to be as naturally tender as for sautés, where it is left in portion-sized pieces. All fat, gristle, and silverskin must be removed from flesh items for the best results in stir-frying.

Because the food product is cooked briefly, nutrients and flavor are better retained in this method of cooking. Production time is quick with the benefit of fresh, healthy cuisine (not overcooked) being served to the dining public. The number of ethnic entreés available through stir-frying make this cooking method particularly attractive to those operations wishing to offer diversity to the menu.

■

INNOVATIVE COOKING METHODS

The first heat used for cooking may have come from burning wood. Gas and oil escaping from the earth and accidentally lit by some means—as well as escaping steam and hot water—were also available sources of heat from the earliest times. Coal undoubtedly became available near the beginning of civilization and was known for its heat-giving properties. Thousands of years ago, the Chinese drilled into the earth to capture natural gas and piped it in bamboo tubes to locales where it could be used. Today, oil, gas, wood, coal, and some other combustible substances are available for heat; and energy from electricity can be used in different forms for heating, as well.

Combustible fuels must have oxygen from the air in order to burn. The gases formed during combustion must be exhausted to keep the air pure for humans to breathe. The loss of heat resulting from this necessary venting has been previously noted, but many combustible fuels are still used and will perhaps be used for a long time to come.

Because electricity can be converted to different forms of energy, it is much used in foodservice work. New ways to use electricity have made it possible to improve some cooking methods and to speed others up. Three such advanced forms of cooking based on electricity are infrared, microwave, and quartz cooking.

Infrared Cooking

If a beam of sunlight falls on a prism, it separates into a spectrum of different colors. The shortest visible wavelengths of light appear as violet, and the longest as red. If a sensitive thermometer is moved across the spectrum, greater heat is found in the red band. But when light is broken up, most of its energy goes into the area just beyond the red band, where no color at all appears. This is known as the *infrared band* of the spectrum.

Infrared heat is used considerably in foodservices. The heat penetrates food directly as a form of radiant energy that does not rely on a carrying or transporting medium (as conducted and convected heat do). The amount of radiant energy reaching the food is directly related to the food's distance from the heat source. Ovens for reconditioning frozen foods, broilers, and modern bake ovens all use infrared energy. Infrared lamps are good for keeping cooked food warm in foodservices, because they maintain a constant temperature and can keep foods warm and tender for up to 24 hours with good moisture retention.

Microwave Cooking

Longer waves than light waves also carry energy. Radio and TV waves are quite long waves of electromagnetic energy; some radio waves, in fact, are as long as a football field. We can produce an energy wave with an amplitude (wavelength) of several inches or more in a magnetron tube. Once generated, the wave travels just as radio or light waves do, strikes hard objects such as metals, and bounces back. This discovery led to the use of these waves in detecting distant objects during World War II. Later it was found that, when these waves penetrated soft food, they disturbed the food's molecular structure, causing the molecules to vibrate at a frequency comparable to that of the waves, 915 to 2450 million times per second. This intense activity creates friction within the food, producing heat. We call this process of vibration *kinetic action*, and it is what happens during microwave cooking. Because water molecules become most active, foods high in moisture cook most rapidly by microwave. A potato, for example, may burst during microwave cooking because of the violent boiling going on within it. Foods also tend to dry out quickly because the hard boiling evaporates moisture rapidly. Cooking times must be measured very precisely in working with a microwave oven, since the heating occurs at so much faster a rate than in conventional cooking.

Microwaves do not brown food, and it is also not possible to give food a fried or broiled appearance or texture in a microwave oven. Steaming more closely resembles microwave cooking in its results. Cooking with microwaves begins at the food's center or about 2 in. deep inside. Small foods such as potatoes, fish, and some vegetables cook better in microwave ovens than larger foods such as roasts. Larger foods need more attention because of this. Bakery goods cook too fast to rise properly when subjected to microwaves.

For various reasons, therefore, microwave cooking cannot replace conventional cooking; still, it is excellent for defrosting foods, warming them, boosting liquid temperatures, heating rolls, and finishing foods during rush hours. Many frozen prepared foods are handled nicely by microwave cooking. Cooking times increase in direct proportion to food quantities. If a single potato takes four minutes, two take eight minutes. Some foods have a higher yield in portions when cooked by microwave, while others have a lower yield. The determining factors are the composition and size of the item, and (of course) the time in the oven. Thinner slices of meat seem to come out juicier, since they are cooked for a shorter period of time than roasts (which tend to dry out). Metal cooking containers cannot be used in microwave ovens, as the metal disrupts the flow of the microwaves within the oven and can result in damage to the magnetrons in the unit.

Quartz Cooking

Quartz cooking may be considered a variation on infrared cooking, since the heat wave is very similar. The quartz tube diffuses infrared energy waves evenly, giving a steady, consistent source of heat. This infrared heat is developed by running an electrical charge through a quartz filament in a vacuum tube. Such tubes used in a broiler can produce up to twenty times the energy concentration possible from other infrared sources, such as glass or metal. A quartz tube can tolerate very high temperatures, and it can withstand a greater thermal shock (a quick extreme change in temperature) without cracking or breaking than can glass. There are now quartz ovens, quartz broilers, and other heating units using the quartz tube. The quartz oven was originally designed to thaw bulk frozen foods, and its success in this regard made possible the cooking-to-order revolution in volume feeding.

Other New Technologies

As a rule rather than an exception, foodservice operators have been somewhat reluctant to embrace new equipment technology. There is new technology on the horizon as well as combinations of old technology, and the smart operator should consider each for its own merit.

One new piece of equipment is the Steam'n'Hold compartment steamer (see Fig. 10-3), which will cook food, then hold it until ready for service. Time and temperature are set and when the cooking is over, the steamer changes automatically to the hold mode. Food products may be held in a ready-to-serve state for several hours. With no boiler, the unit uses water added to generate only enough steam to do the job. Energy is saved, then, as existing compartment steamers with boilers typically generate 10 times the steam necessary to cook the product.

Another new technology is controlled vapor technology (CVAP), a cooking, holding, and thermalizing technology using vapor pressure to maintain food temperature and moisture content. Early acceptance of the technology has been good particularly in holding cabinets. If you compare a pan of fresh-baked biscuits with one held for 45 minutes in a CVAP unit,

one would be hard-pressed to tell the difference. This may prove to be highly beneficial to operators considering home-meal replacement applications.

Although not considered cutting edge, since they were introduced some years ago, equipment using a combination of technologies have come into their own. Combo-ovens, quick-chillers, and cook–chill systems have all revolutionized the cooking and holding of food as convection heat is combined with steam and cooked food is introduced to blast freezing for safer holding.

Even faster cooking and better holding appear to be on the horizon as these and new technologies are introduced and refined. Foodservice operators are wise to view these new technologies at trade shows or to speak with their equipment representatives to stay abreast of new developments.

■

DINING TRENDS

Patrons value taste and nutrition more than price. In other words, quality is remembered long after the price is forgotten. In the Midwest, fat-free foods are extremely popular, but on the east and west coasts, patrons prefer fresh vegetables and grains. The demand for fresh, all-natural, and organic foods may result in lower sales for processed and packaged foods over time. The bottom line sees patrons searching increasingly for nutritionally balanced foods that also offer convenience.

These and other trends can be summarized in the following sections on local, regional, national, and international cooking trends. The smart foodservice operator will ask his or her own patrons what they wish and reflect that in the menu offered.

Local

Depending on their locale, foodservice operators are seeing people take their main meal of the day at various times, largely because the workforce is able to set its own work schedule. Operators should implement a "clock" symbol near menu items quickly served to promote convenience for the patron. Homestyle foods

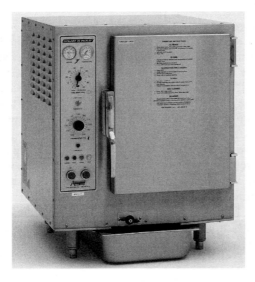

FIGURE 10-3 The Steam'n'Hold compartment steamer is innovative in that it has no boiler. Lower-temperature steam is used to produce a cooked product that is more tender, juicier, healthier, and more nutritious, with less shrinkage. (Courtesy of AccuTemp Products, Inc.)

that are produced fresh suggest that someone is working hard in the kitchen. There is an advantage to letting patrons view the chef at work since a cook on stage is one who will maintain a cleaner work area and take greater pride in the job done.

Locally, the best trend is to visit the competition and see what is selling. People talk about watching their waistline, yet when an operation has taken the time to make its dessert the best in town and to promote it as such, everyone is buying calorie-laden items to consume on-site or off-premises.

Regional

Americans continue to have a growing appetite for ethnic variations, whether Mexican, Asian, or Italian. On the regional scene, each area has it favorites, but the diverse larders of the Americas, including Central and South America as well as the Caribbean, seem to be influencing dishes everywhere. Increasingly, chefs and restaurateurs are finding culinary inspiration from the kitchens around the world.

Fifteen years ago, sun-dried tomatoes and goat cheese were the latest trends. They're now as American as apple pie. Regional American, including Louisiana Cajun and southwestern dishes, is on the scene, with the butter and cream of France gone and the big, brash flavors of Latin America being hot. Foodservice operators need to read of and sample the latest cuisines constantly to offer patrons the best in regional cooking.

National

Speed seems to be the need on the national level. Americans want their food and they want it now. Foodservice operators would be wise to consider the car's dashboard as a placemat and begin to package good-quality cuisine for the ever-growing take-out markets.

Ready-made salads, precut vegetables, presliced fruits, and convenience in preparation and service are the key to high revenues for operators tomorrow. Cooking skills are lacking at home, few have time to sit and enjoy a meal together, and lifestyles have individual family members all wanting different foods to eat.

At the national level, operators should note that there is a slower growth in the population, greater ethnic diversity, an aging population, more women in the workforce, and slower growth in income, with a widening disparity in its distribution. These demographic changes imply that total food sales will not grow very much, a greater variety of ethnic foods will be consumed, and the elderly will increase demand for healthy, nutritious food.

International

The world's population is expected to double in the next 40 years. Minorities will use their increased economic power to influence new choices in foods and services to reflect their own tastes and lifestyles. The bottom line is that foodservice operations offering good value and "something for everyone" will probably flourish in the decade ahead.

International competition increasingly calls for work to be performed during hours that match the business hours in different zones. More night workers will probably create a demand for more 24-hour foodservices.

With the influx of immigrants to the United States, Americans use 68 percent more spices today than a decade ago. The consumption of red pepper rose 105 percent in the same time period, while that of basil, 190 percent. Patrons are increasingly willing to try new flavors and familiar dishes with a twist when dining away from home.

International cooking trends will continue to be something for every foodservice operator to watch. As the world changes, so will those operations aware of what is happening.

■

SUMMARY

Cooking basically involves the application of heat to make food more appetizing, more digestible, safer to eat, and more palatable. Although heat has been used to cook food for centuries, only recently have new methods been developed to improve and hasten the process.

To cook food well, a person must know a lot about what happens to food when it is cooked. Heat changes food. Starches gelatinize or absorb water, thickening foods. Proteins coagulate, changing color and becoming more firm. Other chemical and physical reactions occur, as well. Knowing these principles is an important prerequisite for producing good food.

Standard measures of heat are the calorie and the Btu. The amount of heat available in calories or Btu helps determine whether satisfactory cooking performance can be obtained from a particular piece of equipment. Heat transfer is also important. To cook well, heat must be quickly and evenly transferred to cooking vessels, cooking media, and the food itself. Heat frequently gets into food by conduction—the direct transfer of heat from one particle to another. Another way heat gets into food is by convection; air or a liquid such as fat or water move through an enclosed space, carrying heat with them. Food can also be cooked by means of a form of heat energy called *radiation*, which transfers heat by electromagnetic waves. The amount of heat and the length of time it is applied to food must be controlled if good cooking is to occur.

Cooking usually occurs in one of two ways: cooking with liquid, or cooking without liquid. Boiling is a method of cooking in water or other liquid at a temperature of 212°F (100°C). Blanching (sometimes called scalding) is another way of heating food in water; the food is only partially cooked, but the heat treatment is sufficient to produce the desired result. Parboiling is another method of partial cooking; afterward, the food is usually cooked to completion by another method. Braising is a method of cooking in small amounts of liquid. Pot-roasting, swissing, fricasséeing, and jugging are all forms of braising. Poaching involves cooking in liquid below the simmering point of water, about 195°F (91°C). The liquid should at least half cover the food. Simmering involves cooking food in liquid at a temperature of between 185 and 205°F (85 and 96°C), so that the water is sufficiently hot to cook but does not boil. Stewing is a method of cooking in water or other liquid whereby small cuts, disjointed fowl, cut-up meat, and other products are submerged in liquid and cooked at simmering temperature.

Steaming may be free-vent, pressure, or forced convection. Free-vent cooking uses steam at normal boiling pressure to cook food. Sometimes waterless cooking uses such steam for cooking; in this case, the food itself furnishes the moisture to produce the steam. Pressure steaming may occur in conventional steamers at around 6 or 7 psi of pressure or in high-pressure steamers at around 15 psi. The time of cooking is very short in the latter instance. Forced-steam convection is a relatively recent idea in steam cooking. Steam at a temperature of 212°F (100°C) and under considerable forced convection strikes food. The cooking is very rapid, as occurs under steam pressure.

Baking is a method of cooking by dry heat in an oven. No liquid is used. Barbecuing is a method of cooking over coals or in a pit; a tangy sauce is used to baste the product. Broiling involves cooking food by means of radiated energy. Grilling involves cooking food on a flat, heated surface at moderate temperatures; it sometimes is called griddle-frying, and it is closely akin to sautéing and pan-frying. Griddle-broiling and pan-broiling are two other forms of cooking that resemble sautéing, except that in these the fat is not allowed to accumulate around the item as it cooks. Ovenizing is a process by which items are given a fried or sautéed appearance by being cooked in the oven. Roasting involves cooking by dry, indirect heat in an oven. Baking is a term used interchangeably with roasting, although each describes the cooking of certain distinct (and different) kinds of foods. Both baking and roasting are done in an oven with dry, indirect heat.

Frying includes both deep-frying and sautéing. Deep-frying is a type of frying in which the product is completely immersed in hot fat. Foods for deep-frying are often breaded, batter-dipped, or dough-covered. Pressure frying is a form of frying under pressure. Sautéing involves frying food in shallow fat; the food may be braised after such sautéing. Pan-frying is a form of sautéing.

Three new kinds of electrical energy are being used to create heat for cooking: infrared, microwave, and quartz. They have special advantages but also some disadvantages. Other new cooking technology is also being introduced on a continuous basis.

Local, regional, national, and international dining trends suggest that patrons will value taste and nutrition more than price. In the Midwest, fat-free foods are extremely popular, but on the east and west coasts, patrons prefer fresh vegetables and grains.

The demand for fresh, all-natural, and organic foods may result in lower sales for processed and packaged foods over time. The bottom line sees patrons increasingly searching for nutritionally balanced foods that also offer convenience.

CHAPTER REVIEW QUESTIONS

1. From the material provided in this book as well as from an interview with two different local foodservice operators, define the cooking trends within your area and present them in a written report. Are these local trends similar to regional and national trends? Why or why not?

2. Compare and contrast the three forms of heat transfer. Give examples of each, noting which pieces of equipment transfer heat by which form.

3. Describe the step-by-step procedure for braising. What do the terms *blanc* and *brun* mean? What foods are braised?

4. Discuss the methods of cooking by steam. What are the advantages or disadvantages of each method? What is pressure-fryer cooking?

5. Discuss several new ways in which electricity is being used to cook foods in quantity kitchens.

CASE STUDIES

1. Janet Joplin is the kitchen manager for the Spinnaker Bar and Grill, a restaurant specializing in the simple preparation of fresh seafood. She has been working closely with the owner/manager, Chris Cristopherson, on developing a wide-ranging new menu. Their goal has been to come up with a well-balanced menu that exhibits a variety of cooking styles and presentations. In the past, the restaurant has had a standard range and a deep-fat fryer in the kitchen. The operation is in the process of installing a broiler and a steamer, as well, which will give Janet and Chris considerably more freedom in planning the menu.

Janet has taken particular pride in being able to hire people with little previous cooking experience and train them in a short time to produce a high-quality product. She knows that the new menu will require that her staff be trained in the new types of cooking becoming available to them with the addition of the new equipment. She also wants to strengthen their knowledge of the cooking procedures they have been using heretofore. Janet wants to be able to do this in a short (not more than two pages), concise guide to exactly what each cooking technique is and what its strengths and weaknesses are, particularly with regard to seafood.

Since she knows that you are covering this material in a class you are now taking, she has approached you for help in designing this manual. In return, you get a candlelit dinner for two at the Spinnaker. Draw up a document for her that covers all the major cooking techniques available to her. Bear in mind that she has an oven, range top, deep-fat fryer, broiler, and steamer at her disposal, as well as a wide variety of pots and pans.

2. You are considering the purchase of some new foodservice equipment that uses steam. You have been told to research the Internet and have decided to purchase your equipment from that offered at *www.mfii.com*. What equipment will you purchase? Of those pieces, do any involve new cooking technology? What is that technology?

■

VOCABULARY

Celsius	CVAP
Fahrenheit	Baking
Calorie	Ovenizing
Btu	Frying
Conduction	Sautéing
Convection	Deep-frying
Radiation	Free fatty acid
Mise en place	Smoking point
Boiling	Hydrogenation
Braising	Catalyst
Simmering	Scalding
Steaming	Griddle-broiling
Brun	Infrared
Blond	Microwave
Mirepoix	Broiling
Poaching	Barbecuing
Blanching	Pan-broiling
Parboiling	Grilling
Free-vent	Acrolein
Stewing	Turnover (fat)
Roasting	Fricassée

■

ANNOTATED REFERENCES

Chesser, J. 1992. *The Art and Science of Culinary Preparation.* East Lansing, Mich.: Educational Institute of the American Hotel & Motel Association. (Quality food production techniques and standardized recipes are presented in this text.)

Culinary Institute of America. 1995. *The New Professional Chef,* 5th ed. New York: John Wiley & Sons, Inc. (Complete reference for cooking methods and trends.)

Dornenburg, A., and K. Page. 1996. *Classical Cooking the Modern Way,* 2nd ed. New York: John Wiley & Sons,

Inc. (Recipes, the art of cuisine, and more are presented in this book.)

Escoffier, A. (translated by H. L. Cracknell and R. J. Kaufman). 1995. *The Complete Guide to the Art of Modern Cookery*. New York: John Wiley & Sons, Inc. (This is considered a culinary classic.)

Gisslen, W. 1992. *Advanced Professional Cooking*. New York: John Wiley & Sons, Inc. (Explores contemporary American cuisine and its classical roots.)

Gisslen, W. 1994. *Professional Cooking*, 3rd ed. New York: John Wiley & Sons, Inc. (Cooking theory and applications are presented in this text.)

Labensky, S. R., and A. M. Hause. 1999. *On Cooking*. Upper Saddle River, NJ: Prentice Hall. (Cooking is covered in detail in this book.)

Labensky, S., G. G. Ingram, and S. R. Labensky. 1997. *Webster's New World Dictionary of Culinary Arts*. Upper Saddle River, NJ: Prentice Hall. (Culinary terms are presented from A to Z.)

Laconi, D. V. 1995. *Fundamentals of Food Preparation*. New York: John Wiley & Sons, Inc. (Introductory text on foodservice fundamentals.)

Mizer, D. A., M. Porter, and B. Sonnier. 1987. *Food Preparation for the Professional*, 2nd ed. New York: John Wiley & Sons, Inc. (From food preparation to food selection and presentation, this book covers it all.)

Pauli, E. 1997. *Classical Cooking the Modern Way*, 3rd ed. New York: John Wiley & Sons, Inc. (Cooking techniques are presented in this book.)

Riely, E. 1996. *The Chef's Companion: A Concise Dictionary of Culinary Terms*, 2nd ed. New York: John Wiley & Sons, Inc. (Culinary terms are presented in this book.)

Rubash, J. 1996. *The Master Dictionary of Food and Wine*, 2nd ed. New York: John Wiley & Sons, Inc. (Culinary definitions are presented in this text.)

Salkind, N. J. 1997. *The Online Epicure*. New York: John Wiley & Sons, Inc. (Good cooking via the Internet is the basis for this book.)

Schmidt, A. 1996. *The Chef's Book of Formulas, Yields, and Sizes*, 2nd ed. New York: John Wiley & Sons, Inc. (An abundance of routine calculations and more complex formulas for mastering the efficient purchase of food for the kitchen are presented in this book.)

CHAPTER 11

Sanitation and Safety

I. Sanitation
- A. Microorganisms and Their Classifications
  1. Bacteria
  2. Viruses
  3. Parasites
  4. Molds
  5. Yeasts
- B. Foodborne Illnesses
  1. Chemical Poisons and Poisonous Food
  2. Ptomaines
  3. Foodborne Infectants
  4. Foodborne Intoxicants
- C. Preventing Foodborne Illnesses
  1. Food Protection
     a. Purchasing Food
     b. Storing Food
     c. Preparing Food
     d. Serving Food
  2. Personal Hygiene
  3. Facility and Equipment Sanitation
     a. Cleanliness of Surfaces
     b. Ventilation and Lighting
     c. Garbage Disposal
     d. Comfort Facilities
     e. Equipment Sanitation
     f. Dishwashing Procedures
  4. Rodent and Insect Control
     a. Rats and Mice
     b. Flies
     c. Cockroaches
- D. Responsibilities for Sanitation
- E. Damage Control Impact

II. Safety
- A. Employee's Responsibility
  1. Cuts
  2. Slips and Falls
  3. Burns and Scalds
  4. Improper Lifting and Hauling
- B. Management's Responsibility
- C. Worker's Compensation

III. Americans with Disabilities Act

IV. Hazard Analysis Critical Control Points

V. Computer-Assisted Sanitation and Safety Programs

■

LEARNING OBJECTIVES

By the end of this chapter, the reader will:

1. Know the different types of microorganisms and their classifications.

2. Understand how foodborne illnesses occur and how to prevent them from occurring in the future.

3. Recognize the difference between foodborne infections and foodborne intoxications.

4. Be able to distinguish between the manager's and employee's responsibilities in sanitation and safety.

5. Better understand the role the computer can play in sanitation and safety programs.

6. Learn of HACCP, what it means, and how it affects foodservice operations.

7. Be familiar with the Americans with Disabilities Act and how it affects foodservice operations.

■

SELECTED WEB SITE REFERENCES

American Gas Association
www.aga.org

Ameri Clean Systems, Inc.
www.americleansystems.com

Cooper Instrument Corporation
www.cooperinstrument.com/

DEEZEE Chemical Home Page
www.dzchem.com

Ecolab
www.ecolab.com/

Johnsonite Division
www.johnsonite.com/

Tips on Carpet Care
www.dalton.net/cptcare.html

Water Management Services
www.watermgmt.com/laundry.html

Sanitation and safety in a foodservice operation are the responsibility of every person working in the establishment. According to the Federal Centers for Disease Control, over 9000 deaths are caused from food poisoning alone every year in the United States. The proper discharge of sanitation and safety duties ensures safe food and freedom from accident or injury.

Sanitation and safety regulations established by local, state, and national bodies must be enforced constantly by management. Sanitary regulations are usually outlined in public health codes, while the Occupational Safety and Health Act (OSHA) and other state and local regulations specify standards for safety. Proper sanitation and safety, however, do not just happen because a government agency has established appropriate standards; rather, they are made to happen by the concerned and committed personnel of a foodservice.

Proper sanitation results in healthful, clean, and wholesome food, an orderly environment, and pleasant working conditions. Personnel, as part of the working environment, must also be clean and healthy if sanitary conditions are to be maintained. The term *safety* refers to freedom from risk of accident, injury, or other harm. Clearly, safety and sanitation are related, since unsanitary food can be unsafe and lead to illness. It is best to consider them separately, however, because the basic factors associated with them differ.

As presented in this chapter, the sanitary standards of meat and poultry have been greatly improved since new standards have been implemented through Hazard Analysis Critical Control Points (HACCP). This system calls for identifying points during food handling at which contamination is likely to occur, establishing ways to prevent it, and monitoring the system to ensure the measures are working. In many states or locations today, foodservices are required to have at least one worker on duty who has satisfactorily completed the Educational Foundation of the National Restaurant Association's sanitation course.

■
SANITATION

There is a difference between the words *clean* and *sanitary*, although both can mean the same thing. *Clean* refers to the lack of soil or dirt on an item. *Sanitary* refers to the lack of organisms that may do harm. Thus, a utensil may look clean and yet not be sanitary, because unseen organisms are present. Conversely, something may be sanitary and yet not look clean. An example would be a utensil coming from a dish machine in which it was sanitized by immersion for 10 seconds in rinse water heated to 180°F (82.2°C): soil may still show on it, but the utensil is sanitary because all harmful microorganisms have been destroyed by the hot rinse. In a good system of sanitation, the aim is to make things clean as well as sanitary.

1. UNSAFE FOOD HOLDING TEMPERATURES	**Examples:** Holding prepared, potentially hazardous foods at room temperature; unsafe refrigeration temperatures; unsafe hot holding temperatures.
2. POOR PERSONAL HYGIENE	**Examples:** Failure to wash hands before starting work, after using the toilet, or after touching any soiled object; wearing soiled aprons and outer garments.
3. CROSS CONTAMINATION	**Examples:** Cutting raw foods and cooked or ready-to-serve foods on same cutting board without sanitizing between changed use; use of slicers, graters, choppers and grinders for more than one food product without cleaning between changed use; same for cook's knives.
4. UNSANITARY DISHWARE, UTENSILS AND EQUIPMENT	**Examples:** Improperly cleaned and sanitized tableware, utensils, and cutting equipment; failure to protect sanitized ware from contamination.
5. INFECTED FOOD HANDLERS	**Examples:** Food handlers with infected cuts, burns, or sores; boils or pimples; sore throat; nasal discharge or diarrhea.
6. IMPROPER FOOD HANDLING	**Examples:** Unnecessary use of hands during preparation and serving; thawing of frozen food at room temperature or in warm water.
7. UNSAFE COLD HOLDING AND REHEATING OF DELAYED-USE FOODS AND LEFTOVERS	**Examples:** Slow cooling and reheating of foods; large mass food storage in large-quantity containers; failure to reheat food to safe serving temperature; use of holding or warming units to reheat food.
8. IMPROPER FOOD STORAGE	**Examples:** Uncovered foods on refrigerator shelves; raw foods stored directly on shelves or against refrigerator walls; raw foods stored in direct contact with prepared foods.
9. INSECTS AND RODENTS	**Examples:** Failure to eliminate pest breeding or entry areas; failure to eliminate grime, spilled food, and trash which become food, breeding, and nesting attractions for pests; failure to report and take control action when pests or evidence of pests is noted.
10. CHEMICALS STORED NEAR FOOD	**Examples:** Storage of cleaning and sanitizing compounds, solvents, pesticides and other non-food chemicals near food; use of unlabeled containers in kitchen or serving areas.

FIGURE 11-1 Factors that may lead to an outbreak of foodborne illness.

Microorganisms and Their Classification

Our world is filled with countless living organisms that are so small we cannot see them without using a microscope. For this reason, they are called *microorganisms*. In order to have sanitary conditions, we must understand how microorganisms live, grow, and reproduce and how they are transported. Employees who know how to keep things sanitary understand the "why" as well as the "how." Some states today require foodservice operations to keep a qualified staff sanitarian on duty at all times. Such a person has satisfactorily completed a sanitation course and is qualified to oversee the maintenance of proper sanitary conditions. Laws and regulations governing sanitation are designed to promote and protect health by requiring adoption of good sanitary practices built on sound microbiological principles (see Fig. 11-1). Some of the common "criminals" in foodborne illnesses are listed in Table 11-1.

Microorganisms can be helpful or harmful; as a group, however, microorganisms are dangerous because some can cause serious illness to individuals exposed to them. Microorganisms that cause diphtheria or that develop the powerful poison called *botulinum*—an amount of which small enough to lie on a needle point can kill a person—are examples. Other microorganisms do no harm to humans. Some that live in the soil break down dead organic material, returning the natural substances to the soil so that they are again available to plants. If these microorganisms did not exist, dead trees, leaves, animals, and other matter would accumulate without decomposing. In a remote way, some microorganisms are helpful to human beings; in a sanitary way, they are not.

TABLE 11-1   Harmful Microorganisms

Criminal	Route to the Customer	Disease
Salmonella group	Food contaminated by unwashed hands, rats	Food infections
Staphylococcus	Food contaminated by sores, boils	
Eberthella typhi	Water, milk, shellfish contaminated at the source; food contaminated by unwashed hands or by flies	Typhoid fever
Endamoeba histolytica	Water contaminated at the source; defective plumbing; food contaminated by unwashed hands or by flies	Amoebic dysentery
Shigella group	Water contaminated at the source; dishes or silverware contaminated by a carrier; food contaminated by unwashed hands or by flies	Bacillary dysentery
Trichinella spiralis	Insufficiently cooked pork	Trichinosis
Clostridium botulinum	Home-canned foods improperly prepared	Botulism
Streptococcus group	Raw milk contaminated at the source	Septic sore throat, scarlet fever
Diphtheria bacillus	Dishes or silverware contaminated by a carrier; sneezing, coughing, and spitting	Diphtheria
Campylobacter jejuni	Undercooked poultry and meals, raw milk, and untreated water	Muscle pain, fever, headache, diarrhea
Escherichia coli, E. coli, or *E. coli* 0157:H7	Improper handling of meats, unwashed hands, and contaminated utensils exposed to the bacteria	Hemorrhagic colitis, hemolytic uremic syndrome, or thrombocytopenic purpura
Listeria monocytogenes	Unpasteurized dairy products, meat patés, and processed meats	Listeriosis

Many bacteria and other microorganisms are beneficial. Microorganisms in our intestines manufacture vitamin K. Yeasts help leaven bread, ferment wine and beer, and preserve sauerkraut and pickles. Many of our best cheeses are produced (in part) through the action of microorganisms. The holes and flavor in Swiss cheese come from a special bacteria, while Roquefort relies on the working of a particular mold. Some harmless microorganisms may be good indicators that harmful ones are present. Coliform bacteria, for example, are natural and useful inhabitants of the intestinal tract; but when they appear in food, milk, or water or on solid objects, they indicate that harmful (pathogenic) organisms may also be present.

Scientists have classified many thousands of microorganisms, using a classification system that identifies ancestral connections by general category (kingdom and phylum) and then by increasingly specific groups culminating in categories called *genera*. It also is helpful to classify microorganisms on the practical basis of what they do, since this is important in establishing which ones are harmful and which ones are not. A simple classification suited to our purposes categorizes dangerous microorganisms according to whether they are *toxic* or *infectious*.

A *toxic* organism produces harmful toxins in food. It is the toxin and not the organism itself that causes the problem. The organism may have been destroyed when the food was cooked, but the toxin still remains when the food is eaten and causes the illness. Illnesses caused by toxins are called *foodborne intoxications*. It is important to understand that because of toxins, cooking does not always make food safe.

An *infectious* organism must be taken alive into the body in order to grow and cause harm. It may develop a toxin in the body, or it may just multiply in a way that causes illness. Illnesses caused by infectious organisms are called *foodborne infections*. If food is cooked long enough, infectious agents are destroyed, and no harm occurs.

A third classification of microorganisms (similar in many respects to the first) places them into the following five categories: bacteria, virus, parasite, mold, and yeast. These categories are considered in the ensuing subsections.

Bacteria

Bacteria are a common cause of illness in human beings. They are single-celled microorganisms that resemble plant cells but contain no chlorophyll (the green material found in the leaves or blades of plants). Bacteria are so small that, even when hundreds are clustered together, they cannot be seen. Their average size is 1/25,000 in. in diameter.

Bacteria are usually classified by shape. Round individual cells that group into tiny clusters so that under a microscope they look like a bunch of grapes are called *cocci*. Cells having the shape of rods or elongated tubes are called *bacilli;* they may appear as single rods, in pairs, or in chains that resemble sausage links. Some bacilli can change into spores. In the spore stage, bacilli become dormant within a hard shell that protects them from heat and other outside forces that destroy bacteria. When they are in this tiny shell, they do not grow; but when conditions become good for growth outside, they break out and reproduce living bacteria in the process (see Fig. 11-2). Bacteria in spores may successfully withstand normal sanitary treatment, since the spores can withstand temperatures that kill normal bacteria. *Clostridium botulinum*—the bacterium that makes the deadly botulin toxin mentioned previously—forms spores and thus increases in danger. Cells of bacteria that are spiral-shaped are called *spirilla*. These are always single-celled and not grouped. Spirilla are not known to be pathogenic (harmful to human beings).

Bacteria reproduce by dividing from one into two cells. The two single cells then redivide to form four

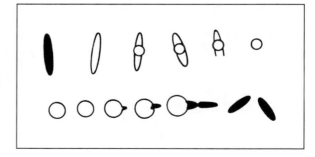

FIGURE 11-2 Reproduction by a bacillus bacterium through an intermediate spore stage.

single cells, and so on. Thus, the number of bacteria that can grow from one cell in a very short time is astronomical. The time between divisions can be as little as 15 to 20 minutes under the right conditions. Poor environmental conditions slow down reproduction. Bacteria in frozen food do not multiply at all, and may gradually die out. However, one should not depend on this happening and think that frozen food has no bacteria. In arctic areas some bacteria have been found that have been frozen for thousands of years.

Bacteria need three things to grow: food, moisture, and temperature. Even if these three are present, other factors may cause bacteria not to grow. First, bacteria thrive on specific kinds of foods. Most pathogens live on the same types of food that people eat. Some, such as *Clostridium perfringens*, are attracted to proteins such as meat, milk, and eggs. Others prefer carbohydrate and protein mixtures such as cream fillings, macaroni and cheese, and gravies. Some can subsist fairly well on foods they are less attracted to, but they do not prosper as well.

Second, bacteria do not do well in dry food; they need moisture. If salt or sugar is present in sufficient quantity in food, the salt or sugar absorbs so much moisture that bacteria cannot grow in it. Foods containing more than 55 percent sugar or more than 10 percent salt will inhibit the growth of microorganisms.

Third, some bacteria thrive in the cold. They are called *psychrophilic* (cold-loving) bacteria. They rot organic material in the cool soil and do other jobs in a cool atmosphere. Other bacteria prefer high temperatures; they may be found in hot water pools or in the water of geysers, and they are called *thermophilic* (heat-loving) bacteria. Bacteria suited to middle-range temperatures (*mesophiles*) include most types of harmful bacteria. They exist in temperature conditions that are neither too hot nor too cold. Their best growth occurs at temperatures of from 98 to 100°F (36.6 to 37.7°C), but public health authorities have a wider danger zone that extends from 45 to 140°F (7.2 to 60°C).

Food preservation discourages the growth of bacteria and other microorganisms. If food is heated and then sealed airtight while still hot (so that no more microorganisms can get to the food), it will not spoil. Dried food discourages organism growth. An acidity level of around pH 4.5 often is enough to deter bacteria from growing, as is an alcohol content of more than 14 percent. Figure 11-3 illustrates the pH scale and gives pH values for representative foods. Creosote in smoke is a preservative, and spices can also be. Some foods are preserved by a combination of antibacterial agents. Thus, sauerkraut and naturally cured pickles keep because of the combined forces of the salt, the acid, and a cool environment. Smoked meat is first cured with salt, saltpeter, and other substances, and then it is smoked; refrigeration also helps preserve it. An understanding of the effects that various processes or substances can have in protecting foods helps encourage better sanitation.

A bacterium eats by absorbing food through the tiny membrane that surrounds the cell. If the membrane is destroyed, the bacterium dies. Some bacteriacides work by destroying the cell membranes of bacteria.

Viruses

The tiniest known living organism is the virus, some of which cannot be seen even under a regular microscope. Viruses are never toxic; all harmful ones are infectious. They are usually transferred through food or water. Hepatitis, a virus infection of the liver, often

Strongest Acidity							Neutral						Strongest Alkalinity
0 pH	1	2	3	4	5	6	7 pH	8	9	10	11	12	13	14 pH

lemon juice	2.0
some vinegars	3.0
tomato juice	4.3
bananas	4.8
fresh bread	5.2

some fruit cakes	6.4
fresh milk	6.8
distilled water	7.0
egg white	7.2
soda cracker	7.8

FIGURE 11-3 Acidity and alkalinity as measured on the pH scale (7 = neutral).

comes from polluted materials in food and water and is contagious. Polio, a virus disease, is transferred in milk or other food substances. Ornithosis is a virus in birds that can cause serious illness in people. Birds and animals should not be allowed into foodservices because they may carry dangerous microorganisms.

Viruses thrive in the same living conditions that support bacterial growth, and they are destroyed by the same things. They seem to be cells or some other self-contained life form that reproduces in organic matter, but how they reproduce and how they live are not known. Viruses are highly feared because of their sometimes fatal effect on infected human beings.

Parasites

A third category of harmful microorganisms consists of parasites. Some, such as *Endamoeba histolytica*, are protozoalike organisms that live in water, in the soil, on fruits or vegetables, or in other foods. The intestinal tracts of people and animals can be infested with this parasite and transfer it to water or food. Food handlers can carry it on their hands. Other parasites include trichinae and tapeworms. Trichinae are usually found in pork but also can be in bear, rabbit or other meat. They are small larvae that grow in the body, lay their eggs in the muscles and cause a disease called *trichinosis* that is characterized by nausea, cramps, diarrhea, aching muscles, fever, and chest pain. We cook pork until it is well done to ensure that we destroy any trichinae present. Eggs of the tapeworm in raw meat are usually the source of its infection in the body.

Molds

Molds, which are types of fungus, usually do not cause illness, but they do spoil food. Molds will attack foods that other organisms cannot invade. They can thrive on cured and smoked meats, jellies and jams, and other preserved foods. More difficult to control by low temperatures alone, molds can be found in refrigerated foods.

Molds are threadlike growths that bear spores or seeds at the end of the thread. They appear fuzzy and cottony and are seen in a variety of colors. Mold spores are constantly circulating in the air, and the best form of control is to keep foods fresh. Molds have difficulty growing on highly acidic foods, and most are destroyed

by heat above 125°F (51.6°C), although some spores may be resistant to even higher temperatures.

Some molds produce toxic substances, but not all molds are harmful or undesirable. The blue mold found in some cheeses give the cheese its characteristic flavor and appearance. A similar blue mold is used in the production of the antibiotic penicillin. Mold found on the surface of aged meat can be trimmed off without affecting the wholesomeness of the underlying meat.

Yeasts

As is the case with molds, yeasts (which are also fungi) do not cause illness, but they can cause spoilage of food. Yeasts are single-celled organisms of the fungus kingdom that are present in abundance. In a vineyard, yeasts will quickly group and grow on a ruptured grape.

Yeasts, acting on carbohydrates, cause fermentation. The yeast acts on the sugar present in a carbohydrate to produce alcohol, carbon dioxide, and water. Some yeasts are used in breadmaking, while others are used in the production of alcoholic beverages such as beer. Yeasts can be cultured to perform specific jobs. Breadmakers, for example, use a cultured yeast to obtain predictable results, rather than using wild yeasts present in the air (which might produce an unwanted reaction).

Foodborne Illnesses

People can become ill from poisons in food, from bacteria or viruses, or from parasites. Poisonous substances can be specially manufactured by chemical companies or they can come from substances that normally are not a threat. If a poisonous substance is consumed, illness or death may result.

Chemical Poisons and Poisonous Food

Danger from chemical poisons is ever present. Toxic agents such as cadmium in graniteware or zinc on galvaneal (galvanized iron) can cause illness. Lead, antimony, and even copper can cause metallic poisoning; acids in some fruit drinks or from other food sources can dissolve these metals. Many cleaning or sanitizing solutions are toxic, and some look like food substances. Trisodium phosphate, a cleaning agent, looks like cornmeal. One powerful roach poison looks like powdered milk. Oxalic acid bleaches well but is very poisonous and should not be used around food. Any poisonous agent that must be used around food should be locked up in a separate area and issued only for a fixed period of time to people who know what they are handling. Cleaning agents and other poisons should be stored separately from foods. All poisons should be clearly marked with the word *POISON* and a skull-and-crossbones label.

Chemical poisons can be introduced at other stages of the food production process, as well. Fruits and vegetables may be covered with a film of insecticide spray or other substances that are harmful if ingested. Consequently, careful washing of all foods is a standard procedure in quantity food production. Some food additives, when present in excessive amounts, may cause poisoning. In large quantities, nitrites, niacin, and cobalt can cause illness. Excessive use of monosodium glutamate (MSG) or of the whitener, sodium sulfite,* can trigger allergic reactions in sensitive individuals. Knowledge of the composition of the food being used in an establishment helps prevent an excessive presence of these chemicals.

Some plants are poisonous, such as some mushrooms, some toadstools, water hemlock, jimsonweed, castor bean seeds, and rhubarb leaves. The green substance that appears in potatoes exposed to sunlight is poisonous, but it takes 17 lb of this green substance (called *solanine*) to kill an adult person. Ergot, a substance in moldy rye, is much more deadly and once killed many people.

Poisons can be found in fish, in shellfish, and even in milk. Some Japanese eat a very poisonous puffer fish that, if prepared correctly, poses no danger. Because mistakes are sometimes made, however, eating it is called playing "Japanese roulette"! A tiny red mite in the ocean can infest shellfish, causing them to become poisonous. If snakeroot forms a significant part of a cow's diet, it makes the milk poisonous.

Purchasing food from sources known to have high standards of purity and sanitation should be a rule in foodservices. Inspection of purveyors' places of business by the buyer is recommended. Public health departments require all commercially offered shellfish to come from beds that it has inspected to ensure sanitation. Vendors can obtain a public health certificate for shellfish taken from such beds. Many original containers are stamped with the health permit number issued by the public health department.

Ptomaines

Some foods are safe to eat on arrival at the facility, but spoilage afterward makes them unsafe. Ptomaines are products of the putrefaction of proteins. Many people complain that they have suffered ptomaine poisoning, but this is usually not true. Indeed, as far as can be determined, ptomaines are not poisonous. They may cause a person to become ill, however, simply because the thought of eating spoiled food is nauseating. (Foods may be poisonous because they have been intentionally sprayed with insecticides or other potentially harmful substances. "Wash all foods thoroughly before using," is a standard saying in quantity food work.)

Foodborne Infectants

Infectious organisms are often the cause of foodborne illnesses. To cause trouble, these must be alive in the food or drink a person consumes. To prevent infection, the microorganism should be kept out of the item; cooking infected foods at high temperatures for extended periods of time will also render the food safe to eat, but the risks of failing to destroy all of the bacteria usually makes this option unsatisfactory. Meat salads, sandwich salad fillings, and other high-protein foods usually are the source of the poisoning. The basic source of *salmonella* is fecal matter. Food on a block on which chickens have been cut up can pick up the *salmonella* present in the interior cavity of the chicken and become contaminated. Poison symptoms from *salmonella* appear in 12 to 24 hours; fever, abdominal pain, diarrhea, vomiting, and chills are characteristic.

Clostridium perfringens is a species of bacteria found in meats and proteins. Incomplete cooking may

---

*Use of sodium sulfite is no longer permitted in foodservices; its use in processed foods is also restricted and must be shown on the label.

leave them in the food. These bacteria form in food and develop a foul odor. Again, the basic source is fecal matter. People become ill in 8 to 22 hours, with acute abdominal pains and diarrhea rather than with nausea, fever, chills, and other usual symptoms of illness. Until recently, these bacteria were unknown.

Shigella, an infection causing diarrhea, is a bacillus that lives in moist foods containing meat, eggs, or other protein. Again, the basic source is fecal matter. Illness is experienced from one to four days after infection.

Different species of *streptococcus* cause strep throat, scarlet fever, and other illnesses. People infected with strep often contaminate food or drink consumed by others, but other causes may bring about infection. Some of these infections are mild; others can be fatal.

Virus infections often come from raw shellfish or other foods that may come into contact with raw sewage. The urine of rats has been known to carry hepatitis and to cause infection. A person may be a carrier—even someone who appears to be healthy but carries germs of the infection that can be spread to others. Most virus infections are dangerous.

Many illnesses are caused by infectants, including the common cold, tuberculosis, scarlet fever, and diphtheria. People who work around food should be in good health. A sneeze or a cough can infect food and spread illness.

This nation is fairly free of foodborne illnesses because of its high sanitation standards. Nevertheless, people still become ill from them because of mistakes and failure to maintain safeguards; and from time to time, outbreaks of poisoning or illness still occur because of bacterial, viral, protozoan, or larval contamination. Clean food, drink, premises, and people are the best safeguards against such contamination.

In areas where municipal drinking water is not used—as in a resort, a rural restaurant or hotel, or a camp—care must be taken to see that the water is potable. Wells and other water sources should be checked by having the public health department examine water samples. If problems exist, chlorination systems should be used. Improper plumbing connections, poor trapping, backwater, blockages, or siphoning can bring sewage water into an area, and contamination can then occur.

Beef, fish, and poultry from uninspected sources may be the source of tapeworms or other larval infections. Raw pork, bear, or rabbit can contain trichinae, a microscopic, wormlike organism that settles in the muscle meat of the host. If diseased meat is processed improperly, consuming it may lead to trichinosis, a serious illness. Federal law requires all sausage or smoked meat that may be eaten uncooked by the consumer to be cooked by the manufacturer during processing to a temperature of 155°F (68.3°C). This is done to ensure that trichinae are destroyed.

Foodborne Intoxicants

Staphylococcus, a genus of cocci, is a common bacterial source of toxic food poisoning. Most people carry staph, but it is also in the air and on most objects. Staph often can be spread by coughing or sneezing, or from infected cuts, boils, pimples, other skin infections, acne, or pus. Foods commonly involved in promoting staph are cooked ham, other cooked meats, cream-filled or custard-filled pastries, potato salad, ham salad, tuna salad, and other carbohydrate–protein foods.

The staph toxin usually causes illness in 1 to 6 hours; symptoms are nausea, vomiting, diarrhea, fever, and abdominal cramps. People in poor physical condition, young children, and old people are relatively susceptible to the poisoning.

It is often found in instances of foodborne illness that some people become ill while others eating the same food do not. Those who remain unaffected may have a higher tolerance to infection or be in better physical condition and thus be able to throw off the illness. On the other hand, some people may escape infection because a food bears the harmful microorganisms in only one area, and only some patrons are served the infected portion. This can be true of either infectious agents or toxins in a batch of food. All disease-producing organisms must be controlled by appropriate sanitation procedures.

Clostridium botulinum has been mentioned as a species of bacteria that makes one of the most toxic substances known, botulin. The bacteria grow without oxygen in a neutral, slightly alkaline, or slightly acidic food. Highly acidic foods such as tomatoes, pickles, and fruit juices will not support the bacteria.

Pickles of low acidity, however, have been found to contain botulin. Outbreaks of botulism have come from canned tunafish and from canned vichyssoise soup. Smoked whitefish placed in plastic wraps that sealed it from the air caused another outbreak. Under-processed canned food may develop botulin, and home-canned foods have been the cause of deaths from botulism.

A small quantity of botulinum is enough to cause death. Earliest symptoms occur 12 to 36 hours after eating. Dizziness, double vision, muscular weakness, and difficulty in swallowing, speaking, or breathing are characteristic. The toxin attacks the nervous system and—unless an antitoxin is administered in a short time—a painful death occurs.

Escherichia coli is a bacterium that usually comes from fecal matter. Several years ago four children died and many others became ill because of undercooked hamburger patties that contained the bacteria. A fifth child died from the same cause because of becoming infected after playing with one of the four who died. Shortly after this a wide number of people became ill from the same cause after drinking unpasteurized fruit juices. The HACCP came about as a result of concern about the safety of our meat and poultry supply. Even after passage of this regulation, deaths and illnesses have been reported from this food poisoning. They came from eating undercooked hamburger, hot dogs, and cold sausage meats.

These continuing and other deaths and illnesses caused by eating contaminated foods emphasize the need for greater care in food preparation to see that the highest possible sanitary standards are maintained. Ground meats should be cooked thoroughly to destroy any bacteria. Eggs and poultry, frequent sources of *Salmonella* poisoning, should also be well cooked. In some cases it may not be possible to cook an egg product, such as in mayonnaise or hollandaise sauce. In that case, the eggs used should be of highest quality and come from approved sources. Some patrons will still want their steak, prime ribs, roast lamb, or lamb chops slightly undercooked and not undergo any danger. If contaminated at all, these whole pieces of meat will be contaminated only on the outside, and the heat in cooking will destroy living organisms there. Meat is fairly sterile inside if nothing has penetrated it to cause

contamination. The reason that ground meats are dangerous is because the outside portions are mixed with the inner ones, making it possible to contaminate the entire product.

Preventing Foodborne Illnesses

Food may already be contaminated when purchased or it may be contaminated later by personnel, equipment, utensils, or vermin. A carrier of contamination is called a *vector;* good sanitation procedures call for stopping the vector at its source. The potential for transfer is always present, and safeguards must be established to prevent this from occurring. These safeguards come under four headings: food protection; personal hygiene; facility and equipment sanitation; and rodent and insect control.

Food Protection

Each of the functions performed in the purchasing, storing, preparing, and serving of food should be isolated in establishing a food sanitation program. Each offers a chance for contamination to occur, unless the food is properly handled. Management is responsible for establishing suitable standards and procedures, and the remaining staff is responsible for seeing that these guidelines are carried out. It often is said that food protection is 95 percent people and 5 percent equipment and facilities.

Purchasing Food. Food should be purchased from purveyors that operate sanitary facilities. Buyers should pay visits to their purveyors to inspect processing conditions. Only food approved for wholesomeness should be purchased. Meat and poultry should carry the round inspected-and-passed stamp. Shellfish should come from beds approved by the public health department, which publishes a semimonthly list of these. Milk should be processed according to prevailing codes. Delivery trucks and delivery procedures should be overseen to ensure that good sanitation is observed. Frozen food should be delivered frozen and refrigerated food should not have an interior temperature that is higher than 40°F (4.5°C). Receiving personnel should be instructed to check the temperatures of incoming perishables, particularly during warm weather.

Operators should ask federal, state, or local agencies about the reliability of food sources. Municipal water is usually considered safe, but private water sources should be checked frequently.

Storing Food. Storage refers to every place where food is kept before preparation, before service, or during service. If contamination occurs at any of these stages, sufficient time may elapse to cause illness in people eating the food. Several rules of thumb can be identified. Rotate stocks, in order to use oldest foods first rather than newer items. Store foods off the floor and away from walls, and maintain correct storage temperatures. Pathogens are most active at about 100°F (38°C) and the danger zone ranges from 40 to 140°F (4 to 60°C). Temperatures below 40°F (4°C) do not kill pathogens, but they do slow down their rate of growth. At temperatures above 40°F (4°C), bacterial growth speeds up, until the optimum of 100°F (38°C) is reached. Bacteria continue to grow at higher temperatures until the temperature reaches 125 to 140°F (52 to 60°C). At temperatures above 125°F (52°C), most bacteria die. Maintaining correct storage temperatures can do much to reduce food deterioration and contamination. In fact, the Food and Drug Administration (FDA) has moved forward with its new Food Code, suggesting that temperatures in refrigerated storage should be lower than required previously. Operators are encouraged to stay abreast of current literature to learn what the government is doing in their state regarding these new regulations.

Fresh fruits and vegetables should be refrigerated. Hot foods from the range or steam table can be cooled quickly if the container is placed in cool water and its contents are stirred. Studies have shown that cooling in this way brings temperatures down as fast as does cooling under refrigeration, until a temperature of around 125°F (52°C) is reached. The item should then be placed under refrigeration to obtain the speediest further cooling.

Factors affecting the rate of cooling include the amount of food in the container, the thickness or density of the food, the velocity with which the cooling medium (movement of air or liquid) strikes the

container, the radius of the mass (distance from its outer edge to its center), and whether the food is stirred during cooling. Small amounts of food cool more quickly than large amounts; and the greater the surface area of a given amount of food, the faster the cooling will be. This is why storing food in shallow pans not more than 2 in. deep is recommended for fast cooling and good sanitation. Microorganisms have a good chance of developing in food that stays warm a long time after being cooked. Storing warm containers close together in a refrigerator also favors the growth of microorganisms. Preferably, storage shelves should be slatted, and containers should be placed so as to allow good circulation of air around them. Containers should be covered so that air-borne bacteria and organisms cannot drop into them. Storage areas should be cleaned regularly, and temperatures should be checked frequently. Good air circulation helps to maintain a sanitary storage place.

Certain foods such as meats, eggs, and milk must be given special handling. Store these at 40°F (4.5°C) or below (meat discolors at 31 to 34°F [0 to 1°C]). Hold warm meat, egg, and casserole dishes for service at temperatures above 140°F (60°C)—except perhaps rare meats. Food held within the danger zone longer than several hours can become dangerous. Holding foods in this zone for as short a time as possible is a good idea. Times in the zone are cumulative. A turkey that is left out to cool for 45 minutes, then refrigerated, then brought back out to pick the meat for turkey salad for another 45 minutes, and then made into salad and left standing at room temperature for 1 hour has accumulated 2.5 hours of danger-zone time. For this reason, frozen foods that have thawed should be used as quickly as possible and not be refrozen.

Preparing Food. Following good sanitary practices during food preparation minimizes the likelihood of infecting patrons with a foodborne illness. Many organisms are present on the skin, nose, hair, or body of the preparer and can be transferred to the food. Coughing or sneezing readily transfers germs, and hand contact with food should be avoided. If the hands are used, they should be scrupulously clean. Alternatively, single-service gloves are a good sanitary measure (see Fig. 11-4). Foods should be handled with clean utensils, such as tongs, scoops, spoons, or

FIGURE 11-4 The use of single-service gloves is a good sanitary measure.

of the parasites. Figure 11-5 identifies some recommended temperatures for achieving good sanitation.

Serving Food. Opportunities for food contamination to occur are particularly numerous during holding for services and during service itself. For this reason,

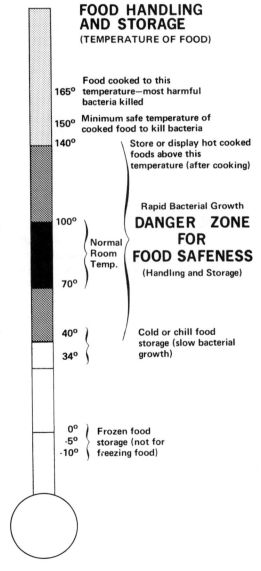

FIGURE 11-5 Recommended temperatures for food safeness (Fahrenheit scale).

forks, as much as possible. Good hygiene and work habits are absolutely necessary.

All fresh fruits and vegetables should be washed thoroughly and, in some cases, scrubbed to remove any organisms or toxic chemicals on their surface. Soaking, followed by thorough washing in fresh water, is usually sufficient to remove most chemical contaminants.

Proper cooking destroys harmful organisms. Solid or viscous food must be stirred during heating to distribute the heat completely throughout the mass. In quantity food preparation, temperatures never get as high as they do in small-batch cooking, and this may cause problems. A food may appear to be boiling when boiling is actually occurring only near the heat source. Trichinae are destroyed at 138°F (59°C), but pork and other meats that can carry it should be cooked to at least 155°F (68°C) to ensure destruction proper holding temperatures must be maintained. The following holding temperatures are recommended:

- *Soups, thin sauces, coffee, tea:* 160°F (71°C) and up
- *Entrées, meats:* 160°F (71°C)
- *Rare meats:* 140°F (60°C)
- *Chilled foods:* 35 to 40°F (2 to 4°C)
- *Frozen foods:* 8 to 12°F (−13 to −11°C)

Protecting food from contamination during holding is essential. If an item is not being served, it should be covered. If customers are serving themselves, a number of steps must be taken to reduce the risk of contamination. Food on a cafeteria counter should be shielded by a sneeze guard. Silverware and dishes to be used by guests should be placed so that contamination is avoided. Silverware should be placed in containers with the handle, not the eating end up. Glasses and cups should be upside down in their holders. All necessary precautions should be taken to see that customers do not contaminate food while serving themselves. In all operations, employees should handle dishes, cups, glasses, and eating utensils in such a way that their hands do not touch the eating surfaces. Packaged foods—such as packets of sugar, crackers, or salad dressings—should be used if possible, and no opened packages should be reused. Some states or localities do not permit open sugar bowls or other open foods on tables. Local sanitation agents can give information on local recommended practices. Employees should be trained to handle food and equipment properly in front of guests (see Fig. 11-6).

Personal Hygiene

Good personal hygiene promotes good health and more effective and efficient job performance. It also helps prevent the spread of disease. Personal hygiene includes getting enough sleep, routinely seeing a doctor or dentist, staying away from work when sick, and avoiding poor personal habits such as picking the nose or biting the fingernails and then handling food or utensils. Unless the hands are clean, food contamination can occur. A clean handkerchief or tissue should be used to cover the nose and mouth when coughing or sneezing occurs. Employees should wash their hands

FIGURE 11-6 Good personal hygiene and proper serving techniques are a must for proper sanitation.

after sneezing or coughing. Scratching the head, wiping the mouth with the fingers, or wetting the thumb with saliva to pick up a paper napkin or to turn pages of an order are unsanitary habits. Smoking a cigarette, cigar, or pipe can transfer organisms from the saliva to the hands or to other items. Smoking should only be permitted in designated areas. Employees should always wash their hands after visiting the rest room, returning from meal breaks, or handling money.

A clean body is the foundation for all the other factors that support a good appearance and proper sanitation. Wearing a cap or hair net is a must to prevent hair from falling into food. Clean hands and nails not only give a good appearance but also protect against contamination. Service personnel make a better impression if they are neat and well-dressed. Makeup may be used sparingly. Nails should be short, well-trimmed, and clean. A well-pressed uniform adds to an employee's appearance of caring about personal hygiene. A healthy, clean person wearing a spotless uniform is more likely to feel like observing good sanitation rules than one who is sloppily dressed, badly groomed, and not particularly clean.

Care is also required in removing soiled dishes and utensils from guests' tables, since these can carry organisms from customers. Table-busing personnel should wash their hands before handling food or clean utensils.

The hands can be one of the main ways in which food becomes contaminated. Good hand-washing sinks should be available at all workplaces. Washing the hands in sinks used to wash dishes or pots and pans is not recommended.

Facility and Equipment Sanitation

Proper cleanliness and sanitation of facilities and equipment is important from both a health standpoint and an aesthetic one. The chances for disease-producing organisms to exist are great in operations in which the housekeeping is poor. Customers who are served food on unclean tables or who see a messy, untidy operation will not wish to continue their patronage. A lack of sanitation and cleanliness is repugnant to customers, who are likely to seek a better atmosphere elsewhere. A clean and appealing place is an economic plus to an operation. Foodservices have an ethical responsibility to serve foods that are clean and safe and to maintain an atmosphere that is in keeping with those food standards (see Fig. 11-7).

FIGURE 11-7 Sanitary working conditions are evident in this kitchen.

An operation should establish a program for achieving proper standards in the facility and equipment. This should cover at least six areas: the cleanliness of floors, walls, and ceilings; proper ventilation and lighting; garbage and refuse disposal; adequate handwashing and comfort facilities; equipment sanitation; and dishwashing procedures. These rules and procedures should be established, and then taught to employees.

Cleanliness of Surfaces. The ability to maintain clean floors, walls, and ceilings depends on their construction. Thus the form of sanitation to pursue is established in the original design of the facility. Periodic cleaning should be instituted, preferably in accordance with a written schedule. Walls and ceilings that are smooth and hard are most easily cleaned. Using light-colored surfaces aids in pinpointing soiled surfaces. The surface structure and composition of floors, walls, and ceilings affect how they should be cleaned. Floor surfaces may vary from carpets in the dining room to concrete, terrazzo, ceramic tile, or polyvinyl chloride in other areas. Each requires unique cleaning materials and procedures (see Table 11-2).

Employees should be taught to maintain a clean environment as they work. When food is spilled on a surface, it should be cleaned up immediately. Spilled food is a safety hazard, and as it sits and dries out, it becomes progressively more difficult to clean. Keeping a well-organized and clean workplace is often called *mise en place.*

Employees should also maintain equipment well. Any equipment needing repair or attention should be reported at once so that it can be put into proper working order.

Ventilation and Lighting. Proper ventilation and lighting are important aspects of facility and equipment sanitation. Accumulations of grease or dirt on walls and ceilings indicate improperly operated ventilation equipment. A well-designed ventilation and exhaust system removes excessive heat, objectionable odors, and noxious gases. This creates an environment in which patrons can enjoy a meal and workers can function efficiently. Filters and ducts in the ventilation system should be inspected and cleaned frequently.

			TABLE 11-2	Characteristics and Cleaning of Various Floors		
Type of Flooring	Hardness	Where Used	Advantages	Disadvantages	Cleaner	
Terrazzo	Hard marble chips in cement	Kitchen, dining rooms	Durable	Damaged by alkalies, stained by grease unless well sealed	Use mild cleaners, scrub, rinse, mop dry	
Ceramic tile	Hard	Kitchen, dining rooms	Resists grease stains if glazed, durable	Unglazed tile easily stained by grease, spaces between tile difficult to clean, slippery when wet	Alkaline cleaners are safe; scrub to clean between tiles, rinse, mop dry	
Concrete	Hard		Durable	Gives off dust, becomes stained by grease unless well sealed	Alkaline cleaners are safe; scrub, rinse (sealing concrete will make it more cleanable)	
Asphalt tile	Soft	Dining rooms	Durable; nonslippery, even when wet	Poor resistance to grease	Mildly alkaline to neutral cleaners; mop, rinse, mop dry	
Rubber tile	Soft		Durable	Damaged by grease	Mild cleaners, mop, rinse, mop dry; never use strong cleaners	
Vinyl	Soft		Nonslippery, even when wet; durable; resistant to grease		Mildly alkaline to neutral cleaner; mop, rinse, mop dry	
Vinyl asbestos	Soft		Durable, resistant to grease		Alkaline cleaners safe; mop, rinse, mop dry	
Cork tile	Soft			Easily damaged by sharp objects, grease will penetrate unless well sealed	Use special cleaners for cork floors (sealing cork tile makes floors more cleanable)	

A dirty exhaust system can pose an extreme fire hazard. A fire igniting in greasy ductwork can spread rapidly and be difficult to control.

Good cleaning cannot occur without proper lighting. Something that is visibly dirty is more likely to be cleaned. It is especially important to have good lighting in the kitchen, in storage rooms, and in hand- and dish-washing areas. Where lighting may be low (to create mood and atmosphere), better lighting should be available for after-hours cleaning. Excess lighting is never desirable. Under energy-saving programs, lighting may be reduced—but never so much as to reduce sanitation or safety.

Garbage Disposal. Garbage and refuse disposal are commonly neglected areas of foodservice sanitation. Battered, leaky refuse cans with covers missing are all too frequently seen. Unless a garbage can is actually in use, it should be tightly covered. Garbage cans or containers should be fly-tight and leakproof. Liquid flowing from a garbage can can coat the ground and furnish a breeding place for flies. Plastic liners may help keep cans clean and facilitate disposal. The time saved in cleaning containers may be well worth the extra cost. Garbage grinders, compactors, and incinerators can also simplify refuse and garbage removal and help improve sanitation.

Comfort Facilities. Adequate handwashing facilities and toilets must be provided if the transfer of microorganisms from person to person is to be prevented. Management must supply adequate facilities. Each employee has the responsibility to see that facilities are used properly and are left in a clean condition for the next person. Facilities that are not properly maintained are open to invasion by insects and rodents; these vermin carry disease and filth from toilets to food. Because employees' hands are probably the most common vehicle for transmitting contamination, good handwashing facilities and practices are important. Such facilities should be located near food preparation areas. Employees should not wash their hands in work sinks. Soap, towels, and a waste container should be provided.

Equipment Sanitation. Equipment sanitation comprises the methods, materials, and times required to keep equipment clean. Good maintenance means good operation. Moreover, it helps prevent the transmission of microorganisms. Properly cleaned equipment lasts longer and costs less to operate than badly maintained equipment.

A cleaning schedule and procedure for each major piece of equipment should be developed (see Fig. 11-8). In establishing a cleaning schedule to be

---

PROCEDURE 1.   HOW TO CLEAN FOOD SLICER

When	How	Use
After each use	1. Turn off machine 2. Remove electric cord from socket 3. Set blade control to zero 4. Remove meat carriage (a) turn knob at bottom of carriage 5. Remove the back blade guard (a) loosen knob on the guard 6. Remove the top blade guard (a) loosen knob at center of blade 7. Take parts to pot-and-pan sink, scrub 8. Rinse	Hot machine detergent solution, gong brush, clean hot water, 170°F for 1 minute; use double S hook to remove parts from hot water
	9. Allow parts to air dry on clean surface. 10. Wash blade and machine shell **(Caution: Proceed with care while blade is exposed)** 11. Rinse 12. Sanitize blade, allow to air dry 13. Replace front blade guard immediately after cleaning shell (a) tighten knob 14. Replace back blade guard (a) tighten knob 15. Replace meat carriage (a) tighten knob 16. Leave blade control at zero 17. Replace electric cord in socket	Use damp bunched cloth* dipped in hot machine detergent solution; clean hot water, clean bunched cloth; clean water, chemical sanitizer, clean bunched cloth

*Fold cloth to several thicknesses.

FIGURE 11-8   Sample cleaning schedule for a major piece of foodservice equipment.

posted in the foodservice, planners should take the following seven steps:

1. Give the name of the equipment to be cleaned.

2. Have three columns headed *when, how*, and *use*.

3. Include all steps (written as briefly as possible) that are necessary to dismantle, clean, and reassemble the equipment properly.

4. Take into consideration the safety precautions necessary to make the equipment safe and clean.

5. Specify the amount of detergent to use and the temperature of water to use for washing, rinsing, and sanitizing.

6. Stress the need for a clean surface for air drying, and for clean hands for reassembling equipment.

7. Provide a separate procedure if daily cleaning differs from weekly or bimonthly cleaning.

Equipment that disassembles should be taken apart, scraped free of major soil, and washed by machine or by hand. All surfaces that food comes into contact with should be washed with hot detergent water, rinsed in warm water, and sanitized in a final rinse. Equipment can be also sanitized by spraying or wetting its surfaces with a chlorinated solution that is about twice as strong as those used for sanitizing by immersion. Different kinds of soil—grease, carbon, egg, and protein, for example—need different treatment. Proteins and eggs must be not be subjected to too high an original temperature or they will be cooked onto the surface. Grease dissolves only at certain washing temperatures.

Water is the primary solvent used in cleaning; a detergent or soap simply makes more things soluble in water. Heat helps to increase solubility, too, and will melt grease. Friction is the main cleaning agent and does about 90 percent of the job; water with detergent does the other 10 percent of the job. After washing, a good rinse in clean water removes detergents and cleaning agents. Sanitizing is the final step. A rinse of 10 seconds in 180°F (82°C) water or of 30 seconds in 170°F (77°C) water is a minimum for sanitizing. Chlorine and other sanitizing agents may be added to the sanitizing rinse to kill any remaining contaminants. Rinse compounds may be used to get better runoff of rinse water, reducing drying time.

Dishwashing Procedures. There are two ways to wash dishes. The first is known as manual or hand dishwashing. The second is mechanical, since a machine is used. Either can give good results, provided that the employee uses proper procedures. The cleaning agents, equipment, work procedures, times, temperatures, and other factors must be correct to get optimum results. Equipment in good operating condition is also required.

Before either manual or mechanical dishwashing can occur, employees must perform the following steps:

1. Separate silverware, glassware, and dishware.

2. Stack trays.

3. Scrape food from utensils.

4. Preflush over a drain. [Using a spray is best; if heated water is used, its temperature should not exceed 120°F (49°C). If preflushing is done into a disposal, use water no hotter than 110°F (43°C). Some machines have a preflush built in. If so, this step can be omitted.]

5. Soak silverware in a mild detergent solution.

6. Stack or rack utensils.

Manual dishwashing may require a slightly different procedure than mechanical dishwashing does. The Ohio Department of Health recommends the following manual dishwashing procedures:

Procedure for a Three-Compartment Sink

1. Wash, rinse, and give a chemical sanitizing.

2. Wash, rinse, and give a hot-water sanitizing.

3. Prerinse, wash, and use a detergent-sanitizer.

4. Prerinse, wash, and use a hot-water sanitizer.

Procedure for a Two-Compartment Sink

1. Wash and use a chemical sanitizer (detergent-sanitizer).

2. Wash and use a hot-water sanitizer.

For mechanical dishwashing, many spray-type machines are available, ranging from the simple single-tank, stationary-rack, stationary-hood, or stationary-door type to the more complex multiple-tank, conveyor type in which dishes move on a conveyor. All are intended to yield a clean and sanitary product, but each must be used correctly to do so.

The temperature of preflush or prerinse water should be below 125°F (52°C) and that of wash water should be between 140 and 160°F (60 and 71°C). Built-in gas, electric, or steam heaters will maintain proper water temperatures. Many machines are built so that the 180°F (82°C) fresh, uncycled final rinse water flows back into the rinse, and water from the rinse area flows back to the wash area, and then back to the prerinse (if the machine has one). This gives a flow of fresh water—from final rinse to rinse, from rinse to wash, and from wash to prerinse—at about the right temperature needed in the machine at each stage.

Detergents are designed for proper cleaning. They are alkaline to help dissolve grease and other products. The alkaline reaction also helps to saponify grease; that is, it helps turn the grease into a soapy substance. Some detergents contain chelating or sequestering agents that hold soil away from utensils once it has been removed. Some detergents contain a sanitizing agent as well, to improve sanitizing. A rinse compound, used to get a better runoff of final rinse water, is usually added separately. The right detergent should be selected for the job, and detergent companies can be helpful in identifying which one is the right one. Sometimes an acid-type detergent may be needed to remove mineral deposits and some stains. Phosphoric acid-containing compounds are used to remove built-up white alkaline salts that form on the surfaces of equipment subjected to hard water. Antispotting compounds or water softeners may be needed in some localities where the water is very hard.

The detergent may be added directly to the wash tank by hand, or it may be dispensed by machine. A light or signal usually indicates when this dispenser is working, and another signal device may warn when the dispenser needs refilling. These dispensers work in different ways but usually do a better job of maintaining a proper detergent-to-water ratio in the wash water than can be achieved by hand addition. Detergents designed for mechanical dispensers may be caustic; they should never be used for manual washing.

Friction in the wash water is produced by a pump forcing water under high pressure through holes in spray arms. The arms may be stationary or moving. Additional friction is generated when spray arms whirl around, throwing the water against the items to be cleaned. The wash water falls, and the soil it carries is collected in strainers or scrap trays. This prevents too much soil from accumulating in the wash water or plugging the machine's spray arms.

After washing is complete, the dishes are given a rinse. The rinse water comes from a tank that is booster-heated to around 190°F (88°C), so it strikes the dishes at about 180°F (82°C). The water should not be heated to 212°F (100°C) because steam produced at that temperature may interfere with the rinsing and cleaning action and because, at such an unnecessarily high temperature, energy is being wasted. In an effort to reduce energy costs and eliminate the need for a booster-heater, some establishments are using low-temperature dishwashers. In these washers, dishes are rinsed in 140°F (60°C) tap water that contains a special chemical sanitizer. The water pressure should be at least 8 psi (pounds per square inch) but not more than 15 psi at the jet, or it should be at least 15 psi but not more than 30 psi in the manifold ahead of the rinse valve. After rinsing is complete, the water flows into the wash section, where it does three things: it helps to keep the wash water hot; it dilutes the wash water, keeping it cleaner; and (as a disadvantage) it dilutes the concentration of the detergent in the wash water.

A drain at the bottom of the wash tank allows for complete draining of the machine. The machine should be cleaned each day. Scrap trays may have to be emptied periodically during the day.

After washing, utensils need to air-dry. This requires space, regardless of whether manual or mechanical washing is done. Sometimes planners set aside one-third of the surface area for soiled dishes and two-thirds of it for clean dishes to dry in. Large operations may need space for five or more drying racks during peak periods. Clean utensils should be removed from racks after drying and stored in dry, clean storage areas above the floor, where they are protected from insects, soil, and other contamination. Employees should take care when handling clean dishes and utensils not to touch eating surfaces.

Rodent and Insect Control

Like people, rodents and insects must have food to live. Because they like much the same foods we do, they seek to feed upon our provisions. In doing so, they contaminate it. The most frequent offenders are rats, mice, flies, and cockroaches. Again, to break the chain of infection or contamination, contact with the vector must be broken. These vermin must be eliminated from the premises.

Even though no single measure completely controls the problems associated with rodents and insects, a good program comprises both basic environmental sanitation and effective chemical or blockage controls of the pests. Different methods must be used for each kind.

Rats and Mice. Rats and mice are highly resourceful, highly adaptable contaminating agents. They are extremely intelligent and often can outwit human beings in gaining their ends. It is necessary to be constantly on guard against their invasion and to adopt severe countermeasures when they attempt to invade. The first step is to be able to identify the enemy (see Table 11-3). Rats are smarter and more aggressive than mice; but mice, because they are smaller, may gain access where rats cannot.

Rats often enter foodservice areas and reach food by gnawing through wood or other material. They need to gnaw to wear down their long incisors, which grow about 5 in. each year. They have been known to gnaw through concrete and even lead pipe. They are agile and can climb wires, pipes, vertical walls, and even metal barricades. They have excellent balance; if they happen to fall, they can drop 50 ft without injury. They have excellent senses of smell and touch. They create rat runs, blackening surfaces with their oily skins, which helps them rediscover their means of entrance. They locate their nesting places in secure areas such as between walls or in burrows. They can get through a ½-in. opening or burrow to a depth of 4 ft to find food. Rats can be vicious and their bite can cause infection. The life span of a rat is from three to five years.

Control of rats and mice starts with the sanitary disposal of refuse and garbage. Rats and mice will not stay where there is no food. Removing trash and other inviting refuse will eliminate hiding and nesting places. Storage areas should be kept clean so as not to harbor them. Poisoning can be an effective method of control, as well. Mice are much more susceptible to poisoning than rats. Some rats can eat strong poisons and show no ill effects; some even seem to thrive on poisons. Slow-killing poisons, which allow the animal

TABLE 11-3 Rodent Pests

Type	Length (Head and Body)	Weight	Coloring	Appearance	Characteristics
Norway rat	7–10 in.	¾ lb (approx.)	Usually gray, may vary from reddish brown to black	Thick body, blunt head, tail shorter than body	Usually infests old structures, places where food and shelter are near
Roof rat	6–8 in.	½ lb (approx.)	Usually gray or black	Slender head and body, long tail, prominent ears, pointed nose	Good climber; may infest upper floors of buildings
House mouse	4 in. (max.)	Less than 1 oz	Gray to brown	Large prominent ears, pointed nose, long very slender tail	Able to enter structures through cracks and holes too small to admit a rat

to leave the premises before dying, should be used. Trapping an also be effective.

Rat- or mouse-proofing performed in a competent manner can do much to eliminate them. This is best undertaken when the building is first built. A thoroughgoing effort will include steps in the following areas:

- *Doors.* All doors should be tight-fitting, with no opening greater than ½ in. between the base of the door and the threshold. The base of all rear doors and other doors where rats can gnaw undisturbed should be flashed with metal, or a U-shaped metal channel should be installed around the bottom of the door. Keep rear doors closed or screened when not in use.

- *Windows.* All basement windows and other windows that are opened within 3 ft of the ground should be screened. Hardware cloth screening should be used for all basement windows and for all lower windows of business buildings. Fly screening may be installed in the same frame with the hardware cloth.

- *Floors.* All basement floors and all ground floors of structures with no basements should be made of concrete and should be joined tightly with the walls. Rats find excellent protection and harborage under wooden floors.

- *Foundations and walls.* Holes in the foundation and the lower walls should be closed with mortar, sheet metal, or other substantial materials. Openings around pipes and wiring should be tightly sealed.

- *Miscellaneous.* Openings on roofs, ventilators, and other building elements should be screened or appropriately closed.

- *New buildings.* All new buildings should be rat-proofed. Most modern, permanent-type constructions can be made rat-proof at no additional cost.

- *Curtain walls.* Concrete floors and shallow foundations should be constructed with a curtain wall around the outer edge extending 36 in. into the ground, or in an L shape 24 in. into the ground with a 12-in. lip extending outward at the bottom.

Flies. Houseflies, persistent and ever-present pests, are major spreaders of filth (which they live and breed on) and disease (which they carry on and in their bodies). When a fly walks over filth, some sticks to its hairy body or to a sticky material on the bottom of its feet. (A fly can walk upside down because of this sticky material.) Harmful microorganisms also can be ingested into the fly's body when it feeds on filth.

A fly contaminates food by walking on it or by eating it. In order to eat, a fly regurgitates liquid from its stomach onto the food. This dissolves the food and allows the fly to suck it back up. Of course, all that is vomited out is not sucked back in. Since the stomach liquid is likely to contain harmful bacteria, it contaminates the area. Fly specks are fecal matter left by the fly wherever it goes. These too contain harmful bacteria. To combat flies effectively, a person should know something of their life cycle.

Flies come from eggs. An adult female may lay as many as 3000 in her lifetime. The eggs hatch into larvae called *maggots*—the little white "worms" often seen in decaying matter. In several days, these change into small, hard, brown capsules called pupa from which, after several days, full-grown flies emerge. A single fly could be the progenitor in forty days of more than 7 million offspring weighing en masse around 140 lb (63 kg), if half of each generation's offspring reached adult fly stage.

Management and workers must work together to control flies. Fly control starts with getting rid of breeding places. This means not allowing food to accumulate and decay, and it means removing garbage, manure, and other favorable factors. Flies often travel considerable distances from where they hatch, so controlling the removal of all breeding places may not be possible. Blocking flies from entrance by using screens on doors and windows is a valuable procedure. Screens must be tight-fitting and in good repair. If flies enter when screen doors are opened, sprays, poisons, and other insecticides present acceptable countermeasures only when used correctly. Used improperly, they can be dangerous to workers and can contaminate food. Such chemical methods of control should only be used by qualified individuals and should only be attempted in locations where food and equipment will not be contaminated. An effective alternative is the electric fly catcher, which draws flies to an electric element that electrocutes them. Keeping the premises scrupulously clean also helps to keep flies away.

Cockroaches. Cockroaches are one of the oldest forms of life on earth, and they have adapted well to human habitations. Cockroaches carry many diseases and also bear a characteristic (and offensive) odor. They need moisture and will be found near it—for instance, in a dirty old wet mop or near hidden moisture that accumulates behind a piece of equipment. Roaches contaminate by crawling on dishes, foodstuffs, and other items. Like flies, they contaminate with their bodies, their manner of eating, and their feces. And once again, they can be controlled better if their life cycle and habits are understood.

Cockroach eggs are deposited in a leathery sack that contains from twenty-five to thirty eggs. They may be dropped anywhere, or they may be glued to walls, shelves, or equipment. It takes about one month for the eggs to hatch and about one year for the young to grow to maturity. Their rate of growth is affected by temperature, moisture, and the amount of food available. They prefer warm, dark, damp places, and they hide in cracks, behind equipment, in boxes, and inside walls during the day, emerging to feed at night when it is dark and quiet. If one or more cockroaches are seen in the daytime, it is certain that swarms of them are nearby, waiting to come out at night.

As in combating flies, cleanliness is one of the most important control measures. Besides eliminating their food, frequent cleaning can pick up the egg sacks and thereby reduce the numbers of new cockroaches. Finding their hiding places helps establish where to place the poisons that will control them. Insecticides of the proper type work well, but they should be used by someone who knows how. Usually a pest control company is best equipped to eliminate rats, mice, and cockroaches, and they can also be helpful in getting rid of ants (see Fig. 11-9). A program of regular treatments is essential. All deliveries should be examined for cockroaches, since this is one of the most common ways they get into operations. Delivery crates and boxes should be unpacked and disposed of as soon as possible.

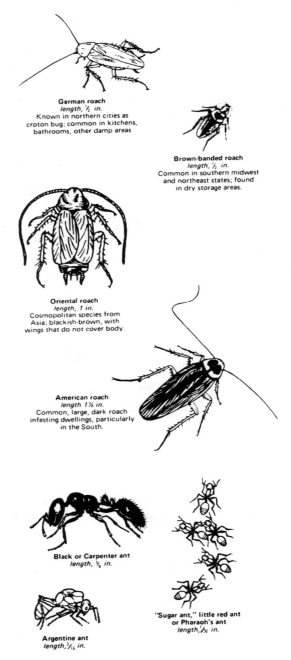

German roach
length, ½ in.
Known in northern cities as croton bug; common in kitchens, bathrooms, other damp areas

Brown-banded roach
length, ½ in.
Common in southern midwest and northeast states; found in dry storage areas.

Oriental roach
length, 1 in.
Cosmopolitan species from Asia; blackish-brown, with wings that do not cover body.

American roach
length 1½ in.
Common, large, dark roach infesting dwellings, particularly in the South.

Black or Carpenter ant
length, ⅜ in.

Argentine ant
length, 3/16 in.

"Sugar ant," little red ant or Pharaoh's ant
length, 1/32 in.

FIGURE 11-9 Crawling insect pests.

Responsibilities for Sanitation

As with other factors in foodservice operations, different personnel have different responsibilities for maintaining satisfactory sanitary standards. Management's responsibilities in implementing a sanitation program include planning the program and training and supervising the staff; the employees' responsibility is to see that sanitary standards are maintained.

Management must provide the proper facilities, equipment, and resources for maintaining the standards. After establishing appropriate policies and standards, management must train the staff in the proper procedures for implementation. Clearly explained instructions and practical guidance are necessary for success. Supervision of the program calls for monitoring the staff to ensure that standards are met. Equipment maintenance and repair policies and procedures should be part of the sanitation program too. Equipment can easily be monitored and inspected during cleaning.

Employees have a responsibility to develop good working habits, to use care in handling foods, and to know what factors promote or impede the growth of bacteria and other organisms that can contaminate food. Employees must know how to control microorganisms and how to practice good hygiene. They should see that equipment and the facility itself are kept in a sanitary condition.

Foodservices must meet certain standards of sanitation to be legally certified to operate. Federal, state, and local regulations require compliance with such standards. Periodically, inspections are made of operations; those that are found to be substandard are cited for violations and may be fined. Correction of the condition is required within a given time. If conditions remain substandard, the operation may be closed. An operation guilty of repeated violations may also be forced to close. Many states today require that a qualified sanitarian (one who has taken and satisfactorily completed a sanitation course) be on duty at all times. The trend is toward placing more responsibility on the foodservice industry itself for the achievement of satisfactory sanitation standards.

Damage Control Impact

Damage control suggests that the foodservice operator is aware of what can go wrong within an operation and be prepared to handle those problems. Studies indicate that those managers who have been mandated to attend a training and certification program in sanitation and safety effectively improve the sanitary conditions in their operations and reduce the spread of foodborne illnesses.

The impact of such training can help to change the fact that foodborne illnesses are a major public health concern today. With 6.5 million to 33 million cases of foodborne illness estimated to occur each year in the United States, costing an estimated $6.7 billion, a majority of the reported cases can be traced to public eating establishments. In an era of widespread travel and international distribution of products, safe food-handling practices are especially important because of the potential for widespread outbreaks of foodborne illness.

Damage control impact has become even more important as increasingly more dollars are spent on foods prepared outside the home. To meet consumer demand, the types of foods served in eating establishments have changed. Formerly, many cooked foods such as soups and stews were offered on a limited menu. Currently, many offerings are cold foods, including raw vegetables and fruits, which require extensive handling by preparers, with greater possibilities for transmission of contaminants.

The bottom line for foodservice operators is to concentrate on food, its preparation and service; on equipment, cleaning, sanitizing, and maintaining standards; on facilities, monitoring water systems, pipes, housekeeping, and ventilation; and on personnel, ensuring proper hygiene and training of staff. The impact of such activity will result in a safer and healthier environment for everyone involved.

SAFETY

In a modern quantity foodservice, a multiplicity of hazards daily confront both operator and employee.

Few occupations, in fact, present so many inherently hazardous features on the job: open flames, hot liquids, high-pressure steam, sharp tools, cutting and grinding machines, and slippery floors are just a few of the unremovable dangers. Accident rates are said to be as high in the foodservice industry as in mining.

Everyone has a responsibility for maintaining good, safe working conditions. Although management is responsible for overall safety guidelines and their enforcement, every person must see that hazards are not created or tolerated out of negligence. Proper observance of safety rules by everyone is essential.

Employee's Responsibility

One of the most important factors in keeping safe is keeping alert. Inattention, failure to anticipate or plan, ignorance of surprise or danger signals, and just plain loafing are factors that bring about accidents. The Industrial Commission of Ohio's Division of Safety and Hygiene identifies six unsafe acts that frequently result in accidents:

1. Failure to look where one is going: kitchen workers carrying pots, pans, food, or supplies collide with other workers. *Result:* bumps, bruises, other injuries.
2. Failure to observe surroundings: cooks at the range stand too near the stove, brush against a hot stove, knock over pots and pans. *Result:* severe burns or other injuries.
3. Failure to handle knives and tools with care: knives are left exposed on open tables instead of being put away in a drawer or other safe place; knives are left protruding from edges of tables and chopping blocks; a knife falls and someone grabs for it. *Result:* cutting or stabbing.
4. Reaching too high or too low, lifting too heavy a weight, or lifting incorrectly. *Result:* strained muscles and sometimes permanent injury. *Rule:* Lift with the leg and thigh muscles, not with the back muscles.
5. Failure to pay attention when using knives, grinders, or other cutting tools: a cook, slicing onions, looks away; a kitchen helper chopping celery in a grinder reaches too far into the grinder. *Result:* cuts or lost fingers.

6. Failure to protect hands from hot pans, pots, and plates. *Result:* burns. *Rule:* Many severe burns result from a failure to use side towels or potholders.

The Industrial Commission of Ohio's Division of Safety and Hygiene and the Ohio State Restaurant Association cite cuts, slips and falls, burns and scalds, and improper lifting and hauling as the major causes of lost employee time in the foodservice industry. Such lost time would be reduced materially if the suggestions given in the subsections that follow were kept in mind.

Cuts

Cuts mean lost time and lost wages to employees and lost profit to the employer. They are among the most common of accidents in the foodservice industry and, unfortunately, are often caused by carelessness. Important preventive measures include the following:

1. Put knives away when you are finished with them. When cleaning a knife, wipe it with the cutting edge away from you. Slice with the sharp end going away from you, preferably toward a cutting board. Pay attention at all times when using a knife.
2. Always use a food stamper to press food into a grinder. Use all slicers properly, with the guard (not your own hand) holding the food to be sliced. Remember, the slicer is power-driven and will not stop just because your finger gets into it.
3. Keep hands away from all moving cutters on ice-cubing machines.
4. Remove all chipped or broken glass immediately, and keep it in a separate container provided for this purpose. Many serious cuts are caused by leaving broken glass and broken dishes in the same area with dirty dishes. As Figure 11-10 suggests, breakage and resulting accidents can be reduced if employees are made aware of the potential costs and dangers of broken glass.
5. Watch for nails and sharp fasteners when unpacking boxes, crates, cartons, and barrels. Use proper tools, and be careful.

FIGURE 11-10 The cost and potential dangers of breakage must be understood by all foodservice employees.

Slips and Falls

Slips and falls are often the result of poor housekeeping. Spilled liquids or foods should be cleaned up immediately. Cluttered aisles and passageways, torn and loose carpeting, and worn stairway present hazards as well. Caution must be exercised while moving about, climbing stairs, and rounding corners. Overheated work areas can lead to fatigue and increase the likelihood of an accident. Inadequate lighting can exacerbate a hazardous situation. By being alert and aware of potential hazards, employees can increase their chances for maintaining personal safety.

Burns and Scalds

Hot materials or equipment are a part of the foodservice business; the best protection against harm from them is constant awareness and caution. Handling hot food with tongs or long-handled spoons or forks and using hot pads to move hot dishes can help avoid injury. Water temperatures at the spigot should seldom be higher than 160°F (71°C). Because of the need to save energy, high water temperatures are being lowered except where essential for a particular job.

Pots filled with hot food should be lifted carefully. The individual carrying them should look ahead to preview the route of travel and the availability of a proper landing place at journey's end. The handles of hot pots should not project over table or range edges. Knowing how to work around hot ovens, griddles, and steam equipment is a part of job competence. An inexperienced worker can receive a severe burn from lifting the lid of a boiling pot or steam-jacketed kettle without using a long fork or other extension tool. Frequently, burns or scalds occur when workers try to do something with hot equipment or food that cannot be done. For example, a worker may reach into an oven at about shoulder height, pull out a roasting pan full of hot meat and juice, and try to lift this down to a table; the weight of the item at that height is too great to permit good leverage, and the result is often injury. Getting help to lift hot items in such cases is the proper procedure to follow.

Improper Lifting and Hauling

Improper lifting or carrying accounts for 25 percent of lost-time accidents. These accidents occur because workers were not trained to carry or lift properly. More injuries occur while putting objects down than while picking them up. Too many people try to lift objects beyond their strength, often suffering back strains or hernias. A person's ability to lift and carry largely depends on physical condition, build, stature, and experience or training. Heavy lifting can be dangerous to a person handicapped with a weak heart, high blood pressure, or a lung disorder.

Simple rules govern proper lifting. People do not realize that leg muscles, because they are bigger and get more exercise, are strong than those of the back and abdomen. Instead, people lift with their legs straight and their back bent over, putting most of the strain on the back and thereby causing injury. Some supervisors erroneously believe that everyone knows how to lift or carry, or they simply do not concern themselves with the question. This is a mistake.

Supervisors should demonstrate to all new employees the following, correct way to lift:

1. Get a good footing.
2. Make certain no foreign object lies on the floor under your feet.
3. Place your feet about shoulder width apart.
4. Bend at the knees to grasp the weight.
5. Keep your back straight.
6. Get a firm hold.
7. Keep your back as upright as possible.
8. Lift gradually by straightening your legs.
9. When the weight is too heavy or bulky for you to lift comfortably, get help.
10. Place heavy weights on dollies or carts before moving them.

Secondary rules include the following:

1. Inspect the object to be lifted for sharp corners, nails, and other things that might cause injury.
2. Distribute the weight evenly by keeping your hands at opposite corners of the object being lifted.
3. If you must shift the position of the load while carrying it, first rest it on something for support.

Many people readily develop good safety habits, but others do not. Studies have shown that some people are accident-prone whereas others seldom have an accident. Older workers can do much to help build good safety habits in newer workers. They should notice things the new person does that might cause injury and, in a helpful way, show the newcomer the correct procedure. Learning and following safety rules can help prevent injury.

Management's Responsibility

Management has primary responsibility for seeing that workers and customers are safe from injury or harm. Customer injuries may translate into claims, lawsuits, or loss of patronage; employee injuries can lead to compensation costs, loss of productivity, and perhaps a lawsuit.

In 1971, the Williams–Steiger Occupational Safety and Health Act became a part of our national labor law. It created the Occupational Safety and Health Administration (OSHA) to administer the act. The purpose of the act was "to assure so far as possible every working man and woman in the nation safe and healthful working conditions and to preserve our human resources." The act performs several functions:

- It encourages employers and employees to reduce hazards in the workplace and to start or improve existing safety and health programs.
- It establishes employer and employee responsibility.
- It authorizes OSHA to set mandatory job safety and health standards.
- It provides an effective enforcement program.
- It encourages the states to assume responsibility for administering and enforcing their own occupational safety and health programs designed to be at least as effective as the federal programs.
- It provides for reporting procedures on job injuries, illnesses, and fatalities.

All employers engaging one or more employees in a business affecting commerce must comply. Certain federal agencies are exempt. Specific standards have been established; in case they do not apply in a particular situation, a "general duty requirement" covers hazards not otherwise mentioned. It reads, in part, "each employer shall furnish to each of his employees employment and a place of employment which are free from recognized hazards that are causing or likely to cause death or serious physical harm to his employees." Any recognized hazard created in part by a condition not covered in the established specific standards may be cited as a violation under this general duty clause.

The various states have been encouraged to set up laws similar to the federal Occupational Safety and Health Act. If provisions of the state law are more strict than those of the federal government, the former hold precedence; if the federal law is more strict, it takes precedence. Upon inspection, a business can be cited for violations if hazardous conditions are found. In the first five months during which the act was in force, OSHA inspectors of foodservices made 10,668 inspections, found 26,771 violations and issued 2141 citations that carried a combined monetary penalty of over $500,000.

The effect of the federal act has been to increase both the responsibility of management for safety and the number of steps management has taken to reduce hazards in foodservices. In some cases the law has brought hardship to a foodservice that was established before the law existed and failed to meet the required standards. Conflict has also developed between foodservices and OSHA inspectors because of interpretations of the law or because of seemingly unreasonable viewpoints of inspectors toward some operational conditions in the industry. So much criticism has been directed against the law and the OSHA administrators by the entire business community of this country that some modification of provisions has occurred. Whatever its shortcomings, the law has brought about a major change in safety in the industries of this nation and has placed upon management a much greater legal responsibility for safety than it ever had before.

It is management's responsibility to see that the plant facilities are safe, to establish rules for safety, and to see that they are enforced. Training programs in safety should be instituted, and meetings should be held to discuss potential hazards and the adequacy of current safety measures. If accidents occur, management should make a careful study of their causes and of possible measures to reduce or eliminate them. A good repair and maintenance schedule should be set up to ensure that equipment and the facility as a whole are kept in top operational shape. All equipment should

bear the stamp of approval of the American Gas Association (AGA) or of Underwriters' Laboratories (UL) (see Fig. 11-11). Publications of the National Safety Council and others can help management develop and carry out good safety programs.

One of the best ways to encourage safe conditions is to maintain a good safety inspection program. OSHA has safety lists that can be used for this purpose, and the forms inspectors use in making the OSHA inspection can be helpful. To be effective, a safety self-inspection program should include the following four points:

1. Inspection of equipment and physical facilities on a department-by-department or function-by-function basis
2. Review of the physical fitness of employees to perform assigned duties safely
3. Review of actual operational procedures and personal practices of employees as they perform their required tasks and of the manner in which employees use hazardous equipment and work in observance of good safety practices
4. Check of compliance with safety and health regulations with respect to displaying required permits, posters, and reports and maintaining required reports

Inspection checks of equipment and facilities should be comprehensive and should include the following fifteen specific areas:

1. Receiving areas for food and supplies
2. Storage facilities for dry and perishable food and for nonfood supplies
3. Food preparation areas and equipment
4. Warewashing and storage facilities
5. Dining rooms and food serving areas
6. Exits, stairways, and landings
7. Customer rest rooms
8. Employee facilities such as toilets, lavatories, locker rooms, and lunchrooms
9. Storage areas for soaps, detergents, solvents, pesticides, and other hazardous supplies

FIGURE 11-11 Symbols of national testing associations: (*a*) American Gas Association; (*b*) Underwriters' Laboratories; (*c*) National Sanitation Foundation.

10. Garbage and trash storage and disposal areas
11. Boiler rooms, compressor installations, and other utilities
12. Exterior surroundings, including parking lots, drive-in service areas, and incinerators
13. Vehicles for transporting food
14. Equipment and systems installed or available for fire protection
15. First-aid supplies

Worker's Compensation

Worker's compensation is the payment by all employers, including foodservice operators, for some part of the cost of injuries, or in some cases of occupational diseases, received by employees in the course of their work. The degree of responsibility varies in different countries and in different states of the United States. Most modern worker's compensation systems consist of legislation requiring the employer to furnish a reasonably safe place to work, suitable equipment, rules and instructions when they are reasonably necessary, and reasonably competent supervisors. The employer is liable for an employee's act of negligence, for the employer's own gross negligence, and for extraordinary risks of work. In most cases the employer is not liable for accidents occurring outside the place of work or for those that have not arisen directly from employment.

Worker's compensation legislation was first passed in Germany, Austria, and Great Britain in the late 1800s. Such legislation came later in the United States, but by 1920 all but six states had passed some form of it; at present all states have some sort of worker's compensation. Private insurance companies offer employer's compensation insurance; some states have made such insurance compulsory, and a few have created state insurance funds to secure payments even when the employer is insolvent.

Most states provide similarly for public employees, although some limit this coverage to workers engaged in dangerous occupations. In Great Britain the payment of compensation is required for almost all industrial accidents. In France all noninsured employers are taxed for a state fund that guarantees compensation payments. In the United States, as well as in other countries, benefits usually cover medical expenses, cash payments in the case of temporary or permanent incapacity, and increasingly, vocational rehabilitation.

The necessity for foodservice operators to pay worker's compensation and provide a safe work environment all add up to the fact that the buck stops with management. An operator is not able to pass the blame for accidents onto others. If negligent, the manager is going to pay dearly, so the best policy is to pay attention to matters of sanitation and safety, as this chapter has so clearly stated.

■

COMPUTER-ASSISTED SANITATION AND SAFETY PROGRAMS

Computers can be of assistance to management in maintaining a safe, sanitary foodservice environment. With the proper software programs, computers can offer help in two primary areas: training and record keeping.

For any sanitation or safety program to be effective, all personnel must be trained in the proper policies and procedures. Personal computers could assist in this training by means of interactive training packages. After reading a booklet or viewing a videotape, the trainee sits at the computer and is quizzed. Right and wrong answers are noted immediately. In the case of a wrong answer, the computer system can backtrack, review material, and lead the student to the correct answer. This type of system can save the manager time by eliminating the need to train each employee personally. In addition, the employee is more receptive to the training because of the immediate feedback on progress and the freedom from any stigma associated with asking "dumb" questions. It allows employees to be trained at their own pace at any time of the day.

As in many other areas of foodservice, sanitation and safety records need to be kept. These necessary records include duct and filter cleaning schedules; pest control schedules related to spraying, fogging, or the placement of traps; employee training records; and accident records. Using a computer to maintain these records renders them readily accessible, easy to update, and space-efficient. The computer can also be programmed to alert the manager to upcoming inspections or periodic cleaning dates.

Clearly, although a computer is not a substitute for a well-planned sanitation and safety program, it can be a useful tool in making an established program more effective.

AMERICANS WITH DISABILITIES ACT

The Americans with Disabilities Act (ADA), passed in 1990, requires covered organizations to take steps to accommodate disabled persons, including employees, and to employ qualified disabled job applicants. It prohibits discrimination against persons with mental or physical disabilities in such areas as employment, public service, and public accommodations, such as foodservices and hospitality businesses.

The ADA's definition of disability is important: "A physical or mental impairment that substantially limits one or more major life activities of such individuals; a record of such an impairment; or being regarded as having such an impairment." The definition includes persons with a wide range of disabling conditions, such as arthritis, amputees, blindness, emphysema, hearing impairments, heart conditions, and shortness of stature, in addition to the more traditional definition involving persons requiring wheelchairs.

In general, public accommodations must remove architectural barriers in existing facilities if such removal is readily achievable: if it is easily accomplishable and able to be carried out without much difficulty or expense. Each public accommodation is to evaluate its facilities and determine what can be done. If a problem exists, the organization typically must provide the goods, services, facilities, and accommodations through alternative methods.

With regard to the facility itself, there are several priorities, including:

- Access to the public accommodation (doorways, entrance ramps, etc.)
- Access to goods and services
- Access to rest-room facilities
- Access to any other services, goods, or facilities offered by the organization

Regulations become applicable when there is new construction or alterations to existing facilities. Enforcement provisions include opportunities for victims of discrimination to take their cases to court. Accordingly, foodservice operators must be aware of the ADA and how it affects their operation and be willing to comply with its regulations.

HAZARD ANALYSIS CRITICAL CONTROL POINTS

Hazard analysis critical control points (HACCP) refers to a food safety system in which one can identify the foods and procedures that are most likely to cause foodborne illnesses. The system encompasses procedures to reduce the risk of foodborne outbreaks and monitors practices throughout a foodservice operation to ensure food safety. As learned in previous chapters, the flow of food consists of receiving, storing, preparing, cooking, holding, serving, cooling, and reheating. Proper time and temperature controls are at the heart of HACCP.

The development of the system was an outcome of the space program through the 1960s, and the Food and Drug Administration (FDA) made it mandatory in the 1970s. By the mid-1990s the government had a HACCP-based inspection program mandated for the seafood industry. As the nation needs better systems to monitor food safety, the principles of HACCP will find their way into all foodservices.

Yet for the time being, foodservice operators may be following the system to some extent without being mandated to do so. HACCP is a science, so to speak, that reduces food-safety risks by prevention at points where contamination is likely to occur. It is formulated to achieve the same result every time. Thus, illness is more likely to be prevented.

By using a simple flowchart, the foodservice operator may diagram the flow of food through the operation, noting critical control points (CCPs). Those recipes involving potentially dangerous foods are considered first, with hazards being assessed. The recipe's ingredients are considered from where they have been received through storage, preparation, cooking, holding, serving, cooling, and reheating, as stated previously.

The CCPs are identified at each step where a preventive or control measure can be applied that would eliminate, prevent, or lessen the risk of hazards. The procedures and standards for each CCP are established and monitored with corrective action steps taken at any point where deviation is noted. A record-keeping system is established with assessment taking place to verify that the system is working. Standards include times, temperatures, and other requirements to keep food items safe.

A system implemented through HACCP takes commitment and time. The operator needs to acknowledge that the move toward such a system is a lengthy process. Eventually, HACCP becomes a living part of the facility as staff and employees become trained and more aware of food safety.

■

SUMMARY

Sanitation and safety are the joint responsibility of everyone in a foodservice. Good sanitation ensures that customers are served wholesome food that will neither make them ill nor do other harm. Good safety ensures that the environment confronting both employees and customers is as free as possible from potential sources of harm.

Four kinds of substances cause food poisoning: organisms that infect; organisms that develop toxins that are ingested with food; chemical poisons; and poisonous foods or animals. The first two kinds of substances are living things that belong to a large family called *microorganisms* because they can only be seen with a microscope. Some are harmful, others do no harm, and still others are beneficial. Harmful microorganisms are called *pathogens*. Time, temperature, availability of food and moisture, cleanliness, and the purchase of noncontaminated food are factors that help control them. Bacteria, viruses, parasites, and molds are types of microorganisms that can be pathogenic. Molds and yeasts tend to spoil food, rather than to cause illness in man; however, some few molds can develop toxic substances.

Foodborne intoxication occurs when certain microorganisms contaminate food and produce a toxin in it that causes illness. Thus, the direct cause of illness is the toxin and not the organism itself. Foodborne infection occurs when microorganisms get into the food or drink and are taken live into the body; their growth in the body does the harm. Destroying infectious agents by cooking food thoroughly at a sufficiently high temperature destroys the microorganisms but may not eliminate the danger from toxins.

Chemical poisoning results from errors in handling harmful substances such as cleaning agents, poisons, or other chemical products used in connection with foodservice operations. Such substances should not be used around food and should only be entrusted to employees who know how to use them safely. Chemical poisons should not be stored in the same area where food is stored. Many plants and animals are poisonous, but most of these have been eliminated from our regular food supplies. They do cause problems occasionally, however. For example, eating oysters or clams made poisonous by tiny red mites can cause serious illness. Purchasing food from approved sources can do much to eliminate a person's risk of consuming poisonous food.

Successful prevention of foodborne diseases depends on the factors of food protection, personal hygiene of workers, facility and equipment sanitation, proper ventilation and light, good garbage and refuse disposal, adequate handwashing and toilet facilities, and good pest control.

Purchasing food from approved sources helps ensure that it is safe and wholesome to eat. Following sanitary practices in preparing food is another way to prevent foodborne disease. Storing food under the right conditions is another crucial step, and serving food properly (including holding it for service properly) is the final step in food protection.

Personal hygiene deals with an individual's habits of cleanliness. A failure to follow good hygienic principles endangers food and may cause a loss of patronage; customers do not like to be served by unkempt, unclean personnel.

The foodservice facility and its equipment must be properly maintained to prevent food-borne diseases. Floors, walls, and ceilings should be easy to clean and should be properly maintained. The maintenance requirements depend largely on the original construction of the building. Proper ventilation and lighting are other aspects of physical plant sanitation. Garbage and refuse disposal must be done in a proper manner.

Good equipment sanitation ensures that foodborne disease organisms will not be transmitted through improperly cleaned and sanitized equipment.

Adequate control of rodents and insects must occur if good sanitation standards are to be achieved. Elimination of breeding places is an important preventive measure in all cases. Rats and mice are the most resourceful of pests. Keeping them out of the building is one of the best ways to control them. Poisoning and trapping can also be effective. Houseflies carry many diseases. Measures for keeping them out of a facility should include screens and other blockage devices, cleanliness, and elimination if they get inside. Cockroaches, too, can carry disease organisms, and they have a very offensive odor. As with flies, an understanding of their life cycle is an important step in the proper control of these pests.

Both management and employees are responsible for seeing that good sanitation is practiced in an operation. Local, state, and federal sanitary codes must be followed.

Damage control suggests that the foodservice operator is aware of what can go wrong within an operation and be prepared to handle those problems. Studies indicate that those managers who have been mandated to attend a training and certification program in sanitation and safety effectively improve the sanitary conditions in their operations and reduce the spread of foodborne illnesses.

Safety is achieved in a foodservice when both employees and management properly perform their duties in this area. Employees should be alert, follow safety rules, work safely, and think safety. Experienced workers can help train newer ones in good safety procedures. Employees should learn to handle tools and materials correctly. In the foodservice industry, the four most common causes of major injuries are cuts, slips and falls, burns and scalds, and improper lifting and hauling. Cuts mean lost time and lost wages to the employee and lost profit to the employer. Slips

and falls usually result from poor housekeeping or from inattention. Burns and scalds usually result from carelessness or from a failure to observe good safety rules. Improper lifting and hauling account for the final 25 percent of lost-time accidents. They occur when proper principles of lifting and carrying are not followed.

Management is responsible for seeing that a facility is safe by providing a safe place to work, by keeping equipment in good operating order, by publishing rules for good safety, by making sure that these rules are carried out, and by training employees in good safety principles. The Occupational Safety and Health Act is a law that has established high standards for safety in U.S. industry.

Worker's compensation is payment by all employers, including foodservice operators, for some part of the cost of injuries, or in some cases of occupational diseases, received by employees in the course of their work.

The Americans with Disabilities Act (ADA) passed in 1990, requires covered organizations to take steps to accommodate disabled persons, including employees, and to employ qualified disabled job applicants. Years after its passage, the ADA has not fulfilled the greatest fears of its critics, nor the greatest hopes of its supporters.

Hazard Analysis Critical Control Points (HACCP) refers to a food safety system in which one can identify the foods and procedures that are most likely to cause foodborne illnesses. By using a simple flowchart, the foodservice operator may diagram the flow of food through the operation, noting critical control points (CCPs).

Computer assistance in implementing sanitation and safety programs can be quite helpful. Computers can be used effectively by management for training employees and for maintaining safety and sanitation records.

■

CHAPTER REVIEW QUESTIONS

1. What is HACCP, and how can a foodservice operation set up its own HACCP system?

2. List the different kinds of microorganisms that can cause food-borne infections and intoxications. Indicate: (a) how infection and intoxication types differ; (b) the necessary factors that must be present for their growth; and (c) how they can be controlled.

3. Discuss foodborne poisoning from chemicals, plants, and animals. How does such contamination occur? What control measures can be taken to see they do not occur?

4. Food protection, personal hygiene, facility and equipment sanitation, and pest control are important areas in the prevention of foodborne disease.

Select one of these and describe how it could be appropriately implemented in an operation of your choice. For example, how would you see that food protection assured good standards if you operated a foodservice? Be specific.

5. Discuss and divide the responsibilities for achieving good sanitation standards between employees, management, and governmental authorities.

6. What are the responsibilities of management and employees in achieving high safety standards? Can you contribute others you feel are important? Share these in an oral report before the rest of the class.

7. What is OSHA?

■

CASE STUDIES

1. John Smith is a newly appointed regional manager for Barrt's, a chain of family-style restaurants. In the third week on his new job, John receives a frantic call from Dave Schultz, who is the manager of one of John's units. It seems that Dave's restaurant has just been visited by a county health inspector, and the restaurant failed to pass the inspection. The health inspector has given Dave one week to make corrections, after which the restaurant will be reinspected. After calming Dave down, John asks him about the inspection and learns that there are no major problems—just numerous small ones. John asks Dave about his sanitation and safety programs and is not too surprised when Dave informs him that there is no formal program. John tells Dave to begin addressing the immediate problems and says that he will get back in touch with him in the next few days. John then makes calls to the managers of the other units in his region and discovers that, while some managers have established their own

programs, there is no company-directed program. These managers assert, however, that their establishments are clean, safe, and well maintained. John, knowing how important cleanliness is to Barrt's customers, decides that some action must be taken. What would you suggest he do and how should he do it?

2. Being a man of hospitality, Ken Jordan has decided he wishes to be a humanitarian as well. Accordingly, he has announced that he will be the first *hospitarian*, a term he coined himself. He will open a Community Feast Food Bank in his city to cater to the less fortunate. He wishes to obtain information about who to call regarding the Americans with Disabilities Act (ADA) to obtain answers to general and technical questions about the ADA. He has asked you to search the Web site *www.usdoj.gov* or *www.yahoo.com* to find the phone number. What is that number? Will he be able to order technical assistance material as well?

■
VOCABULARY

Clean

Microorganism

Coliform

Cocci

Mesophiles

Hepatitis

Danger zone

Salmonella

OSHA

Infection

Parasite

Trichinosis

Pupa

AGA

Sanitary

Virus

Intoxication

Botulism

Amoeba

Spore

Japanese roulette

UL

Bacteria

Protozoa

Staphylococcus

Vector

pH

HACCP

■
ANNOTATED REFERENCES

Educational Foundation of the National Restaurant Association. 1995. *Applied Foodservice Sanitation*, 4th ed. Chicago: National Restaurant Association Educational Foundation. (A standard for all books on the subject.)

Loken, J. K. 1998. *The HACCP Food Safety Manual*. New York: John Wiley & Sons, Inc. (Total food safety is presented in this book.)

Longree, K., and G. Armbruster. 1996. *Quantity Food Sanitation*, 5th ed. New York: John Wiley & Sons, Inc. (A classic text on the subject of sanitation and safety.)

McSwane, D., N. Rue, and R. Linton. 1998. *Essentials of Food Safety and Sanitation*. Upper Saddle River, N.J.: Prentice Hall. (A complete text on the topics of sanitation and safety from education and training to HACCP, facilities, and environmental considerations.)

Mill, R. C. 1998. *Restaurant Management: Customers, Operations, and Employees*. Upper Saddle River, N.J.: Prentice Hall, pp. 240–260. (Sanitation and food safety are presented in one chapter.)

National Assessment Institute. 1994. *Handbook for Safe Food Service Management*, 2nd ed. Upper Saddle River, N.J.: Prentice Hall. (HACCP to managerial responsibilities are included in this second edition.)

CHAPTER 12

Pantry Products

I. Appetizers
    A. Canapés
    B. Cocktails
    C. Dips
    D. Hors d'Oeuvres
    E. Relishes
    F. Salads
    G. Soups
    H. Management's Responsibility

II. Sandwiches
    A. Planning Production
    B. Supplies
        1. Breads
        2. Spreads
        3. Fillings
        4. Garnishes
    C. Sandwich Care
    D. Sandwich Presentation
    E. Sandwich Preparation
        1. By Order
        2. In Quantity
        3. Advance Preparation
    F. Management's Responsibility

III. Salads and Salad Dressings
    A. Salad Planning
    B. Salad Menu
        1. Appetizer Salads
        2. Accompaniment Salads
        3. Main-Dish Salads
        4. Dessert Salads
    C. Salad Equipment
    D. Salad Ingredients
        1. Salad Greens
        2. Fruits and Vegetables
        3. Other Salad Ingredients

    E. Salad Presentation
        1. Individually Served Salads
        2. Buffet Salads and Salad Bars
    F. Salad Preparation
        1. Base or Underliner
        2. Body
            a. Fruits and Vegetables
            b. Gelatin Mixtures
            c. Protein Foods
            d. Rice and Macaroni
            e. Salad Greens
        3. Salad Dressings
            a. French Dressing
            b. Mayonnaise
            c. Cooked or Boiled Dressings
            d. Other Basic Dressings
        4. Garnish
    G. Management's Responsibility
        1. Standards for Salads
        2. Cost Controls

IV. Beverages
    A. Coffee
        1. Coffee Equipment
        2. The Product
        3. Brewing Coffee
            a. Vacuum Coffeemaker
            b. Urn Brewer
            c. Semiautomatic Coffeemaker
            d. Automatic Coffeemaker
        4. Sanitation and Quality Control
        5. Serving and Merchandising Coffee
    B. Tea
        1. Preparing Tea
        2. Tea Quality
    C. Cocoa

LEARNING OBJECTIVES

By the end of this chapter, the reader will:

1. Learn about the production of various pantry products: from appetizers, sandwiches, and salads to beverages, including coffee, tea, and cocoa.

2. Be able to distinguish among various types of appetizers.

3. Know the various types of salads: their ingredients, presentation, and preparation, and service with various dressings.

4. Understand beverages: their consumption, quality standards, and service.

SELECTED WEB SITE REFERENCES

Anchor Foodservice
www.anchorfoods.com/

Blue Diamond Almonds
www.bluediamondgrowers.com/

De Choix
www.dechoix.com

Dressings & Sauces
www.dressings-sauces.org/

ED & F Man Cocoa Products
www.edfman.com/

Frito-Lay
www.fritolay.com/

Harney & Sons
www.harney.com

McCormick & Company, Inc.
www.mccormick.com/

Millstone Coffee
www.millstone.com/

Mr. Coffee
www.mrcoffeeconcepts.com/

Starbucks Coffee
www.starbucks.com/

Terra Chips
www.terrachips.com/

Thomas J. Lipton Company
www.lipton.com/

A number of occasional or partial meal foods are prepared in the pantry, including appetizers, sandwiches, salads and salad dressings, beverages, and fountain items. Desserts may also be dished here. In some operations, the pantry prepares breakfast items. In a large continental kitchen, much of this work is taken over by the garde-manger section, and the pantry may be a part of it. In this chapter, however, only appetizers, sandwiches, salads and their dressings, and beverages are considered.

Pantry work is distinguished by the wide variety of foods prepared there. Much preparation is required to get foods ready for final service. Vegetables, fillings for sandwiches, butters and toppings for appetizers, syrups and dessert toppings, garnishes, and a host of other items must be readied for service, and handling this takes advance preparation. In addition, work must be fast, neat, and accurate. A salad—even one with everything ready for assembly—can take twenty or more manipulations, and a sandwich can take just as many. Foods must be very attractive. Since many cold foods are served from the pantry, sanitation there must be very good. Workers must be dexterous and skilled and must be able to work under pressure.

Keeping a clean and organized work area is paramount to good-quality products and fast production. The pantry is often a place where workers get their first job, before moving up to more sophisticated sections and more responsible positions. It is excellent training for the exacting work required for more complex tasks in the kitchen.

■

APPETIZERS

An appetizer is a food used to whet the appetite. Appetizers should be zesty, often tangy food that encourages hunger by stimulating the taste buds and salivary glands, starting the flow of gastric juices in the stomach. Because of their fairly high salt content and piquant flavor, they go well with alcoholic cocktails. This description may not seem to fit all items classed as appetizers (such as salads and soups), but they all are served to help whet the appetite.

Appetizers should look appealing. Contrasts in color and shape are important; attractive service also enhances their appeal. If well-prepared, appetizers can be an extremely important drawing card for an operation.

Appetizers usually require a lot of labor, so good organization and planning in their production are essential. Ingredients used include seafoods, meats, poultry, fruits, vegetables, juices, and dairy products. Finished products include canapés, cocktails, dips, hors d'oeuvres, relishes, salads, and soups. A soup or salad may also be considered a course in a meal or a food proper, rather than an appetizer.

Canapés

Canapés are small finger foods often served as tiny open-face sandwiches cut into fancy shapes and served either hot or cold. Size and thickness depend on the items used. A base of crackers, toasted bread, rye bread, pumpernickel, pastry shells, puffs, or something similar can be topped with a rich, savory paste or seasoned butter and artistically decorated. Imaginative shapes can be made with a variety of tools, but special decorative equipment is needed, including the following items:

- Pastry bags and small pastry tubes
- Vegetable cutters
- French knives
- Serrated bread knife
- Small pair of scissors
- Large and small cookie cutters
- Channel knife
- Decorating tools
- Oval food cutters and vegetable cutters
- Shallow oval vegetable timbales
- Small and large parisienne cutters
- Small timbales with high sides for aspics and the like
- Apple (rose) cutter
- Special design cutters
- Small vegetable knives and other knives

Canapés can be topped with good-quality carry-over foods such as meat, fish, hard-cooked eggs, and seafoods. Many foods can be made into pastes and worked into butters. A tiny shrimp with a caper nestling inside it, resting on top of a small round of bread spread with a tangy shrimp butter is but one example of an attractive canapé. The garnish is important, although it can be simple or ornate. Possible garnishes include slices of stuffed olives, mushrooms, caviar, parsley, pimiento, or truffles. Labor costs must be contained, since labor can be intensive. Mixing a few fancy canapés with relatively many simple ones reduces labor costs.

To mass-produce canapés with a minimum of labor, slice trimmed two-day-old bread loaves lengthwise, ¼-in. thick, and arrange them in rows so that they cover a flat surface; then apply spreads with a spatula or pastry bag. Arrange the toppings for each canapé, garnish, and then cut for service. Refrigerating (to give more firmness) may help cutting. A layer of seasoned butter can prevent the garnish or other food on top from soaking into the base of the canapé. Wrapping to prevent drying before service and refrigerating to reduce deterioration are recommended practices.

When other appetizers are to be served alongside canapés, it is advisable to plan to run out of canapés and let the guests switch to eating dips and other more simply prepared foods toward the end of the function. The nature of the product allows prices to be set rather high, with a good markup on food and labor.

Cocktails

Beverage cocktails may be alcoholic or nonalcoholic. At most occasions, both should be offered. Patrons are beginning to prefer wines or light alcoholic drinks to the heavier ones. Too much alcohol dulls the taste buds, but a small amount stimulates the flow of gastric juices in the stomach.

Cocktails should be tangy, highly palatable, and not too sweet, since sweetness dulls the appetite. Sweet cocktails are usually served with light foods; light, tart, stimulating cocktails best complement heavy or rich meals. Alcoholic cocktails consumed as apéritifs include martinis, manhattans, and other mixed drinks. Light dry wines, Dubonnet, sherry, and Madeira are usually offered as well.

Nonalcoholic cocktails served as a first course may include shrimp, crab, oysters or clams on the half shell, fruit cups, lobster, fish, and fruit or vegetable juice. Meat or fish cocktails are usually served with a tangy sauce. An unusual cocktail could be made with fruit juice and a bit of soda water, with a small ball of sherbet swimming in the center.

All cocktails should be ready to eat when served. Fruits should be free of seeds, inedible pulp, and membranes. Seafoods should be shell-free, except for oysters or clams in the half shell and minced lobster in its shell. About 10 lb of seafood is needed for 100 1½-oz portions; about 18 lb of fruit or other solid food is needed for 100 3-oz servings (about ⅓ cup each). Creative use of garnishes can increase the appeal of a cocktail.

Dips

Dips are appetizers that can be scooped up with a chip, cracker, vegetable stick, or other crisp food. They are low in labor cost and can be made from many carry-over foods. The variety of dips is almost unlimited. Bacon and onions, clams, shrimp, or brown beans stirred into sour cream or cream cheese or blended

with tiny-curd cottage cheese can be offered. Fondues and other hot dips can also be used.

A dip's consistency is as important as its flavor. If the dip is too thick, the dipping food item may break; if it is too thin, the patron will be unable to scoop enough onto the dipper, and what is scooped may dribble onto the patron's chin or clothes. Refrigerating cold dips until about 15 minutes before service helps keep them thickened to the right consistency. Hot dips are usually served in chafing dishes. The amount of dip required depends on what else is served and on the duration of the function. If each dip occupies a volume of 1 tsp, it follows that 1 qt holds about 200 dips. At a rate of 5 dips per person, 1 qt of dip serves about forty people. This can vary. A popular dip can run out very quickly, and a less popular one may receive scant attention.

Hors d'Oeuvres

Once hors d'oeuvres were small portions of food served during a multiple-course meal and eaten with a fork. Today, they are bite-sized morsels that are often picked up with the fingers. For this reason, they should be small. Toothpicks can be used to facilitate picking them up. Hors d'oeuvres should be tangy and stimulating to the appetite. They are often placed on trays (as are canapés) and passed to guests or left on a buffet to be picked up as desired (see Fig. 12-1). Hot hors d'oeuvres

FIGURE 12-1 Hors d'oeuvres are often plated and passed among guests.

are usually served in a chafing dish on a buffet table. Tiny plates and napkins may be provided there so that several of these foods can be picked up at one time. Hors d'oeuvres may also be placed on a small dish and served as a first course to a seated guest. Oblong or square glass dishes, china, earthenware, wood, or plastic service can be used. If the food is highly acidic, some metals should not be used. Silver and copper resist most normal food action, but copper may form a poisonous compound when it reacts with some acids.

Like dips, hors d'oeuvres come in almost endless varieties. The type used should be suited to the other foods served with the meal. Usually about five or six selections are sufficient. Some hors d'oeuvres can be easily made for fifteen to twenty people but not for several hundred. Labor cost can be a big factor in the selections and quantities made.

Typical hors d'oeuvres include shellfish (especially oysters, clams, and mussels); fish such as trout, eel, herring, whitefish, and sturgeon; caviar; meats such as veal tongue, smoked beef, sausages, and various hams and pickled meats; vegetables that have been stuffed, marinated, pickled, or cooked; and poultry and dairy products such as chicken wings, turkey, deviled eggs, and cheese tidbits.

Carryover foods can be used. Color and flavor contrasts, along with attractive presentation, are desirable. Simplicity reduces cost and improves the product, since overgarnished hors d'oeuvres do not look appetizing.

The number of hors d'oeuvres needed depends on the time of day, duration, and nature of the function. Some foodservices prefer to use premade products that are purchased frozen.

Relishes

Relishes are either crisp, fresh vegetables or pickled items. Many may be purchased ready to serve. They should be attractively arranged and presented.

Fresh vegetable relishes include radish roses, celery hearts, carrot curls, cauliflower buds, turnip or green pepper strips, and small scallions. Pickled items include pickles, corn, sweet onions, cranberries, watermelon rind, green tomatoes, spiced fruits, and stuffed olives.

Crushed ice is often served over relishes when they accompany a meal.

Salads

Some salads, such as a seafood salad, are offered as appetizers. Many operations serve a small chilled tossed salad as a first course; this is called *California service*—usually on a glass plate or wooden bowl. Pâté, mousses, terrines, chopped chicken livers, pickled herring, and other items are also frequently served in this way and thus are considered appetizer salads.

Soups

Most soups can be appetizers, except the heavy ones that are almost meals in themselves. A consommé, bouillon, turtle soup, light cream soup or bisque, split-pea or lentil soup, or delicate French onion soup can be a good appetizer. The meat juices in meat soups stimulate the taste buds and start the flow of gastric juices.

Management's Responsibility

The formula for successful appetizers is to offer a diverse selection varying in eye appeal, form and shape, and price. Presentation is critical as many appetizers will attract the eye of another patron, stimulating an order. Proper management will ensure the correct temperature of product served with the correct utensil for consumption. There is nothing more inappropriate than a well-presented hot bowl of soup arriving without a soup spoon.

Many fine convenience appetizers are available on the market today. These afford the foodservice operator an opportunity to save on labor while offering a greater selection. The key is to follow the recommended preparation and serving instructions that accompany these fine products.

In total, management must have the tools and equipment necessary to produce appetizers. Workers must know how to use these tools and equipment under proper sanitary conditions. Again, proper presentation of appetizers is a must, so management control is especially important.

<h1 style="text-align:center">■ SANDWICHES</h1>

A sandwich consists of bread, a spread, a filling, and a garnish; the garnish may be omitted. Sandwiches can be close- or open-face, and they can be broiled, grilled, deep-fried, baked, or toasted. A sandwich can be a simple snack, a main entrée, or a hearty meal in itself. A toasted open-face sandwich with a filling of butter, brown sugar, and pecans and topped with a scoop of ice cream is a dessert; a toasted open-face tangy cheese sandwich can be served as an appetizer. Sandwiches can be simple or elaborate, hot or cold, big or as small as finger foods. They can account for a significant portion of an operation's income, especially for lunch. Some operations specialize in sandwiches and serve nothing else.

Planning Production

Sandwiches require much labor. All items to be used should be ready at final assembly. Tools and dishes must be close at hand, and simplification of motions is essential.

Each kind of sandwich requires special equipment and tools. The work center should be comfortable for the worker. Short workers may prefer a surface that is 34 in. high, while taller ones may be best-served by a surface that is 36 to 42 in. high. A worker should have about 16 in. of free space (measured from the elbow) on either side to move about in. Containers for fillings and spreads should be tilted toward the worker for easy access.

A free work area in front can act as a cutting board. A scrap opening or chute should be located on the right for right-handed workers. Plate and bread dispensers can help a worker reach what is needed. A toaster, grill, or microwave oven may be needed; a sink is essential. Necessary tools include a sharp knife, a two-tine fork for picking up small items, scoops for fillings, a short spatula (about $1\frac{1}{2}$ in. wide and 4 in.

long) for spreading, and tongs for picking up items. An electric knife is good for cutting sandwiches after they have been made. A portion scale may be needed to check fillings. Each tool should have its own storage area within easy rich.

Supplies

Sandwich supplies include breads, spreads, fillings, and garnishes. All should be of high quality.

Breads

Fit the kind of bread to the sandwich; a Reuben sandwich, for example, needs good pumpernickel or rye bread. Breads used may be cracked wheat, whole wheat graham, white, rye, pumpernickel, French, Italian, Boston brown, or the various quick breads such as nut, date, or banana. Buns, hard or soft rolls, and specially shaped breads are used for special kinds of sandwiches. All bread should be fresh and firm enough to take spreads easily.

Rolled sandwiches are made using moist, freshly baked bread no older than one day. The loaf is sliced lengthwise about $\frac{1}{4}$ in. thick and then the slice is lightly flattened with a rolling pin to give it firmness. The spread is applied and the bread is rolled up and stored in a refrigerator, tightly wrapped in a damp cloth or plastic wrap. Later, the wrap is removed and the roll is sliced so the rounds show the roll. Carryover bread makes good toasted or grilled sandwiches. Sandwich bread is sliced from $\frac{1}{8}$ to $\frac{3}{4}$ in. thick. The number of slices per loaf for different types of bread is given in Table 12-1.

Spreads

Butter and margarine are the usual sandwich spreads. They add richness and keep moist fillings from soaking into the bread. They should be softened for easy spreading, but no extra moisture should be added. Spices, prepared mustard, chopped chives, parsley, or pimiento can be added to give special flavor. Melted spreads do not prevent soaking as well as unmelted ones do. Other spreads include peanut butter, cream cheese, and some types of whipped cheese. Jelly, mayonnaise, and salad dressings are apt to soak.

TABLE 12-1 Bread Slices per Loaf, Excluding End Crusts

Bread Type	Weight of Loaf (lb)	Thickness of Slice (in.)	Number of Slices
White	1¼	⅝	19
	1½	⅝	24
Sandwich white	2	½	28
	2	⅜	36
	3	½	44
	3	⅜	56
Whole wheat	1	⅝	16
	2	½	28
	3	½	44
	3	⅜	56
Rye	1	¾	23
	2	¾	33

About 1 tsp of spread is usually sufficient for 1 slice of bread; 100 sandwiches (200 slices) are spread with about 2 lb of butter, which equals 1 qt or 192 tsp of spread. (See Appendix D for information on how to translate these quantities into metric units.) With peanut butter, 1 lb is 1¾ cups; so if 1½ Tbsp are used per sandwich, 5⅓ lb (2⅓ qt) are used for 100 sandwiches. Many other spreads, such as cream cheese, take the same quantity as butter.

Fillings

Sandwich fillings may be spreads, or they may be cuts of cheese, meat, poultry, or sausage. A thinly sliced or well-chopped filling makes the sandwich easier to eat; 1 to 2 oz of thinly sliced meat gives greater production volume per pound and may be more tender than a thick slice. Chopping should not be so fine that it results in a paste.

Portion control is important. Appropriate scoops should be used to ensure uniform measures of spreadable fillings, and a close check should be made of sliced portion size. If a No. 20 scoop (1½ oz) of spreadable filling is used, 100 sandwiches take 5 qt of filling.

The average weight of sliced filling used per sandwich is 1½ to 2 oz. Table 12-2 lists equivalent measures of various fillings.

Garnishes

Garnishes should be planned to complement sandwiches. They should be edible and eye-appealing. Lettuce or other salad greens, relishes, cheese, peppers,

TABLE 12-2 Equivalent Measures of Various Fillings

Food Item	Quantity	Equivalent Volume	Other Measure
Bacon	1 lb	1½ cups, cooked and chopped	25 strips
Celery	1 lb	1 qt, chopped	
Cheese, grated	1 lb	1 qt	
Chicken	1 lb	1 qt, cooked and diced	
Cottage cheese	1 lb	1 pt	
Eggs, hard-cooked	1 dozen	3½ cups, chopped	
Jelly	1 pt		16 sandwiches
Lettuce	1 head		16 leaves
Mayonnaise	1 pt		50 bread slices
Meat, ground	1 lb	3 cups, cooked	
Nut meats	1 lb	3¾ cups, chopped	
Olives, drained	1 qt	3 cups, chopped	
Onions	1 lb	1 pt, chopped	
Pickles, drained	1 qt	3 cups, chopped	
Salmon	1 lb	1 pt, flaked	
Tomatoes	1 lb		3–4 fruits
Tunafish	1 lb	1 pt, flaked	

pickles, pimiento, and olives are commonly used garnishes. Other possible garnishes include potato chips, nuts, and shoestring potatoes.

Sandwich Care

Sandwiches are quite perishable and must be stored properly if they are being held for service. To prevent bread from becoming stale or developing mold, store it at 75 to 85°F (24 to 29°C). Bread stales most quickly under refrigeration and least quickly when frozen. In a warm, moist climate, bread may have to be refrigerated to prevent mold. The bread storage area should be situated well above floor level and should be dry. Fresh bread absorbs flavors easily, so it should not be kept near fish, onions, cabbage, garlic, or other strong-smelling foods. To avoid crushing breads, stack them properly.

Storage for fillings and other sandwich products is also necessary. Usually, a refrigerator is set aside for such supplies, but a lower part of the work area can be refrigerated to keep them there.

If sandwiches are to stand for several hours, they should be refrigerated. Place them on trays not more than two deep, and cover the tray with plastic or other wrap to prevent drying. Most sandwiches should not be stored for more than 12 hours at 40°F (4.5°C). Some fairly dry sandwiches with cheese or meat fillings can be stored for 24 hours. Depending on their fillings, some sandwiches can be frozen for from several days to several weeks at 0°F (−18°C) or lower.

Filling ingredients that freeze well include the following:

- Canned, ready-to-serve, or cooked meat
- Chopped and sliced cooked or canned poultry and fish
- Dried or ground cooked beef and ham
- Cooked egg yolk but not egg white
- Baked beans
- Peanut butter
- Butter or margarine
- Applesauce
- Canned crushed or tidbit pineapple
- Mashed apricots

- Grapefruit, lemon, orange, or pineapple juice
- Dairy sour cream
- Milk or cream
- Horseradish
- Chopped parsley, chives, or onions

Filling ingredients that do not freeze well include the following:

- Chopped, sliced, or grated raw vegetables and fruit
- Meats such as bacon, frankfurters, and liver-sausage
- Cooked egg white
- Nuts
- Jelly, jam, or preserves
- Honey
- Olives
- Pickles (dill or sweet)
- Piccalilli
- Chili sauce or catsup
- Pimiento
- Prepared mustard
- Mayonnaise or salad dressing
- Strong spices such as garlic, cloves, and cinnamon

Sandwiches to be frozen should be individually wrapped. Frozen sandwiches thaw in about 1 to 2 hours at room temperature; they should be consumed soon after thawing.

Sandwich Presentation

Sandwiches should be presented in good arrangement with suitable garnishes. The plate should fit the size of the sandwich and garnish. If other foods (such as a cup of soup) are served with the sandwich, a larger plate should be used. Crowding is not recommended. Figure 12-2 shows different ways to cut and arrange sandwiches.

The sandwich and its size should fit the occasion. For an outdoor picnic, sandwiches can be larger to suit hearty appetites; but at a tea, sandwiches should be much smaller. At such functions, contrasts in fillings, shapes, and garnishes add interest.

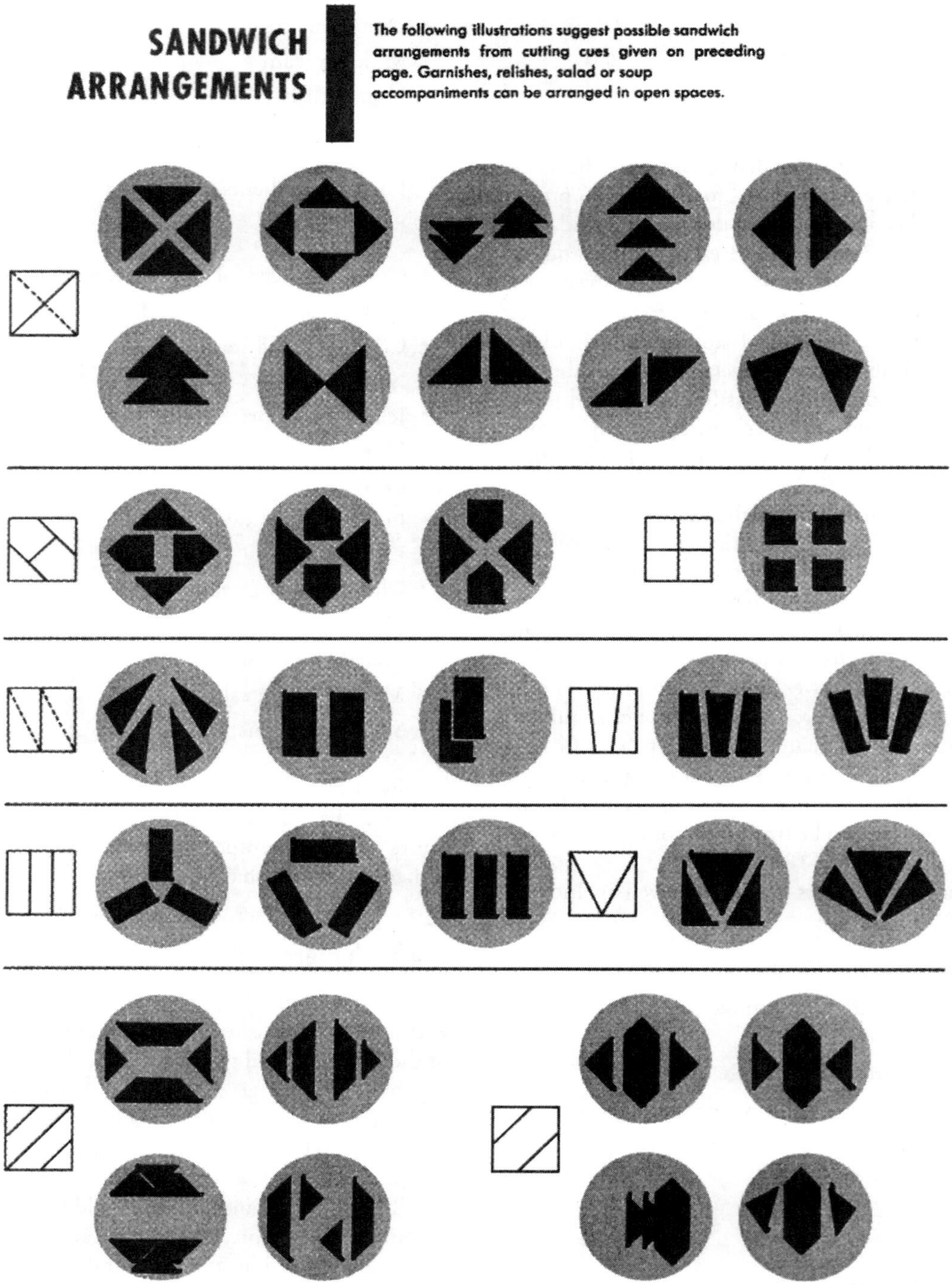

SANDWICH ARRANGEMENTS

The following illustrations suggest possible sandwich arrangements from cutting cues given on preceding page. Garnishes, relishes, salad or soup accompaniments can be arranged in open spaces.

FIGURE 12-2 Sandwich cuts and arrangements.

Sandwich Preparation

Unorganized sandwich production wastes time, causes confusion, and yields a poor product. A completed sandwich reflects the skill of its creator and the character of the work center. Sandwiches are prepared in three different ways: by order, in quantity, or in advance.

By Order

The freshest sandwich is the one prepared to order. A well-planned sandwich center should be able to fill each individual order as it comes in. Figure 12-3 identifies manipulations to use in producing such sandwiches.

In Quantity

When many sandwiches are to be made at one time, all fillings, spreads, breads, and garnishes should be prepared and assembled. To produce from thirty-two to forty-eight sandwiches at one time, pick up four slices of bread with each hand and place them on the work counter. Repeating this process creates a 4-by-4 square (16 slices). Place the spread on the individual slices with one motion of the spatula across each surface. Then measure and deposit the filling on each piece of bread with a scoop, or place a sliced filling on each piece of bread by hand. Using a short spatula, distribute the spreadable filling evenly, moving from the upper right to the upper left and continuing in an S motion to the

JOB: board method for sandwiches made to order (1 operator)

What to do	How to do it
1. Place 2 matching slices of bread on board.	1. Use left hand. 2. Place bread slices in front of you. 3. Open them as you lay them down.
2. Spread softened butter or margarine on bread slices.	1. Use right hand. 2. Use spatula to spread butter to edges of both bread slices.
3. Place filling on one slice of buttered bread.	1. Use right hand. 2. Transfer sandwich filling portion with scoop or spatula to center of bread. If filling is sliced cheese, tomatoes or meat, transfer with fork. 3. Use one stroke of the spatula away from you and one stroke toward you to spread filling evenly to edges of bread.
4. Place lettuce for sandwich.	1. Use left hand. 2. Place 1 portion of lettuce on sandwich filling.
5. Close sandwich.	1. Use left hand. 2. Place other slice of buttered bread over lettuce leaf lining up edges of bread.
6. Cut sandwich.	1. Use both hands. 2. Hold sandwich together with thumb and first finger of left hand. 3. Cut sandwich into desired sections with a sharp knife held in right hand in a straight downward motion.
7. Serve sandwich.	1. Use both hands. 2. Transfer completed cut sandwich to cold plate by sliding cutting knife under it and supporting sandwich with other hand. Don't push completed sandwich from the board to the plate. 3. Garnish as desired. 4. Place check with order after completion.

FIGURE 12-3 Step-by-step method for sandwich preparation by the order.

lower right and finally the lower left portion of the bread. Garnishes may be placed on each slice at this point. With four slices of bread in each hand once again, place two on each sandwich, one to finish a layer of sandwiches (16 total) and another to begin a second layer (32 total) on top of the first. To make 48 sandwiches, repeat the process a third time. After filling, cut the two or three layers of sandwiches, wrap

them, and store them in a refrigerator, if they are to be held. Figure 12-4 lays out an alternative, assembly-line approach to making sandwiches in quantity.

Advance Preparation

Sandwiches sometimes must be made ahead. If so, they must be properly made and stored to retain their

JOB: assembly line method for uncut sandwiches	
What to do	How to do it
1. Station workers.	**1.** Station 6 workers in a sitting or standing position along one side of a long work table, forming a single assembly line that flows toward the left.
2. Number 1 worker cuts wrapping material and places 2 matching slices of bread.	**1.** Use both hands. **2.** Cut sandwich wrapping material to proper size (8-inch × 12-inch). **3.** Place 2 matching slices of bread, side by side, on cut wrapper, sliding it along to number 2 worker.
3. Number 2 worker spreads softened butter or margarine on bread slices.	**1.** Use both hands. **2.** Use spatula to spread butter to edges of both bread slices, sliding it along to number 3 worker.
4. Number 3 worker places filling on one slice of buttered bread.	**1.** Use both hands. **2.** Transfer sandwich filling portion with scoop to center of bread. If filling is sliced cheese, tomatoes, or meat, transfer with fork, sliding it along to number 4 worker.
5. Number 4 worker spreads filling and closes sandwich.	**1.** Use both hands. **2.** Use one stroke of spatula away from you and one stroke towards you to spread filling evenly to edges of bread. **3.** Place lettuce on filling if it is used. **4.** Place other slice of buttered bread over lettuce or filling, lining up edges of bread. **5.** Slide closed sandwich along to number 5 worker.
6. Number 5 worker wraps sandwich.	**1.** Use both hands. **2.** Bring long ends of wrapping paper together in a pharmacist's fold onto center of sandwich. **3.** Follow fold with both hands, running left hand to left edge of sandwich and right hand to right edge of sandwich. **4.** Tuck paper in securely at each end with left and right hands, folding top edge in first, then bottom edge. **5.** Slide wrapped sandwich along to number 6 worker.
7. Number 6 worker stacks wrapped sandwiches and transfers them to refrigerator.	**1.** Use both hands. **2.** Stack wrapped sandwiches in wire baskets, on bun pans or trays; allow some space for air circulation between each row for quick chilling. **3.** Label sandwich containers with variety they contain. **4.** Place filled sandwich containers in refrigerator on shelves or a movable rack during the entire holding period.

FIGURE 12-4 Step-by-step method for sandwich preparation in quantity.

quality. Their ingredients should be ones that do not deteriorate during refrigerated or frozen storage. Do not prepare in advance sandwiches that are apt to soak badly.

Management's Responsibility

To produce good sandwiches, management must provide well-planned work centers equipped with the proper tools and equipment and the right kind and quality of ingredients. Workers should be trained to use these tools and ingredients correctly to make good sandwiches. Sanitation must be carefully watched. Management must also guide workers in how to present products. Pricing is a special management concern. Since the labor cost is high in relation to food cost, inappropriate pricing can result if the amount of labor involved is not considered. Sandwiches should carry a good markup.

■
SALADS AND SALAD DRESSINGS

A salad should be made creatively so that its various parts contrast in color, texture, flavor, and form (see Fig. 12-5). Originality must be employed to ensure that salads are different from the competition's. In total, eye appeal, arrangement, simplicity, and variety in the end product are keys to successful salads. Dressings should

FIGURE 12-5 This salad contrasts in color, texture, flavor, and form.

complement the salad. A fruit salad is improved when a slightly tart, honey-fruit dressing is used with it. Bland salads require relatively tangy dressings, while zesty ones may use milder dressings.

Salad Planning

Salads generally consist of chilled ingredients placed on an underliner and topped with a dressing and garnish, but other foods are called salads that do not fit this definition, including hot salads, frozen salads, salads without underliners, and salads without garnishes. Salads are especially popular because of today's emphasis on healthful foods. People realize that salads are low in calories and high in nutrition for the calories they carry. Salad bars have become so popular that even fastfood operations feature them. Nonetheless, some people find that salad bars get very boring when they endlessly repeat the same offerings. Giving a salad bar special distinction by accenting it with different offerings or with a slightly different manner of presentation can do much to heighten patron approval of the salad bar.

Salad Menu

A wide variety of salads can be offered on a menu. They should suit the taste of patrons, meet cost restraints, and be planned around items that will be available. They are often high in labor cost compared to material cost. A salad may be used as an appetizer, as an accompaniment to a meal, as a main dish, or as a dessert.

Appetizer Salads

The appetizer salad selections should be light and small and should be designed to stimulate the appetite.

Accompaniment Salads

Accompaniment salads, whether served with the main course or offered as a separate course, should provide a contrast in flavor, texture, color, form, and temperature with the main course food. Moderate portions are served. Hearts of lettuce, cabbage, pickled beets, savory vegetables, fruits, fruits with cottage cheese, and molded or frozen salads are proper. A heavy meal calls for a lighter salad, and vice versa. A salad featuring meats, beans, potatoes, macaroni, or other heavy items should accompany a light meal, such as a sandwich or a soup and a glass of milk. A lighter, more delicate salad should be offered with a meal consisting of roast beef, potatoes, and vegetables.

Main-Dish Salads

Some salads are served as entrées. They often have a base of meat, fish, seafood, or cottage cheese, extended by chopped celery or other vegetables. Cooked macaroni products, rice, potatoes, or other starchy items can be used as well. The pasta salad, for example, is quite popular. A main-dish salad—such as a chicken salad with pineapple and tomatoes, or a tomato stuffed with tunafish salad—can be served as the main dish for a luncheon or even for a dinner.

Dessert Salads

A dessert salad is usually sweet and can be quite rich. It may be served as a single item accompanied by a beverage at parties, or it may be served as a complete meal. After a bridge or club session, a dessert salad can be served with finger sandwiches. Fresh salads made of nuts, fruits, or dairy products, gelatin salads, and frozen salads can all be used as dessert salads. A small salad of some sweet mixture served in a bite-sized creampuff shell or in a small lettuce cup, a strawberry mousse, and jellied fruit are other dessert salads. The dressing is usually sweet.

Salad Equipment

Mise en place is organizing beforehand, getting all materials needed, and then keeping things organized, cleaned up, and moving toward the final goal and is vital to efficient salad production. It must be accompanied by the use of good tools and equipment. The work center must be planned thoughtfully and arranged for proper placement of foods, tools, and equipment, to save time and motion. Sanitation is important, since many foods are perishable and cold. Storage is another important consideration, since so many items are prepared ahead of service.

Small tools include sharp knives, zesters (often used to peel orange or lemon rind), parisienne scoops

(used in making melon balls), peelers, corers, wire whips, appropriate cutting tools, and suitable trays and containers. A good cutting board is needed, preferably of plastic, for better sanitation. Larger equipment includes a vertical cutter/mixer (used to prepare both salads and salad dressings), a food slicer, a vegetable peeler, a food cutter (with an attachment hub for grating, dicing, and slicing), and a food mixer. The tools and equipment should be sufficient to enable the pantry section to produce high-quality products with a minimum expenditure of labor. Adequate refrigeration space should be available, providing a constant 40°F (4.5°C) holding temperature.

Salad Ingredients

Salads are often considerably more than greens, sliced tomatoes, or other simple products served with a dressing; they can be complex combinations of many ingredients. Salad greens are, however, the base of most salads. They can serve as both the underliner and the body. Fruits, vegetables, and other ingredients are also used.

Salad Greens

Salad greens contribute body and bulk to a salad, as well as variety in color, texture, and flavor. Only fresh young greens that are free of blemishes should be used. These should be trimmed, washed, and chilled before using to ensure that they are well drained, cold, and crisp at service. Salad greens are best dipped into water, shaken free of excess moisture, and then placed into wet cloths, plastic sacks with holes in them, or some other suitable container in which they can cool and crisp. *Do not use sodium sulfite to keep the greens fresh in appearance.* This chemical sometimes produces severe allergic reactions in people who ingest it. Do not put greens into water for holding, since air and water are both needed to crisp. Soaking also causes a loss in nutrients.

Fruits and Vegetables

Fruits and vegetables are the main ingredients of salads. They augment the greens by adding flavor, texture, and color. Such a salad may use (for example) the greens of torn romaine and spinach, mixed with fresh, crisp bean sprouts, chopped hard-cooked egg, and fresh bacon bits, and topped with a sweet-and-sour dressing. Some fruit salads may contain only celery or one other crisp item for flavor and texture contrasts to the fruit. The blend of fruits and vegetables with greens should be suitable. It is also important to use ingredients that are available year-round.

A number of fresh, frozen, or canned fruits are available for salads throughout the year, including the following:

Apples	Lemons
Apricots	Nuts
Avocados	Oranges
Bananas	Peaches
Cantaloupe	Pears
Cherries	Persimmons
Cranberries	Pineapple
Dates	Raisins
Figs	Raspberries
Grapefruit	Strawberries
Grapes	Tangerines
Honeydew melon	Watermelon
Kumquats	

Many vegetables for salads are plentiful throughout the year, as well, including the following:

Artichoke	Olives
Asparagus	Onions
Beans, green	Parsley
Beets	Parsnips
Broccoli	Peas
Brussels sprouts	Peppers
Cabbage	Potatoes
Celery	Radishes
Corn	Squash
Cucumbers	Swiss chard
Garbanzo beans	Tomatoes
Garlic	Turnips
Mushrooms	Watercress

Other Salad Ingredients

Other items such as croutons, nuts, and seeds, plus seasonings such as basil, chives, chervil, rosemary, tarragon, and thyme, can be used to give salads special texture, body, or flavor. Crisp foods such as crackers or even rolls or hot breads may be served as an accompaniment. A chapon (a dry piece of bread flavored with garlic) can be added to the salad for flavor.

Salad Presentation

Not all salads are served as individual units. Salads may be dished into bowls and served from a buffet, or salad materials can be put into containers on a salad bar where patrons are invited to make their own combinations. As little handling as possible is usually best. Salads served immediately after preparation or after chilling look and taste the best.

Individually Served Salads

Small or appetizer salads are usually served on a plate about 5½ in. in diameter. A regular dinner salad is served on a 7-in. plate. The salad should not come to the plate's edge; instead, some space should appear between the margin of the salad and the edge of the plate, so that the food is framed by the plate. Salad height is important. A flat salad appears small and insignificant. By using chopped greens, chunks of firm lettuce, and similar items, preparers can give a salad the appearance of having better form and of being lighter. Colors in a salad should be mixed; various greens can give a good color contrast. Messiness is to be avoided. Salads made of soft materials should be mixed as little as possible. Garnishes should be appropriate, selected to suit the flavor, texture, and appearance of other salad components. Showmanship can be exercised by using form and color to give striking effects; salads should not, however, present garish contrasts; they should look good through careful attention to the principles of proportion and color matching.

Preparing and presenting individual salads at the tableside can do much to merchandise salads. This can be done from a mobile cart on which various salad materials are assembled, including the dressings and garnishes. The server doing the preparation should be deft and should set up attractive salads with good proportions and form. Good measurement should be achieved by eye and practice, just as it is when a good bartender mixes a drink.

Buffet Salads and Salad Bars

Salads can be prepared in large bowls, attractively garnished, and placed on buffets, from which guests serve themselves or are served. Simplicity makes salads attractive and reduces labor costs. Salads can be served from attractive ceramic bowls, silver platters or other platters, or ice bowls; a fruit salad can be placed in a half scooped-out watermelon. It is important to offer different kinds of salads on a buffet. Thus, a fruit salad, a potato or pasta salad, a seafood salad, and a vegetable salad could be there.

Salad bars, too, should hold a variety of salad materials in containers set on a refrigerated counter. Chopped greens, sliced or sectioned tomatoes, chopped sweet onion or thinly sliced onion rings, green and red pepper strips, carrot cubes, crisp green vegetables, olives, celery, sliced mushrooms, cottage cheese, spinach leaves, garlic-seasoned croutons, chopped chives, and other items can be offered. At a slightly greater cost, marinated quartered artichoke hearts, pickled pepperoncini, molded fruit, thinly sliced sausage, and chopped cheese can be offered. Four to six dressings should be presented.

Salads and the plates onto which they are dished should be chilled. Some localities require that sneeze guards be installed and other sanitary precautions be taken when customers are allowed to dish their own foods.

Salad Preparation

Fresh, crisp ingredients, skillfully and imaginatively prepared, make good salads. Eye appeal plays a large part in salad acceptance.

Base or Underliner

The salad base usually is a salad green that holds the salad and leaves a border in the bowl or plate on which it is served. Limp lettuce hanging over the plate's edge can ruin appearance. The base may contribute to color, texture, or both. It should be clean and edible, even though guests usually do not eat it. To encourage consumption, the base can be sliced or chopped.

When the salad is made of large, loose particles of food, a cup or good base helps hold the salad's shape.

A base can also help give the salad height. The base should be in proportion to the salad, with its highest point at the back of the plate, away from the guest.

Bibb, Boston, iceberg, leaf, or romaine lettuce, as well as chicory, curly endive, escarole, and other greens such as spinach leaves, may be used for underliners. Large outside lettuce leaves do not cup well, but they can be used flat on trays or in bowls for buffet salad underliners. Unusual underliners might include thinly sliced apple latticed under a molded fruit salad or sliced tomatoes laid under a mixed bean salad. Cabbage is not used as a base because of its tough texture.

Body

The body of a salad comprises its main ingredients and usually gives the salad its name. The arrangement of the body on the base or in the serving dish should be neat and appealing. Simplicity can enhance the salad's attractiveness. Individual ingredients (particularly main ones) should remain distinct; they should not be chopped, minced, or mashed to the point where desirable appearance and crispness are lost. Variety of form produced when cutting or tearing salad materials adds interest. Many kinds of foods may be used as the main body, and all require different treatment.

Fruits and Vegetables. The size of the fruits or vegetables in salads should be established in purchase specifications. Thus, tomatoes or artichokes of a specific size may be needed.

Fresh, dried, frozen, and canned fruits and vegetables are used effectively. All inedible parts—including seeds, skins, and stems on fresh items—should be removed. Particle size can be varied for interest, and contrasts in color, flavor, and texture should be sought. Soft-textured items such as bananas, pastas and artichoke hearts combine well with crunchy items such as apples, celery, nuts, and crisp greens. Some fruits darken when cut and exposed to air. Cutting these with a stainless steel or nonmetal knife and dipping the pieces into an acid such as pineapple or lemon juice, a special oxidizing solution, or even salted water helps prevent darkening.

Firm vegetables such as cucumbers and radishes should be cut thin enough to be easily chewed. Cutting vegetables into julienne, diced, sliced, or round shapes adds to their interest, as can random chopping, cutting, or grating. Cooked items should be firm, not mushy. Carryovers can be used, but they should not be inferior products.

Gelatin Mixtures. Gelatin is often used for salads—either plain or as a prepared, sweetened, and flavored mixture. Gelatin adds a different flavor, texture, and color to salads. Soak dry gelatin in cold water and then bring it into solution at 170°F (77°C) or higher in more liquid. As it cools, it absorbs water or liquid, forming a soft gel. A good gelatin makes a strong yet tender gel; a poor gelatin produces a weak and watery gel or a tough and leathery one with poor flavor. Gelatin has no grades; it should be purchased on the basis of the quality of the product or brand.

Gelatin strength is decided by the following six factors: the gelatin's quality; the amount of liquid; the amount of sugar and acid (sugar strengthens and acid weakens); the amount of added materials; the temperature at which the gelatin sets; and the temperature of the gelatin mixture. Typically, 3½ oz of unflavored gelatin forms a good gel in 1 gallon of liquid; 1½ lb of powdered gelatin dessert mix is required per gallon. Fruit or vegetable juices add flavor, as do syrups from canned fruits, meat stocks, and even carbonated beverages. Sweetness in plain gelatin mixtures can come from the ingredients or from added sugar. Solid materials can be added as soon as the gelatin is put into solution. Ingredients that tend to float to the top should be stirred in just before the mixture sets. If a lot of ingredients are to be added, less liquid should be added.

Gelatin powders are added to hot liquid to bring them into solution; water is usually the liquid used, but drained juice or other liquids also work well. The sweetness of canned fruit juice can produce too sweet a mixture, however. Gelatin products usually hold their quality for several days under good refrigeration. A number of gelatin products can be produced.

Basic Gelatins. Whether unflavored or flavored, basic gelatin is a simple combination of gelatin and liquid, with the possible addition of seasonings, solids, or a sweetener. Sugar increases the firmness of a gel, as does milk, while acids and any solids added tend to weaken the gel. Gelatin in any form is derived from the hides, skins, connective tissues, and bones of all types of animals.

Cubed and Flaked Gelatins. Variations in the cutting of basic gelatins result in interesting salads. Cubed gelatin is made by cutting gelatin ½ to ¾ in. deep into cubes and serving it alone, with different-flavored cubes, or with mixed fruits. Flaked gelatin is prepared by running firmly set gelatin through a food grinder at high speed and serving it with other flaked gelatins (for color contrast), with coconut, with fruit, or with ice cream.

Gelatin Foams. Because gelatins have a high protein content, they retain air well and easily foam when beaten. *Whips* are basic gelatins blended or beaten until they double in volume. Such whipping occurs when the gelatin mixture has the consistency of thick, raw egg white. Foaming takes place best if the partially thickened gelatin is whipped in a chilled bowl or over ice water at high speed. *Snows* or *sponges* are similar to whips, except that they have egg white blended in at the point of whipping. The result is a gelatin product whose volume is tripled.

Layered Gelatins. Two or more basic gelatins can be layered on top of another, creating different colors and flavors for each layer. The first layer must be completely jelled when the second layer is poured—already partially thickened—on top of it. Each succeeding layer is then poured, partially thickened, after the preceding layer poured has completely jelled.

Gelatin Molds. Decorative molds offer attractive shapes to gelatin salads. The mold may be lightly oiled first to prevent gelatin adhesion. If oil is not used, the jelled mold must be dipped in warm water up to its edge and then inverted, allowing it to slip onto a serving platter. Molds frequently have arranged fruit in them. This is accomplished by taking a chilled mold, coating a section with partially thickened gelatin, arranging the fruit as desired on the gelatin, allowing this area to jell quickly, and proceeding one section at a time until the mold is done. Of course, many times the fruit or solid food is simply stirred into partially thickened gelatin, the mixture is poured into a mold, and the salad is unmolded when it has jelled, with food particles suspended throughout the mold. (*Note:* Raw pineapple, papaya, figs, and some other fruits containing certain enzymes cannot be used in gelatin production because the enzymes they contain break down the gelatin's structure and do not allow the gelatin to jell.)

Other Gelatin Products. By adding various ingredients and performing certain preparations, other gelatin products as popular as salads are created.

- *Aspics:* similar to basic gelatins and may be clear or seasoned with meat stock, tomato, spices, or other foods.
- *Bavarian creams:* combine basic gelatins and egg yolks cooked a few minutes to form a custard; when chilled, the mixture is folded into a stiff meringue, and whipped cream is added.
- *Mousses:* rich desserts made by combining cream thickened with gelatin and whipped cream.
- *Chiffons and Spanish creams:* basically the same as bavarian creams, except that whipped cream is not added.
- *Charlottes:* mousses enclosed in or trimmed with sponge cake or ladyfingers.

Quick Gelatin Preparation. Dissolve the gelatin in up to one-half of the liquid called for, and measure the rest of the required liquid as crushed ice, covered with cold water, in the measuring container. Combine the crushed ice and water with the gelatin mixture, and stir until the ice melts and the gelatin thickens. A quickly formed gel is more likely to lose its structure when removed to higher temperatures than is one that solidifies at higher temperatures.

Protein Foods. Main-dish salads usually contain meats, fish, or poultry as the major ingredient—roast beef, ham, tongue, crab, lobster, shrimp, salmon, tunafish, chicken, or turkey—along with cheese or eggs. The pieces should be bite-size. Protein foods can be cut in a food chopper or mechanical cutter. Meats and poultry are usually diced or cut into julienne strips; fish is often coarsely flaked with a fork. It is difficult to cut some foods into pieces of the proper size. Do not mince foods, as this often gives them a pasty consistency when mixed with a salad dressing.

Rice and Macaroni. Rice and macaroni products are frequently used in salad bodies because they are neutral in color and taste and because they accept seasoning and garnish well. Rice and macaroni products should be cooked *al dente* (firm and chewy), rather than mushy or overcooked. Rice must be dry and fluffy, and pastas must be thoroughly drained. Rice and pastas can replace vegetables in some salads, or they can be mixed with protein foods.

Salad Greens. Salad greens give a variety of textures, flavors, and shapes to a salad. When they are used as the main ingredient and garnished with vegetables and other foods, the result is a tossed or green salad. To prepare, simply toss the salad greens with other ingredients or by themselves to make the body. Wash all of the greens thoroughly. The core in an iceberg lettuce should be pushed in with the palm of the hand or be removed by being pressed on a hard surface. The head should be placed in cold water and then thoroughly drained, with the open-end surface down. Other greens should be washed in cold water rather than in a colander under running water. Colander washing does not free dirt as well as does putting the greens into a cold-water bath and letting soil sink to the bottom. A small bit of salt added to the water draws out worms and insects, but delicate greens may wilt from such treatment. For such greens, carefully scrutinize the leaves instead of using salt. Place washed clean greens in plastic bags or covered containers, and allow them to crisp under refrigeration.

To use, shake the greens free of water, and cut them with a sharp knife or tear them to avoid bruising. When time allows, tearing is preferred. Pieces should be bite-size. Heads of solid materials may be quartered, their cores removed, and the heads roughly chopped or evenly sliced or grated.

Salad Dressings

Salad dressings are flavorful sauces applied to salads to enhance their flavor or appearance and sometimes to add moistness. They are usually emulsions, mixtures containing two liquids that do not combine. Thus, when oil and vinegar (two common salad dressing ingredients) are shaken together, they mix for a few minutes, but the oil soon rises to the top of the vinegar. Such a mixture is known as an *unstable* or *temporary emulsion*. If as a third ingredient an emulsifying agent is added, the mixture can become a *stable* or *permanent emulsion*. One such emulsifying agent is egg, as in mayonnaise, which coats the droplets of oil and holds them in a water-oil emulsion. French dressing is a temporary emulsion, while mayonnaise and cooked or boiled dressings are permanent emulsions. A relative few basic dressings provide the foundation from which most dressings are prepared.

French Dressing. French dressing is a 3:1 blend of oil to an acid such as lemon juice or vinegar. Seasonings such as salt, pepper, paprika, mustard, sugar, honey, and garlic are added for flavor and to help stabilize the emulsion and give some thickness. Tomato paste also can be added. The oils most commonly used are corn, cottonseed, olive, peanut, safflower, or soybean. Vinegars may be cider (apple), distilled (white or brown), malt, or wine.

Mayonnaise. Mayonnaise is a permanent emulsion of acidic liquid (such as vinegar or lemon juice), oil, and egg yolk or whole eggs, with added seasonings. Usually the acidic liquid, seasonings, and egg are mixed together, and the oil is then added very slowly while the mixture is beaten. Gradually, a thickened emulsion forms.

Cooked or Boiled Dressings. Cooked or boiled dressings are thickened with starch and perhaps eggs; they contain less oil than French dressing or mayonnaise. Flour is often the thickener, although cornstarch, arrowroot, or another starch can be used. The liquid is milk, water, vinegar, juice, or some blend of these. When the mixture is cooked and thickened, eggs are added, as in making a pudding or sauce. The oil, margarine, butter, or other fat is then added; and vinegar or lemon juice and seasonings are added last. Acids tend to break down starch pastes, so they are added last to minimize this possibility. Care must be taken to ensure that the dressing does not become lumpy or get scorched.

Other Basic Dressings. Other popular dressings are based on sour cream, cream cheese, whipped sweet cream, and other liquids and semisolids. These are often seasoned with vinegar, herbs, spices, and salt. Among the variations on these are sour cream with chives, cream cheese dressing with chopped Spanish olives and honey, and sour cream with fruit and nuts. A hot dressing used for wilted salads is a French dressing variation that sometimes is lightly thickened.

As patrons wish to watch their weight, healthy modifications and low-fat substitutes to traditional salad dressings have occurred. The replacement of oil with fruit and vegetable juices or stocks has become popular. Vinegar or just plain lemon juice have been used by some wishing to cut all fat from the salad dressing. Continued use of salad dressings low in calories will dictate that most foodservices operations will need to feature these alongside the more traditional salad dressings discussed previously.

A marinade is a mixture, such as oil and vinegar plus seasonings, in which foods are soaked or marinated to gain flavor. A five-bean salad is often marinated. Other marinades are used to flavor meats, fish, and some vegetables before they are used in a salad. Leafy vegetables are not marinated because they wilt. Any excess marinade should be drained off when the salad is dished for service.

Garnish

The garnish of a salad corresponds to the icing on a cake; it should offer contrasts in color, flavor, and texture, as well as harmony with the salad and dressing. Garnishes should be edible, eye-appealing, and simple. The following garnishes are among the many possibilities:

Anchovies

Apple—slices, balls, sticks

Aspic cubes

Bananas—scored with fork and sliced

Beet—rings, shreds

Cabbage—shredded, slaw

Carrots

Celery—curls, hearts, stuffed

Cheese—grated, balls, slivers

Cherries (green and red) stuffed with almonds or cheese, cut into petals, or whole with stem

Chives—chopped

Coconut—white or faintly tinted

Cucumber—slices, fingers, rings, sticks, boats, curls, fluted

Dates, figs, or prunes—stuffed with peanut butter, cheese, or nuts

Eggs—hard-cooked, sliced, chopped, riced, deviled

Endive

Fruits—glazed, candied

Grapes—plain, frosted, in clusters

Jelly cubes

Kumquats—fresh or preserved

Lettuce—edges dipped in paprika

Melon—balls, cubes, pickled, rings, sticks

Mint—fresh sprigs or chopped

Nutmeats

Olives—ripe or green, stuffed, sliced, or whole

Onion—rings

Oranges or grapefruit—segments, slices

Paprika

Parsley—fresh sprigs or chopped

Peppers—green or red, rings, chopped, or strips

Pickles—dill or sweet, sliced, stuffed, fans, strips

Pimiento—strips, stars, chopped

Pineapple—fingers, cones, fans, dipped in paprika or edge rolled in finely chopped parsley or nuts

Pomegranate seeds

Preserved or spiced fruit

Radishes—slices, roses, fans

Raisins

Relish—raw fruit and vegetables

Rhubarb—curls, raw

Scallions

Sherbet

Spinach—small leaves

Strawberries—sliced, dipped in sugar, whole

Tomatoes—sliced, diced, quartered

Truffles—fancy shapes

Turnip—strips, thin slices

Watercress

Management's Responsibility

Management must plan to coordinate salads and salad dressings into the total production picture. Standards must be set and followed, and costs must be controlled. Quality can be controlled by proper purchasing and handling procedures; quantities must be controlled by good recipes and portion control.

Standards for Salads

Production standards can be established by using good recipes, by referring to pictures of salads, and by good training. Employees should understand that good salads are light, appetizing blends of ingredients with a proper dressing. They must be arranged with symmetry and uniformity of design. Creative blends of salad ingredients with dressings and artistic presentation should be incorporated into the recipe to achieve consistency in how products are served. Fresh, high-quality ingredients that offer complementary contrasts of textures, colors, and flavors should be used. Salads should not appear tired from having been overhandled. No inedible portions or wilted or rusted particles should appear. Portions should fit the plate.

Cost Controls

Labor can be significant in making salads and dressings because much of the preparation is by hand. Purchasing preprepared materials, such as diced vegetables, chopped lettuce, and salad dressings, reduces costs and still can maintain high standards.

Food costs can be controlled by controlling portions. Tossed or green salads served as meal accompaniments should be approximately 1 cup (2½ oz) each; about 16 lb European plan or 20 lb American plan of greens is needed for 100 salads. A 3-oz scoop (No. 12) is used for potato, pasta, and fruit salads. A good portion of gelatin salad is ½ cup; thus, slightly more than 3 gal gives 100 portions. Comparing portion costs between convenience items and items made from scratch can indicate which should be used, provided that quality is equal.

Labor is an important consideration in salad making. Space should be provided so that many salads can be prepared at one time. Plates can be spread out on 18 in. × 26 in. sheets, filled, and then loaded into refrigerators to be held until needed. Enough refrigerator space must be planned. Making one set of motions for many salads reduces labor time. Thus, all bases can be set down for twenty-four salads, followed by the body, the garnish, and perhaps the dressing. Sometimes, however, it is preferable to have serving personnel add the dressing and garnish just before serving.

The salad buffet and salad bar are labor savers and customer pleasers. The hand labor required to place underliner, body, garnish, and dressing on each individual plate is saved. Even though larger servings may be taken, which raises costs, they take the edge off appetites with relatively inexpensive food, which again saves the operation money.

■

BEVERAGES

Beverages are an important part of meals, partial meals, and snacks, and may even be taken on their own. Many foodservice operators feel that coffee, for example, is something guests remember and come back especially to have. It is commonplace in the foodservice industry that a good cup of coffee or tea draws customers. While seemingly simple to make, beverages must be made with care. Proper attention must be paid to see that ingredient proportions are correct, that equipment is clean, that the making is right, and that the service ensures preservation of the care taken in making the beverage.

Coffee

Coffee has been a symbol of hospitality for generations. The Arabs use it to greet guests and start business meetings. The coffee break is an important part of a worker's day.

Coffee Equipment

The right size of coffeemaker must be used. The equipment must be flexible enough to meet coffee needs at both peak and slow periods. A cup of coffee is 6 oz; if free refills are offered, every patron can drink 1 pint or more of coffee. Small units are used during low-demand periods, and often the number of units in service is increased or larger equipment is used when demand is high.

The coffee brewer should be constructed of glass, porcelain, or stainless steel. Bitterness, astringency, or metallic taste results if the equipment contains tin plate, aluminum, copper, or nickel.

Another concern is whether an automatic or a manual brewer is best. Labor costs, skilled workers, and the availability of service personnel for maintenance and repair are factors that must be considered. Manual brewers are less expensive and have fewer mechanical devices that fail, but they require more attention from personnel and may, therefore, increase costs. Semiautomatic brewers are also available. The filtering devices and the water feed into the grinds are also important to the purchase decision. Their design should provide adequate waterflow over and around the grinds, sufficient capacity to prevent backup and overflow, sufficient support of the grounds so that they do not touch the beverage, and easy access to all areas for cleaning. The mechanical reliability of each unit also should be evaluated before the final selection is made. Automatic brewers can save labor and simplify the process so that inexperienced workers can make good coffee.

Urns are large units that make coffee by the drip method. The coffee is placed evenly in a basket and boiling or very hot water is put over it to drip through. Vacuum coffeemakers work by forcing boiling water (by steam) into an upper chamber, where it comes in contact with finely ground coffee. The water stays in contact with the grounds for a short time, and then the steam pressure is released—often by removing the boiling lower part from the heat. When the steam condenses in the lower part of the coffeemaker, a vacuum is formed that sucks the liquid in the top down through a filter, leaving the grounds in the upper chamber to be removed and discarded.

New types of drip equipment can make 9-cup, 12-cup, or larger batches. The coffee is put into a small cartridge or container, a button is pressed, and the correct amount of very hot water flows through the coffee. The old cartridge is then replaced with a new one to make a fresh batch. Coffee is also made by boiling, although this is seldom done in today's foodservices. Special coffeemakers such as an espresso machine are also used. An *ibrik* is used to make Turkish coffee.

The Product

Coffee is a small bean that comes from a red, cherrylike fruit. Many different kinds exist. The Arabica plant is the one most often grown. It makes a rich mellow coffee if grown under the right conditions. Coffee flavor and quality vary with the season of the year, the processing method, and the growing area. Colombian coffee is rich and aromatic in flavor, with a bouquet quite different from the full and mellow flavor of the Santos coffee of Brazil. The Robusta plant is also grown and is the primary source of African coffee. Its flavor is stronger than the Arabica's but not quite as full. It is often blended with the Arabica.

Before roasting, coffee is a whitish green bean. The raw beans are often blended to produce a coffee of particular flavor. Rio coffee is coffee grown in the lowlands; it has a somewhat harsh and bitter flavor. It is blended into coffees to give them more bitterness or acidity. Coffee beans are roasted to develop their flavor and color. After roasting, coffee beans are highly perishable. Roasting can be light, medium, or dark. The Italians and the French like a dark or heavily roasted coffee.

The next step in the coffee-making process is the grinding of the beans. The grind may range from coarse to very fine, but it must be suited to the type of equipment on hand. A drip machine needs a fine drip grind. An urn coffee is ground more coarsely. The contact time with the water and the size of the grind are related: fine grinds stay in contact with the water for less time (see Table 12-3). If the grounds and

water are in contact too long, overextraction occurs, and bitter flavors are released. If the water is not in contact with the grounds long enough, the coffee is underextracted and weak in flavor.

Coffee loses flavor less rapidly in the whole bean than when ground. After grinding, flavor loss is very rapid; in three days, the loss can be 20 percent. Coffee held over a week should be either vacuum- or gas-packed. It should be stored at room temperature only when necessary. A cool, dry place is needed. Coffee that is not to be used within a day should be stored in a freezer, since freezing or refrigerating delays deterioration.

Coffee stocks should be rotated so that the oldest coffee is used first. Packaging should be the exact size needed to produce the desired amount. Some operations purchase bean coffee and grind it as needed to get the freshest coffee possible.

Coffee best releases its flavors in water heated to about 205°F (96°C). The contact time is controlled to suit the grind. Tastes or odors in some water spoil coffee. Some waters that are high in iron make a very dark coffee that turns green when cream is added. Odors, flavors, and iron content can be filtered out. Hard water makes acceptable coffee, provided that it is not too hard. Water containing too many bicarbonates or other hard materials may cause a sediment to form that delays water flow through the grounds, resulting in overextraction. Hard water salts or lime can build up in the equipment, destroying temperature controls or otherwise interfering with proper water delivery and ultimately reducing the life of the equipment. Using water softeners under these circumstances is recommended, but they must not be allowed to deliver too many bicarbonates. Polyphosphate substances can be introduced to reduce bicarbonates. The water used should be freshly boiled, since overheated water can be stale and can lack vitality.

The right amount of coffee of the right grind and blend should be used. The correct ratio is usually 2½ gal of water to 1 lb of coffee. Some experts recommend 3 gal of water per lb. Management must decide on the strength of the brew it wants to serve and should set standards accordingly.

Home measurements are 2 Tbsp per 6 oz or ¾ c of water. For a strong demitasse, however, 4 oz of water are used. Coffee purveyors usually indicate the correct

TABLE 12-3 Proper Exposure Times to Water of Various Coffee Grinds

Type of Grind	Use	Exposure Time (min)
Fine	Vacuum coffeemaker	1–3
Drip	Urn brewer	4–6
Regular	Electric percolator	6–8

proportions for their product. Coffees designed for the home do not hold up well in commercial use.

Brewing Coffee

The right amount of water heated to the right temperature and added to the right amount of properly ground coffee for the right contact time gives a good coffee brew. In an urn, the grounds are usually spread evenly on the filter to a depth of 1 to 2 in. Evenly spread coffee promotes uniform extraction. Paper or cloth filters are most commonly used, although filters can also be made of woven wire screens, perforated plates, or metal disks, which do not allow the finest particles of coffee to pass through. The water should thoroughly saturate the coffee. If only a part of the desirable flavor is extracted, a weak coffee results; too much saturation produces an overextracted coffee that is strong and bitter.

If all of the water does not pass through in the prescribed time, it may be necessary to run less water through the grounds and to add hot water directly to the new brew or to place some water in the urn before brewing. Coffee should not be extended by pouring too much water through it, since this extracts bitter and unflavorful products. An acceptable brew extension only results if the flavorful and good fractions alone are extracted; then these are diluted to the smallest extent possible.

The filtering device should be removed immediately after the water passes through. Leaving the grounds over the brewed coffee harms the brew. Pouring fresh coffee through extracted grounds increases its bitterness. In an urn, most of the flavor comes off first; bitterness and less flavor come off last. For this reason, it is wise to mix the coffee in the urn by removing the grounds and then repouring some of the coffee from the bottom of the urn back in at the top, to give a more balanced strength and flavor. The coffee can also be mixed by stirring. In some new urns, automatic agitation does this.

Brewed coffee should be held at 185 to 190°F (85 to 88°C) for up to an hour. Coffee cannot be reheated without a distinct flavor and color loss. Each category of coffeemaker—vacuum, urn, semiautomatic, and automatic—has its own brewing procedure. The guidelines that follow were drawn up by the Coffee Brewing Center.

Vacuum Coffeemaker

1. Accurate measurement is the most important step. Use 1 oz of coffee to each 16 to 20 fl oz of water. Measure cold, fresh water into the lower bowl, leaving at least 1 in. at the top for water expansion. Place the coffeemaker on heat and bring to a boil.
2. Place a clean filter in the upper bowl. If a cloth filter is used, rinse it in hot water.
3. When the water boils, reduce the heat or turn off the electricity.
4. Measure fine-grind coffee into the upper bowl, and insert the upper bowl with a slight twist to ensure a tight seal. Return the maker to reduced heat.
5. Let water rise into the upper bowl, and allow it to mix with the ground coffee for 1 minute, stirring thoroughly in a zigzag motion for the first 20 seconds.
6. Remove the maker from the heat. The brew should return to the lower bowl in no more than 2 minutes.
7. Remove the upper bowl, rinse it and the filter with hot water, and place the rinsed filter cloth in cold water until next use.
8. Hold the coffee at 185 to 190°F (85 to 88°C), and never allow it to boil. Never reheat coffee, and always serve it fresh. Brewed coffee should be discarded after 1 hour.

Urn Brewer

1. Accurate measurement is the most important step. Use 2 (never more than 2½) gallons of water to 1 lb of coffee.
2. Spread fresh drip-grind coffee evenly in the filter. To obtain good extraction, spread the layer of coffee evenly (1 to 2 in. in depth). If a new urn bag is to be used, rinse it thoroughly in hot water before placing it on the urn bag ring. If the urn is not equipped with a brewing basket, use a gridded riser to ensure proper support of the urn bag.
3. Use fresh boiling water. The urn should be attached only to a cold-water line. Be sure that the water temperature is 200°F (93.5°C) when it comes into contact with the coffee.

4. Pour the water in a slow, circular motion. This helps ensure even extraction. Make sure that you wet all the grounds evenly. Total contact time must be between 4 and 6 minutes. Brewing time starts when the first water touches the coffee and ends when the last water has passed through the grounds and the brewing device is removed. Replace the urn cover between pours, to preserve the aroma.

5. Remove the filter device as soon as the water has dripped through. If the brewing device is left in the urn, steam from the coffee rises, condenses into water, and passes back through the grounds, releasing astringent and bitter materials. Rinse the cloth filter in clear hot water, and then store it in cold water until next use.

6. If all of the water cannot be passed through the grounds within the prescribed brewing time, it may be necessary to place some of the water (up to 40 percent) in the bottom of the urn before brewing. This prevents overextraction and bitterness.

7. Mix the brew. Draw off the heavy coffee from the bottom of the urn and pour it back into the brew to ensure uniform mixing. Mix at least 1 gallon for each 1 lb of coffee used.

8. *Never* repour brewed coffee back through spent grounds. It only makes the coffee bitter, and is a waste of time.

Semiautomatic Coffeemaker

1. Accurate measurement is the most important step. Use 1 oz of coffee to each 16 to 20 oz of water. For 48 oz of water (8 cups), use from 2.4 to 3 oz of coffee. For 60 oz of water (10 cups), use from 3 to 3.75 oz of coffee.

2. Place filter paper in the brewing cartridge. Spread fresh fine-grind coffee evenly in the filter paper; an even coffee bed is important for even extraction.

3. Check the water temperature. The coffee coming from the brewing cartridge must be at least 190°F (88°C) to ensure that water was 200°F (93.5°C) when it was filtered through the grounds.

4. Check the brewing time. All pour-over brewers should deliver a bowl of coffee in from 3 minutes 20 seconds to 4 minutes 20 seconds. Remove the grounds as soon as the water has dripped through.

5. Rinse the cartridge; wipe the underside of the spray head area on the coffeemaker at least once a day.

Automatic Coffeemaker

1. Use 4 oz (never less than 3⅕ oz) of coffee to each ½ gallon of water. If a 3-oz package is used, reduce the beverage yield to 54 oz or less.

2. Place a filter in the brewing chamber. If a cloth filter is to be used, rinse it thoroughly in hot water before its first use.

3. Spread fresh coffee grounds evenly in the brewing chamber. An even coffee bed is important for even extraction. Use either a drip grind or a fine grind, depending on the machine's brewing cycle.

4. Check the water temperature. The coffee coming from the brewing chamber must be at least 190°F (88°C) to ensure that the water was 200°F (93.5°C) when it filtered through the grounds.

5. Remove the grounds immediately after the water has dripped through. If a cloth filter was used, rinse it in clear hot water, and then store it in cold water until next use.

Sanitation and Quality Control

Coffee flavors are carried in volatile or essential oils that can break down easily, become rancid, or otherwise react unfavorably with substances. Long holding or improper making leads to major flavor loss. Coffee that smells good may not taste good because much of what is good is escaping into the air. High temperatures encourage rapid flavor loss, and yet for best flavor coffee must be served hot.

A rancid odor or taste in coffee can be caused by a dirty filter, uncleaned equipment, or a clogged gridded riser that contains old grounds. Urn bags absorb coffee flavors that deteriorate and unpleasantly flavor new brew. It is best to use paper filters and discard these after each use. A new cloth bag should be rinsed well, since starch filters can give off-flavors to the brew. All surfaces should be thoroughly cleaned; rinsing is not enough. Serving cups should also be clean.

Urns should be rinsed after each use and a brush should be used to clear away sediment or coffee oils. Before a new batch is made, clear water should be run through the urn to rinse it. At the end of the day, equipment should be disassembled and thoroughly cleaned. The urn is reassembled and partly filled with water, which remains in it overnight. Leave the urn top ajar so air can circulate through. Twice a week, the urn should be filled with water and an urn-cleaning compound added, in accordance with the manufacturer's directions. The compound is left in the water overnight, and the following morning the unit is thoroughly cleaned. If this overnight standing is not possible, allow the urn to soak for at least 30 minutes before cleaning. Check to see that all parts are working. Do not use bleaches, soaps, or detergents on coffee equipment unless these are thoroughly rinsed off before use. Small coffee equipment can be soaked frequently in cleaning compound, thoroughly scrubbed, and rinsed. Good sanitation is one essential step in producing good coffee.

A cup of coffee should be judged for its flavor, odor, color, body, and clarity. The flavor should be sweet (but not too sweet) and slightly bitter. No saltiness should be evident. Some acidity should be present, but the effect should be mellow, not sharp. Coffee odors should be fragrant and aromatic, with a rich, true coffee smell; harsh or strong odors indicate poor quality. The color should be a blackish brown, not pale or weak. Coffee should have some body or density, since good making extracts about 22 percent solids from the grounds. Body is judged by using 18 percent milkfat cream (not half-and-half or whipping cream): the cream should flow into the coffee and feather; that is, it should stay in a separate layer and not mix. Cream or milk quickly blend with tea because tea has no body. Clarity is essential. A cloudy cup of coffee almost always has a poor flavor. When the coffee is consumed, no sediment or grittiness should be evident. A silver spoon lowered into the coffee shows up clearly until it is hidden on the bottom.

Serving and Merchandising Coffee

A good cup of coffee sells itself by exuding the characteristic rich odor of coffee so many people find appealing. Coffee tastes best right after it is made. In formal service, coffee is served after all dishes are removed and the table is crumbed. The coffee is poured from the container into a cup sitting on a saucer directly in front of the patron. The handle of the cup should be to the right and at an angle toward the patron. Less formal service allows the cup to be lifted with the saucer for pouring and then replaced in front of the patron.

Special coffees can add interest to a menu. Espresso is made by using steam pressure to force water over very finely ground coffee that has a dark roast. The brew has a heavy flavor with some bitterness and a heavy body. Cappuccino consists of espresso and hot milk poured simultaneously from two different pots into a cup containing nutmeg, cinnamon, powdered chocolate, or whipped cream. Turkish coffee is made from sugar and heavily roasted, finely powdered fresh coffee; it is boiled at tableside in a small brass pot called an *ibrik*. Café diablo is espresso poured into a silver bowl at the table and heated (usually over a guéridon), liqueurs are added, and the drink is flamed. Demitasse coffee is strong dark-roasted coffee often served with liqueurs, lemon, or spices in small cups; sugar should be offered with it. Viennese coffee is similar to demitasse coffee, but it is always topped with whipped cream; hot milk may be used to thin it. French coffee (also called *café au lait*) is a combination of strong, dark-roasted coffee and milk. Irish coffee is a blend of black coffee, Irish whiskey, and heavy cream. By mixing coffee and liqueurs of its own choosing, an operation can create its own specialty coffee.

Few operations today offer instant coffee. The delicate nature of coffee flavors and their high volatility make it difficult to capture them, and much escapes during processing. Instant coffee processed by the freeze-dry method has the most flavor of the instant types, and some brands do a good job of approaching good coffee flavors.

Tea

Tea is a popular beverage that is less expensive than coffee. The stimulant (theine) in a cup of tea contains about half the caffeine contained in a cup of coffee. Tea can be served hot or cold. If hot, it should be very hot; if cold, it should be iced. A good tea is brisk in flavor. Black tea produces a beverage the color of a bright new copper penny. Tea prepared with hard water is dark, and a slight film may cover the surface because of tannin precipitates. Using soft water corrects this. Lemon juice also lightens tea.

Tea comes from the young leaves of a tropical bush. Normally, two leaves and a bud are picked from the end of each stem. The highest-grade teas come from the tiniest leaves. Tea grown at high elevations and picked at the start of the season is best. Climate and the making of tea influence quality, too.

After being picked, tea is withered, rolled, and fired. Black tea is fully fermented between the rolling and firing processes; fermentation consists of an oxidation of the tannins. Oolong tea is partially fermented, and green tea is not fermented at all. Firing stops fermentation and preserves the tea.

Preparing Tea

Teas may be blended from as many as thirty different varieties of tea. Oolong tea is a somewhat tan-colored beverage, and green tea brew is a yellowish green. Green tea contains the most tannins, oolong has less, and black tea has the least. Tea should be made by what is called *wet service*. The tea is put into a preheated pot (1 tsp for every 2 cups of water), and boiling water is poured over it and left to steep for about 3 to 5 minutes. A poor cup of tea results when boiling water is put into a cup or pot and the tea bag is put on the saucer beside it, for the patron to immerse. Tea is not as perishable as coffee, but it can deteriorate, so fresh supplies should be maintained.

Tea also clouds easily, particularly under refrigeration. This clouding is caused by the precipitation of tannins. The Tea Council suggests making iced tea in quantity by the 1,2,3 method:

- 1 qt of boiling water over
- 2 oz of tea; steep for 6 minutes; stir; pour off the brew into
- 3 qt of cold tap water

This tea is then added to ice cubes as needed. Finished tea can be held for about 4 hours for making iced tea. Mint or lemon is usually served with iced tea. The best sugar to use is fine or fruit sugar, because it goes into solution more quickly than regular granulated sugar. Often, tea for icing is made quite strong, and the hot brew is poured over ice cubes. Some operations use instant tea for iced tea.

To make tea in quantity, use 5 to 8 oz of tea and 4 to 7 gallons of water for 100 to 150 cups. One cup (1 oz) of dry loose tea, makes 50 cups, or a little more than 2 gallons of beverage. Freshly boiling water makes the best tea. Tea can be made in a large pot on the range or in a steam-jacketed kettle. The water is brought to a boil, the tea added, the heat shut off, and the tea allowed to steep for about 3 to 5 minutes. Leaving tea leaves in the brew longer than this extracts too much tannin, giving a bitter flavor and an excessively dark tea.

Tea can be served with milk, but cream masks its delicate flavor. Lemon adds a bit of briskness to tea; sometimes a clove is served.

Some teas contain special flavors. Jasmine tea contains jasmine flowers, which give the tea a fruity flavor and a highly perfumed aroma. Teas can be blended with fruit juices and other liquids for special occasions. A spiced tea is made by steeping tea with orange or lemon rind, whole cloves, cinnamon sticks, and sugar. The Russians make a tea that is sweetened and seasoned with orange, lemon, pineapple juice, and a bit of cinnamon.

Tea Quality

A good cup of tea should be judged by its flavor, strength, clarity, and color. Body is not a factor, since tea has little or no body. The tea should have a slightly bitter flavor. Green tea is the most bitter (again, because of its tannin content), oolong next, and black tea least. Some teas have a distinct sweetness, and others have a distinct acidity; still others have blends that fall in between these. The taster should know what to look for in the particular tea. Teas should be aromatic with a fruity, fragrant odor. Flavor and odor are best obtained by swishing the tea inside the mouth, and then breathing in to catch the aroma. The mouth is then opened so that the taste buds can give their signal on what they find. A brisk tea should have a zesty stimulating quality. Weak teas are described as thin or weathery. All teas should be clear and have no oiliness.

Cocoa

Where cocoa service is infrequent, it is best to use individually packaged mixes of good quality. When service is regular or large, cocoa can be made from dry cocoa, sugar, and either liquid milk or dry milk and water. Breakfast cocoa is 22 percent cocoa butter. Some cocoas contain less fat. Cocoa that is too high in fat solidifies. Butter or margarine can be added if nonfat dry milk is used. Richer cocoas can be made by adding chocolate and by using half-and-half or cream in place of some of the milk.

Cocoa should have a rich, chocolate color. A gray or muddy color indicates poor quality. A definite chocolate flavor should be discernible. A poor cocoa powder makes a cocoa that has a flat flavor and produces much sediment. Wateriness or syrupy quality is undesirable.

There should be no scum on top or in the cocoa. Putting foam on the cocoa by whipping it or by adding marshmallows or whipped cream stops scum formation. Covering the kettle also accomplishes this. Cocoa contains starch, so some slight thickening occurs when it is heated, giving it some body. Adding starch to improve body is not recommended.

∎

SUMMARY

The pantry is responsible for the production of a number of food items that require a lot of labor and are often served cold. These include appetizers, sandwiches, salads, salad dressings, and beverages. Because of the number of motions they must perform, employees should be skilled and well-trained, and their work centers should be efficiently planned.

Appetizers and sandwiches are related products requiring somewhat similar production techniques. Sanitation is important. Appetizers are foods designed to whet the appetite, such as canapés, cocktails, dips, hors d'oeuvres, relishes, salads, and soups. A sandwich is a food item made of bread, a spread, a filling, and a garnish. Sandwiches may be double, open-face, or even tiered, and they may be hot or cold. Sandwiches can be served as snacks, entrées, desserts, and luncheon offerings.

A salad is usually a crisp green, vegetable, or fruit medley combined with other foods and served either hot or cold. Salads may be used as appetizers, meal accompaniments, full meals, or desserts. A salad usually consists of a base or underliner, a body, a garnish, and a dressing. Salad greens are the most common ingredient and contribute color, texture, and flavor. Fruits and vegetables are used to accent the greens and form a major portion of the body. Other salad ingredients are added for interest, variety, and flavor. The success of a salad depends on its freshness, on its crispness, and on the adequacy of the various parts. The body of a salad may consist of greens, meat, fish, poultry, gelatin mixtures, cheese, pastas, or combinations of these.

A dressing is a flavorful sauce used to add flavor, texture, and perhaps even color to a salad. The two basic types are French dressing and mayonnaise or stable-emulsion dressing. French dressing separates into vinegar and oil components, so it is called an *unstable emulsion*. Mayonnaise and boiled dressings form a fixed phase of oil in water and are called *stable* or *permanent emulsions*.

Garnishes for salads should be edible and should offer contrasts in color, flavor, texture, and form.

Beverages are liquids served before, during, or after meals to provide a flavorful, stimulating drink. Coffee, tea, and cocoa all contain stimulants related to caffeine. Coffee should be made with special care for the product, the equipment, and the brewing method used to produce it. A fresh product, scrupulously clean equipment, and the right production method can yield a desirable cup of coffee. The right roast and right grind are also crucial. Coffee water should remain in contact with the grounds for a set time and be of the right temperature—around 205°F (96°C). Some alkalinity in the water is acceptable, but if the water contains too many bicarbonates, the contact time may have to be increased because a sediment will form that delays passage of the water through the grounds. Approximately 2½ gallons of water should be used for each 1 lb of coffee.

Tea comes from the last new leaves on the stems of a tropical bush. Quality also depends on the variety of tea, on growing conditions (including elevation and climate), on processing, and on brewing technique. Black tea is fully fermented, a process that removes tannin. Oolong tea is partially fermented, and green tea is not fermented. After plucking, tea is withered, rolled, fermented (or not), and then dried.

Iced tea can be made by the 1,2,3 method recommended by the Tea Council. Iced tea is best made from freshly brewed tea that is poured over ice. Storing tea in refrigerators encourages clouding.

Cocoa is made from cacao beans, which are also the source of chocolate. In processing, more than half the cocoa butter is removed, and the remainder of the bean is ground. Breakfast cocoa is high in cocoa butter, containing 22 percent or more. When service is infrequent, cocoa should be made from packaged mixes. There are several ways to prepare cocoa in quantity from cocoa powder and other ingredients.

CHAPTER REVIEW QUESTIONS

1. You have been retained as an instructor in a junior college to teach a quantity food preparation course along with two other food-related courses. The junior college lacks a good quantity food preparation kitchen but does have a space in which most of the major pieces of quantity production equipment are available. You are able to set up a pantry work center; but lacking a real production center, you decide to put together a set of slides that demonstrate the production of appetizers, sandwiches, salads, salad dressings, and beverages. Set up a list of the slides that will be shown, and write a commentary to accompany each slide as it is shown.

2. Compare the different kinds of appetizers, and indicate the nature of each. Why are color, contrast, and arrangement important in preparing these?

3. What are some of the methods workers should use in making sandwiches?

4. Identify the amount of finished salad, by weight or measure, that is required for 100 portions of the following salads: (a) potato salad, (b) salad of mixed greens, and (c) gelatin salad.

5. What are the four parts of a salad?

6. What factors strengthen gelatin mixtures, and what factors weaken them?

7. How is French dressing prepared? What is a stable emulsion? What is an unstable emulsion? How is mayonnaise made?

CASE STUDIES

1. Jill's Bistro has decided to enhance its offering of appetizers. Jill has always featured specialty main dishes and desserts and needs your help in suggesting five or six good-quality appetizers. She has heard good things about Anchor Food Products and recommends that you check out their Web site at *www.anchorfoods.com*. Which appetizers would you suggest for her medium-priced operation? Would you suggest other appetizers on the menu, too? Why?

2. Mr. T has decided to open a coffee bar and he is interested in serving tea as well. He knows little about the product and has asked you to research the subject by looking on the Internet. He recommends that you consider *www.harney.com*. List the various teas available and describe each briefly. Which teas would you recommend, and why?

VOCABULARY

Tannin

Oolong tea

Appetizer

Canapé

Arabica

Urn

Boston lettuce

Mayonnaise

Salad bar

Sodium sulfite

Hors d'oeuvres

Garde-manger

S motion

Salad body

Stable emulsion

Cocktail

Pasta

Salad

Escarole

Gel

Dessert salad

Base or underliner

Curly endive

Green tea

Permanent emulsion

Automatic brewer

Espresso

Rancidity

Wet service

Robusta

Salad dressing

Bibb lettuce

Iceberg lettuce

Mise en place

Appetizer salad

Temporary emulsion

Garnish

Main-dish salad

Open-face

Romaine

1,2,3 method

Accompaniment salad

Black tea

Salad greens

Boiled dressing

Brisk flavor

Chapon

Breakfast cocoa

Relish

ANNOTATED REFERENCES

Chesser, J. M. 1992. *The Art and Science of Culinary Preparation.* East Lansing, Mich.: Educational Institute of the American Hotel & Motel Association. (Appetizers are considered in this book as well as other sections on food preparation.)

Donovan, M. (ed.). 1996. *The New Professional Chef*, 6th ed. New York: John Wiley & Sons, Inc., pp. 353–374 and 917–1024. (Garde manger and pantry production are considered in these chapters.)

Donovan, M. 1997. *Cooking Essentials for the New Professional Chef.* New York: John Wiley & Sons, Inc., pp. 399–424 and 669–708. (Garde-manger, salad preparation, and pantry production are considered in these chapters.)

Gielisse, V., M. E. Kimbrough, and K. G. Gielisse. 1999. *In Good Taste.* Upper Saddle River, N.J.: Prentice Hall, pp. 81–144. (Pantry products are discussed in this chapter.)

Gisslen, W. 1995. *Professional Cooking*, 3rd ed. New York: John Wiley & Sons, Inc., pp. 525–550. (Sandwiches and hors d'oeuvres are considered here.)

Janericco, T. 1990. *The Book of Great Hors d'Oeuvres.* New York: John Wiley & Sons, Inc. (Appetizers in all shapes and forms are discussed in this text.)

Kapoor, S. 1996. *Healthy and Delicious: 400 Professional Recipes.* New York: John Wiley & Sons, Inc., pp. 269–320. (Healthy recipes of salad and salad dressings are presented in this chapter.)

Knox, K., and J. S. Huffaker. 1997. *Coffee Basics.* New York: John Wiley & Sons, Inc. (Coffee, its history and preparation, is discussed in this book.)

Labensky, S. R., and A. M. Hause. 1999. *On Cooking: A Textbook of Culinary Fundamentals*, 2nd ed. Upper Saddle River, N.J.: Prentice Hall, pp. 828–860. (Hors d'oeuvres and canapés are considered in this chapter.)

Larousse, D. P. 1996. *The Professional Garde Manger.* New York: John Wiley & Sons, Inc. (The art of the buffet is discussed in this book.)

Matsuo, Y. 1992. *Ice Sculpture.* New York: John Wiley & Sons, Inc. (Guide to the art and skill of ice-carving.)

Monaghan, J., and J. S. Huffaker. 1995. *Espresso!* New York: John Wiley & Sons, Inc. (Running one's own coffee business is discussed in this book.)

Nam, I., and A. Schmidt. 1993. *Art of Garnishing.* New York: John Wiley & Sons, Inc. (Creating flowers from food is pictured and discussed in this text.)

Pauli, P. 1997. *Classical Cooking the Modern Way*, 3rd ed. New York: John Wiley & Sons, Inc., pp. 89–148. (Appetizer recipes are offered here.)

Schmidt, A., and I. Nam. 1996. *The Book of Hors d'Oeuvres and Canapes.* New York: John Wiley & Sons, Inc. (This book features appetizers from start to finish.)

CHAPTER 13

Stocks, Soups, and Sauces

I. Stocks
- A. Ingredients
- B. Preparation Methods
    - 1. Procedures
        - a. Brown Stock
        - b. White Stock
        - c. Chicken Stock
        - d. Fish Stock
    - 2. Stock Products
- C. Care After Preparation
- D. Prepared Bases

II. Soups
- A. Soup Preparation
    - 1. Consistency
    - 2. Seasoning and Garnish
    - 3. Special Procedures
- B. Soup Service
- C. Soup Care
- D. Portion Control

III. Sauces
- A. Planning
    - 1. Thickening Agents
        - a. Roux
        - b. Eggs
        - c. Beurre Manié
        - d. Starches
    - 2. Other Ingredients
- B. Preparation
    - 1. Brown (Spanish or Espagnole) Sauce
    - 2. Velouté Sauce
    - 3. Béchamel Sauce
    - 4. Tomato Sauce
    - 5. Hollandaise Sauce
    - 6. Miscellaneous Sauces
- C. Care After Preparation

■

LEARNING OBJECTIVES

By the end of this chapter, the reader will:

1. Know that quality stocks are the foundation for excellent soups and sauces.

2. Learn about stocks: their ingredients, preparation, care, and use.

3. Be able to distinguish between the various types of soups: their preparation, seasoning and garnish, care, and service.

4. Be able to identify the various types of sauces: their preparation, care, and service.

■

SELECTED WEB SITE REFERENCES

Campbell's Soup Company
www.campbellsoup.com/
Cookshack Barbecue
www.cookshack.com/

Nestle Foods
www.nestle.com/
Stouffers
www.stouffers.com/index.html

Stocks form the base of many soups, sauces, and prepared dishes. The difference among individual soups and sauces is often a matter of seasoning, ingredients, and consistency of a stock.

■

STOCKS

Stocks are thin, flavored liquids made from simmered meat, fish, or poultry flesh or bones, vegetables, seasonings, and a liquid such as water, milk, or tomato juice. Simmering extracts flavors, making a rich, flavorful stock. Proper ingredients, preparation methods, and care after preparation are necessary to make a good stock.

Ingredients

Quality stocks result only when clean, wholesome ingredients in the right quantities are used (see Fig. 13-1). The right equipment must also be available. A gallon of stock is made from 4 lb of bones, 5 qt of water, 1 lb of mirepoix (8 oz of leeks or onions, 4 oz of celery, and 4 oz of carrots), and seasonings. The best bones to use are knuckle, shank, or neck bones, in that order. Chicken carcasses and parts make an excellent stock. The bones, head, and skin of lean white fish make a light, delicate fish stock. Veal bones may be added to produce gelatin; beef for a rich flavor; and pork or ham for special flavor. Bones should be cut into 4-in. lengths and cracked. Meat may be simmered with the stock, and then removed for other uses.

A mirepoix is a mixture of roughly cut vegetables of high quality. Besides the primary ingredients (leeks or onions, celery, and carrots), mirepoix mixtures can contain parsley, turnips, cabbage, and other vegetables. Kitchens save vegetable parts such as the tops of fresh tomatoes, pieces of parsley, and water from canned vegetables; however, the stock pot is not a garbage can for disposing of undesirable wastes. Seasonings and spices such as bay leaf, thyme, clove, peppercorns, garlic, and parsley stems are added to stocks. The spices are usually wrapped in a small cloth or bag called a *sachet* or *bouquet garni*. Stocks should only be lightly salted because they are often concentrated or combined with other foods, and too much salt here may oversalt the finished product.

Equipment for making stock can be very simple. Bones are cut and cracked in the meat section, where the bandsaw, cleavers, and meat blocks are found.

FIGURE 13-1 Quality ingredients help ensure a fine stock.

A stock pot for top-of-the-range cooking or a steam-jacketed kettle can be used to cook the stock. The stock pot should have a spigot so that stock can be withdrawn from the bottom. A cutting board, a French knife for preparing vegetables, a ladle, a skimmer, and a china cap are necessary tools.

Preparation Methods

Most stocks are simmered for 4 to 8 hours, but fish stocks should not be simmered for more than forty-five minutes and chicken stocks for more than 3 hours, to prevent clouding.

Some operations maintain continuous stock kettles; the French call them *pots-au-feu*. Stock is taken as needed, and additional bones, water, vegetables and vegetable water are added. After four or five days, the stock becomes weak or cloudy and must be started anew. Recent evidence indicates that adding a bit of vinegar (not enough to flavor the stock) helps dissolve calcium from the bones, producing stock that is a richer source of calcium than milk.

Procedures

Certain common procedures must be followed to prepare good stocks. Place the bones in a stock pot, add cold water to cover them, and bring the liquid to a boil, skimming off scum as it rises. When the stock begins to boil, lower the heat to a simmer. Some chefs start stocks from hot water, saying that it produces a clearer, more flavorful stock. Cold water dissolves some proteins in bones, which then coagulate to form a cloudy scum. Some chefs discard the first water just as it comes to a boil, to get a clearer stock, but some flavor and nutrition are lost thereby.

Mirepoix and seasonings may be added to the bones at the start or later, depending on the type of stock to be made. When the stock is done, it is drawn or poured from the pot through a china cap lined with cheesecloth. It is then ready to use or store. Four major types of stock are made: brown, white, chicken, and fish.

Brown Stock.
A brown stock is usually made with beef and veal bones. The beef bones give the stock a rich flavor, and the veal bones provide gelatin that give it body. To prepare, oil the bones lightly, place them in a roasting pan, and brown well in an oven. Over-browning results in a bitter stock; underbrowning gives a weak flavor and a poorly colored stock. After browning, put the bones into a stock pot. Pour off the fat, and, over heat, deglaze the roasting pan by removing the fond (browned portion) with water. Add this to the pot, and cover the bones with water. Bring to a complete boil, skim, and simmer until the stock is full-flavored. Add the mirepoix and sachet about midway through the cooking. Some chefs add tomatoes to enrich the stock's color and add flavor; others say this should not be done to a base stock.

White Stock.
White stock is more delicately flavored than brown. Veal bones are best for it, but a combination of beef and veal bones is usually used. Do not brown the bones. Instead, cut them, wash them, and place them in cold water; then bring the liquid to a boil. Change the water if the stock appears cloudy at the start. Simmer for from 4 to 6 hours. Add the mirepoix and sachet about halfway through. Strain the finished stock, and use or store it.

Chicken Stock.
Chicken stock is sometimes also called *white stock*. Prepare it as you would white stock, but simmer it for only 2½ to 3 hours. Add the mirepoix and seasonings after the first hour of simmering. Some chefs omit carrots from the mirepoix for this stock, but the decision is a matter of personal preference. End by straining the chicken stock.

Fish Stock.
Bones, heads, skins, and trimmings from lean, white deep-sea fish are often used for fish stock. Rich, fatty fish such as salmon, mackerel, and trout give a strongly flavored and slightly dark stock. Do not wash the bones before cooking. Start with cold water; bring the liquid to a boil, skim, and then simmer. Add the mirepoix and spices immediately after skimming. An acid such as lemon juice may be added. Simmer for 45 minutes.

Adding vinegar, lemon juice, or some other acid to a stock helps give it clarity and adds flavor. It also helps to leach calcium out of the bones in the stock pot. Add only a small amount of acid; most of it boils away in the cooking, so little flavor is left.

Stock Products

Many items made from stocks are used to enrich foods. A glaze is stock reduced by one-fourth in volume. The mixture is syrupy and coats a spoon when warm; it produces a solid gel when cold. A stock simmered to one-half its original volume is called a *demiglaze*. Stocks reduced to a paste consistency are called *glacés*; thus we have *glacé de poulet* or *glacé de volaille* (chicken), *glacé de viande* (from brown or white stock), and *glacé de poisson* (fish). Essences are rich stocks simmered with wine, vegetables, and herbs. They are used to flavor and enrich items. After simmering, essences are strained and reduced to the desired consistency, usually that of a demiglaze. A fumet is a rich essence reduced almost to a glacé and thinned with sherry or Madeira wine.

Care After Preparation

Cool stocks rapidly to below 40°F (4°C), to prevent them from souring, as they are an excellent medium for microorganism growth. This is best managed by dividing the stock into 2-gallon lots, placing the containers in a sink, and circulating cold water around them. Then refrigerate or freeze each lot. Leave the congealed fat on top to prevent things from falling into the stock. When reheating a stock, bring it to a boil as quickly as possible. Continual recooling and reheating lowers a stock's quality.

Prepared Bases

Many operations today use prepared bases to make stocks because they lack the bones and meat required to make their own or because they find the prepared bases' laborsaving aspects desirable and their product quality adequate. These bases may also be used to enrich existing stock and other items. To ascertain the ingredients in a prepared base, read the list of ingredients; these are listed by quantity, from greatest to least. A base that reads "Salt, monosodium glutamate, fat, sugar, flavoring, spices, and coloring" contains no meat product and is probably not a high-quality product. Poor-quality bases make poor products. One base prized by many chefs has a label listing "Chicken meat including natural chicken juices, salt, monosodium glutamate, chicken fat, sugar, flavoring, and turmeric," indicating that chicken meat is greatest in quantity, and turmeric least.

■
SOUPS

One way of classifying soups is to separate them by their consistency, which reflects their ingredients and indicates their use. Thus, one simple classification has three categories: thin soups, thick soups, and national or specialty soups.

Thin soups are generally clear—such as broths, bouillons, and consommés—or begin with clear stocks that are made opaque by only a few ingredients. A broth is a rich, flavorful stock. A bouillon is a soup made from stock, extra meat, and seasonings; it may or may not have been clarified. A bouillon is often a rich broth that carries the flavor of beef. A consommé is a rich, delicate, clear soup made by clarifying a rich stock or broth after enriching it by simmering it with meat and other flavoring ingredients. Thin opaque soups may include cream soups, bisques, purées, light vegetable soups, and some cold soups such as vichyssoise.

The borderline between thin and thick soups is difficult to draw. A purée can be thin or quite heavy and thick. Some vegetable soups are very heavy because they contain a lot of ingredients, while others are almost clear. Many heavy soups are thickened with rice, potatoes, pastas, starch, or eggs. A chowder or gumbo is thick because of its many ingredients.

A thick purée is made from vegetables such as potatoes, tomatoes, and asparagus; from legumes such as lentils, split peas, and lima beans; or from fruit such as prunes and apples that are pulped, strained, and added to the desired base—either plain or thickened stock. Starch may be added to thicken the mixture and hold the ingredients evenly in the soup. Neither milk nor cream is added to a purée, since that would make the product a cream soup.

The word *chowder* comes from the French word *chaudron*, which was the name of a large cooking kettle used by French Canadian fishermen of the Maritime Provinces. Chowders originally contained seafood or fish as their main ingredient, but they may now be made primarily from corn, lima beans, or mushrooms. Diced potato, chopped onion, and bacon fat or salt pork are added after the vegetables have been sautéed in fat. Manhattan (Philadelphia or Long Island) chowder contains fish stock and tomatoes; Boston or New England chowder contains milk or cream.

Bisques can be thin or thick. The base is usually some type of shellfish, such as lobster or abalone. A gumbo is quite heavy and often contains seafood, chicken, vegetables, and rice; it must contain okra—a vegetable of distinct flavor and some thickening power. Gumbo can be thickened with cornstarch. Potato soups are heavy soups.

National or specialty soups may be thick or thin. They fall into this category because of their ingredients, their method of making, or their origin. French onion soup, mulligatawny from India, Scotch mutton broth with its typical barley, olla podrida of Spain, borsch from Russia, and minestrone from Italy fall into this category. Many specialty soups are served cold, including a delicate avocado soup seasoned with lemon and sherry, a jellied madrilène, a gazpacho, and many chilled fruit soups.

Soup Preparation

High-quality ingredients, the right production techniques, and proper service are all important in soup preparation. Attention must also be given to the soup's consistency, seasoning, and garnish, and to special production needs.

Consistency

Many thin soups are watery because the stock lacks suitable body. Good body results from using a fairly large amount of gelatin that occurs naturally in the bones of young animals. Additional body can come from thickening agents, meat, fish, poultry, and vegetables. When a roux (equal parts fat and flour, by weight, cooked together) is added to a stock, a velouté results. The amount of roux used determines whether the velouté is thin or thick. Adding the velouté to a soup such as a bisque or a cream soup determines its thickness.

When cream soups are made using a velouté, 1 pint of the vegetable purée (for flavoring) is added for each gallon to be prepared. A small portion of this hot soup is then gradually added to warmed milk or cream, and the milk or cream mixture is stirred vigorously back into the hot soup. Such precautions must be taken to prevent curdling. This method is often called the *velouté method* of making a cream soup. High temperatures and excessive salting, as well as the presence of any acid, may cause curdling. A cream soup that is too thick for service can be thinned with warmed milk or cream. The consistency desired in cream soups and most thick soups is similar to that of moderately heavy cream.

Seasoning and Garnish

Seasoning a soup successfully involves achieving a blend of flavors, with no one flavor predominating. Adding spices and other seasonings toward the end of preparation ensures maximum flavor. Delicacy of seasoning should be sought. Just a tiny pinch of ground cloves added to a cream of tomato soup can make a great difference in its flavor. A touch of sage has a similarly beneficial effect on clam chowder. Monosodium glutamate may help to smooth out and elicit delicate flavors. Subtlety in flavoring can add much to a soup's quality.

Many thin soups are enhanced by an appropriately chosen garnish. A cream of asparagus soup with a tiny asparagus spear floating in it, lobster bisque graced with chopped lobster, cream of corn soup with popcorn croutons, and other garnishes of this kind make soups stand out (see Fig. 13-2). Croutons made

FIGURE 13-2 This uniquely presented cream of squash soup is garnished with shelled sunflower seeds.

from toast or pastry, small-grained cereals such as rice or barley, grated cheese, or a daub of sour cream add a distinctive touch when used sensitively. Garnishes used for consommés and bouillons sometimes give the soup its name. Thus, a consommé Florentine has a few tiny cooked leaves of spinach in it, and a bouillon vermicelli has a few cooked vermicelli in it and a tiny slice of truffle.

Special Procedures

The following procedures for making a good soup are standard:

1. Skim the surface of the stock to remove fat and scum. This gives a clearer stock and a better soup.

2. Strain stocks and soups. All stocks should be strained after completion to remove impurities and give a clearer soup. Particularly thin soups should be strained through a china cap covered with cheesecloth. Bouillons, consommés, and other clear soups should be so clear that they sparkle.

3. Add spices with care. Subtle seasoning is the key. A stock and a soup should be a blend of flavors. Some chefs remove the little round part of a clove because it can give a slight bitterness. Others crack whole peppercorns to extract more flavor. Spices should not be added except in a sachet bag, and they should be kept in only long enough to contribute the flavor they should.

4. Sauté some vegetable garnishes to improve their flavor.

Cream soups often have problems with curdling. Asparagus, potatoes, tomatoes, and other vegetables or ingredients may contain tannins, acids, or other substances that cause curdling. Small curds separate out as a result of curdling, damaging the soup's appearance. One precaution that can be taken is to blend a small quantity of the curdling product *into* some milk or cream to temper it. This mixture can then be blended back into the soup. If all of the soup is to be added at once, take two precautions: add the curdling product *into* the milk or cream slowly; simultaneously, give the mixture rapid agitation with a wire whip. Do not add cold milk or cream; warm them first. If the velouté method is used, the curdling product is bound in with

the starch, which lessens chances for curdling. Reducing the heat and cutting back on salt help reduce curdling as well. An old stock of cream soup should never be mixed with a new one. Preparing only the quantity of soup needed for a short serving time is a good practice.

Some cream soups (and others) may get extra flavor, smoothness, and body from a liaison. This is a mixture of one part egg yolk and three parts rich cream. Some of the hot soup is blended into the liaison, and then this blended mixture is stirred carefully into the hot soup mixture. The heat is raised as the soup is stirred well, and cooking proceeds only until thickening occurs. The soup should be served shortly after it thickens.

Some soups evaporate when they stand in the steam table. As this occurs, the soup may become too salty or the seasonings too heavy. To counteract this, watch the soup closely, and add liquid to replace any lost to evaporation.

Soup Service

Soup temperatures at service should be 160°F (71°C) or more. Many soups are held at 190°F (88°C). Chilled soups should be served at between 40° and 45°F (4.5° and 7°C). Containers into which hot soup is to be served should be preheated. Menus should offer both thin and heavy soups, plus others. The soups served should contrast in appearance, presentation, and accompaniments (such as bread or crackers). A kettle of bouillabaise can be served with garlic toast, and a hearty bowl of clam chowder with the traditional oyster crackers. Fat globules should not be visible, as a good sheen is important (see Fig. 13-3).

Soup Care

Carryover soups have many uses. They can be added to the stock pot, if they are of the right kind, or they can be used as a base for other dishes. Often they can be combined with other soups to make new soups. For example, cream of tomato soup blended with split-pea soup makes a Mongole soup. most soups sour easily and must be handled as stocks are. Cool

FIGURE 13-3 Fat globules such as those in this cup of soup are a disgrace to any foodservice operation.

them quickly in 2-gallon (or smaller) containers in a sink, and refrigerate them as soon as possible. Thick soups cool more quickly when they are stirred.

A runout time for soups can be planned, to reduce carryovers. For backup, you can use a convenience soup, many of which are of good quality.

Portion Control

The right quantity of soup must be served to meet management's cost estimates. A standard portion for a first-course soup is about 6 oz (¾ cup or 160 cc). A big bowl holds about 10 oz (1¼ cups or 300 cc) and often suffices for a light meal. Because soups are not high-cost items, a tureen or crock of soup may be put on the table, and guests invited to help themselves. Alternatively, a soup bar may be offered, along the lines of a salad bar.

■

SAUCES

Sauces are used to enhance the flavor, appearance, nutritional value, and moistness of food. They were originally created to increase foods' flavor and palatability. Sauces should not be designed to hide poor food quality but to complement the flavor of items served. A sauce's flavor should never overwhelm the flavor of

the food. Sauces should be of correct consistency so that they flow readily but are not soupy. All sauces must be based on a rich stock or liquid to enhance the end product. Good planning, careful preparation, and proper handling result in good-quality sauces.

Planning

Planning for sauce production includes obtaining good-quality ingredients and using proper production agents.

Thickening Agents

Sauces can be thickened with roux, eggs, butter and flour (beurre manié), and starches. A sauce's consistency is important to its function (see Table 13-1).

Roux. Roux is often used to thicken sauces. A slack roux has more fat than flour, by weight; a lean roux has more flour than fat. Cooking roux is done to give sauces color and to remove any starchy flavor that might be present. A white roux is cooked just long enough to

TABLE 13-1 Items to Add to Different Sauces to Achieve a Desired Consistency		
Sauce Prepared	*Purée Added*	*Liquid to Finish With*
Velouté sauce (stock thickened with roux)	Vegetable purée	Milk or cream
Béchamel sauce (stock thickened with roux and thinned with milk or cream)	Thickened vegetable purée	Milk or cream (if the sauce is too thick)
White or cream sauce (milk or cream thickened with roux)	Vegetable purée	Milk or cream (if the sauce is too thick)

blend the fat and flavor properly. A blond roux is cooked slowly until a fawn color is obtained; this type of roux is used to make velouté sauce. A brown roux is cooked to a rich brown color.

Stir all roux while cooking them. Brown roux is often finished slowly in an oven; slow browning gives a finer product. To speed browning, the flour may be prebrowned in an oven. Browning destroys some thickening power: a heavily browned roux has one-third of white roux's thickening ability.

Add cool roux to hot liquids, and add warm or hot roux to cool liquids. Continue stirring until the mixture boils, and then cook for 10 minutes.

Roux can be refrigerated and held. A roux makes a smooth sauce, if properly handled, by separating starch granules with fat. The starch swells or gelatinizes in the presence of hot liquid, and these granules then crowd each other and thicken the entire mixture. Acids such as vinegar and tomato juice reduce the thickening power of starch granules by reducing their swelling.

Eggs. Eggs—both whites and yolks—can be used to thicken. Their proteins hold moisture loosely, giving the sauce a creamy consistency. Yolks have more thickening power than whites and produce a more stable thickened mixture. When eggs coagulate with liquid, they bind it in and thicken it. If the mixture is stirred and not heated too fast, a smooth thickening occurs. Thickening occurs in most mixtures at about 185°F (85°C).

The use of liaisons has already been mentioned. In this process, eggs are blended with milk, cream, or stocks before being added to a sauce. To add the eggs, first remove the sauce from the heat; next add a ladleful of hot sauce into well-blended eggs, stirring rapidly with a wire whip; then stir this back into the sauce, and continue stirring until thickening occurs. Use low heat, and cook only until thickening occurs. Curdling can occur if too much heat is used. A slightly curdled mixture can be saved by being strained through a cheesecloth.

Eggs are often used to thicken blanquettes, white ragoûts, fricassées and some special soups. A hollandaise sauce is thickened with eggs, but the result is more an emulsion than a cooked product.

Beurre Manié. Beurre manié consists of butter and flour that have been kneaded together to form a soft paste. It is usually composed of equal parts butter and flour. Small portions are added to hot sauces to thicken them. Vigorous stirring is necessary. The butter melts as heat is applied, and the flour thickens the mixture. No lumping occurs. Kneaded butter and flour are sometimes used in place of roux, but they may leave a slight taste of uncooked flour if overused.

Starches. Many starches are used as sauce thickeners. Each has a different thickening power, clarity, and viscosity (resistance to flow). Waxy maize or converted starch is the most powerful thickener. Corn, potato, rice, and arrowroot starches come next and thicken about equally. Tapioca and wheat flours are weakest. Cake and pastry flours thicken better than bread flour because they contain more starch. Waxy maize or converted starch has the most clarity, followed by arrowroot, tapioca, and cornstarch; rice starch, potato starch, and wheat flours are the least clear. Cornstarch, wheat flours, and rice starch are the most viscous; each makes a firm gel. Potato starch, arrowroot, and tapioca have low viscosity and form somewhat fluid gels. Waxy maize or converted starch forms the least viscous gel of all. A firm gel is desirable for puddings and cream pies, because it allows a piece of cream pie to hold its shape when cut. Waxy maize or converted starch is used for fruit pie fillings since it has good clarity and is not too viscous when cool, giving a soft, pasty texture that is desirable in such a pie. Arrowroot, tapioca, and waxy maize or converted starch can be frozen without breaking (curdling) upon thawing.

Starch in very hot water swells and absorbs moisture; maximum thickness is attained at 194°F (90°C). This process is called *gelatinization*. The swelled starch granules begin to crowd each other, and soon the mixture becomes thick. Overlong exposure to high temperatures or excessive stirring can cause a starch to lose some of its thickening power. Acids break down starches. Even the slight amount of acid contained in meat juices necessitates using a bit more flour to get thickening equal to what would be achieved in milk or water. For this reason, acidic ingredients are sometimes withheld until cooking is complete and the product is about to be cooled, since the action of the acid is less at lower temperatures.

A starch blended with a liquid and then added for thickening is called a *slurry*. To add a slurry, take a bit of the hot mixture to be thickened, add it to the slurry, and blend well; then add this mixture to the hot mixture, giving good agitation to work the starch in quickly. Use a heavy pan that gives good heat distribution.

Other Ingredients

Other ingredients used in sauces are wines, spices, seasonings, vinegar, butter, fats and oils, mirepoix, and sachet bags. To *finish* a sauce means to add the final, completing touch. An ingredient such as heavy cream, butter, or liaison is added to give a smoother, mellower, and subtler flavor. Wines may be used. Many sauces take a garnish such as chopped mushrooms, truffles, the grated rind (zest) of a lemon or orange, raisins, or pineapple.

Preparation

The sauce cook is called a *saucier*. Sauce making is an art that must be carefully learned and practiced. Most sauces are derived from one of a few basic sauces that are called *major, grand,* or *mother* sauces. The sauces derived from them are called *small* or *secondary* sauces.

The major sauces are brown (Spanish or Espagnole), Béchamel, velouté, tomato, and hollandaise. The first three have a meat base; tomato sauce may or may not have a meat base; and hollandaise sauce does not have a meat base. Sauces that lack a meat base are called *neutral*.

Brown (Spanish or Espagnole) Sauce

The most commonly used sauce is brown sauce. A common preparation is to sauté 2 lb of mirepoix in ½ lb of fat until the onions are transparent. Then ½ lb of flour is added, and the mixture is cooked for 10 minutes, until it acquires a tan color. Then 1 gal of cold brown stock is stirred into the hot roux. The cook then adds ½ cup of tomato purée, a bay leaf, and salt and pepper to taste, and simmers the mixture for 1½ hours. When strained, the mixture yields 1 gal of sauce. Some cooks add a ham bone or lean bacon during simmering to give a distinctive flavor, although others say this is not a true brown sauce. Brown sauce is the base of many small sauces.

Velouté Sauce

A velouté is neither a white nor a cream sauce. To make it, add 1 gal of white stock to 20 oz of blond roux, and simmer until proper thickness occurs. Fish sauce (*vin blanc*) is a velouté with a white fish stock and dry white wine as an addition; this might also be used for a Newburg sauce. Small sauces based on velouté include *allemande, bonne femme* caper, homard, horseradish, and *suprême*.

Béchamel Sauce

Count Béchamel was a famous French gourmet whose chef invented the Béchamel sauce. Some chefs say a Béchamel sauce is a velouté sauce because it is made from a meat base and is a white sauce. Foodservice professionals feel that a white or cream sauce should be called a Béchamel. A white sauce has a milk base thickened with a white roux; a cream sauce is made of rich milk or is a white sauce to which cream is added. In this text Béchamel is considered a basic sauce and white and cream sauces are considered variations of Béchamel. The great chef Escoffier's recipe for Béchamel calls for veal poached in milk as the basic liquid, with white roux added. Today, we use a rich veal or chicken stock, thicken it with roux, and add rich milk or cream. This way of making the thickened stock actually produces a velouté.

To make 1 gal of Béchamel, reduce 3 qt of white stock to 2 qt. Add ½ cup of chopped onions, ½ cup of carrots, two bay leaves, and 1 Tbsp of crushed peppercorns. Blend 1 lb of white roux with the stock. Strain, and then add 2 qt of coffee cream. Béchamel is an excellent sauce for chicken, cheese, white fish, or other delicate preparations that need a light, white sauce. Small sauces based on Béchamel include Albert, à la king, cardinal, cheese, egg, mornay, mustard, and Newburg.

Tomato Sauce

Stock, a tomato source (tomatoes, tomato purée, or tomato paste), seasonings, and roux make tomato sauce. It is served with many fried items, such as breaded veal cutlets, pork chops, and fish. It is the base of many other sauces, as well. To make 1 gal of tomato sauce, sauté 1 Tbsp of minced garlic in 12 oz of butter.

Add 1 lb of finely chopped onions and 8 oz of finely chopped celery, and sauté until soft. Then add 6 oz of flour and cook for 5 minutes. Add 3 qt of hot stock (brown, white, or ham), and stir until thickened and smooth. Next add 1 qt of tomatoes, 2 qt of tomato purée, 1 tsp of ground thyme, two bay leaves, 1 tsp of crushed peppercorns, and three whole cloves. Reduce the mixture's volume by a third, and strain through a china cap. Check for seasoning, and serve.

Hollandaise Sauce

Hollandaise is often called the queen of sauces. It is an emulsion of egg yolks and melted butter. One yolk can absorb about 3 oz of butter. Melted butter is gradually beaten into the yolks, which have been placed in a pan over hot water. Adding butter too fast does not result in a good emulsion. If overcooked, the sauce curdles.

Hollandaise is not difficult to make if the proper techniques are used. To make 2 qt, melt 2 lb of butter, and clarify it (melt the butter, and remove the scum on top and the milk solids on the bottom, leaving clear butter). In a separate stainless steel bowl, whip sixteen egg yolks and 3 oz of water. Place the bowl above simmering water, and whip the mixture steadily over slow heat until it is thick and creamy. Remove the bowl, and add the clarified butter very slowly to the yolks while whipping. When all the butter has been added, season with the juice of two small lemons, with cayenne pepper, and with salt to taste. Serve with a vegetable such as broccoli or asparagus, with eggs, or with fish, or use as a base for other sauces. It is the sauce used for eggs Benedict.

Miscellaneous Sauces

Butter sauces can be made from butter or margarine. The basic butter sauce is butter blended with lemon juice and cayenne pepper. From this, compounded butters are made by adding minced or pulped ingredients. Compounded butter sauces include anchovy, caper, garlic, and mustard, maître d'hôtel butter, beurre meunière, noisette beurre, and beurre noir. The last four butters mentioned are melted with a bit of lemon juice and chopped parsley. The first is served just that way; the second is lightly browned (with almonds, it makes amandine sauce); the third is browned to a nut-brown color; and the last is almost blackened, but not burned. Burned butter tastes bitter.

Basic and compounded butters are made into fancy shapes with a pastry tube, and then are chilled before being served. Butters are used for canapés and can also be served on hot foods. For example, a rosette of anchovy butter might be served on top of a broiled steak; it finally melts but makes a nice presentation. Melted butters are served over fish, steaks, chops, vegetables, and other foods. *Beurre noir* is often served over sautéed calves' brains.

Cold sauces include cocktail sauce, tartar sauce, and dill sauce. A sauce used to decorate cold meats, called *chaudfroid*, is made by hydrating 12 oz of unflavored, granulated gelatin in cold water and dissolving it into 2 cups of boiling water. This is then added to 1 gal of warm velouté sauce or mayonnaise. Next, 1 qt of coffee cream is added, and the whole mixture is seasoned with salt and white pepper. This is cooled until syrupy, and then ladled over a meat item to give it a white cover. The item is chilled to allow the sauce to set, and the pouring of the sauce over the meat is repeated until a thick white coating covers the meat. The item can then be decorated and glazed.

Simple sauces are gravies or sauces made from meat drippings. *Au jus* is an unthickened browned meat juice made by deglazing a pan in which meat has cooked. After extraction, this fond is strained through a cheesecloth. Rich brown stock may be added, if needed. *Jus lié* is very slightly thickened *au jus* and is used in much the same way.

Gravy differs from other sauces in that its flavor is dominated by the flavor of the meat from which it came. Gravy is usually made from meat drippings to which some good stock is added. The mixture is usually thickened with a roux. Different gravies include country, giblet, roast beef, and roast lamb.

Sweet sauces include sweet butters, sweet purées, and creams. The consistency may range from fluid to quite stiff. Starch and eggs frequently are used as thickening agents. Some sweet sauces may consist of fruits in their own juice. Sweet sauces served with meat include raisin sauce (with ham) and pineapple sauce (with Hawaiian chicken). Many flavored syrups are served over desserts. Sweet sauces may be hot or cold.

Care After Preparation

Some sauces take great skill to produce. They also require proper handling after preparation. A good sauce cook can resort to many tricks to save a sauce if something goes wrong. If a sauce is too thick, a liquid such as stock, wine, milk, or water can be blended in, with a wire whip. If the sauce is too thin, a bit of beurre manié, roux, or slurry may be added. A lumpy sauce can be strained or even vigorously whipped to remove lumps. A curdled sauce from a broken emulsion can be repaired by adding a small amount of liquid to a bowl and then vigorously whipping the broken emulsion into the liquid. If the emulsion is mayonnaise or hollandaise sauce, egg or egg yolk is used in place of liquid. Since too much heat can cause a rich gravy or sauce to break, these are only given moderate heat.

To avoid having a crust or skin form on sauces, cover a hot sauce with a tight lid, or put at thin film of melted fat or butter over the top. Buttered paper, wax paper, or plastic film can also be put over the surface and kept on the sauce until it is to be used. Whenever a crust does form, do not stir it into the sauce; instead, remove it, or strain the sauce to remove it.

Sauces should be quickly cooled to room temperature and then properly stored. The same procedures and precautions should be taken for sauces as for stocks and soups. Some sauces do not store as well as soups and stocks. These should be used at once. Sauces can be frozen only if thickened with specific kinds of starches.

A good sauce should have a high sheen from the fat and should be rich in color—whether creamy white or dark brown. Many sauces should have a varnishlike brilliance. A dull, heavy, pasty product gives evidence of poor quality. Consistency should be very thin to thick, depending on the particular sauce. Thin sauces flow readily; thicker sauces cling to the product. In a sauce, the many flavors should blend subtly and agreeably, with no particular flavor predominating. A standard portion of sauce or gravy is 2 to 3 oz.

SUMMARY

Stocks are thin, flavored liquids made by simmering meat, bones, vegetables, spices, and other substances. Some stocks can be just a plain liquid. The equipment required is simple and includes a stock pot, cutting board, French knife, ladle, skimmer, and china cap. Scrap ingredients can be used to make a stock, if they are of good quality. Seasonings include bay leaf, thyme, cloves, peppercorns, garlic, parsley, and bouquet garni. Stocks are lightly salted and may be started in either cold or hot water.

Major stocks are classified according to their ingredients and color. Brown stock is made from well-browned beef and veal bones. White stock ideally is made from veal or chicken bones, but often is based on unbrowned beef and veal bones. Chicken stock is made from chicken carcasses or chickens that are cooked and then used for other purposes. Fish stock is made from the bones, heads, skins, and trimmings of white, lean, deep-sea fish.

Stocks can be used to make a glaze, a demiglaze, a glacé, an essence, or a fumet. Stock care is important. Stocks should be cooled rapidly, kept refrigerated, and used soon after making. Convenience stocks are widely used today. Buyers should read the ingredient list to see what they contain, before buying.

Soup categories include thin, thick, and national or special. Thin soups are served with hearty meals. The meat flavor in a consommé, bouillon, or similar soup aids in stimulating the appetite. Thick soups are used with lighter meals or constitute meals in themselves. There are many national or special soups. Soups to be carried over should be cooled immediately and refrigerated.

Sauces are used with other foods to enhance their flavor, appearance, nutritional value, and moistness. Thickening agents include roux, eggs, beurre manié, and starches. They give body and consistency. Other ingredients are stock, seasonings, wine, and sometimes a garnish. Grand sauces are the foundation of many small sauces; they include brown or Espagnole sauce, velouté, Béchamel, tomato sauce, and hollandaise sauce. Miscellaneous sauces and their derivatives include butter sauces, cold sauces, simple sauces, and sweet sauces. Achieving proper consistency, preventing curdling, and obtaining proper flavor blends take skill on the part of the saucier. Standards in sauce preparation dictate the appearance, consistency, flavor, and overall eating quality of the finished product.

■

CHAPTER REVIEW QUESTIONS

1. What are the ingredients of a good stock? What are the reasons for starting a stock in cold water versus hot water? Why would only a small amount of salt be used? Describe the preparation of a standard mirepoix and of a sachet bag.

2. How are brown, white, fish, and chicken stocks made? Name at least one soup and one sauce derived from each stock. Describe what a high-quality stock should be like.

3. Name the different classes of soups, and identify the consistency of each. What seasonings and garnishes are used for soups?

4. List the common thickening agents, and explain in detail how each is used. Give examples of products in which each agent might be used.

5. Identify the grand sauces, and describe how each is prepared. What ingredients are used for each? Give some major points in the care of sauces. Are these points also applicable to stocks and soups? List at least two small sauces derived from each grand sauce.

■

CASE STUDIES

1. You have been hired as the manager of a new club that is still in the planning stage. A meeting is to be held with the house-building committee on planning the kitchen. No menu has been set. The club is not intended to appeal to a deluxe membership but to individuals of above-average income; therefore, the food must be very good but not elaborate. Some members of the committee feel that in keeping with the *nouvelle cuisine*, the menu should not feature sauces and other rich foods. Other members do not agree. Opinion is also divided as to whether cooks are needed who have the ability to produce a wide variety of sauces or whether some of the modern food bases could be used instead for what is needed in stocks, soups, and sauces.

   You do not want to take sides and want the members to make the final decision, but you feel personally that making stocks, soups, and sauces from scratch is the better policy. You jot down some pros and cons that you will present to the members when they meet on this matter. Set up this list, covering such things as product quality, variety, equipment needs, personnel needs, and

cost differences. What about *nouvelle cuisine?* Are stocks, soups, and sauces eliminated, or are they merely changed in nature?

2. You are the manager of a restaurant catering to a fairly good clientele. You are not satisfied with the soups and sauces being produced and are planning to talk with the chef and steward about the matter. You want to go into the discussion well-prepared to deal with these two professionals, so you set up a list of questions you want to ask, inquiring into why these products are not meeting a sufficiently high standard. What are these questions, and how should you handle this discussion?

3. The preparation of stocks, soups, and sauces is a time-consuming process. Accordingly, you are asked to review convenience items in the processing of recommending soups ready made. The Internet site *www.stockpot.com* has been suggested as a source of information. List the soups they offer. Do they also provide sauces to the foodservice industry? Would you buy products from this company? Why?

■ VOCABULARY

Stock

Bouquet garni

Sachet

Mirepoix

Brown stock

White stock

Chicken stock

Fish stock

Glaze

Demiglaze

Glacé de viande

Glacé de poulet

Glacé de volaille

Glacé de poisson

Fumet

Essence

Convenience stock

Bouillon

Consommé

Bisque

Chowder

Gumbo

Purée

Boston chowder

Coney Island chowder

Olla podrida

Borsch

Mulligatawny

Velouté

Liaison

Roux

Lean roux

Brown roux

Blond roux

Slack roux

Beurre manié

Arrowroot

Waxy maize starch

Slurry

Saucier

Brown sauce

Espagnole or Spanish sauce

Béchamel sauce

Escoffier

Hollandaise

Butter sauce

Maître d'hôtel butter

Beurre noisette

Beurre meunière

Beurre noir

Chaudfroid

Au jus

Jus lié

■ ANNOTATED REFERENCES

Bridge, T. 1995. *200 Classic Sauces: Guaranteed Recipes for Every Occasion.* New York: John Wiley & Sons, Inc. (From basic to dessert sauces, this book presents 200 recipes.)

Donovan, M. (ed.). 1996. *The New Professional Chef,* 6th ed. New York: John Wiley & Sons, Inc., pp. 259–300 and 419–554. (Stocks, soups, and sauces are considered in these chapters.)

Donovan, M. 1997. *Cooking Essentials for the New Professional Chef.* New York: John Wiley & Sons, Inc., pp. 295–342. (Stocks, soups, and sauces are considered in these chapters.)

Gisslen, W. 1995. *Professional Cooking*, 3rd ed. New York: John Wiley & Sons, Inc., pp. 121–194. (Stocks, soups, and sauces are considered in these chapters.)

Kapoor, S. 1996. *Healthy and Delicious: 400 Professional Recipes.* New York: John Wiley & Sons, Inc., pp. 1–62. (Healthy recipes for stocks, soups, and sauces are presented in this chapter.)

Labensky, S. R., and A. M. Hause. 1999. *On Cooking: A Textbook of Culinary Fundamentals*, 2nd ed. Upper Saddle River, N.J.: Prentice Hall, pp. 178–261. (Stocks, soups, and sauces are considered in these chapters.)

Larousse, D. P. 1993. *The Sauce Bible.* New York: John Wiley & Sons, Inc. (Recipes and more about sauces are presented in this book.)

Larousse, D. P. 1997. *The Soup Bible.* New York: John Wiley & Sons, Inc. (Recipes and more about soups are presented in this book.)

Pauli, P. 1997. *Classical Cooking the Modern Way*, 5th ed. New York: John Wiley & Sons, Inc., pp. 1–88. (A discussion on stocks, soups, and sauces is offered here.)

Peterson, J. 1998. *Classical and Contemporary Sauce Making*, 2nd ed. New York: John Wiley & Sons, Inc. (From equipment needed to stocks used, this book presents what is needed to know about sauces.)

CHAPTER 14

Fruits, Vegetables, and Cereals

■

LEARNING OBJECTIVES

By the end of this chapter, the reader will:

1. Understand the importance of proper purchasing, receiving, storing, and issuing of fruits and vegetables.

2. Learn the various cooking methods of fruits and vegetables as well as the preparation techniques for quality presentation and service.

3. Recognize the various market forms of fruits and vegetables, including canned, dried, fresh, and frozen and be able to distinguish the quality among them.

4. Know how to prepare and serve cereal, pasta, and rice.

■

SELECTED WEB SITE REFERENCES

American Fruit Processors
www.americanfruit.com/

American Mushroom Institute
www.americanmushroominst.org/

California Artichoke Advisory Board
www.artichokes.net/

California Avocado Commission
www.avoinfo.com/

California Olive Industry
www.calolive.org/

California Prune Board
www.prunes.org/

California Raisin Advisory Board
www.net-asset.com/raisins/

California Strawberry Commission
www.calstrawberry.com/

California Walnut Marketing Board
www.webcom.com/walnut/

Cherry Marketing Institute Incorporated
www.cherrymkt.org/index.html

Dean Foods Vegetable Company
retail.deanfoods.com/

GeniSoy Products Co.
www.genisoy.com/

Hazelnut Growers of Oregon
www.hazelnut.com/growers/welcome.htm

Idaho Potatoes
www.idahopotato.com/

Kellogg's
www.kelloggs.com

Michigan Apple Committee
michiganapples.com/

Produce Marketing Association
www.pma.com/

Quaker Oats Company
www.quakeroats.com/

Washington Red Raspberry Commission
www.red-raspberry.com/

■

FRUITS AND VEGETABLES

Fruits and vegetables are good sources of vitamins and minerals, while often contributing a fairly low number of calories. They are also important sources of bulk in the diet and enhance many meals with their color, flavor, and texture.

Planning Production

Public interest in healthful foods has made fruits and vegetables more popular on the menu. At one time, many menus stinted on these food items, but the popularity of salad bars and the addition of vegetables and fruits with main entrées have brought them back. Proper planning for service demands a knowledge of the seasonal availability and cost of various fruits and vegetables, plus a knowledge of how best to merchandise and prepare them. Properly prepared fruits and vegetables can be as good as any other food in a delectable meal.

The menu should offer a variety of fruits and vegetables. Fresh fruits can be featured as cocktails, fruit juices, breakfast items, and (with or without cheese) as desserts. Similar presentation opportunities at lunch and dinner exist for raw and cooked vegetables.

Fruits are the seed-bearing part of a plant. Thus, cucumbers, squash, green beans, snow peas, and tomatoes are fruits although we class them as vegetables. In food preparation, fruits are usually fleshy or pulpy, often juicy, and usually sweet, with aromatic flavors. Cereal grains and nuts are dry seeds and thus are prepared differently for various dishes. Vegetables come from various parts of a plant—the bulb, part of the root, the flower, the fruit, the leaves, the stems, or even part of the stalk (see Table 14-1).

Improved handling, packaging, and processing have extended the season in which fruits and vegetables are available. Menus should be planned with a view toward seasonality, since prices tend to be high during the off-season (see Table 14-2). Canned, frozen, and dried fruits and vegetables can add variety and compensate for a lack of fresh items during some periods of the year. Some fruits and vegetables store very well and so are available for much longer periods than others are.

Purchasing

Many fruits and vegetables are highly perishable, and their quality and cost can vary. Thus, purchasing must be done with care. The market can fluctuate quickly. Prices must be balanced against freshness, tenderness, shape, appearance, trim loss, total weight, and cooking loss. Best buys do not always involve the lowest-cost item. Specifications should be made in such a way as to ensure that the right vegetable or fruit is purchased for the production need.

Many fresh fruits and vegetables are available now in a ready-to-use form. They reduce storage space,

labor, equipment, and preparation area space. Their use often increases food cost while reducing labor and other costs.

Grades

Most fruits and vegetables are sold by grades that indicate their quality. When grades are not available, brands may give a quality clue. Grades are set by the U.S. Department of Agriculture (USDA) on the basis of color, shape, uniformity of size, ripeness, and freedom from defects. U.S. Grade No. 1 is a high-quality grade suitable for use in foodservices. It represents most of the crop. U.S. No. 2 is marketable but is generally lower in quality. A few U.S. No. 3 grades exist; and combination and field-run grades are sometimes found. U.S. Fancy and Extra Fancy are grades above U.S. No. 1 and represent better-than-average quality. Some fruits and vegetables may be available for only part of the year; off-season, the above-average grade must be purchased at times. Sometimes buyers have to purchase in consumer grades, which are U.S. A, B, and C. The grades for processed vegetables and fruits (canned, frozen, and dried) are as follows: U.S. Grade A or Fancy (first grade), U.S. Grade B or Extra Standard or Choice (second grade), and U.S. Grade C or Standard.

Other Purchase Factors

The product's size, trim, variety, ripeness, packaging, and place of origin, among other factors, should be considered. Production needs dictate which factors are especially important.

The amount to purchase depends on storage available, product perishability, and amount needed. Potatoes, cabbages, yams, and some other items keep well, whereas others keep poorly. Processed items can be purchased in quantities suitable to a foodservice's needs and available storage. Dried fruits and vegetables can spoil quickly and become infested with weevils and other insects.

Purchase quantities should be calculated on the basis of the number of times within a purchase period that the item will be served multiplied by the number of portions per service day multiplied by the portion size. Thus, if frozen peas are offered five times in a

TABLE 14-1 Classification of Vegetables by Part of Plant	
Category	*Vegetable*
Bulb	Garlic, leeks, onions, shallots
Flower	Broccoli, cauliflower, artichoke (French or global)
Fruit	Cucumbers, eggplant, okra, peppers, squash, tomatoes
Leaf	Brussels sprouts, cabbage, greens
Root	Beets, carrots, radishes, potatoes (sweet), turnips
Seed	Beans, corn, peas
Stem	Asparagus, celery
Tuber	Jerusalem artichokes, potatoes (white)

TABLE 14-2   Availability of Fruits and Vegetables, by Month

Monthly Availability of Fruits, Expressed as a Percentage of Total Annual Supply

COMMODITY	% Jan	% Feb	% Mar	% Apr	% May	% June	% July	% Aug	% Sept	% Oct	% Nov	% Dec
Apples	11	10	10	8	6	3	2	3	10	15	11	11
Apricots					5	62	31	2				
Avocados	11	11	11	11	9	7	6	6	5	6	8	9
Bananas	7	8	9	9	10	10	8	8	7	8	8	8
Blackberries					13	56	19	12				
Blueberries					2	32	39	23	4			
Cantaloupes	*	1	1	3	8	24	24	22	12	4	1	
Casabas						1	5	16	29	29	18	2
Cherries	*				14	39	42	4				1
Coconuts	8	6	8	5	3	3	4	4	11	14	18	16
Cranberries	1	1							6	21	49	22
Crenshaws				*	3	8	17	27	27	15	2	*
Figs, Fresh					1	15	8	29	24	19	4	*
Grapefruit	12	12	13	12	10	6	3	2	2	8	10	10
Grapes	3	3	3	3	2	6	10	18	19	15	11	7
Honeydews	2	5	7	6	3	7	14	22	21	12	1	*
Lemons	7	6	7	8	10	11	11	10	8	7	7	8
Limes	6	4	4	4	6	15	16	13	10	7	6	9

COMMODITY	% Jan	% Feb	% Mar	% Apr	% May	% June	% July	% Aug	% Sept	% Oct	% Nov	% Dec
Mangoes			2	19	39	29	10	1				
Nectarines	2	5				16	35	34	8			
Oranges, all	12	11	11	10	9	7	5	5	5	6	8	11
Oranges, West	9	9	10	10	9	7	7	7	7	8	6	11
Oranges, Fla.	15	15	14	11	10	6	2	1	*	3	10	13
Peaches	*	*	*		2	26	31	27	13	1		
Pears	6	6	7	6	4	1	6	15	16	15	10	8
Persians				*	5	15	29	31	19	1	*
Persimmons	1								2	41	39	17
Pineapples	8	9	12	14	15	17	7	4	2	3	4	5
Plums-Prunes	1	1	1		2	18	25	25	24	3		
Pomegranates									20	63	15	2
Raspberries					1	21	55	5	6	7	4	1
Strawberries	1	2	5	15	31	26	10	5	3	1	*	*
Tangelos	10									7	46	37
Tangerines	24	8	3	1							20	44
Watermelons		*	1	2	11	27	33	21	5	*		

*Less than 0.5 of 1% of annual total

The table is based on unloads of fresh fruits in 41 cities as reported by the U.S. Department of Agriculture, and on import figures.

Monthly Availability of Vegetables, Expressed as a Percentage of Total Annual Supply

COMMODITY	Jan %	Feb %	Mar %	Apr %	May %	June %	July %	Aug %	Sept %	Oct %	Nov %	Dec %
Artichokes	6	7	12	23	15	5	4	3	4	6	9	8
Asparagus	*	2	20	33	28	15	1	*	*	1	*	*
Beans, Snap	5	5	6	8	9	13	12	11	10	8	7	6
Beets	4	4	5	5	7	13	15	13	12	11	7	4
Broccoli	10	10	12	11	9	5	4	3	6	10	11	9
Brussels Sprouts	14	9	5	2	1	1	1	2	11	19	20	16
Cabbage	9	8	10	9	9	10	8	7	8	8	8	8
Carrots	10	8	10	9	8	8	8	7	8	8	8	8
Cauliflower	9	7	7	5	5	5	4	5	11	19	14	8
Celery	9	8	9	9	8	7	7	7	8	9	10	9
Chinese Cabbage	9	8	7	7	6	8	9	8	9	10	11	10
Corn, Sweet	1	1	2	6	15	16	20	18	11	4	3	2
Cucumbers	4	4	5	9	12	14	14	10	8	8	8	6
Eggplant	8	6	7	8	7	8	9	12	12	8	8	8
Escarole-Endive	7	7	9	8	7	9	9	9	9	10	8	7
Endive, Belgian	5	15	15	14	11	4			4	10	9	15
Greens (misc.)	10	10	11	11	8	7	6	6	6	8	8	10
Lettuce	8	7	8	9	9	9	9	9	8	8	8	8
Mushrooms	10	9	10	10	8	7	5	5	6	8	10	11
Okra	*	*	1	4	11	19	23	21	12	7	3	1
Onions, dry, all	8	7	8	9	9	10	9	8	9	8	8	8
Onions, dry, Texas	*	*	6	32	30	14	11	6	1	*	*	*

COMMODITY	Jan %	Feb %	Mar %	Apr %	May %	June %	July %	Aug %	Sept %	Oct %	Nov %	Dec %
Onions, dry, New York	10	9	10	6	2	1	2	12	14	13	11	11
Onions, dry, California	2	2	1	1	9	21	23	14	9	8	7	3
Onions, Green	6	6	8	10	10	11	11	10	8	7	7	7
Parsley and Herbs	6	6	8	8	7	8	8	8	8	10	13	11
Parsnips	13	11	11	9	6	5	3	3	8	14	11	9
Peas, Green	7	8	9	15	13	15	13	9	5	3	2	2
Peppers, Sweet	7	6	7	7	8	10	10	10	10	9	9	7
Potatoes, All	9	7	9	9	9	9	8	8	8	8	8	8
Potatoes, California	4	4	4	3	10	26	24	11	5	3	3	4
Potatoes, Maine	11	11	16	19	16	9	2	*	*	1	5	9
Potatoes, Idaho	14	12	14	14	9	2	*	2	3	8	12	12
Radishes	6	6	8	9	11	12	11	9	8	7	7	7
Rhubarb	6	12	15	22	27	13	3	1	*	*	*	1
Spinach	9	8	10	10	10	8	6	5	7	9	9	8
Squash	7	5	6	7	8	9	10	9	10	11	11	8
Sweetpotatoes	9	8	9	7	4	2	3	6	10	12	17	13
Tomatoes, All	6	6	8	8	11	11	12	10	8	8	6	6
Tomatoes, Florida	12	10	14	16	23	8	*				4	13
Tomatoes, California	2	1	*	*	2	8	19	17	16	19	12	4
Tomatoes, Mexico	13	20	24	22	13	3	*	*	*	*	1	4
Turnips-Rutabagas	12	11	10	6	4	4	4	4	9	12	13	11

*Less than 0.5 of 1% of annual total.

The table is based on unloads of vegetables in 41 cities as reported by the U.S. Department of Agriculture.

three-week period, with eighty 3-oz portions needed each time, the purchase amount is $5 \times 80 \times 3 \div 16$ oz = 75 lb. If an additional reserve of 25 lb is desired, the purchase amount should be 100 lb.

Quantities needed should be stated precisely. Thus, if fresh spinach by the bushel is to be purchased, the specification should say "Spinach, bushel, minimum net weight 18 lb"; other appropriately precise specifications might say "32-size grapefruit, minimum weight per carton 42 lb" or "50-lb carton Idaho baking potato, 100 count." Grade and other quantity and quality factors should be added as necessary.

Storage and Handling Care

Fruits and vegetables are living things, and even after harvest they can change quality. Refrigeration and proper humidity retard the natural processes of deterioration and drying out. All items should be inspected on delivery to ensure that they meet specifications.

Management should establish procedures for handling and storing fruits and vegetables. Crates and packages should never be dropped, thrown, or pushed roughly. Many products bruise easily, and some crates of produce are as delicate as a crate of eggs.

Suitable handling during pre-preparation can avoid damage. Do not allow food to stand in unfavorable temperatures. Do not stack items so high that items underneath are crushed. Rotate stock, using older items first; dating may be desirable. Fruits and vegetables need air to hold well. Piling them up without access to air can harm them. Items that give off odors should be stored separately. Canned fruits and vegetables with a high acid content (such as plums, sauerkraut, and tomatoes) are best held for a maximum of six months. As mentioned previously, it is now possible to install equipment that absorbs ethylene gas that causes ripening in fruits and deterioration of vegetables and thus give them a longer shelf life.

Proper temperature, humidity, and ventilation are important for fresh items as well as for frozen, dried, and canned foods. Hold fresh produce under refrigeration at 45°F (7°C) or lower, but do not freeze it. Store frozen products at 0°F (–18°C) or lower. Most canned foods keep for about a year. Optimal storage for all forms of food should be known and adopted.

Cooking Methods

Cooking can change the texture, color, flavor, and nutritional value of fruits and vegetables. Personnel should be informed about what these changes are and how best to control them to obtain only desirable changes.

Baking

Baked potatoes are common on many menus. Numerous other fruits and vegetables can be baked with equally favorable results, including apples, carrots, cucumbers, eggplant, green peppers, onions, stuffed mushrooms, squash, tomatoes, turnips, and parsnips.

Sufficient moisture must be present to soften the cellulose of the baking item. Low-moisture items can be parboiled before being baked; covering items while they bake also helps. Baking may act to caramelize sugar and to brown items, producing a desirable flavor. Nutrients are retained well in baking, since there is no water-leaching. Baking in covered casseroles that contain only a small amount of water also reduces nutrient loss.

Baking temperatures range from 350 to 450°F (177 to 232°C); squash and potatoes should be baked at the higher temperatures. A baking time of 30 to 60 minutes is usually sufficient. Overbaking high-starch products or failing to prick and open the skin immediately after baking can result in a soggy product.

Boiling

Fruits and vegetables should not be boiled at a hard boil, since this can break them up. Bring the cooking liquid to a rapid boil, then add the product, and bring the liquid back to a boil as fast as possible; then reduce the heat to a simmer. Water is never hotter than 212°F (100°C), regardless of the size of the bubbles in it. Cook delicate vegetables and fruits for as short a time as possible. It is common today to serve vegetables with some crunch still in them. Some vegetables, such as onions, must be cooked for a long time to remove their strong flavor. Others, such as members of the cabbage family, develop strong flavors with long cooking. Cooking in a small amount of water is often recommended for delicate vegetables; but if they are dropped into a large amount of boiling water, the cooking proceeds faster. It is also desirable to cook strongly flavored vegetables in plenty of water, to reduce the flavor somewhat. Water should be salted at the rate of 1 tsp of salt for every 1 qt of water.

Fruit cooked in plain water breaks up more easily than fruit cooked in a syrup, because sugar strengthens the fruit's fiber. The higher the sugar content of the boiling liquid, the less the breakup of the fruit.

Covering the cooking pot saves heat, but some vegetables and fruits should be cooked uncovered (see Table 14-3). Most fruits can be cooked covered; but

delicate green vegetables should be cooked uncovered for at least the first 5 minutes of boiling so that they retain their greenness.

Batch-cooking is recommended for vegetables and some fruits. In each batch, cook about the quantity needed for 20 minutes of serving. Never mix old and new batches. If demand suddenly increases, use a high-pressure steamer to get a batch to service quickly. Vegetables can be parboiled in advance, plunged into cold water, and refrigerated. When needed, they can be cooked to doneness in their original cooking water.

Metals and some chemicals can change the texture, color, and flavor of fruits and vegetables as they cook. Hard water preserves the green color in vegetables; very hard water destroys thiamine and ascorbic acid. Acids destroy the green color. An alkaline reaction (from hard water or soda added to water) helps green vegetables retain their green color but also softens the product. Legumes cook more quickly if soda is added to the cooking water. Soda destroys thiamine and ascorbic acid, however, and gives some vegetables a poor texture. Iron reacts with red vegetables and fruits, causing them to turn a muddy color. Acids make red fruits and vegetables redder, while alkalies turn them a muddy red. Acids preserve the color of white vegetables and fruits (see Table 14-4).

TABLE 14-3 Boiling of Various Vegetables, Covered or Uncovered

Vegetable	Covered or Uncovered	Exceptions
Bulb	Uncovered	
Flower	Uncovered	
Fruit	Covered	Okras, pepper
Leaf	Uncovered	
Root	Covered	Rutabagas, turnips
Seed	Uncovered	Corn
Stem	Uncovered	
Tuber	Covered	

TABLE 14-4 Effect of Boiling Pigments in Acidic or Alkaline Liquid

Name of Pigment	Color	Example Where Pigment Is Found	Effect of Acid on Product	Effect of Alkali on Product
Anthoxanthins[a]	White	Onion	White	Creamy white to yellow with more alkalinity
Chlorophyll[b]	Green	Green vegetables	Olive green	Intensified green, loss of nutrients, slimy texture
Carotenoids[b]				
Carotenes	Orange	Carrots	Little effect	Little effect
Lycopene	Red	Tomatoes	Little effect	Little effect
Xanthophylls	Yellow	Corn	Little effect	Little effect
Flavonoids[a]				
Anthocyanins	Red to blue/purple	Red cabbage	Little effect or intensified red	Bluish green

[a]This pigment dissolves in water.
[b]This pigment dissolves in fat.

Braising

Vegetables are braised more frequently than are fruits. Celery, onions, cabbage, lettuce, and Belgian (French) endive are among the vegetables that can be braised. To braise a vegetable, sauté it in a small amount of fat until it is almost tender. Then toss, with quick stirring, and serve partially cooked (this is the stir-fry method). Other vegetables should continue cooking while a thickening agent (usually starch or flour slurry) is added. If more moisture is desired, add hot stock to the pot, and simmer it. Continue cooking the product until it is tender, turning and stirring it.

Broiling

Fruits and vegetables are usually not broiled, since a lot of moisture evaporates in broiling. Only very tender, moist produce broils well. For example, some fruits that are first topped with butter, sugar, and seasonings can be broiled successfully.

Many items are parboiled first to make them tender and moist. An eggplant that has been parboiled almost to tenderness can be covered with mayonnaise and broiled. Tomatoes, because of their high moisture, can be broiled. Other items put onto shish kebabs broil well after parboiling. The lack of moisture in many broiled items leaves them crisp and slightly charred, which may add flavor but may also leave a tough texture. Many broiled items are served with sauces or butters, for added moistness.

Deep-Frying

Fried foods have rich flavor and good texture. Potatoes deep-fry well. A one- or two-step process can be used. Potatoes cut into strips fry crisply in 5½ to 7 minutes in fat heated to 350°F (177°C). The two-step method fries potatoes at 260°F (127°C) for 6 minutes, and then finishes them at 325°F (163°C) for 2½ minutes. Shake the strips well after the first step of frying, and set them aside until orders come in. The potatoes can then be finished quickly. Refrigerate the strips if they are to be held for any length of time. Frozen strips are prefried in this manner, so only a short cooking time is needed to prepare them fully.

Vegetables such as parsnips and some fruits can be parboiled and then deep-fried. Breaded bananas and other fruits can be deep-fried without requiring any previous cooking. Some breaded vegetables such as eggplant and mushrooms deep-fry well. Even slices of dill pickles, batter-dipped and deep-fried, make a delicious accompaniment to roast ham, corned beef, or a pastrami sandwich. Table 14-5 lists ways to deep-fry selected items.

Sautéing.

A shallow sauté pan, a griddle, or a well-greased baking sheet can be used to sauté vegetables. Proper cooking temperatures range from 335 to 425°F (168 to 223°C). Most vegetables and a few fruits sauté well, including apple rings, tomato slices, mushrooms, zucchini, eggplant, onions, and all forms of potatoes.

Steaming

Vegetables that do not pack tightly around each other tend to steam well. Vegetables that must be put into water to steam are usually better when boiled. Spinach, peas, and some other vegetables pack so tightly that the centers undercook. Steaming retains nutrients well because of the absence of water-leaching, but the high temperatures can destroy some vitamins.

Few fruits are steamed. They are better when stewed or boiled with sugar added, except in certain special preparations. Vegetables that are usually covered when boiled come out better when steamed. Steaming (especially pressure steaming) reduces cooking time. Timing must be watched closely, since temperatures are much higher, and cooking is consequently faster. Pressure at 15 psi cooks vegetables very rapidly. If a pressure of 5 to 7 psi is used, quantity production is best. Pressure cookers, unlike convection steamers, cannot be opened at any time during cooking without releasing steam—sometimes very dangerously.

Many vegetables and some fruits can be steamed in perforated pans that allow the steam to filter around the product. Tough parts such as asparagus or broccoli stalks may come out best when cooked first in hot water with only the stalk immersed and then tipped over with the entire vegetable (including the tender head) immersed. When steaming vegetables in pans of water or in perforated pans, do not fill the pans with water to a height of more than 2 in.

TABLE 14-5 Deep-Frying Methods for Various Vegetables

Vegetable	Blanch[a]	Breading	Batter	325°F (163°C)	350°F (177.5°C)	375°F (191°C)
Asparagus tips	×	×				1
Broccoli flowers	×		×			1½
Brussels sprouts	×	×			1½	
Carrots, peeled halves				1½		
Cauliflowerettes, plain				1½		
breaded	×	×				1
Corn, whole ears				1½		
Cucumbers, thin slices			×		1½	
Eggplant, ¼ in. slices, peeled					1	
thin slices			×		1½	
Snap beans, whole					1½	
Squash, 2 in. strips	×		×			1½
Sweet potato, ¼ in. slices					2	
Zucchini, 2 in. slices				4		

[a]Items to be blanched are precooked, breaded, or battered, and finished at the indicated fryer temperature just before service.

Stewing

Fruits and some vegetables stew well. Fruits to be stewed should be fresh, sound, and not overripe; they should be stewed in water, if a sauce is desired, or in sugar or a sugar syrup, if the fruit is to retain its shape or if a fruit compote is desired. To obtain a good compote syrup, add 1 cup of sugar for every 2 cups of water. Adding more sugar results in a preserve or a jam. Too much sugar causes fruit to shrink and harden.

Preparing Fruits and Vegetables

Fruits and vegetables are often prepared according to whether they are fresh, frozen, canned, or dried. Each kind of food takes a different kind of preparation.

Canned Fruits and Vegetables

Canned fruits are used because of their nature, convenience, cost, or because fresh fruits are not available.

A canned pear or pineapple and cottage cheese salad is not at all inferior to a fresh fruit served in the same way. Because fresh Kadota figs are seldom on the market, the canned product usually must be used. Canned ready-to-use pie fillings are labor-saving good-quality items.

Selecting a canned product should be based on the production need. Higher-grade fruits are more uniform in size and color, are free from defects, and have a heavier syrup than lower grades have. On the other hand, the lower grades are suitable for many purposes. High-grade peaches can be served as a sauce; a lower water-pack grade can be used for pies.

When purchasing canned items, the buyer should distinguish between styles offered. Fruits may be whole, halved, sliced, cut, diced, chunks, spears, tidbits, or crushed. Can sizes vary from the large No. 10 (3-qt and larger) size to No. 303 (16 oz) or even smaller. Smaller cans may be used at the end of a serving period, when a smaller quantity is needed.

Canned items should be stored at temperatures lower than 70°F (21°C) but above freezing. Good ventilation is important. Rusted, leaking, bulging, or badly dented cans should be discarded. Often they can be sent back to purveyors for credit.

When fresh items are not available, or when high labor costs militate against the daily preparation of fresh items, many operations rely on canned fruits and vegetables. The nutritive value of canned items is nearly the same as that of any fresh produce, there is little waste, and the cost per serving is easy to calculate.

The syrup accompanying canned or frozen fruit can be used to make fruit drinks, sweetened juices, or molded salads, or it can be used as the base for cold fruit soups or served with the fruit. Carryover fruits can be used in salads, in fruit soups, and in puddings.

Fruit drinks or juices made from natural or synthetic fruit juices are served as breakfast beverages, as appetizers for lunch, or as cool drinks in the afternoon. A 4- or 5-oz glass is an adequate portion. A No. 3 (46-oz) cylinder can yield from 9 to 11 portions.

About two-thirds of a can of vegetables consists of the drained item, and the remainder is juice. Pour the brine from the vegetables into a pot, and heat it until it is reduced by about half. Then return the vegetables to it, season, and serve. Unused brine can go into the stock pot. Vegetables served with some of their brine have added nutritional value. Canned vegetables should be served as soon as possible after reheating.

No. 10, No. 2½, No. 2, and No. 303 cans are used for vegetables. They hold, respectively, 106 oz, 24 oz, 20 oz, and 16 oz. Vegetables may be garnished at service with fried bacon, pimiento, browned almonds, cooked chopped green pepper, cubed carrot bits, chopped parsley, or various cream or cheese sauces.

Dried Fruits and Vegetables

A dried fruit such as a prune contains about 25 percent moisture. Dried vegetables contain much less. Fruit may be dried naturally or vacuum-dried. The latter method reduces moisture to about 2 to 5 percent. Some vegetables such as beans, peas, and other legumes dry naturally. In vacuum-drying, the product

is placed into a chamber and warmed, and the air is evacuated, creating a partial vacuum into which the moisture in the vegetable is quickly drawn. Most products dried in this way have a rather porous texture and rehydrate well. Vegetables are rarely freeze-dried, although a few items such as mushrooms and chives might be. In freeze-drying, the item is frozen and placed into a chamber where a vacuum is created. A small amount of heat is applied and the moisture goes from ice to water vapor without ever turning to water, a process called *sublimation.*

Apples, apricots, figs, dates, peaches, pears, prunes, and raisins are the most commonly used dried fruits. They go into breads, pies, puddings, cookies, cakes, candies, and desserts. Some are eaten in their dried form.

Dried fruits should be washed well before using. To rehydrate a fruit, place it into water; the dried fruit absorbs liquid and plumps up. Just 1 lb of dried prunes yields 3 lb of cooked prunes. Some types of dried fruit should be soaked for a short time, while others can be simmered immediately. Use 1 qt of water for each 1 lb of fruit. Add sugar as necessary, depending on fruit sweetness.

Dehydrated potatoes are the most commonly used dried vegetables, with dried onions following. Just 1 lb of dried onions equals about 8 lb of fresh EP. Dried parsley, peppers, tomatoes, and others are used occasionally.

Legumes are dried beans, peas, and other seeds of leguminous plants. Chick-peas (garbanzos), cow-peas (black-eyed peas), split peas, and lentils belong to this family. Legumes are used in soups, salads, and main dishes. All foreign matter should be removed from the legumes. Small rocks of about the same weight as a legume can cause a patron to break a tooth. After sorting, the legumes are washed. They may be soaked in cold water, although split peas and lentils need no soaking. After soaking, the legumes are simmered for several hours until tender; water is added as needed. About 2 tsp of salt are added for every 1 lb of legumes. Cooking legumes in steamers after soaking them requires less time and less water.

A quantity of 5 to 6 lb of dried legumes yields 100 3-oz portions—or more, if other ingredients are added. Because acid interferes with the softening of legumes, brown sugar (slightly acidic), molasses, catsup, and tomatoes should not be added until the legumes are tender.

Fresh Fruits and Vegetables

Fresh fruits can be served raw and eaten in hand. They also can be cut up for salads, appetizers, prepared dishes, and desserts. All fruits should be washed thoroughly before being used, since chemicals and sprays may be on them. Many fresh fruits are pared, cored, pitted, or trimmed before serving. Table 14-6 describes ways to prepare different fruits.

Fruits such as peaches, pears, and apples tarnish (turn brown) when their cut surface is exposed to air. Dipping the cut pieces into acidic juices (such as lemon or pineapple juice) or into an antioxidant solution retards or eliminates the browning. Putting cut fruit into a gelatin mixture also stops browning to some extent.

TABLE 14-6   Preparation Techniques for Fresh Fruit

			Quantity Needed for 100 4-oz Portions	
Fruit[a]	Preparation Technique	Trim Loss Percent	AP (lb)	EP (lb)
Apples[b]	Wash under running water; peel and core as required; cut as desired.	10 peeled 25 peeled and cored	28 33.5	25 25
Apricots[b]	Wash under running water; cut in two; remove pit; slice as desired.	6 pitted	27	25
Avocados[b]	Peel; cut in two lengthwise; remove pit; cut as desired.	25 pitted	33.5	25
Bananas[b]	Peel and cut as desired.	30 peeled	36	25
Berries (except strawberries)	Sort and remove stems; wash under spray; drain; cut as desired.	5 topped	26.5	25
Cherries[b]	Sort; remove stems and pits; wash under spray.	12 topped and pitted	28.5	25
Citrus fruit	Peel and cut into sections. (Grapefruit sections may be replaced into the peel.)	40 pared	42	25
Grapes	Remove stems; wash grapes; cut in halves; remove seeds.	10 seeded	28	25
Melons	Cut in half; remove seeds before or while preparing.	50 pared	50	25
Peaches[b]	Blanch to remove skin; cut to remove pit; serve as desired.	15 pitted	29.5	25
Pears	Wash under running water; peel and core as required; cut as desired.	10 peeled 25 peeled and cored	28 33.5	25 25
Pineapples	Remove tops and bottoms; pare; cut as desired.	45 pared	45.5	25
Plums	Wash thoroughly in containers of water; cut in two to pit.	5 pitted	26.5	25
Rhubarb	Remove leaf and hard area at base of stem; wash; cut across stem into desired width.	10 prepared	28	25
Strawberries	Remove stem; wash under spray; cut if desired.	10 topped	28	25

[a]All fruit listed is of medium size.

[b]Place in solution to retard browning.

Fresh fruit can be stewed or baked and served in the form of fruit sauces, compotes, preserves, or baked fruit. Some fruits, including apples, bananas, and pineapple, can be sautéed. Grapefruit halves, bananas, pineapple slices, and some others can be broiled. Fresh fruit can be seasoned with brown or maple sugar, nuts, raisins, nutmeg, cinnamon, or cloves.

Fresh vegetables are served raw or cooked. Many raw vegetables are served as relishes or in salads. Some are used raw as garnishes. Vegetables may be cooked by boiling, simmering, steaming, baking, or broiling. Table 14-7 describes ways to prepare different vegetables.

Clean vegetables thoroughly before using them. Lukewarm water cleans better than cold water does. Some thin-skinned vegetables, such as potatoes, can be cooked and then peeled. Others, such as turnips and rutabagas, should be pared. Thin paring or peeling after cooking saves nutrients. Some vegetables tend to discolor when pared or cut. Dipping them in

TABLE 14-7　Preparation Techniques for Fresh Vegetables

Vegetable[a]	AP (lb)	Trim Loss (%)	Cooking Weight Gain (+) or Loss (−)	EP (lb)	Preparation and Possible Seasonings[b]	Boiling[c] 212°F (100°C)	Baking 350°F (177°C)	Baking 450°F (232°C)	Steaming 0 psi 212°F (100°C)	Steaming 6 psi 230°F (110°C)	Steaming 15 psi 250°F (121°C)
Artichokes French	50	20	0	8⅓ doz	Wash; cut top and tips; cook with lemon; serve whole (6.5 oz).	40	—	—	—	20	10
Jerusalem	25	25	0	19	Pare thin; serve whole, quartered, or sliced.	30	45	—	35	25	15
Asparagus	38	35	−10	21	Wash, removing sand, dirt, and scales from stem; break woody base from stem and tip; cook and season with 17, 24, or 25.	15 stems in water	—	—	18	13	1½
Beans Lima	50	60	−5	20	Shell; wash; drain; cook and season with 15, 19, 22, 23, 25, or 26.	20	—	—	30	12	1½

(continues)

TABLE 14-7 Continued

	Approximate Quantities Needed for 100 3-oz Portions					Approximate Average Time (min) and Temperature (in 10-lb Lots)					
						Boiling*c*	Baking		Steaming		
Vegetable*a*	AP (lb)	Trim Loss (%)	Cooking Weight Gain (+) or Loss (−)	EP (lb)	Preparation and Possible Seasonings*b*	212°F (100°C)	350°F (177°C)	450°F (232°C)	0 psi 212°F (100°C)	6 psi 230°F (110°C)	15 psi 250°F (121°C)
Beans (*cont.*)											
Snap	23.5	15	−5	20	Wash, remove ends; cut as desired; cook and season with 2, 14, 15, 16, 17, 19, 23, 25, or 26.	20	—	—	25	20	2½
Beets											
New	27	25	−5	20	Remove roots and tops; wash; cook; peel; cut as desired and season with 1, 3, 4, 9, 11, 13, 17, 23, or 26.	35	50	—	50	40	7
Old	27	25	−5	20	Same as for new.	70	60	—	75	65	14
Broccoli	35	40	−10	21	Remove woody stems and coarse leaves; wash; drain; split stem for uniformity; cook and season with 4, 11, 17, or 25.	9	—	—	15	8	2
Brussels sprouts	22	17	+5	18	Remove leaves and stalks; wash; soak briefly in salt water; cook and season with 2, 4, 11, 17, 22, or 26.	13	—	—	15	11	2
Cabbage*d*	25	20	−5	20	Trim as needed; wash; quarter; cook and season with 4, 6, 11, 16, 17, 18, 23, or 25.	12	—	—	15	10	2

TABLE 14-7 Continued

| Vegetable[a] | Approximate Quantities Needed for 100 3-oz Portions | | | | Preparation and Possible Seasonings[b] | Approximate Average Time (min) and Temperature (in 10-lb Lots) | | | | | |
	AP (lb)	Trim Loss (%)	Cooking Weight Gain (+) or Loss (−)	EP (lb)		Boiling[c] 212°F (100°C)	Baking 350°F (177°C)	Baking 450°F (232°C)	Steaming 0 psi 212°F (100°C)	Steaming 6 psi 230°F (110°C)	Steaming 15 psi 250°F (121°C)
Carrots	27	22	−10	21	Remove tops; peel; cut as desired; cook and season with 1, 3, 6, 11, 12, 13, 14, 15, 16, 18, or 26.	15	35	—	20	13	3
Cauliflower	45	50	−15	22.5	Remove outer leaves and stalk; wash; break into flowerets; cook and season with 4, 6, 11, 14, 15, or 25.	9	—	—	11	8	2½
Corn-on-the-cob	60	15	+2	8½ doz	Husk, removing all silk; cook and season with butter and salt.	10	30	—	13	9	3½
Eggplant	24	20	0	19	Pare; cut before cooking; cook and season with 15 or 19.	14	45	—	15	7	1
Onions	26	12	−15	22.5	Pare; cook and season with 4, 17, 18, 19, 22, or 26.	25	45	—	30	20	5
Peas	50	60	−5	20	Shell; wash; cook and season with 2, 11, 15, 16, 19, 21, 22, or 23.	7	—	—	10	4	1
Potatoes Sweet	25	23	0	19	Scrub and wash; cook; pare; cut (unless baked); season with 1, 5, 8, 9, or 18.	20	—	50	25	15	6

(continues)

TABLE 14-7 Continued

	Approximate Quantities Needed for 100 3-oz Portions					Approximate Average Time (min) and Temperature (in 10-lb Lots)					
			Cooking Weight				Boiling[c]	Baking		Steaming		
										0 psi	6 psi	15 psi
	AP	Trim Loss	Gain (+) or Loss	EP	Preparation and Possible	212°F	350°F	450°F	212°F	230°F	250°F
Vegetable[a]	(lb)	(%)	(−)	(lb)	Seasonings[b]	(100°C)	(177°C)	(232°C)	(100°C)	(110°C)	(121°C)
Potatoes (cont.)											
White	25.5	25	0	19	Scrub and wash; cook; pare (unless baked); cut; season with 2, 3, 4, 6, 7, 11, 17, 19, or 26.	25	—	50	35	20	8
Spinach	25.5	27	−5	20	Remove coarse stems; wash; cook and season with 2, 14, 15, 18, or 19.	5	—	—	8	4	1

[a] All vegetables listed are of medium size.

[b] The numbers listed in this column correspond to the following:

1. Allspice	8. Cinnamon	15. Marjoram	21. Rosemary
2. Basil	9. Clove	16. Mint	22. Sage
3. Bay leaves	10. Curry powder	17. Mustard seed	23. Savory
4. Caraway seed	11. Dill	18. Nutmeg	24. Sesame seed
5. Cardamom	12. Fennel	19. Oregano	25. Tarragon
6. Celery seed	13. Ginger	20. Poppyseed	26. Thyme
7. Chives	14. Mace		

[c] Boiling times indicated are from the point after the vegetables are added and the water has returned to a boil.

[d] Shredded cabbage is cooked approximately half the time of quartered cabbage.

an antioxidant or holding them in water can prevent discoloration. Dipping leaches fewer nutrients from the product; any soaking should be minimal. All personnel who prepare vegetables should know cooking procedures and times.

Frozen Fruits and Vegetables

Freezing can retain the nutrients, flavor, color, and texture of many fruits and vegetables and can reduce spoilage and storage space. Frozen items can be purchased in 5-, 10-, 20-, or 30-lb packages. Some 2- and 2½-lb packages are also available. Store frozen foods at 0 to −10°F (−18 to −23°C). Items are best thawed under refrigeration.

Cook vegetables directly from the frozen state by dropping them into salted boiling water. Frozen vegetables cook in less time than fresh ones do because they have been blanched before freezing to hold their color and flavor; they may also contain salt. Do not refreeze items once they have been thawed.

Management's Responsibility

Management is responsible for establishing an maintaining proper standards for vegetables and fruits.

Grading can be used to help select good-quality products. Yield and comparative tests are used to check quality and select the best items. Standards should cover flavor, color, texture, form, and other relevant attributes. Flavors should be natural to the product and should blend with the flavors of other foods. Colors should be bright and natural. Shape and size should favor easy eating and should contrast with other foods. Textures should complement those of other foods.

The choice among canned, dried, fresh, or frozen fruits or vegetables should be based on availability, storage space, labor, and cost. Fresh items should be used if they are available at a good price and if labor is available to process them at a satisfactory cost. Menus should feature fresh or processed fruits and vegetables because of the variety they give and because they carry a good markup.

Vegetables and fruits should be served in ways that bring out their true flavors. Good seasoning and buttering are important in serving vegetables. Fruits should be plump and well-filled.

Fruits and vegetables should be merchandised by having menus emphasize an interesting range of cooking styles, such as scalloped, au gratin, souffléed, glazed, and stuffed. Offering an attractive salad bar can also encourage produce consumption and patron satisfaction.

■ CEREALS

Over 50 percent of our calories should normally come from carbohydrates. Foods such as potatoes, rice, wheat, legumes, sugars, and syrups contain a liberal supply of carbohydrates, and these are frequently included as one of the foods in a meal. People commonly eat a cooked or dry cereal for breakfast, and rice, potatoes, pasta, hominy, or even couscous may be eaten at lunch or dinner.

Cereals require special treatment in cooking. The starch swells and absorbs considerable moisture. Usually, about 1 gal of lightly salted water is used for every 1 lb of cereal or pasta (noodles or spaghetti). Fine breakfast cereals such as cornmeal, farina, and grits make a desirable thickened mixture in these ratios, but coarser ones such as oatmeal and cracked wheat require about 1¼ lb for each 1 gal. Yields of various cereal

TABLE 14-8 Yields in Weight and Volume of 1-lb Batches of Dry Cereal Products		
Cereal Product	Cooked Weight	Cooked Volume
Rice	3–4 lb	1½–1¾ qt
Pasta	4 lb	2¼–2½ qt
Noodles	3 lb	2 qt
Fine cereals	9 lb	1⅛ gal
Coarse cereals	9¼ lb	1¼ gal

products are identified in Table 14-8. The salted water is brought to a boil, and the cereal is added while stirring. Some items (e.g., pasta and rice) have enough water to obviate the need for further stirring, but breakfast cereals do require some stirring while cooking to a thick paste. Fine cereals can be mixed with a bit of water and then added to the main volume of water, to prevent lumping. The more a cereal is stirred during thickening, the more slick and pasty it becomes.

Pasta, rice, and other such products should be *just* cooked: cooks describe the resulting product as *al dente*, which means "to the tooth" or still having some firmness. Selecting a good-quality long-grain rice or a pasta that is made from semolina paste gives a firmer product and one that holds up better in the steam table. Poorer-quality products get flabby and soft and are less pleasant to eat. Breakfast cereals are cooked to complete doneness but should not be overcooked. After cooking rice, noodles, or pastas in a large quantity of water, drain, blanch in cold water, drain, and then use.

Some people cook pasta items with 3 qt of water and 2 Tbsp of salt for each 1 lb of product. The water should be boiling when the product is added. Stir, cover tightly, remove from direct heat, and let stand for 10 minutes or until the product is tender. A small amount of oil added to the cooking water can reduce the amount of foam produced in boiling. Then drain and blanch in cold water. Some thick-walled pasta products must be left on low heat to cook properly by this method. If cooking is done in a steam-jacketed kettle, add the cereal products to the boiling water, stir, cover, turn off the steam, and allow to stand until cooked. For pasta shapes, see Figure 14-1.

Alphabets—This favorite kids' shape is usually used in soups for a fun meal anytime.

Angel Hair, Capellini ("Fine Hairs")—Thin, delicate strands are best if used with thinner, delicate sauces. Other uses: break in half and put in soup; use in salads or stir-fry meals.

Bow Ties, Farfalle ("Butterflies")—Bow Ties brighten any meal with their interesting shape. Thick enough for any sauce, or make into a salad or soup.

Ditalini ("Little Thimbles")—This versatile shape can be used as the base of any dish. Bake it, stir it into soups, or create great salads and stir-fry dishes.

Fettuccine ("Small Ribbons")—Perfect for heavier sauces like cheese, meat and tomato sauces. For variety, try breaking in half and putting in soups, or use for a salad.

Fusilli ("Twisted Spaghetti")—This long, spiraled shape can be topped with any sauce, broken in half and added to soups, or turned into a beautiful salad. Fusilli also bakes well in casseroles.

Jumbo Shells—Best when stuffed with your favorite mixtures of cheese, meat and vegetables. Stuff with meat flavored with taco seasoning, top with salsa and bake for a delicious Mexican dish, or create your own stuffed treat.

Lasagne (From "lasanum," Latin for pot)—Create new Lasagne casseroles by using chopped vegetables, cheeses and any kind of sauce. You can also assemble your casserole and freeze it for later.

FIGURE 14-1 Pasta shapes. When making delicious pasta dishes, be sure to choose a pasta shape and sauce that complement each other. Thin, delicate pasta such as angel hair or thin spaghetti should be served with light, thin sauces. Thicker pasta shapes, such as fettuccine, work well with heavier sauces. Pasta shapes with holes or ridges, such as mostacciola or radiatore, are perfect for chunkier sauces. (Courtesy of the National Pasta Association.)

Linguine ("Little Tongues")—A great shape for all sauces. Also a good choice for salads and stir-fry dishes.

Macaroni ("Dumpling")—A highly versatile shape that can be topped with any sauce, baked, or put in soups, salads and stir-fry dishes.

Manicotti ("Small Muffs")—Stuff Manicotti with a mixture of meat, cheese and vegetables, top with your favorite sauce, and bake. Or stuff and freeze for a later time.

Medium Egg Noodles (from "Nudel," German meaning paste with egg)—This size of Egg Noodle can be baked, tossed in soups or salads, or topped with cream, tomato, cheese or meat sauces for a delicious meal.

Medium Shells, Conchiglie ("Shells")—Shells make a great addition to soups or as the base of a wonderful salad. Try remaking your favorite Macaroni and Cheese using Shells, for a fun twist on a time-honored tradition.

Orzo ("Barley")—This small, grain-shaped pasta can be topped with any sauce, added to soups, or baked as a casserole. Perfect as a side dish as well as a main course.

Penne, Mostaccioli ("Quills" and "Small Mustaches," respectively)—This tubular pasta goes well with sauce, used in salads, baked in casseroles, or made into stir-fry dishes.

Radiatore ("Radiators")—This ruffled, ridged shape adds elegant interest to any sauce. It also works well baked in casseroles, or used in salads and soups.

FIGURE 14-1 Pasta shapes (*continued*)

Rigatoni ("Large Grooved")—Rigatoni's ridges and holes are perfect with any sauce, from cream or cheese to the chunkiest meat sauces.

Rotini ("Spirals" or "Twists")—Rotini's twisted shape holds bits of meat, vegetables and cheese, so it works well with any sauce, or you can use it to create fun salads, baked casseroles, or stir-fry meals.

Spaghetti ("A Length of Cord")—America's favorite shape, Spaghetti is the perfect choice for nearly any sauce, or it can be used to make casseroles or stir-fry dishes. Go beyond tomato sauce and see what your favorite becomes.

Vermicelli ("Little Worms")—Slightly thinner than Spaghetti, Vermicelli is good topped with any sauce, or as a salad or stir-fry ingredient.

Wagon Wheels, Ruote ("Wheels")—Wagon Wheels make interesting salads, casseroles and stir-fry dishes. Add to soups, or simply top with sauce and enjoy.

Wide Egg Noodles (From "Nudel," German meaning paste with egg)—Go beyond the traditional Stroganoff and use, Wide Egg Noodles to create soups, salads and casseroles. Or, top with any sauce and serve hot.

Ziti ("Bridegrooms")— A medium-sized, tubular pasta shape, Ziti is perfect for chunky sauces and meat dishes. It also makes wonderful salads, baked dishes and stir-fry meals.

FIGURE 14-1   Pasta shapes (*continued*)

In the Orient, rice is cooked in less water than it is in the United States, but the result is still a flaky product. The Oriental method has the advantage of yielding a more nutritious product, since the process of cooking in a lot of water, draining this cooking water off, blanching in cold water, and then draining again results in a loss of nutrients. In the Oriental method, all the cooking water is absorbed. Some people, however, advocate washing the rice several times before cooking it, to get rid of excess starch. The following formula is used:

1 lb long-grain rice 2 Tbsp (1 oz) oil 1¾ qt cold water 2 Tbsp salt	1. Wash the rice, if desired; drain well, and add the oil. Mix the oil in well to coat the rice. 2. Add the water and the salt; cover tightly, and bring to a boil. Stir once, employing the following procedure: cover, and lower heat to a very low level. Cook tightly covered for about 20 minutes. During the last 5 minutes uncover so that the rice steams dry. If the rice absorbs all the water but is still uncooked, add more water and cover again tightly and cook until tender. Too much water gives too soft a product.

A steam-jacketed kettle or heavy pot on the fire can be used for this method. Often, the cooked rice is fluffed up with a fork just after the cover is lifted and the rice is cooked. This allows steam to escape and gives the grains more separation.

Beans and other legumes are usually presoaked in 1½ gal of water for each 1 lb of product. After soaking (usually overnight), the product is put into plenty of lightly salted boiling water and allowed to cook until tender. Press between the fingers or against the side of the cooking vessel to check for doneness.

Breaking of the legume's skin indicates overcooking. If legumes, rice, or pastas are to be cooked further after boiling in water, the product is usually slightly undercooked at boiling.

SUMMARY

Fruits, vegetables, and cereals add variety and appeal to menus. Production procedures should protect nutritive values and produce the best-quality product. Fruits and vegetables furnish minerals and vitamins. Fruits are low in protein and fat, and some contain substantial amounts of carbohydrate. Vegetables are good sources of minerals and vitamins and are also low in fat, and some contain significant quantities of protein and carbohydrate. Cereals are an important source of carbohydrates, plus other nutrients. These items should be placed on menus for their nutritive yield and also because of their appeal to patrons.

Classification of fruits and vegetables for culinary purposes usually follows the cooking method used and the use of the product in the menu. Thus, some fruits are treated as vegetables in cookery, and vice versa.

To ensure quality, specify a grade or brand when purchasing fruits and vegetables. Variety should also be considered, since different varieties have different flavors or other qualities. Size or count should be specified, and the amount ordered should be based on storage facilities, quantity needed, and product perishability. Temperature, humidity, and ventilation are important factors in the care of fruits and vegetables.

Cooking methods vary according to the final product desired and according to how particular fruits or vegetables respond to handling, preparation, cooking, and service. Boiling and steaming are two often-used methods. Braising is used more for vegetables than for fruits. Some fruits and vegetables can be broiled to give a different presentation. Frying fruits and vegetables can also add distinction. Frying methods include sautéing, stir-frying, and deep-frying. Stewing is normally used for fruits.

Canned fruits and vegetables have been processed by being sterilized in their moisture- and vapor-proof containers. Fruits are usually packed in syrup, and vegetables are usually packed in a lightly salted brine. The syrup or brine can be used in other foods. Canned vegetables need only reheating and seasoning before service.

Dried fruits are used in breads, pies, puddings, cookies, cakes, and other desserts. Legumes are the seeds of leguminous plants. Other dried vegetables that may be used are potatoes, onions, and peppers. Fruits and vegetables should be soaked, usually before being cooked.

Fresh fruit can be served whole to be eaten in-hand, such as in a fruit basket, or can be cut and served. Fresh berries are popular. Fresh vegetables are served either raw or cooked. Frozen fruits and vegetables are handled in much the same manner as their fresh counterparts. Cooking frozen vegetables takes less time because they have already been blanched during first processing. Special care must be taken to prepare and cook vegetables and fruits so as to gain maximum nutrients, flavor, color, texture, and form.

Processed fruits and vegetables save storage space and labor and reduce the chances of spoilage.

Management must set good standards for the production of fruit and vegetable items on the menu. Costs must be controlled, since preparing these products involves considerable labor. Perishable items deteriorate or spoil and thus cause monetary loss. Only the quantities needed for a limited time should be purchased. Fruits and vegetables should be properly stored and handled.

Cereals require special cookery because of their high content of starch, which swells on cooking and absorbs moisture. About 1 lb of rice, cornmeal, pasta, or the like is added to 1 gal of boiling water and cooked to tenderness. Coarse breakfast cereals require just a bit more product per gallon. Rice, pastas, and other such foods are blanched after cooking and before use. If the product is to receive further cooking, it may be undercooked to allow for the further cooking.

■

CHAPTER REVIEW QUESTIONS

1. You are the manager of a good restaurant and want to feature fruits and vegetables on the menu. What will you do?

2. Customers say that most of the food in a foodservice is good but that the fruits and vegetables are flavorless and lack good color and texture. What should be done by management to improve the quality of the produce they serve?

3. What factors must be considered in purchasing fruits and vegetables? What are the grades for fresh, frozen, and canned fruits and vegetables? How does a buyer compute the quantity needed?

4. Summarize the various cooking methods for cooking fruits and vegetables, giving times, temperatures, cooking methods, and cooking procedures.

5. What kind of vegetables can be baked? Why would a cook parboil an item such as winter squash before baking it?

6. Correctly fill in the spaces left blank in the following table:

Color Pigment	Reaction to Heat	Reaction in Acid	Reaction in Alkali
Green (spinach)	Turns olive green	Turns olive green	
Red (beet)	None		
White (onion)	None		
Orange (carrot)	None		

7. What does *al dente* mean?

8. What are the approximate amounts of salt and water used to cook 1 lb of rice the regular way (in which the rice is blanched after cooking)?

9. If a 5-oz portion of noodles is served, how many pounds of noodles must be cooked to give 100 such portions?

CASE STUDIES

1. J. D. "The Baby Boomer" Carlson has moved forward with his plans to open assisted-care living facilities for the growing population that is aging. He is interested in serving pasta to the captive audience in his foodservice operations, as he believes that it is nutritious. By considering the Web site *www.ilovepasta.org*, research the nutrient content of pasta and report it to "The Baby Boomer." Do you agree with him that pasta would be good to serve? Why?

2. J. D. "The Baby Boomer" Carlson has considered vegetarian meals also as being healthy for his clientele and is considering the value of serving a meatless hamburger on a regular basis. By considering the Web site *www.bocaburger.com*, research the nutrient content of this vegetarian item and report it to "The Baby Boomer." What is soy protein? Do you feel this meatless burger would be good for aging baby boomers? Why?

VOCABULARY

Drained weight
U.S. No. 1
U.S. Fancy or Extra Fancy
U.S. Grades A, B, and C
Stir-fry
Legumes
Rotini
Ziti
Batch cooking

Braising
Sautéing
Long-grain rice
Portion
Sublimation
Vacuum-dry
Freeze-dry
Stewing
Fettuccine

ANNOTATED REFERENCES

Donovan, M. (ed.). 1996. *The New Professional Chef*, 6th ed. New York: John Wiley & Sons, Inc., pp. 791–856. (Vegetables, potatoes, grains, and pastas are considered in these chapters.)

Donovan, M. 1997. *Cooking Essentials for the New Professional Chef*. New York: John Wiley & Sons, Inc., pp. 621–652. (Vegetables, potatoes, grains, and pastas are considered in these chapters.)

Gisslen, W. 1995. *Professional Cooking*, 3rd ed. New York: John Wiley & Sons, Inc., pp. 379–472. (Vegetables, potatoes, and other starches are considered in these chapters.)

Kapoor, S. 1996. *Healthy and Delicious: 400 Professional Recipes*. New York: John Wiley & Sons, Inc., pp. 171–268. (Healthy recipes for dried beans, peas, lentils, vegetables, potatoes, grains, and rice are presented in this chapter.)

Labensky, S. R., and A. M. Hause. 1999. *On Cooking: A Textbook of Culinary Fundamentals*, 2nd ed. Upper Saddle River, N.J.: Prentice Hall, pp. 572–678 and 718–759. (Vegetables, potatoes, grains, pastas, and fruits are presented in these chapters.)

Pauli, P. 1997. *Classical Cooking the Modern Way*, 3rd ed. New York: John Wiley & Sons, Inc., pp. 281–338. (A discussion on vegetables, potatoes, pasta, rice, and grains is offered here.)

CHAPTER 15

Meat, Poultry, and Seafood

■

LEARNING OBJECTIVES

By the end of this chapter, the reader will:

1. Understand the nutritional impact that meats, fish, and poultry have on one's diet.

2. Know the market forms of meats, fish, and poultry as well as how each are inspected, graded, and packaged for distribution to foodservices.

3. Learn how to properly purchase, receive, store, and issue meats, fish, and poultry.

4. Be able to recognize which cooking method is best to use in the preparation of meats, fish, and poultry to enhance flavor and tenderness and ensure proper presentation and service.

SELECTED WEB SITE REFERENCES

American Lamb Council
www.lambchef.com/

American Ostrich Association
www.ostriches.org/

Koegel Meats
www.koegelmeats.com

National Pork Producers Council
www.nppc.org/

Oscar Mayer & Company
www.kraftfoods.com/oscar-mayer/om_index.html

Plantation Turkey
www.plantation-foods.com/

Tyson
www.tyson.com/

Viking Seafood
www.vikingseafoods.com

Meat, poultry, or seafood often constitutes the main item of a meal, as well as being the item prominently featured on a menu. Because these foods are the center of most meals, the other foods served revolve around them.

The term *meat* commonly refers to the edible portions of cattle, swine, and sheep. Beef is the most popular variety of meat in this country, followed by pork, veal, lamb, and mutton, in that order. Poultry comprises chickens, turkeys, ducks, geese, guineas, and squabs. Game birds include grouse, pheasants, partridges, quail, peacocks, wild duck, and wild turkey. Most poultry is relatively cheap and has good menu appeal because of its inexpensiveness and low content of fat and saturated fat. Chicken is the most popular type of poultry, with turkey far behind in second place. The approximate age of a bird is told by moving the keel bone; if this bone is flexible, it indicates a relatively gelatinous content, signifying that the bird is still growing. In this chapter, the word *fish* refers to gilled animals with fins, while *shellfish* refers to crustaceans, mollusks, and other shelled water creatures. The word *seafood* includes both. Seafood is popular today because of its nutritional qualities and its low levels of fat and saturated fat. Most shellfish contain higher levels of cholesterol than meat or poultry.

NUTRITION

Meats, poultry, and seafood are highly nutritious. They contain complete proteins and are highly digestible.

Lean meat, poultry, and seafood are about 20 percent protein. All are low in carbohydrate. Lean flesh often contains about 10 to 13 percent fat. Meat or poultry containing heavily marbled or layered fat has a substantially higher fat percentage. Lean pork has no more fat than other lean meats. All are an excellent source of the B vitamins—especially organ meats. Lean pork is high in thiamine. Most of the fat on poultry is found along the back. Geese and ducks are quite fat, but much of their fat is cooked off. Both meat and poultry liver is high in vitamin A. Seafood items (especially fish liver) are high in vitamins A and D. Meat contributes some potassium, sodium, zinc, and other minerals to the diet; in fact, minerals constitute about 1 percent of the total composition of lean meat. Poultry furnishes good phosphorus, some potassium, zinc, and some trace minerals. Seafood supplies good quantities of minerals and is especially high in iodine, if it comes from saltwater.

PURCHASING

Flesh foods are delicate substances that must be properly selected, handled, and cooked to end up as acceptable items. It is necessary to know what each item offered on the market is, because each kind requires special preparation. Following is a short summary of the types of animals whose flesh can be selected on the market.

Beef

- *Steer:* young male cattle castrated before sexual maturity
- *Heifer:* young female cattle that has had no calf
- *Cow:* older female cattle that has had a calf
- *Bull:* sexually mature male cattle
- *Stag:* sexually mature male cattle that has been castrated
- *Bullock:* young uncastrated bull

Veal and Calf

- *Veal:* young calf 6 to 14 weeks old
- *Calf:* young calf 14 to 52 weeks old

Pork

- *Barrow:* young male hog that has been castrated
- *Gilt:* young female hog that has not had a litter
- *Sow:* older female hog that has had a litter
- *Boar:* mature sexually active male hog

Lamb and Mutton

- *Lamb:* young sheep under a year old
- *Yearling:* young sheep about a year old
- *Mutton:* sheep over a year old, usually castrated
- *Ewe:* mature female sheep
- *Ram:* mature sexually active male sheep

Chicken

- *Broiler:* young chicken of either sex under 9 weeks old
- *Fryer:* chicken of either sex 8 to 12 weeks old
- *Roaster:* chicken about 12 weeks old weighing around 5 lb
- *Capon:* desexed male chicken weighing 4 to 7 lb
- *Hen (fowl):* older female chicken
- *Rooster:* mature male chicken
- *Rock Cornish hen:* classed as a chicken but smaller than one and usually purchased quite young, weighing 1 to 2 lb

Turkeys

- *Fryer-roaster:* turkey about 16 weeks old weighing 4 to 8 lb
- *Hen, young:* female turkey under a year old weighing 12 to 16 lb
- *Tom, young:* male turkey under a year old weighing 12 to 30 lb
- *Hen, mature:* female turkey over a year old
- *Tom, mature:* male turkey over a year old

Ducks

- *Broiler-fryer:* duck less than 8 weeks old weighing about 2½ lb
- *Duckling:* young duck about 16 weeks old, weighing 3 to 7 lb
- *Mature duck:* older duck of either sex

Geese

- *Young:* goose under 6 months old, weighing 4 to 10 lb
- *Mature:* over 6 months old, weighing 10 to 18 lb

Pigeon

- *Squab:* pigeon 3 to 4 weeks old, weighing 8 to 14 oz
- *Pigeon:* older bird

Fish

- *Salmon:* little Atlantic salmon is available and most goes into lox; purchase Pacific salmon, 8 to 12 lb, for slicing or baking
- *Cod:* the best weigh 5 to 8 lb; scrod is a young cod weighing 1½ to 2½ lb; many fish, including haddock, and cusk, are classed as cod; finnan haddie is lightly smoked and salted cod
- *Sole:* member of the flounder family; lemon sole, petrale, and Dover sole (from the Pacific—not true Dover sole) are all sold as sole
- *Mackerel:* fish weighing ¾ to 1½ lb; quite fatty and often broiled
- *Snapper:* red snapper comes from the Gulf and southern waters; other species are caught in the Atlantic and Pacific; should weigh 1½ to 4 lb

- *Trout:* purchase Idaho trout; should weigh about ¾ lb
- *Halibut:* delicate, lean, white-fleshed fish; best ones come from small fish weighing 12 to 20 lb
- *Yellow perch:* freshwater fish caught in considerable numbers in Canada and shipped to this country; fillets weigh 2 to 3 oz
- *Whitefish:* fat fish caught in the Great Lakes; prized as a delicacy
- *Bullhead:* fish now grown in considerable quantities in some southern states

Shellfish

- *Crab:* blue crab comes from north Atlantic and Gulf waters, dungeness from Puget Sound and Pacific waters, king from Alaskan waters, and stone from southern waters of the Atlantic; the body of blue crab is eaten (soft shell crab are blue crab that are molting); only the legs of king and stone crab are eaten; and the legs and body of the dungeness are eaten
- *Crayfish:* small lobsterlike shellfish; they run about 8 to 12 per lb
- *Clams:* mollusks; littleneck and cherrystone clams are eaten raw; most others are eaten cooked, either deep-fried or in chowders
- *Lobster:* true lobster have claws; most come from New England and Maritime Province waters; laguna lobster caught off California is a spiny lobster without claws; chicken lobster weighs just under 1 lb
- *Lobster tails:* spiny lobster with the head removed; most are imported
- *Oysters:* Bluepoints and Chincoteagues from Maryland and Chesapeake Bay waters are well known; Pacific oysters are the large Japanese variety and the tiny prized Olympias
- *Scallops:* mollusks whose inner muscle only is eaten; bay scallops are considered superior to sea scallops; New Bedford scallops are sea scallops of high quality
- *Shrimp:* a small crustacean of the lagoste family; the best come from Gulf and Caribbean waters; sold by count per pound; pink and white shrimp varieties are considered best, but brown and red are marketed; some aquaculture shrimp are available from Taiwan; if purchased glazed, glazing should be only 6 percent of total weight, but it can run 20 percent or more, raising the cost considerably

Market Forms

Meats are available in carcass, wholesale-cut, or portion-cut form. Much beef is sold in boxed form, which is meat that has been boned and wrapped in air-exhausted plastic. Some beef is also now aged in air-exhausted plastic rather than by dry aging. Pork is seldom purchased in the carcass. Most operations purchase legs (hams) and loins as wholesale cuts, but most pork is actually used in the portion-cut form. Cooked and canned products are also available. Most foodservices prefer to buy unfrozen meat, but occasionally it may be desirable to purchase it frozen.

Poultry is available whole, cut up, in steaks, or in cutlets. Some rolls (flesh that is boned, rolled, and tied) are available. Some turkey and chicken today are processed so as to imitate ham or to fill wieners or other sausages.

Market forms of fish are given in Table 15-1. Most foodservices prefer to purchase fresh fish, but the quality of frozen fish has improved so much that more operations are purchasing fish in this form. For purposes of recognition, whole fresh fish may be categorized as either flat or round (see Fig. 15-1). The process of preparing a fresh fish for cooking is outlined in Table 15-2.

Shellfish such as crab, lobster, mussels, oysters, and clams are often sold alive. Live mollusks can be identified by their tightly closed shells; if the shell is gaping open, the animal is dead. Crab and lobster show more active signs of life. A *weak* is a lobster that is dying. A *chicken lobster* is one that weighs about 1 lb. Clams, mussels, and oysters may be sold shucked. Shucked clams and oysters should not have more than 8 percent free liquid on them. Some shrimp is marketed alive but most is frozen—often, headless but unshelled. PDQ (peeled, deveined, and quick-frozen)

TABLE 15-1   Market Forms of Fish

Form	Definition	Edible Portion (%)	Best Ways to Cook
Fin fish (whole or round, fresh)	Fish just as they were caught. They must be cleaned (by taking the insides out), and the scales must be taken off before they can be cooked. The head, tail, and fins may also be cut off before cooking.	45	Bake, poach, broil, fry, steam
Drawn (fresh or frozen)	A drawn fish has been cleaned. It must be scaled and have the fins cut off before it can be cooked.	48	Bake, poach, broil,
Dressed or pan-dressed *a* (fresh or frozen	A dressed fish has been cleaned and scaled. The head, tail, and fins are usually cut off, too.	67	Bake, poach, broil, fry, steam
Steaks (fresh or frozen)	Steaks are slices of fish about ¾ to 1 in. thick. They have been cut across the fish. A part of the backbone is the only bone in a fish steak. Steaks are ready to cook.	84	Bake, poach, broil, fry, steam
Fillets (fresh or frozen)	Fillets are the sides of a fish. They have been cut away from the backbone and are ready to cook.	100	Bake, poach, broil, fry, steam
Butterfly fillets (fresh or frozen)	Butterfly fillets are the two sides of the fish cut lengthwise away from the backbone and held together by the uncut flesh and skin of the belly.	100	Bake, poach, broil, fry, steam
Single fillets (fresh or frozen)	A fillet cut from one side of a fish is called a *single fillet*. This is the type most generally available on the market. The fillets may or may not be skinless.	100	Bake, poach, broil, fry, steam

TABLE 15-1 Continued

Form	Definition	Edible Portion (%)	Best Ways to Cook
Chunks (fresh or frozen)	Chunks are pieces cut across the fish. They are larger than steaks and are ready to cook.	100	Bake, poach, broil, fry, steam
Breaded portions (frozen)	Portions are pieces of fish that have been cut from blocks of frozen fillets. They are covered with batter and all of the portions in a package are the same size. Portions sold raw must be cooked. Precooked portions are cooked only a little bit more before being served. The package label should be read for cooking instructions.	100	Ovenize, deep-fry
Fish sticks (frozen)	Sticks are made the same way portions are, but sticks are smaller. All sticks are cooked a little bit before being frozen; cooking must be completed before they can be served.	100	Ovenize, deep-fry

*Pan-dressed is smaller size to fit pan for cooking.

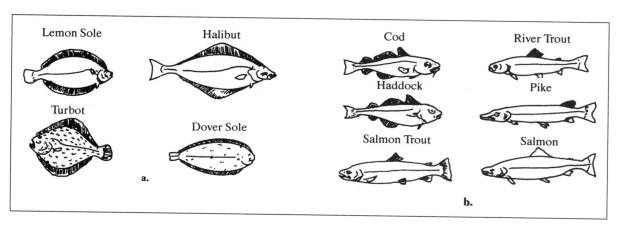

FIGURE 15-1 Categories of fish: (*a*) flat fish; (*b*) round fish.

TABLE 15-2   Preparing Whole Fresh Fish for Cooking

Process		*Procedure*
Scaling		Wash the fish. Place the fish on a cutting board and hold the fish firmly by the head with one hand. Holding a knife almost vertical, scrape off the scales, starting at the tail and scraping toward the head. Be sure to remove all the scales around the fins and head.
Cleaning		With a sharp knife, cut the entire length of the belly, from the vent to the head. Remove the intestines. Next, cut around the pelvic fins, and remove them.
Removing the pectoral fin		Remove the pectoral fins by cutting just in back of the collarbone. If the backbone is large, cut down to it on each side of the fish.
Removing the head and tail		Then place the fish on the edge of the cutting board so that the head hangs over, and snap the backbone by bending the head down. Cut any remaining flesh that holds the head to the body. Cut off the tail.
Removing the dorsal fin		Next remove the dorsal fin (the large fin on the back of the fish) by cutting along each side of the fin. Then give a quick pull forward toward the head, and remove the fin with the root bones attached. Never trim the fins off with shears or a knife because this will leave the root bones at the base of the fins in the fish. Wash the fish thoroughly in cold running water. The fish is now dressed or pan-dressed, depending on its size.
Cutting steaks		Large dressed fish may be cut crosswise into steaks, about 1 in. thick each.
Filleting		With a sharp knife cut along the back of the fish from the tail to the head. Then cut down to the backbone just in back of the collarbone.

TABLE 15-2   Continued

Process	Procedure
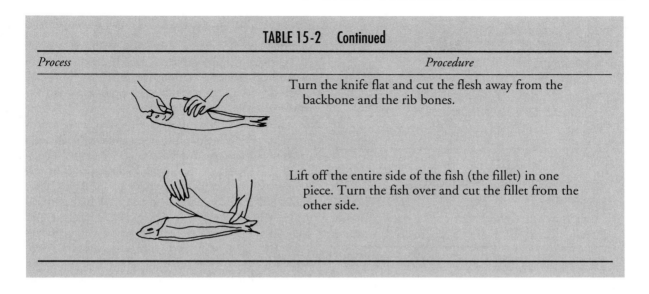	Turn the knife flat and cut the flesh away from the backbone and the rib bones.
	Lift off the entire side of the fish (the fillet) in one piece. Turn the fish over and cut the fillet from the other side.

shrimp are available. Cooked shrimp, either in the shell or peeled and deveined can be purchased as well. King crab legs are often cooked and frozen before sale, and picked crab meat is also available.

Today, many meats, poultry, fish, and seafood items are available in a highly prepared state for cooking or are even cooked so they need only be rewarmed to be ready for service. Thus, many raw breaded items such as veal cutlets, chicken breasts, shrimp, and fish only need cooking. Cooked and deveined shrimp are available. Kinds of cooked roasts, steaks, stews, poultry, fish, and seafood abound. All of these items come either chilled, cook–chilled, or frozen. When these products are purchased, food costs go up and labor costs should go down.

Grading

Quality grading of beef, veal, calf, lamb, and mutton is much the same. Grades are Prime, Choice, Select, Standard, and Commercial for beef; they are Prime, Choice, and Select for calf, veal, lamb, and mutton (see Fig. 15-2). Other grades are available, but foodservices seldom purchase these (see Table 15-3). Pork is graded U.S. No. 1, 2, or 3. Bulls and cows cannot make the Prime grade; bullocks (young bulls) can, but they are graded on a slightly different basis from the one on which steers and heifers make Prime grade. Mutton cannot make Prime. Sows cannot make U.S. No. 1 for pork. Some foodservices find that pork graded U.S. No. 2 is as satisfactory as pork graded U.S. No. 1.

Poultry is graded U.S. Grade A, B, or C. Few operations purchase anything other than U.S. Grade A. Fish and shellfish are not graded except in some processed forms. The grades for the processed product are U.S. Grade A, B, and C. Buyers must therefore rely on their own judgment to identify high-quality seafood. The best insurance for obtaining good seafood is to pick a good, honest purveyor. Most of the quality characteristics of fish depend on

FIGURE 15-2 USDA grade mark for highest-quality meats.

		Lamb and Yearling Mutton	
Beef	Veal and Calf		Mutton
Prime	Prime	Prime	Choice
Choice	Choice	Choice	Good
Select	Good	Good	Utility
Standard	Standard	Utility	Cull
Commercial	Utility	Cull	
Utility			
Cutter			
Canner			

TABLE 15-3 USDA Quality Grading for Meats

freshness. To check quality, look for the following characteristics:

- *Flesh:* firm, not separating from the bones.
- *Odor:* fresh and mild. Fish from the water has practically no fishy odor, but the odor is more pronounced with time. Fish and shellfish should not have a strong odor nor any ammonia smell. Open fish that has been eviscerated should not have a fishy smell.
- *Eyes:* bright, clear, and full. As fish stales, the eyes cloud and turn pink. Eyes protrude on fresh fish, but sink as the fish ages.
- *Gills:* red, bright, and free from slime. Gill color fades with age to a light pink, to gray, and then to a brownish or greenish color.
- *Skin:* shiny, with color unfaded. Iridescence may be present when fish are first caught. Look for pronounced markings and colors. Fresh fish have scales that are tight and difficult to remove.

Frozen fish should be solidly frozen when received. Some may be water-sprayed to prevent dehydration. Such spray should not be over 6 percent of total weight. Good storage temperatures for fish are 0°F (−18°C) or below. Frozen fish should have the following characteristics:

- *Flesh:* no discoloration or freezer burn. If fish is thawed and refrozen, quality is diminished. Brownish edges indicate thawing and refreezing.

- *Odor:* little or no odor. A strong odor usually signals poor quality.

Most frozen fish is wrapped, either individually or in pieces or packages of various weights. Wrappings are composed of moisture- and vapor-proof material, air-exhausted.

Beef, lamb, and mutton are given yield grades based on the ratio of flesh to bone and fat. No. 1 yield indicates highest yield, while No. 5 represents lowest (see Table 15-4). Each yield grade gives approximately 4.6 percent more edible retail cuts than the next lower grade. Prime grades cannot qualify for highest yield grades, since too much fat is required in achieving so high a quality. Foodservices do not find yield grades as helpful as do supermarkets and other buyers. Consequently, purchase is usually by IMPS (Institutional Meat Purchase Specifications) number, which limits the amount of surface fat, specifies the cut size, and so on. The IMPS also allow a foodservice to specify the range from which it wants the carcass or cut to come. For example, there are five ranges for beef, covering carcass weights of 500 to 1000 lb and over. Indicating the desired range of the cut—say, of a No. 109 oven-ready rib—thus indicates the desired weight, as well.

IMPS numbers are standard specifications used nationally to indicate what portion of the carcass is wanted. A No. 103 rib, for example, is a full rib just as it is cut from the carcass; but a No. 109 rib excludes the lower parts of the rib (shortribs), the chine bone, the upper muscle over the chine bone, and other parts, and the remaining rib must be ready to be put into the oven. Purchasing by IMPS number simplifies ordering and ensures getting what is needed. Beef cuts have

TABLE 15-4 Yield Grades for Gauging the Relative Amount of Edible Meat on a Cut or Carcass

Species	Yield Grades
Beef	1–5
Lamb, and yearling mutton	1–5
Mutton	1–5
Veal and calf	Not applicable
Barrows and gilts	Not applicable
Sows	Not applicable

numbers in the 100 range; lamb and mutton, the 200 range; veal and calf, the 300 range; and pork, the 400 range. Portion cuts are numbers in the 1000s. Thus, a lamb loin chop would be No. 1232.

Quality meat grades for beef, veal, calf, lamb, and mutton are based on the conformation or yield of the meat as judged by the size of the rib eye and other conformation factors, the amount of flesh marbling, the color of the lean, the texture and appearance of the flesh, age, and so on. Hog grades are based on carcass weight, depth of fat back, fullness of muscles, and carcass length. Table 15-5 lists quality factors related to color and texture. Poultry is graded on

the basis of conformation, fat cover, freedom from defects, freedom from pin or vestigial feathers, and so on. Again, seafood is usually ungraded, except for certain processed items.

Cuts and Parts

Beef is available in carcass, side, quarter, wholesale, or portion cuts. Wholesale cuts include round, sirloin (loin end), short loin, rib, flank, plate, brisket, shank, chuck, and neck (see Fig. 15-3). Calf, veal, lamb, and mutton are sold by carcass or by breast, shoulder, rib, leg, or portion cuts (see Figs. 15-4 and 15-5). If ribs,

TABLE 15-5   Meat Quality Factors: Color and Texture

Animal	High Quality	Poor Quality
Beef		
Lean	Bright red after cutting, and smooth-grained	Darker red, rough-grained
Fat	Creamy white, firm, and brittle	Yellow or orange, scant, and unevenly distributed
Bone (backbone)	Red, soft, and spongy, with abundant cartilage	White, hard, and brittle
Veal		
Lean	Grayish pink, fine-grained, and smooth	Pale or dark
Fat	Grayish or pinkish white, firm, and brittle	Little or none present
Bone	Red, soft, and spongy, with abundant cartilage	
Lamb		
Lean	Pinkish red, fine-grained, and smooth	Darker and stronger-flavored
Fat	White or pinkish, firm, flaky, and brittle	Heavier layers
Bone	Red, soft, and spongy, with abundant cartilage	
Pork		
Lean	Grayish pink and fine-grained	Darker and coarser
Fat	White, very firm, but not brittle	Excess both internally in lean and externally
Bone	Red, soft, and spongy	Less red

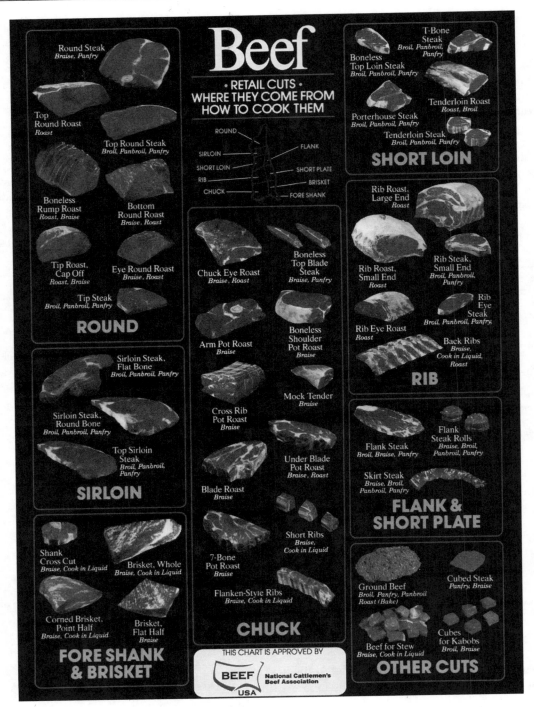

FIGURE 15-3 Wholesale and retail cuts of beef.
(Reproduction courtesy of the National Cattlemen's Beef Association, © 1986.)

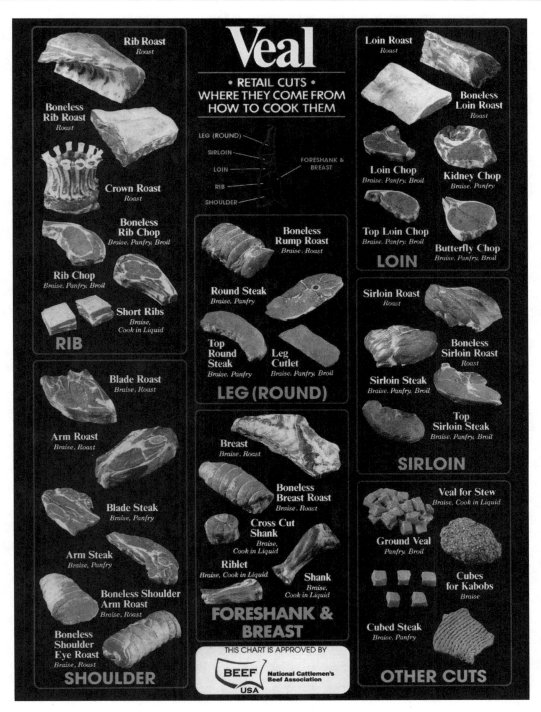

FIGURE 15-4 Wholesale and retail cuts of veal.
(Reproduction courtesy of the National Cattlemen's Beef Association, © 1986.)

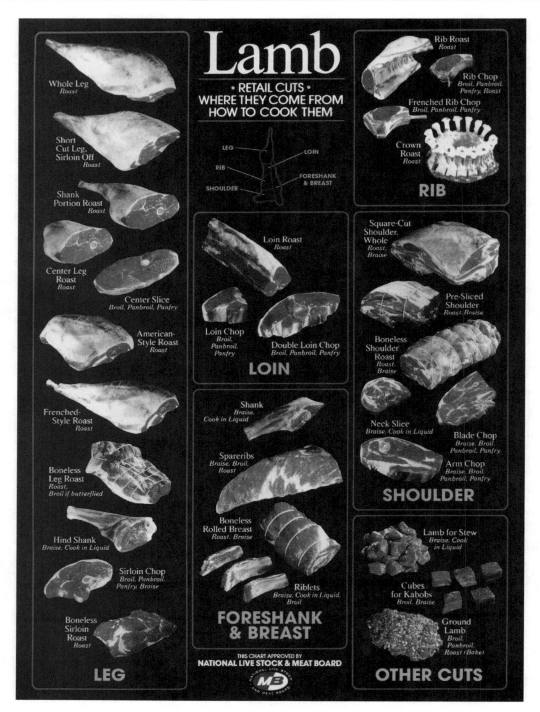

FIGURE 15-5 Wholesale and retail cuts of lamb.
(Reproduction courtesy of the National Cattlemen's Beef Association, © 1986.)

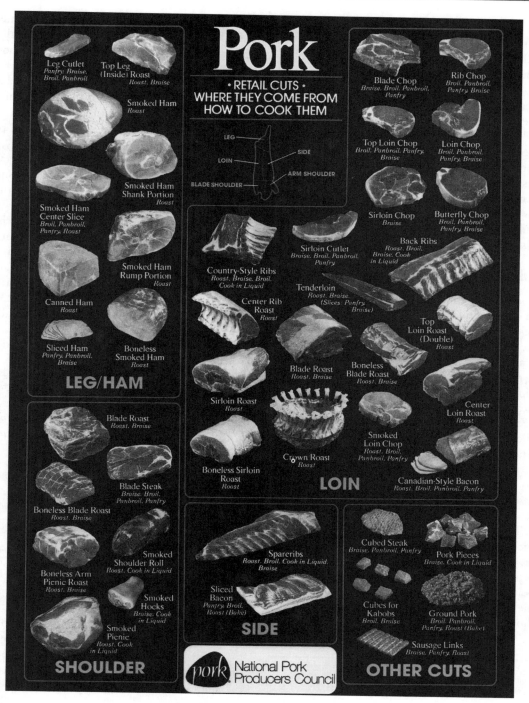

FIGURE 15-6 Wholesale and retail cuts of pork.
(Reproduction courtesy of the National Pork Producers Council, © 1986.)

legs, or other cuts of these meats are purchased, they may come double; that is, they may be joined by the spine (chine) bone into left and right units. Large calves are sometimes cut much as beef is. Pork is purchased as picnics, shoulders (Boston butts), loins, spareribs, side pork (bacon), legs, and portion cuts (see Fig. 15-6). Some few operations may purchase pork jowls, heads, ears, tails, or hocks. Variety meats such as kidney, liver, and sweetbreads from all animals are available.

Poultry is available whole, by sides, in quarters, or in parts; ground turkey or chicken is also available. Chicken parts are breasts (boneless or bone-in, and skinless or skin on), legs (drumsticks and thighs), drumsticks, thighs, wings, and backs. Livers, hearts, gizzards, and necks are also available.

Inspection for Wholesomeness

Inspection considers the fitness of meat and poultry for human consumption; grading considers its palatability and yield. Meat and poultry inspection is required under law in every plant. Animals must be inspected before and after slaughter. The packinghouse or processing plant must be sanitary and meet rigid standards in equipment, procedures, and other categories. Inspection is carried out under municipal, state, or federal auspices. State and local standards must equal or exceed federal ones.

Grading is not mandatory, but inspection for wholesomeness is. Some operations purchase meats and poultry by brand name. Pork is especially often purchased in nongraded form. Meat quality and yield grades are indicated by the grade appearing in the shield. The stamp indicating wholesomeness is a circle in which the words "Inspected and Passed, USDA" appear. A number is also included to identify the particular processing plant. By this means the place where the meat was inspected and passed can be traced. Under the shield on a grade stamp, the initials of the individual who graded the meat may appear.

■

RECEIVING, STORING, AND ISSUING

Flesh products are usually the most expensive food items purchased by a foodservice, and receivers must make sure that the proper quantity and quality are delivered. Temperatures of refrigerated items should not be higher than 40°F (4.5°C), and frozen items should be solidly frozen. All items should be clean, sanitary, and well-packaged or well-wrapped. If a receiver is unable to identify grade or other quality factors, someone in the organization should be called to assess these factors on what has been delivered. Meat, seafood, and poultry receipts are often tagged with the date of delivery, grade, weight, and other delivery data on the tag so FIFO procedures can be followed. When the product is issued, the tag is removed and sent to the accounting department to indicate that it is no longer in inventory.

After delivery, items must be stored in such a way as to prevent waste through pilferage, deterioration, or infestation. Older stocks should be moved to the front, so new deliveries can be stored behind and used last. Refrigerated products should be held at about 35°F (1.6°C) and frozen items at −10°F (−23°C) or lower. Storage areas should be clean, with all stock stored off the floor. Table 15-6 gives some recommended holding times for meat storage. Poultry can

TABLE 15-6   Holding Times for Meat Storage

Meat	Holding Period
Chilled	
Solid, fresh cuts of beef, pork, lamb, and veal	2–4 days
Ground items and variety meats (liver, etc.)	1–2 days
Most cooked, processed meat items, and cooked leftovers	3–7 days
Frozen	
Beef	6–12 months
Lamb and veal	6–9 months
Pork	3–6 months
Ground beef	3–4 months
Ground lamb, ground veal, and variety meats	3–4 months
Ground pork	1–3 months

be refrigerated for about three days and held frozen about six months. Fish may be held refrigerated for about three days. Fat fish (such as mackerel, salmon, and tuna) may be held frozen for about three months, while lean fish (such as haddock, cod, and swordfish) may be held frozen for about six months.

As noted, the issuing of meats, seafood, and poultry is simplified if they are tagged when received. Tagged items need not be reweighed or otherwise identified when they are issued, and tags are also helpful in reordering. The bottom half of the tag is returned to the purchasing agent when the item is issued, thus facilitating inventory control. All items issued should be recorded on properly signed requisitions that are sent to food and beverage accounting (along with records of direct deliveries) for compilation of the food cost of the day.

■

COOKING PRINCIPLES

Cooking meat, poultry, or fish makes it more palatable to most people, changing its texture, flavor, and appearance. Cooking also sterilizes the product.

Color

The red color of flesh comes from a red pigment called myoglobin. Beef, lamb, and mutton flesh are high in it; veal and pork have very little. Chicken has some, especially in the legs, but fish has almost none. Myoglobin changes to hematin when heat strikes it. This accounts for the color change we see in meat when it goes from uncooked to well done. Some red meats become quite red when exposed to air because the myoglobin oxidizes and changes into oxymyoglobin. If myoglobin joins with nitrogen to form nitrosomyoglobin, it will not change to a done color when cooked but will remain red, as happens with cured ham or in cured brisket of beef. A ham loaf containing ham, veal, and pork is red because of the nitrogen in the salts used to cure the ham. Onions dried in inert nitrogen gas or root vegetables grown in a highly nitrogenized soil may contain enough nitrogen to turn meat red. Sometimes, enough nitrogen curing salts get rubbed onto a cutting block to contaminate meat and make it

look red. This can happen if ham steaks are cut, and then the meat block is not cleaned before some braising steaks are cut. The braising steaks become red permanently and do not look done when cooked. The color change to a brownish color when meat spoils is due to a change of myoglobin into metamyoglobin.

Doneness

When flesh reaches about 160°F (71°C), its color begins to change and the protein coagulates; coagulation is complete at around 175°F (78.6°C). Just before coagulation, denaturation (a moisture loss) occurs. The higher the cooking temperature is and the longer the flesh is exposed to it, the greater the moisture loss is. This loss is called *shrinkage*. The degree of doneness often dictates the amount of shrinkage (see Table 15-7). Thus, flesh cooked rare may shrink 10 percent; flesh cooked medium done, 15 percent; and flesh cooked well done, 25 percent. High heat does not seal in the juices and reduce shrinkage. A constant low temperature is best for cooking flesh foods.

High heat develops browning in meat and an accompanying flavor change—a sort of caramelization. Salt delays browning; and since its penetration is small (from ¼ to ½ inch) it is frequently omitted until service.

Flesh Composition

Flesh is about 25 percent solids and 75 percent moisture. Of the solids, 20 percent consists of fat, fatlike compounds, ash, and other substances; the remaining 80 percent is protein. Fatty tissue is from 15 to 50 percent moisture. This is why we lay fatty meat over meat during cooking to help baste it or add moisture. We also lard meat to add moisture. Rib and strip sirloin roasts are roasted fat-side-up so that, when the fat melts in cooking, the released moisture bastes the meat.

Flesh is bound together by connective tissue that winds through the meat fibers and holds them together. They also join together and make a tendon that attaches to the bone, holding a muscle in place. There are two main kinds of connective tissue: collagen, or white connective tissue; and elastin, or yellow connective tissue. Collagen changes into gelatin and water when exposed to moist heat. Acids speed the

TABLE 15-7   Stages of Doneness in Flesh

Temperature			
°F	°C	Doneness	Description
140	60	Rare	Flesh is quite red; there is a good brown color on the outside, with some pink next to it, and then redness. Appearance is full and plump, and the flesh yields under pressure; juices are red but not bloody.
160	71	Medium	Interior is a modified rose; pink juices are apparent but less plentiful than in rare; exterior is well browned; surface is not as plump or full; there is definite resistance to pressure.
175	78.6	Well	Meat is completely cooked; little or no juice appears; flesh is hard and flinty to the touch and has a shrunken appearance; surface is brown and dry.

transformation, and that is why we add tomato juice or other acidic liquids to some meat that is to be braised. Elastin is not changed by heat. It must be broken up by grinding, pounding, cubing, or some other mechanical treatment.

The lean flesh of meat largely consists of meat fibers—thin, long, tiny rods that contain moisture, salts, vitamins, fats, and flavor. Fine, silky fibers are a sign of tender, moist meat. Coarse, dry ones indicate meat that may be tough when cooked. As meat cooks, the fibers shrink and lose moisture. The fibers also pack closer together, and this results in a drier, tougher piece of meat. The longer the heat and the higher the temperature, the more the fibers pack together and toughen. We can see this happen when we compare the tenderness of a rare or medium-done steak with that of a well-done steak. Flesh from the tougher cuts is cooked at low braising, simmering, or other moist-heat low temperatures to allow the collagen to dissolve slowly and leave the meat fibers unbound. Although the individual fibers are themselves tough, they are so tiny and separated, the meat is tender. Seafood flesh, having little connective tissue, is usually always tender. However, overcooking can dry out the flesh and leave it dry because of the loss of moisture.

When the protein of eggs and high-protein dairy products is overheated or cooked too long, the protein hardens and bunches together, expressing moisture

(in cheese, fat is also sometimes lost). Pure egg retains its moisture but hardens, but in a custard the egg protein shrinks, losing moisture, and the solid custard now is broken with free moisture in it. Lightly cooked custards are solid. Lightly cooked cheese is soft and spreadable.

Most meat proteins are soluble in cold water, but coagulation makes them insoluble. We can see soluble proteins coagulating when scum floats to the surface of boiling meat.

Tenderness

Age is perhaps the greatest factor in making meat tough. As an animal ages, its connective tissue content increases, the fibers get coarser, and the moisture content of the flesh decreases. Muscles that receive a lot of exercise are tougher than ones that receive little exercise. Thus, in beef, the flesh of the leg is tougher than the flesh of the rib or loin area. Flesh from the rarely exercised muscles of a young animal is apt to be exceedingly tender. The composition of poultry flesh is much like that of meat, but fish contains little connective tissue and therefore is not tough.

Flesh can be tenderized by breaking up the connective tissue or by cooking it in moist heat. Actually, flesh is most tender when it is uncooked; rare flesh is next in tenderness, medium done is next, and well done is least tender. We can also use an enzyme such as papain to digest connective tissue and make the flesh more tender. This is frequently accomplished by injecting an animal with a papain solution just before slaughter, allowing the heart to pump the papain throughout the tissues, and then slaughtering the animal. When the meat from such an animal is cooked, the papain goes to work digesting some of the connective tissue. We can also dip a cut of meat into a papain product, but this is not as effective. Making meat into what are called *formed products* is another way to tenderize meat. First the meat is frozen and thinly shaved; then it is thawed and mixed well for about 8 minutes with a bit of salt to draw out collagen from the meat. The meat can then be shaped—into steak form, for example—frozen in this shape, and cut into steak cuts. The shaving of the meat cuts through the connective tissue so that it is no longer tendinous, making the flesh tender. Putting meat into a marinade can make the meat tenderer, but only slightly so. Heavy electric shocks on flesh can shatter connective tissue and thereby tenderize it.

Aging flesh increases its tenderness, as well as adding moistness and flavor. Enzymes in flesh work slowly to break down connective tissue. Flesh is dry-aged by being held at 35 to 45°F (1.6 to 4.4°C). Aging can also occur when meat is placed into air-exhausted plastic wraps and left to stand at a temperature of about 35°F (1.6°C). Flesh aged in this way needs no fat cover to protect it from molds and bacteria. Aging for about 14 days is usual, but flesh can be aged for 21 days with good results. After 21 days, the aging benefits drop significantly. Sometimes the process of aging meat is called *ripening*.

Flavor

The flavor of flesh comes from nitrogenous and non-nitrogenous extractives and waste products, plus aromatic, essential oils. The oils are extremely important in giving meat flavor. Flavor develops in flesh during cooking for up to about 3 hours; then it declines.

The animal's diet influences its flavor because many essential oils the fat contains come from the foods it eats. A milk–mash diet for chickens produces mildly flavored flesh. Cattle fattened on distiller's mash have a sharply acidic fermented-flavored meat. Hogs fed on peanuts—as is done with some Virginia pork—yield flesh that has a slightly peanutty flavor. Wild ducks feeding in sea marshes acquire a fishy flavor, but those feeding on wild rice or on grains from farmers' fields in the Midwest develop a marvelous flavor.

A large proportion of the essential oils in flesh can be found in the fat. This is because these oils are soluble in fat but not in water. Fat appears as surface fat, as marbling (tiny flecks of fat between fibers), and as fats or fatlike substances in meat juices. Thus, the most flavorful meat is that having some fat.

Poultry—except, at times, duck—is usually cooked well done. Pork is always cooked well done to destroy any trichinae that may be present. Seafood is *just* cooked to doneness; overcooking leaves it dry and toughened.

Cooking Methods

Meat, poultry, and seafood are cooked by the following dry-heat and moist-heat methods:

Dry Heat
- Broiling
- Roasting or baking
- Barbecuing
- Sautéing, pan-frying, or grilling
- Deep-frying
- Ovenizing

Moist Heat
- Braising (pot-roasting, fricasséeing, casseroling, swissing, or stewing)
- Simmering (poaching or stewing)
- Steaming
- Blanching

The cuts of meat for which each of these cooking methods is appropriate are listed in Table 15-8.

Cooking by radiant heat is called *broiling*. The term *grilling* once meant the same thing as *broiling*, but now it means sautéing or frying. Only tender flesh

TABLE 15-8 Preferred Cooking Methods for Various Cuts of Meat

Beef	Lamb	Fresh Pork	Smoked Pork	Veal	Variety Meats
Roasting					
Rib	Leg	Ham (leg)	Ham	Leg (round)	Liver
Sirloin butt	Loin	Loin	Picnic	Loin	
Loin strip	Rack	Boston butt	Shoulder butt	Rib	
Tenderloin	Rolled shoulder	Picnic	Canadian-style bacon	Rolled shoulder	
Rump butt[a]	Cushion shoulder	Rolled shoulder	Loaf (ground ham)	Loaf (ground veal)	
Round[a]	Loaf (ground lamb)	Cushion shoulder			
Inside (top)		Spareribs			
Outside (bottom)					
Knuckle (tip)					
Inside chuck[a]					
Shoulder clod[a]					
Loaf (ground beef)					
Broiling, Griddle-Broiling, or Pan-Broiling					
Rib steak	Shoulder chops	Fresh pork is not broiled or griddle-broiled	Sliced ham	Veal is not broiled unless it is fairly mature and well-fatted, and then only loin chops or steaks	Sliced veal or lamb liver
Club steak	Rib chops		Sliced bacon		Lamb kidneys
T-bone steak	Loin chops		Sliced Canadian-style bacon		Brains
Porterhouse steak	Leg steaks		Shoulder butt		Sweetbreads
Sirloin steak (cut from loin strip)	Lamb patties (ground lamb)				
Butt steak					
Tip steak (knuckle)					
Inside chuck steak[a]					
Top round steak[a]					
Beef patties (hamburger)					
Flank[a]					

Braising

Chuck (all cuts)	Breast	Chops	Breast	Heart (all kinds)
Brisket	Neck	Loin	Chops	Liver
Plate	Shank	Rib	Loin	Beef
Flank	Shoulder	Shoulder	Rib	Pork
Neck		Feet	Shoulder	Kidney
Outside (bottom) round		Hocks	Cutlets (leg)	Tripe
Heel of round		Spareribs	Neck	
Rump butt		Fresh ham steaks	Shoulder	
Shank		Tenderloin	Shank	
Short ribs				
Skirt steak				
Ox tails (joints)				

Simmering

Fresh	Neck	Spareribs	Ham	Kidney
Chuck (all cuts)	Breast	Backbones	Shoulder	Heart
Rump butt	Flank	Pigs feet	Picnic	Tongue
Shank	Shanks	Hocks	Shoulder butt	For precooking:
Heel of round	(Large cuts of lamb		Hocks	Brains
Brisket	are usually			Sweetbreads
Plate	not cooked		Neck	
Flank	in water)		Shank	
Short ribs			Breast	
Corned			Flank	
Brisket			(Large cuts of veal	
Rump			are usually not	
Plate			cooked in water)	
Round				

ªUsually, only Prime to Choice quality in these should be cooked in this manner.

is broiled. To broil well, the flesh should have some fat content. Lean sole or veal does not broil well. Leaner items should be broiled a bit faster and basted with butter or other fat. Thin pieces should be positioned close to the broiler. The broiler temperature should be lowered for cooking thicker items such as turkey breasts and chateaubrands. Too high a temperature produces a hard, dry crust that slows heat penetration, giving a charred outside and an undercooked inside.

To broil, dip the item into oil and place it gently on the grid. Cook to half doneness, and then turn and season. For larger pieces, two turns may be needed. Otherwise, the two sides can be completely cooked with one turn. Do not allow drippings to flame; the smoke and soot can harm flavor. Grease fires are dangerous. Put fires out with water. Place fish, shellfish, and small pieces of meat or poultry onto grids that hold the product firmly and allow turning without requiring the product itself to be lifted up and flipped.

Pan-broiling is done in fairly heavy skillets suited to the quantity, so drippings do not burn. Rub the pan bottom lightly with fat. Preheat the pan almost to smoking. Cook the item on one side and then on the other. Pour off fat as it gathers. Do not cover and do not add water. To griddle-broil, place the item onto a lightly greased 325°F (161°C) griddle, and cook with no fat.

Items for roasting should fill the pan and leave no space for drippings to burn. A trivet or perforated underliner is placed under the roast so that drippings can collect underneath, away from the flesh. A mirepoix may be introduced to give added flavor. Salt is not added until browning is complete. Do not pierce the flesh of a roasting or broiling item; instead, roll it over or insert the fork into fatty tissue. Use no liquids. Place the fat side of the item up; this means that poultry should be roasted with its breast side down until the last stage of roasting. Among seafood, only fat fish roasts or bakes well. Use thermometers to monitor the temperature for correctness. Low-heat roasting gives a tenderer and more flavorful product. Cook large pieces to just under their doneness temperature (about 10 to 25°F under it), because the heat inside the item will continue to cook it after withdrawal from the oven (see Fig. 15-7). Foodservices commonly have special ovens for roasting.

Barbecuing involves roasting or broiling meat while basting it with a tangy sauce. The sauce is more clearly the indicator of the method than the cooking technique. Barbecued items are often cooked before, under, or beside an open fire or in a covered pit.

Frying is cooking in a thin layer of fat, the same technique that is used in sautéing. Pan-frying involves cooking in shallow fat, and grilling is the same except that the item is cooked on a griddle.

Deep-fat cooking is a method of cooking foods by immersing them in hot fat. The quality of the finished product depends on the frying temperature, the quality of the fat, and the quality of the food item itself. Many breaded or battered meat, poultry, or fish items are deep-fried. Lean fish responds especially well

FIGURE 15-7 These roasts have been slightly undercooked as the correct doneness will be reached as they rest out of the oven.

to deep-frying because it lacks fat, so the enriching by fat during frying helps give it palatability.

Ovenizing consists of placing items onto well-greased baking sheets or pans and then putting them into an oven to fry or sauté. The items should be drizzled with fat after being placed on well-greased pans. Oven temperatures should be 325 to 425°F (161 to 212°C). Ovenizing is most appropriate when large quantities are needed.

Flesh foods can be cooked by any of a number of moist-heat methods in which the quantity of moistness is controlled. Sometimes just enough moisture is used to tenderize the collagen, while in other cases the amount of moisture is almost enough to constitute stewing. Braising may also be called *stewing, pot-roasting, fricasséeing*, or *swissing*. Often a product is sautéed before it is braised. If this occurs, the product is called *brun;* if not, the product is called *blond*. Alternatively, marinating may occur before braising. If a flesh food is served in an earthenware casserole without vegetables, it is listed on the menu as *en casserole*. If it is served with vegetables, it is described as *en cocotte*. *Jugged* means braised in a pot. The word *poêler* means to brown well and then to braise in juices while basting with butter; the temperatures used in this process are high.

In browning, containers should be well filled with meat so that moisture is built up inside. Add fat to cover the bottom, and heat; then add the product. Brown if the item is to be *brun*. Add moisture if needed, and add vegetables if desired. Simmer. When it is done, remove the meat and thicken the gravy. Put all together again, heat, and serve.

Many items in quantity are best braised in shallow steamed-jacketed kettles. To do this, heat the kettle with its cover down and steam on full; add some melted fat, and drop in evenly sized pieces of the meat or poultry (fish breaks up too much when treated this way). Turn the pieces every 15 to 25 minutes. Then add a small quantity of water or liquid, cover, reduce heat, and simmer.

Cooking items in water at just below the boiling point is called *simmering* or *poaching*. Few flesh foods are boiled, since this tends to toughen them. Simmering gives a more tender, moist, and flavorful product. To simmer, bring the water to a boil, add the product, and maintain the temperature at between 185 and 205°F (85 and 94.6°C). Test for doneness by inserting a fork about ½ inch deep and twisting. Do not over-cook, as this gives a dry, shreddy meat. Skim and add liquid as necessary. Add vegetables 30 minutes before serving, or cook them separately and add at service. Meat may be sliced and kept warm in stock.

Meats, poultry, or fish may be steamed over rich stock, a process called *free-venting*. Steam pressure cooking reduces steaming time, making it easy to over-cook products. Many items are given preliminary pre-browning and then are finished in a steam chamber. A court bouillon (flavored fish stock) is used for steaming fish; the aromatic steam rises and flavors the fish.

Dropping products momentarily into boiling or simmering water to give them partial cooking is called *blanching*. It is more a preparation than a method of cooking. Some cooks presoak items in cold, salted water before blanching them. Thus, sweetbreads would be soaked in this manner, and then their outer membranes would be removed, before the actual cooking began. Brains and sweetbreads are blanched in slightly acidulated water to whiten them.

Frozen Meat Cookery

Meats, fish, and poultry are frozen rapidly to produce small crystal growth and to protect the meat fibers from rupture. Slow freezing causes a breakdown of the fibers and a greater loss of drip upon thawing. Frozen items can generally be cooked in the same ways as non-frozen ones. Better flavor and higher nutrient values result if meat is cooked directly from the frozen or partially frozen state rather than from the thawed state, but sometimes this is not possible. Partial thawing, followed by cooking, reduces cooking time. Hard-frozen products take about three times longer to roast than refrigerated products do.

For roasting, put frozen roasts fat-side-up and roast in a 325°F (161°C) oven. Cook until thawed; then insert a thermometer, and roast at 300°F (147°C) until done. Thin frozen items take only slightly longer to cook than do regular unfrozen ones, but thicker frozen items need two to three times as much cooking time.

To broil, place thin or medium-thick frozen pieces 4 in. from the heat source; place thicker pieces

8 to 12 in. from the heat. Reduce the heat to thaw, and then cook as for regular items. To panbroil, place frozen items in an ungreased pan on fairly low heat, and cover until thawed or almost thawed; then cook them in the same way as regular items are cooked. If thick frozen items are dropped into 325°F (161°C) fat, the fat speeds thawing without overcooking the outside. Sautéing may be preferred to pan-broiling because it thaws foods more rapidly, but lower temperatures must be used at the start. To sauté frozen breaded items, partially thaw them first. Braising small frozen pieces is done the same as with unfrozen items. If frozen pieces can be separated, they may be dredged with flour before being browned. Otherwise they must be thawed first. Simmering frozen items differs little from simmering unfrozen items.

To thaw frozen food, leave it wrapped under refrigeration; thawing takes 1½ hours per pound. Thawing at room temperature takes about 1 hour per pound. If the food is put in front of an electric fan, time is reduced. Items in waterproof wraps can be thawed in a basin of cold water or under running cold water, but this cannot be done if the food is unwrapped. Thawing in water takes about 1 hour per pound.

Poultry

Unfrozen poultry should be received at a temperature below 40°F (4.4°C) and stored at a temperature below 34°F (1.2°C). Hold frozen poultry at 0°F (–17.6°C); it holds well for about 8 months at this temperature, but at 20°F (–6.6°C) it holds for only 2 to 4 months. To thaw, place the frozen poultry in a 40°F (4.4°c) area. Small birds thaw in 1 to 2 days, but larger ones take 2 to 4 days. Thaw poultry in moisture-proof wraps under water—1 to 2 hours for chickens, and 2 to 6 hours for turkeys. To speed thawing, remove the giblets and neck as soon as possible, and spread the legs and wings. Cut-up poultry can be separated as soon as it is crinkly. Wash thawed birds in cold water, drain, and prepare them for cooking. Avoid refreezing.

Poultry can contain *Salmonella* bacteria, especially in their intestinal areas. Thus, poultry cut up on a meat block or other surface can contaminate whatever it touches with the bacteria. Other foods then touching such surfaces, equipment, or tools can become contaminated. It is therefore extremely important that

good cleaning and sanitizing occur after handling any poultry.

Poultry should be cooked in the same way as meat. Low-temperature cooking at 250°F (120°C) results in less shrinkage and a moister, better-looking finished bird. Cook old birds in moist heat and young ones in dry heat. Most poultry is cooked to well done except (sometimes) duck. Small birds such as squab and quail are best baked at temperatures around 400°F (202.4°C), to reduce drying out. Rabbit should be cooked just as poultry is.

Turkeys weighing over 18 lb have a skeletal structure that accounts for 9 percent of their weight, and they have about the same size abdominal cavity as 18-lb birds, so purchasing birds weighing more than 18 lb gives a greater yield of servable meat per lb purchased (see Table 15-9). Four portions, with bone and skin, are obtained from a 3½-lb duckling. A 2½-lb chicken, without neck or giblets, gives four cooked portions of 6 to 7 oz each. About 12 oz AP of turkey provides a good portion.

To roast poultry, truss the legs and wings of small birds, but not of large ones. Season the birds inside and out. Place the birds breast-side-down until they are almost done; then turn them and brown the breasts, if they are to be carved in front of guests. Geese and ducks can be roasted breast up. Place chickens and ducks in a preheated 325°F (161°C) oven. When they are done enough to allow it, break away the legs so that heat can penetrate around the thighs; this speeds roasting. Roasting times in an oven heated to 325°F (161°C) are listed in Table 15-10. Roast large birds at 250°F (120°C). Doneness is signaled by a temperature of about 180°F (81.4°C) in the thigh or breast center. Some cooks roast large, tender birds until only slightly underdone, and then cool and slice them. The slices are placed over mounds of dressing— dark meat first, covered with broad slices of white meat. The platter is covered with a moist cloth and refrigerated. When ordered, the sliced meat is warmed in an oven or steamer and served, one mount per plate, covered with very hot gravy. Roasting and then chilling before carving gives a greater yield. Figure 15-8 illustrates how to cut up a cooked bird.

Poach-roasting is a method in which poultry parts are placed into pans skin-side up and then covered with water to about two-thirds of their height. The

TABLE 15-9	Portion Yields of Turkeys of Different Weights					
Dressed Weight (lb)	Loss to Ready-to-Cook (%)	Ready-to-Cook Weight (lb)	Sliceable Meat Cooked (lb)	Number of Portions		
				2-oz	3-oz	4-oz
12–14	15	10–12	4½	36	22	12
14–16	15	12–14	4½	42	26	15
16–18	15	14–16	6	48	30	18
18–20	15	16–17	6⅔	54	32	20
20–22	15	17–18	7⅓	59	37	23
22–24	15	18–20	8	64	40	25
30–35	13	26–30	13½	108	67	40

procedure for cutting up a turkey for poach roasting is illustrated in Figure 15-9. Bones are left in. The parts are seasoned lightly and roasted in a 325°F (161°C) oven until tender. The product is allowed to cool in the stock. Alternatively, poultry may be poached in water sufficient to cover it. Large birds are sectioned and boned before poaching. To poach poultry, place parts in a stock pot, roasting pan, or steam-jacketed kettle, and cover with hot, salted water or stock, using

1 Tbsp of salt and 1 tsp of white pepper for every 6 lb of bird. Simmer at low heat for about 2 to 2½ hours, or until fork tender. Let it cool in the stock. Remove, and chill covered. Then slice (on a machine, preferably). Use the meat and stock within three days. If desired, add some vegetables for seasoning.

Older birds may be steamed. The bird, whole or in parts, is placed into a pan with a small quantity of water and steamed until tender. Like red meat, poultry

TABLE 15-10 Timetable for Roasting Poultry in a 325°F Oven		
Ready-to-Cook Weight[a] (lb)	Approximate Roasting Time (min/lb)	Total (hr)
1⅓–2½	30–40	1–1¾
2½–3½	30–40	1½–2½
3½–4¾	30–40	2–3
4¾–6	30–40	3–3½
6–8	30–40	3½–4
8–10	25–30	4–4½
10–14	18–20	3½–4
14–18	15–18	4–4½
20–30	12–15	5–6

[a]Weights are for unstuffed birds.

a.   b.

c.

FIGURE 15-8 Basic chicken cuts: (a) wing cut; (b) leg cut; (c) breast cut.

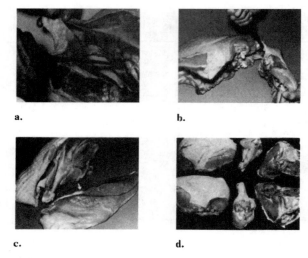

a. b.

c. d.

FIGURE 15-9 Preparing a turkey for poach-roasting: (*a*) remove the legs and wings; (*b*) remove the back; (*c*) separate the breast from the breastbone; (*d*) retain legs, thighs, and breast for poach-roasting.

can also be braised. Fricasséed or stewed poultry is braised poultry. Both tender and tough birds can be prepared in this way. Frying or sautéing poultry proceeds in much the same way as for meat.

Seafood

The flesh of most seafood contains little connective tissue, so it cooks to tenderness quickly. Extended cooking toughens some shellfish, such as clams, oysters, and shrimp. Overcooked fish is dry and breaks up easily.

Seafood is a highly perishable product. It keeps well under refrigeration for about three to five days. Use it as soon as possible after thawing. Sometimes products can be cooked directly from the frozen state, which saves nutrients by avoiding drip loss and also retains more flavor.

About 4 to 5 oz of fish is enough for a portion, if clear meat is used (see Table 15-11); but if there is much bone and skin, larger portions are required. Fish flesh contains from 1 to 20 percent oil, which is spread

TABLE 15-11   Portion Size and Amounts for 100 Portions of Seafood

Species	Serving Size (EP)	Approximate Weight for 100 Portions (EP)
Fish		
Dressed	8 oz	50 lb
Fillets	5 oz	32 lb
Pan-dressed	8 oz	50 lb
Portions	5 oz	32 lb
Steaks	6½ oz	40 lb
Sticks	5 oz	32 lb
Shellfish		
Clams—live, shucked	3 oz	19 lb
Crabs—live, soft-shell	5 oz	32 lb
Lobsters—live	1½ lb	100 lobsters
Lobster tails, spiny—frozen	6 oz	100 tails
Oysters—live, shucked	3 oz	19 lb
Scallops—shucked	4 oz	25 lb
Shrimp—headless, raw; headless, raw, peeled	5 oz	32 lb

throughout. Fat fish such as salmon, mackerel, shad, and trout bake, broil, or poach well (see Table 15-12). Lean fish such as flounder, halibut, and cod are best poached, deep-fried, sautéed, or simmered; sauces are often added to these to increase their richness and moistness. Lean fish is frequently breaded or batter-dipped for sautéing or deep-frying. Higher baking temperatures help keep small fish from drying out. If lean fish is baked, either it or the baking pan must be covered with fat. Lean fish can be broiled if dipped into flour and basted frequently. Cooking in a grid helps keep the fish from breaking up. Seafood is sautéed in slightly more fat than are meats and poultry, especially lean fish, which is often floured lightly. Serve sautéed fish as soon as possible, since quality is lost in standing. To simmer or poach, barely cover the fish with liquid and cook gently.

Any large fish should be cooked in a cloth so that it can be raised easily; a lightly greased trivet with handles can also be used for this purpose. If the fish is to be served cold or decorated, allow it to cool in its own stock. This means removing the fish from the heat before cooking is complete, because otherwise overcooking will occur during standing. Large fish and pieces are best started in cold water. Strong-flavored fish is also best started cold. Put small pieces into barely boiling liquid, reduce the heat to a simmer, and cook until done, about 8 to 10 minutes. Some types of white-fleshed fish are baked in milk. Fish is steamed until just done, usually over a court bouillon. Steamed, simmered, or poached fish is good served cold, especially a fat fish such as salmon.

A *matelote* is a fish braised with vegetables and lightly thickened stock. *Bouillabaisse* and *chioppino* are fish stews served with the stock, together or in separate bowls. Few seafoods are braised, since fish are so tender that they break up under this treatment, and since shellfish get tough with the longer cooking given in braising. When seafood is braised, wine, court bouillon, or some other liquid is added to it. Because of the short cooking time, the cover is left off so that as much evaporation as possible takes place.

Shellfish can be cooked in a number of ways with good results (see Table 15-13). If cooked at too high a temperature or for too long, however, they become tough and rubbery. Simmering at 190°F (87°C) is preferable to boiling; a temperature of 140°F (59.4°C)

immediately kills shellfish. Simmer lobsters in 205°F (95.2°C) water to get a tender and flavorful item. *Do not overcook.*

Shrimp can be cooked either shelled or unshelled. Prepare unshelled cooked shrimp by peeling off the shell and cutting along the back to remove the vein; then wash. To cook shrimp, use ¼ cup of salt to 1 gal of water and 5 lb of raw, headless shrimp. The yield after cooking, peeling, and deveining is 50 percent. *PDQ* shrimp stands for *peeled, deveined,* and *quick-cooked* shrimp. They may be purchased fresh or frozen; the federal government allows a 6 percent water glaze to be added, but no more should be present.

Wash clams and oysters in the shell thoroughly, using a stiff brush. To rid clams of sand, add ⅓ cup of salt per gallon of tap water; soak for an hour or so, and wash. Repeat until they are sand-free. Some cooks put clams into salted water, add cornmeal, and leave them overnight under refrigeration. The live clams eat the cornmeal and disgorge the sand. To steam clams, use six to eight clams (¾ lb) per portion. Place them into a container with a perforated bottom containing ½ cup of water per pound of clams. Steam for 5 to 10 minutes or until partially opened. Serve on a plate, with the clam nectar in a cup and a side dish of melted clarified butter. To roast clams or oysters, place them in a 450°F (230°C) oven for 15 minutes or until they are open. Bake or barbecue in the half-shell. Poach lobster tails for 5 minutes in salted water; then split them and broil for 5 to 6 minutes. To reduce curling of the tail, thaw and split; then remove the thin undershell, and bend the tail back firmly, breaking the connective tissue that causes curling.

MANAGEMENT'S RESPONSIBILITY

Since meat, seafood, and poultry items make up the most important part of menu offerings, much care must be given to their selection, handling, storage, preparation, service, and merchandising. It is management's responsibility to see that the desired standards are achieved in these areas. Management should select competent workers to bring the goals to a satisfactory conclusion. It should recognize and select items popular with patrons, as well as those it can produce well

TABLE 15-12   Time and Temperature[a] for Various Methods of Cooking Seafood

Baking		Broiling			Boiling, Poaching, or Steaming	
Temperature (°F)	Time (min)	Distance from Heat (in.)	Time (min)		Method	Time (min)
350	40–60				Poach	10/lb
350	25–30	3	10–15		Poach	10
350	25–35	3	10–15		Poach	10
350	25–35	3	8–15		Poach	10
350	30–40					
400	15–20					
450	12–15	4	5–8		Steam	5–10
		4	8–10		Boil	10–15
400	15–20	4	12–15		Boil	15–20
450	20–30	4	8–12		Boil	10–15
450	12–15	4	5–8		Steam	5–10
350	25–30	3	6–8		Boil	3–4
					Boil	3–5
350	20–25	3	8–10		Boil	3–5

[a]To convert to degrees Celsius, remember that 350°F = 177°C, 375°F = 191°C, 400°F = 204°C, 450°F = 232°C, and 500°F = 260°C.

and profitably. Specifications should follow through on management's thinking, and receiving procedures should ensure that the right goods are obtained and that management's interests are protected. Storage, handling, and issuing accountability must be present. Good recipes should be used and properly executed. Portion control must be established. Forecasts of quantities needed should be accurately calculated so that waste or carryover is minimized. Service should be in keeping with the high standards established in the other areas. Good sanitation is essential and should be carried to a point where patrons and employees are assured of adequate protection. Meat, seafood, and poultry are highly perishable products and require special attention if they are to be good and safe to eat.

Pricing should be based on good recipe yields or yield tests (see Fig. 15-10). Good accounting procedures should give management knowledge of the cost and profitability of items. Menu offerings should

ITEM: DATE: CARD NO.

                                              %
                                    Quantity   100

Gross Weight

Preparation Loss
(Trimming Waste)

Cooking Loss
(Shrinkage)

Serving Loss
(Slicing Loss
& Tasting)

Net Yield

FIGURE 15-10   Sample yield test card.

TABLE 15-12 Continued					
Deep-Fat Frying		Pan Frying		Ovenizing	
Temperature (°F)	Time (min)	Temperature (°F)	Time (min)	Temperature (°F)	Time (min)
325–350	4–6				
350–375	2–4	Moderate	10–15	500	15–20
350–375	2–4	Moderate	10–15	500	10–15
350–375	2–4	Moderate	8–10	500	10–15
350	4	Moderate	8–10		
350	3	Moderate	8–10		
350	2–3	Moderate	4–5		
375	2–4	Moderate	8–10		
350	2–4	Moderate	8–10		
350	3–5	Moderate	8–10		
350	2–3	Moderate	4–5		
350	2–3	Moderate	4–6		
350	2–3	Moderate	8–10		

be properly balanced among meat, seafood, and poultry items. The program designed to offer these items should extend year-round. They are too important to the success of an operation for the program to be any less comprehensive.

Poultry is an item that is growing in popularity. It is low in cost and provides good protein, vitamins, and minerals. It is usually low in calories, cholesterol, and saturated fats. Ways should be found to serve the item in a manner that differs from those usually presented in the home, since poultry is today a frequent item on the family menu. Seafood is also popular, for reasons of health and taste. Fish is low in saturated fats and cholesterol. Shellfish is low in saturated fats but has fairly large amounts of cholesterol. The caloric yield is not high if the seafood is prepared in a manner that does not add a lot of calories. Many high-quality frozen items are now on the market, so foodservices far from fresh supplies can still serve high-quality items. Meats are still the most frequently selected menu items. They provide an excellent source of protein and yield a good supply of important minerals and vitamins. In the future, while patrons may avoid red meats in general in their diets, they may throw their dietary restrictions out the window when they go out to eat, and select what they miss; thus, meat may still continue to be the most selected item on the menu.

■

SUMMARY

Meat, poultry, and seafood are prepared and cooked by similar methods. All are usually the center of a meal and supply valuable nutrients. Meat and poultry animals are separated by age and sex. Selection by class is necessary to suit the cooking process. Fish and shellfish are only selected by kind, except some few such as

TABLE 15-13   Preferred Ways to Cook Shellfish

	From	Description	Best Ways to Cook
	Clams, shucked (fresh or frozen)	Removed from shell	Bake, deep-fry, pan-fry, poach, steam
	Crabs, in-shell (fresh or frozen)	Raw or cooked	Pan-fry, poach, steam
	Crabs, out-of shell (fresh, frozen, or canned)	Raw or cooked	Pan-fry, poach, steam (thaw for cooking)
	Lobster tail (fresh or frozen)	Meat from the tail of the lobster; usually sold in shell	Bake, broil, poach, steam
	Mussels, shucked (fresh or frozen)	Removed from shell	Deep-fry, pan-fry, poach, steam
	Oysters, shucked (fresh or frozen)	Removed from shell	Bake, deep-fry, pan-fry, steam
	Scallops, shucked (fresh or frozen)	Removed from shell	Bake, deep-fry, pan-fry, poach, steam
	Shrimp, headless (fresh or frozen)	Raw, in-shell shrimps	Bake, broil, deep-fry, ovenize, pan-fry, poach
	Shrimp, peeled (fresh or frozen)	Used to describe shrimp from which shell has been removed	Bake, broil, deep-fry, ovenize, pan-fry, poach
	Shrimp, deveined (fresh or frozen)	Peeled shrimp from which the black sand vein has been removed	Bake, broil, deep-fry, ovenize, pan-fry, steam
	Shrimp, cooked, peeled, and deveined (fresh or frozen)	Cooked	Thaw for cocktail

littlenecks, cherrystones, or chowder (quahog) clams of the New England variety; their classification is by size. Cod is sometimes separated by size, too, into scrod and cod; and lobsters are as well. A chicken lobster weighs around 1 lb and makes a good portion.

Meats and poultry are graded. Some meats are graded both for quality and for yield. It is important to know the cuts of meat in order to cook them properly. Some meats from the less extensively used muscles are especially tender; meat from young animals is also apt to be more tender. Tenderness depends on the cooking method, the age and cut of the animal, the tenderizer used (such as papain), and the mechanical treatment of the item (if any), such as grinding, cubing, or pounding. Meats and poultry must be inspected for wholesomeness. Grading for quality or yield is not mandatory, but inspection for wholesomeness is.

Meat, seafood, and poultry are highly perishable products and need care in handling, storage, and cooking. Cooking makes these items more palatable by causing flavor, texture, and color changes. The color of flesh changes in cooking. Myoglobin, the pigment in red meat, changes into hematin; other flesh—such as the breast of chicken or the flesh of fish—also changes color, but not from red to the gray color seen in cooked meat. Cooking can tenderize flesh but it can also toughen it, especially if the item is cooked too long or at too high a temperature. Myoglobin reacts differently depending on handling and cooking, so some color changes in meat may not follow the pattern of cooking to a gray doneness. Flesh heated to 160°F (71°C) changes color, and the protein it contains coagulates and denatures. Coagulation is a firming process, and denaturation is the loss of water. High cooking temperatures cause greater denaturation than lower ones.

Connective tissue in flesh is the major cause of toughness. Flesh that has little connective tissue (such as that in young animals, birds, and fish) is more tender. Older animals and muscles that get quite a bit of exercise contain more connective tissue and therefore are apt to be more tough, requiring a cooking method that tenderizes. The connective tissue known as *collagen* is tenderized by long, moist cooking. The connective tissue called *elastin* is not affected by cooking and must be broken down by mechanical means or by

using a tenderizer. Flesh gets more tender with aging. The flavor of meat comes from nitrogenous and non-nitrogenous fractions. The fat in flesh also carries much flavor, since flavor esters are absorbed into it. Fat contains from 15 to 50 percent moisture, which helps it keep meat moist in cooking.

Stages of doneness in most flesh are as follows: rare, 140°F (59.4°C); medium, 160°F (70.4°C); and well done, 175°F (78.6°C). Poultry and pork are usually cooked to well done. Seafood is just cooked to doneness. Some meats are cooked rare or medium.

Meat and poultry are cooked by dry-heat and moist-heat methods. Dry-heat methods include broiling, roasting or baking, barbecuing, sautéing, pan-frying or grilling, deep-frying, and ovenizing. Moist-heat methods include braising (pot-roasting, fricasséeing, casseroling, swissing, or stewing), simmering (poaching or stewing), steaming, and blanching. Moist-heat methods usually require a long cooking time to allow the moisture and heat to break down connective tissue. Frozen meat may be cooked from either the frozen state or the thawed state. If flesh is cooked after thawing, the cooking procedure is much the same as for regular products.

Meat is broiled by radiant heat. Only tender and fairly well-marbled meats broil well. Lean meats must be basted well during broiling to prevent them from drying out. Pan-broiling is a method of frying in a hot pan or griddle without any fat. Sautéing involves cooking in shallow fat. Deep-fat-fried meats are usually breaded or batter-dipped. Roasting or baking is done in dry heat. Some products such as veal can be roasted covered to help retain moistness (since veal is quite lean). Barbecuing often consists of broiling over coals or a flame while basting with a tangy sauce. It can also be done in an oven. Ovenizing is a method in which items are placed onto well-greased pans, drizzled with fat, and then baked in an oven at fairly high heat. The product comes out much as if it had been sautéed. The method can be used to produce a lot of items at one time.

Meat may be prebrowned before braising, or it may be braised without browning. The meat is allowed to form its own juices, and liquid may also be added. The item is cooked for a long time covered. Pot-roasting, fricasséeing, casseroling, swissing, and stewing are all forms of braising. Blanching, poaching, and stewing all involve cooking in a small quantity of water. The difference is that blanching occurs for a very short time, stewing occurs for a long time, and poaching is cooking only to doneness. To boil is to cook in boiling water—a method infrequently used in meat cookery. Steaming is just what it says: cooking in steam.

Poultry is cooked in much the same way as meat. Poach-roasting is often used for turkey. The bird is divided into pieces, placed into water that nearly covers it, and cooked in an oven. Turkey can also be simmered, sliced, and served as roasted turkey. Birds cooked in water are often allowed to cool in the stock to prevent them from drying out. Tough poultry is cooked by moist-heat methods.

The methods used to cook seafood depend on the amount of oil in the flesh. Fatty fish broil, steam, poach, or bake well. Lean fish are often sautéed, deep-fried, poached, or steamed. Steaming is done over a court bouillon. All fish are cooked only to doneness. Shellfish are toughened by extended cooking, especially at high heat.

Meat, poultry, and seafood must be well merchandized to meet patron favor. Good merchandising must be backed up with good menu planning and presentation, good purchasing and storage methods, good preparation, and good service.

■

CHAPTER REVIEW QUESTIONS

1. In recent years, the serving of rare hamburgers in foodservice operations has been discouraged. What is the correct temperature at which this meat should be served if it is to be well done? Do you eat rare meat? Why or why not?

2. Why are seafoods considered healthful? Can a person on a low-saturated-fat diet eat both shellfish and fish? Can one on a low-cholesterol diet?

3. What quality factors should be checked when receiving fish and shellfish? What are the various market forms of fish? What are the various market forms of shellfish? Are all seafood products inspected and passed for wholesomeness?

4. What are the different cooking methods suited to lean fish? to fat fish?

5. Why are cooking times and temperatures so important in cooking fish and shellfish?

6. Why should meat be in a person's diet? Name some of the important nutrients meat supplies.

7. Define what each of the following is on the market: veal, calf, steer, heifer, cow, lamb, barrow, gilt, yearling, mutton, and ewe.

8. Should a foodservice heart of the house today be designed with a complete butcher shop? Explain.

9. Where are the most tender cuts on an animal? Are muscles that get a lot of exercise apt to be tougher or tenderer than those that get little exercise? What steps can be taken to make meat tender in handling or cooking?

10. (a) What is meant by *moist-heat methods?* Name four of them. (b) What is meant by *dry-heat cookery?* Name five dry-heat methods.

11. What are the interior temperatures when meat is rare, medium done, and well done?

12. What is aged or ripened meat?

13. What are some of the guidelines for handling and cooking frozen meat? frozen poultry?

14. How can portion control be achieved in the production and service of meat, seafoods, and poultry?

15. A steak dinner costs $9.80, and a chicken dinner costs $7.50. The food costs are, respectively, $3.10 and $2.00. If the desired food cost is 30 percent of menu price, why has management priced these foods as it has? Which item offers the best gross profit, in dollars? Would management wish to sell steak or chicken, based on this difference in gross profit?

16. How do meat, seafoods, and poultry compare in nutritive value?

17. What are the kinds and classes of poultry?

18. What type of chicken should be used to make a roast chicken?

19. What is the top grade for poultry?

20. (a) What are the various methods used in food-services to cook poultry? (b) How should an old tom turkey be prepared?

21. What are the market forms of fresh poultry?

22. What is management's responsibility in seeing that good meat, seafoods, and poultry are put before patrons?

CASE STUDIES

1. You are the owner of a very popular restaurant. You have a chef who is well known for his excellent food. The instructor of the quantity food class at the local university has invited you and the chef to come to the school and tell the class how you achieve such high standards in the preparation of meat, poultry, and seafood. The talk is to be for two school class hours (100 minutes), and you and the chef decide to split the time, leaving some time after each talk for questions. Make an outline of what management presents as its way of achieving high standards in the production and service of these foods, and make another outline of what the chef feels are the essentials in achieving high standards. Since time is limited, there can be little or no demonstration—just a straight lecture.

2. The grading and certification of meat is managed by the U.S. Department of Agriculture. In researching the Web site at *www.ams.usda.gov*, you are asked to provide information on how to buy meats. Your answer should include a discussion on wholesomeness, grading, labeling, and nutrition.

VOCABULARY

Cholesterol
Saturated fats
Complete protein
Seafood
Shellfish
Fish
Crustacean
Mollusk
Chicken lobster
Fillet
PDQ
Flounder

Sole
Salmon
Fat fish
Lean fish
Round (whole) fish
Steak
Butterfly
Deveined
Cherrystone
Littleneck
Bluepoint
Rooster

Tom
Ready-to-cook
Eviscerated
U.S. Grade A
Chincoteague
Olympia
Bay scallop
Sea scallop
Court bouillon
Trivet
Dungeness
Blue crab
King (Alaskan) crab
Stone crab
IMPS
Marbling
Variety meats
Bullock
Steer
Heifer
Veal
Calf
Stag
Cow
Lamb
Roasts
Yearling
Mutton
Ram
Ewe
Barrow
Gilt
Sow
Boar
Prime
Choice
Select

Standard
Yield grading
Quality grading
Rib
Round
Sirloin tip
Flank
Chuck
Shank
Short loin
Plate
Pork
Fresh ham
Ham hock
Jowl
Pork loin
Ripening (aging)
Connective tissue
Elastin
Collagen
Papain
Marinating
Shrinkage
Myoglobin
Nitrosomyoglobin
Oxymyoglobin
Metamyoglobin
Duck
Poultry
Goose
Fryer
Roaster
Capon
Stewing hen
Broiler
Beef

■
ANNOTATED REFERENCES

Donovan, M. (ed.). 1996. *The New Professional Chef*, 6th ed. New York: John Wiley & Sons, Inc., pp. 555–708. (Meat, poultry, and seafood are considered in these chapters.)

Donovan, M. 1997. *Cooking Essentials for the New Professional Chef*. New York: John Wiley & Sons, Inc., pp. 527–586. (Meat, poultry, and seafood are considered in these chapters.)

Gisslen, W. 1995. *Professional Cooking*, 3rd ed. New York: John Wiley & Sons, Inc., pp. 195–378. (Meat, poultry, and seafood are considered in these chapters.)

Kapoor, S. 1996. *Healthy and Delicious: 400 Professional Recipes*. New York: John Wiley & Sons, Inc., pp. 63–170. (Healthy recipes for meat, poultry, and seafood are presented in this chapter.)

Labensky, S. R., and A. M. Hause. 1999. *On Cooking: A Textbook of Culinary Fundamentals*, 2nd ed. Upper Saddle River, N.J.: Prentice Hall, pp. 262–531. (Meat, poultry, game, and seafood are presented in these chapters.)

Pauli, P. 1997. *Classical Cooking the Modern Way*, 3rd ed. New York: John Wiley & Sons, Inc., pp. 157–258. (A discussion on meat, poultry, and seafood is offered here.)

Bakeshop Production

■

LEARNING OBJECTIVES

By the end of this chapter, the reader will:

1. Be able to distinguish between the various functions ingredients play in baked products.

2. Learn about the types of flour as well as the types of leavening agents and their use in baked products.

3. Understand the difference between yeast and quick breads, how they are prepared, and how they might be served.

4. Be able to prepare various desserts and serve them in appealing ways within food service operations.

SELECTED WEB SITE REFERENCES

Archway Cookies
www.archwaycookies.com/

Ben & Jerry's Ice Cream
www.benjerry.com/

Dunkin Donuts
www.dunkindonuts.com/

Fleischmann's Yeast
www.breadworld.com/

Hershey Foods
www.hersheys.com/

RICH's Products
www.richs.com/

Good hot breads and desserts are appreciated by patrons and can do much to draw business. Sweet breads, muffins, or rolls enhance a breakfast; and a baked dessert provides an excellent conclusion to a meal. Many operations try to emphasize bakery products because of their appeal and profitability.

INGREDIENTS

The ingredients used in baking are critical to product quality. Bakers must know what ingredients do and how they interact to produce high-quality goods. Handling the proper ingredients in the proper manner brings baking success.

Flour

For baking purposes, *flour* can be defined as a ground starch product that is used to give structure and body to bakery products. Different flours are needed to make the many products produced, and different wheats are blended to produce the kind of flour needed. Regular flour contains from 63 to 73 percent starch and from 7 to 15 percent protein. The rest consists of moisture and small amounts of fat, sugar, and ash (minerals). Varying the relative amounts of starch and protein dramatically affects how flour acts in baking.

When wheat is milled to make flour, patent flours come off first, then middlings, and then clears. Patent flours are of the highest quality, middlings are next,

and clears are the poorest. Straight flour is all the flour from the kernel. Whole-wheat (graham) flour is straight flour plus some of the bran. Good patents are mixed with rye, straight flour, or poor flours to make them work properly.

Gluten

Flour protein is called *gluten*. It can absorb twice its own weight in moisture. Moistened gluten is sticky, and in a batter of dough it forms a thin, gummy, stringy mass that can stretch out, giving elasticity to the mixture. Working a dough or batter forms a network of gluten throughout the mixture. Baking firms this network and helps support the baked item. Some products, such as yeast bread, need a strong network; others, such as cake, need a weaker one. Mixing a dough or batter helps gluten develop, which is desirable in bread but not in muffins. Bread or hard-wheat flour is called *strong* because it forms strong structures. Pastry or soft-wheat flour is called *soft* because it produces more tender and delicate structures.

Bread (Hard-Wheat) Flour

Hard or spring wheat contains from 65 to 70 percent starch and from 12 to 16 percent gluten. The high protein content (plus good-quality protein) allows it to make strong structures that yeast breads and some pastry products need. Most bread flours contain from 12.5 to 12.8 percent gluten. Hard-crusted breads and hearth breads need a 13 percent gluten flour; a strong patent flour containing about 15.5 percent protein is blended with rye and other flours that need gluten.

Pastry (Soft-Wheat) Flour

Wheat that is planted in the fall and matures in early summer is called *winter wheat*. It develops less gluten and more starch than hard wheat, and its gluten is not as strong as that of spring or hard wheat. Cake flour contains from 7 to 8 percent protein. The patents of soft wheat make the best pastry and cake flours. A weaker gluten is desirable for making cakes, cookies, hot bread, and pies. A cake flour contains enough gluten to give a weak structure that suffices to support the sugar, fat, and other ingredients blended with it. A fine cake flour gives a tender, velvety crumb.

A soft flour holds together when squeezed in the hand, but hard-wheat flour crumbles. Hard-wheat flour is yellowish because gluten is a faintly yellow color. If 2½ oz of cornstarch is added to 14 oz of hard-wheat flour, the mixture approximates a soft wheat flour in its structural characteristics.

All-Purpose Flour

All-purpose flour falls between soft- and hard-wheat flour in gluten content. It makes fair bread, cakes, and pastries but is not really suited to quantity work. It is designed for use in the home.

Shortening

Shortening is a fat used in baking to lubricate ingredients and prevent products from sticking. Shortening interferes with gluten particles' sticking together and thus makes a short dough. It also provides slip, which is needed so that batters or doughs can move when leavening gases start to push them. Different shortenings are used in baking: a short one makes pie crusts and other products tender, while an elastic fat makes a good puff paste. Cakes need a fat that spreads easily and quickly in a batter; emulsifiers such as monoglycerides and diglycerides are used to help do this in cake shortenings. Special margarines or shortenings are used for frostings, icings, and special products needing butter.

The development of hydrogenation was an important step in shortening development, since it made oils solid, thus increasing their plasticity and firmness and giving them moldability and workability over a wider range of temperatures. A plastic shortening should be medium firm—not brittle or oily—and must hold texture over a 55 to 90°F (13 to 32.2°C) range. Most hydrogenated shortenings have a relatively short plastic range but are especially made to be of superb plasticity. A waxy, tough, extensible, plastic, firm fat is needed to produce a flaky pie crust so that, when the dough is rolled out, layers of fat remain between layers of the dough (see Fig. 16-1). Lard is an excellent fat to use for making pie crusts; its waxiness and plasticity spread it out in pieces or layers, giving good shortness. It remains sufficiently firm at mixing temperature so that it does not work completely into the dough. Icings need a flavorless, plastic fat that creams (aerates) well, has no greasiness, and holds a lot of sugar. For some cakes, plasticity and waxiness are needed so that air is incorporated, increasing volume and creating a light batter. Plasticity and shortness are related, since fats lacking plasticity may not spread well and therefore may not shorten the gluten. Tallow (beef suet) and oleo are too firm and brittle to spread well. Chicken fat has excellent shortening power but is too soft to use in many products.

Sugar

Sugar is used to sweeten and also to tenderize products. It gives color and texture to baked goods and

FIGURE 16-1 This piece of pie has a flaky crust, making it a superior product.

adds calories. The bakeshop uses many different sugars. Powdered sugar (ground sugar) is used for dusting, icing, or coating items. The "X" on a sugar label indicates the grind fineness, with 4X to 10X sugars being the ones most often used; 10X is the finest. Coarse sanding sugars are used for coating doughnuts, candies, orange peel, and so on. Regular granulated sugar has a moderate-sized crystal. Fruit sugar is fine granulated sugar used for making cakes, meringues, and other bakery products; it goes into solution rapidly. Angel and sponge cakes are best made with fruit sugar. Fondants and other products high in sugar content are also available on the market.

Brown sugar is less refined than white sugar. It ranges in color from very light to very dark and is used in different ways to achieve different results. The darker brown sugars give a stronger, richer flavor and a moister product. Molasses is a syrup that is a by-product of white cane sugar. Like brown sugar, molasses may range from light to dark; a darker molasses gives a moister product with a rich, heavy flavor. Honey and some other sweeteners are sometimes used for special needs.

Leavening Agents

A leavening agent aerates bakery goods, giving them an open, light texture and greater volume. The kind used, the amount, and the other ingredients in the product all influence how much rise occurs. Insufficient amounts of leavening in a product give a poor, heavy, course crumb and poor volume. As a result, the finished product is too compact to be of good eating quality. Too much leavening gives an open, loose, crumbly grain; the flavor is poor, and the product may fall because the excess gas has broken down the gluten structure.

Getting the right leavening action to accomplish proper aeration takes skill. A leavening agent creates gas in a product, which is captured when protein coagulates and starch gelatinizes, firming the product. The texture is then light and open. Steam alone is the only leavening agent used in popovers and cream puffs; the shells of these baked items are strong enough to hold the gas inside, creating a large cavity. Yeast feeding on sugars and other carbohydrates creates carbon dioxide gas, which leavens breads. Soda, ammonia, and baking powder spread in doughs or batters and, during baking, create carbon dioxide, which leavens the mixture. Sometimes, as in overdeveloped muffins, the gas collects and forms tunnels that work up to the top as gas tries to escape. Thus, many products cannot be worked too much and remain tender enough to allow the leavening to work. The sugar and shortening in cakes and cookies counteract such toughening, so these batters and doughs can be worked more. A double-acting baking powder is one that makes a small quantity of gas in the cold stage but makes a far greater quantity of gas in the hot stage. Single-acting baking powders, which work completely in the cold stage or in the hot stage, are little used in quantity work.

Air

Air is incorporated into products when eggs are whipped to a foam, as in angel cake. Air is also worked into a batter in creaming. Air leavens a pound cake. When heated, air swells, causing the product to rise and develop a fine texture. Once air is in a batter or dough, it must not be lost. Thus, a sponge cake should be folded only until the dry ingredients are incorporated into the egg foam; overmixing loses the air in the foam.

Steam

Water swells to 1600 times its liquid volume when it converts into steam; thus, it can be a powerful leavening agent. Steam leavens many bakery products. A cake is leavened one-third by air incorporated during creaming, one-third by steam, and one-third by baking powder. Cream puffs, éclairs, popovers, crackers, and pie crust depend entirely on steam for leavening. About 1 qt of liquid per lb of flour gives adequate leavening by steam.

Chemical Leaveners

Soda was perhaps the first chemical leavener used. It produces carbon dioxide gas. If the product contains an acid ingredient, such as sour milk, vinegar, or chocolate, the acid reacts with the soda, releasing carbon dioxide (see Table 16-1). Heat also causes this reaction; thus, when a product is baked, the soda gives

	Liquid Required for Complete Reaction		Soda Required for Complete Reaction	
Type	*Weight*	*Measure*	*Weight*	*Measure*
Cream of tartar	5 oz	1 cup	2 oz	4½ T
Honey	11.3 lb	1 gal	½ to 1 oz	1 to 2¼ T
Molasses	11.5 lb	1 gal	1¾ oz	3 to 4 T
Orange juice	4 oz	½ cup	1¼ oz	2⅔ T
Sorghum	11.3 lb	1 gal	1 oz	2¼ T
Sour milk or buttermilk	8.3 oz	1 gal	¼ oz	2⅔ T
Vinegar, 40 to 50 grain	3 oz	⅓ cup	1¼ oz	2⅔ T

TABLE 16-1   Soda–Liquid Combinations for Leavening

off the necessary gas for leavening. Cream of tartar can be used with soda to make a single-acting baking powder, with the cream of tartar acting as an acid. If too much soda is added, a slightly disagreeable flavor results. Soda also causes the pigments in some foods to darken. Bananas, spices, chocolate, and other ingredients change color when soda is used, the excess soda causing an alkaline reaction. Such a product may fall because alkalines tenderize gluten. This sometimes happens to gingerbread.

Two dry ingredients can be combined to make a leavening agent. The first baking powder was cream of tartar and soda, but it reacted entirely in the cold stage, so products had to be completely mixed, the baking powder added, and the mixture *immediately* panned and baked; otherwise, the gas would be lost in standing. Then other acids were found that could be combined with soda and that reacted only in the hot stage. Thus, double-acting baking powders became possible. A popular one today contains two acid reactors—one that reacts in the cold stage, and another that reacts in the hot stage. Another popular baking powder in quantity work contains only one reactor, but it reacts very slowly in the cold stage and releases most of its gas in the hot stage. The federal government requires a double-acting baking powder to yield not less than 12 percent of its weight in carbon dioxide; most commercial powers yield 17 percent. Baking powders contain some cornstarch, which helps absorb moisture and prevents the powders from reacting in their containers.

Baking ammonia gives off carbon dioxide and ammonia gas when heat strikes it in a batter or in a dough. If all of the gas works off during baking, there is no aftertaste, although a distinct odor of ammonia is evident during baking. Thick products such as cakes capture some of the ammonia, so it is not used for them. For thin-shelled products such as cookies or cream puffs, however, baking ammonia works well. It gives good spread to cookies; that is, it causes them to spread out in baking.

The leavening powers of ammonia, soda, double-acting baking powder, and single-acting baking powder are, respectively, 1; 0.75; 2.5; 5. Thus, if a recipe called for 1 oz of soda, you could probably substitute $1\frac{1}{3}$ oz of ammonia ($1.00/0.75 \times 1$), or $3\frac{1}{3}$ oz of double-acting powder ($2.5/0.75 \times 1$), or $6\frac{2}{3}$ oz ($5/0.75 \times 1$) of single-acting baking powder. If the recipe called for 3 oz of double-acting baking powder, the equivalent amount of ammonia, soda, or single-acting powder would be, respectively, 2 oz ($1/2.5 \times 5$), 0.90 oz ($0.75/2.5 \times 3$), and 6 oz ($5/2.5 \times 3$). The rule is to divide the ratio of the leavening agent in the recipe into the ratio of the substitute leavening agent and then to multiply the result by the amount of the first leavening agent (the one called for in the recipe).

Yeast

Yeast is a tiny fungus that thrives on carbohydrates and creates, as by-products, carbon dioxide gas and alcohol. The gas acts as a leavener. Yeast grows best at temperatures of 78 to 90°F (25.5 to 32°C). Its growth slows at around 98°F (37°C), and it is killed at 140°F (60°C). Yeast needs warmth, moisture, and carbohydrate in order to grow. Salt slows its growth. Yeast grows best when certain other substances are present, and bakers add conditioners to provide these in a dough. They also add malt, which contains glucose—the ideal food for yeasts.

Compressed, dry regular, or instant dry yeast is used in foodservices. Compressed yeast should be held at 45°F (7°C). Freezing may do some harm to compressed yeast, but not much. Thaw it at 40°F (4.5°C). Add compressed yeast crumbled into 110°F (43°C) water. Sift dry regular yeast into water heated to the same temperature, and allow the particles of yeast to hydrate thoroughly. Instant dry yeast is mixed with the dry ingredients. Bakers weigh yeast, rather than measure it. The amount of yeast required is less in warm weather than in cold weather (see Table 16-2).

Eggs

Eggs coagulate and give bakery goods body; they also add flavor, color, moisture, nutrients, volume, and tenderness. Eggs help bind ingredients together and also emulsify fats. Bakers treat eggs as moisture, so they count them as a part of the liquid. Thus, a cake should contain about 170 percent liquid to 100 percent flour, with other liquids accounting for 100 percent and eggs 70 percent. Actually, egg whites contribute about three-fourths of their weight in liquid, yolks one-half, and whole eggs about two-thirds.

Liquids

Liquid can act as a leavening agent by turning to steam. It also provides the base into which many ingredients are dissolved, bringing them together in the batter or dough. Liquids contribute moistness, flavor, and tenderness and help retard staling. Moisture is absorbed by starch and gluten, which give the product structure. If moisture were not present leavening agents would not work.

Flavorings and Spices

Spices and flavorings add flavor. Natural oils or esters used as flavorings include lemon oil, clove oil, and vanilla. Other flavorings may be imitation. Some

TABLE 16-2   Yeast–Water Combinations for Proper Mixing

Compressed Yeast	Active Dry Yeast	Water[a] Weight[b]	Approximate Measure
1 oz	½ oz	2 oz	¼ cup
2 oz	¾ oz	3 oz	½ cup
4 oz	1½ oz	6 oz	1 cup
8 oz	3¼ oz	13 oz	1 pt
12 oz	4¾ oz	1 lb 3 oz	1¼ pt
1 lb	6½ oz	1 lb 10 oz	1¾ pt
2 lb	12¾ oz	3 lb 3 oz	1 qt 1¼ pt
3 lb	1 lb 3¼ oz	4 lb 14 oz	2 qt 1 pt
4 lb	1 lb 9½ oz	6 lb 6 oz	3 qt 1 cup
5 lb	2 lb	8 lb	4 qt

[a]Water temperatures should be 90 to 100°F for compressed yeast and 5 to 10°F higher for dry active yeast.
[b]Corrected to nearest ¼ oz.

flavorings are best when they come from the natural product, while others are better in the imitation product. High-quality spices should be used, since the quantity is small but the flavor contribution is large. Spices should be stored in tight containers, away from heat and moisture. Only enough to last about a month should be purchased. Spices are usually weighed in baking; flavorings are measured, if they are liquid.

BREADS

Most operations today buy their loaf breads but bake their own rolls and hot breads to ensure top-quality products (see Fig. 16-2). The prepared mixes on the market produce a variety of yeast and quick breads. The cost and quality of the mix must be evaluated against the merits of making the product from scratch.

Yeast Breads

Yeast breads are evaluated on the basis of their volume, crust, symmetry, evenness of baking, grain (internal appearance), texture, and flavor. White bread is made from good-quality white (bleached) hard flour. Whole-wheat (graham) bread is usually 40 to 60 percent whole-wheat flour, with the rest being a good patent white flour. Whole-wheat bread dough is mixed less because of the weaker gluten in the whole-wheat flour.

FIGURE 16-2 High-quality baked products distinguish one foodservice operation from the next.

Regular rye bread consists of 20 percent rye and 80 percent strong white patent; the ratio can go up to 100 percent rye. Pumpernickel used all of the rye kernel; it is somewhat heavy and compact because it lacks gluten. Rye bread is mixed less and is made with less moisture because rye flour cannot absorb as much moisture as hard-wheat flour can. A loaf of rye bread can burst in baking because of its weak gluten.

Hard, French, Italian, and hearth breads are made from quite lean (nonrich) doughs. The heavy crust develops because steam remains in the oven during the early stages of baking. Sourdough bread is made by adding a fermented starter or old dough to a new one. Sweet doughs vary in richness and may have shortening rolled into them, as puff paste does. Rich doughs can be retarded (refrigerated) or frozen better than lean ones can.

Basic yeast bread requires flour, yeast, liquid, and salt. Fat can also be added. Sweet doughs and other rich types contain sugar, eggs, more fat, and added flavoring. Ingredient proportions for breads and many other bakery products are always given in their percentage relative to flour. Thus, a bread recipe may list the liquid as 60 percent, which means it is present in a weight equivalent to 60 percent of the flour (see Table 16-3). Graham breads, rye breads, and those

TABLE 16-3   Bread Ingredients and Proportions

Ingredient	Lean (%)	Rich (%)
Conditioner	¼–½	¼–½
Eggs, whole	—	10–45
Flavoring	—	¼–½
Flour, bread	100	65–100
cake	—	0–35
Liquid	58–60	40–60
Milk, nonfat, dry	0–6	3–8
Salt	1–2½	1½–2½
Shortening	1–12	8–40[a]
Spices	—	¼–½
Sugar, granulated	2–3	6–25
Yeast, compressed	2–3	2–8

[a]Use additional shortening (20 to 50% of the flour) for rolling in.

high in milk and shortening need more salt. Salt controls yeast growth; thus, more salt must be used in summer than in winter. Salt should be added with the flour rather than mixed directly with the yeast.

The straight-dough method of making dough combines all ingredients at one time. The following steps (and the approximate time required for each) are involved in the straight-dough method:

1. Weighing and mixing (12 minutes)
2. Fermentation (45 to 60 minutes)
3. Punching (3 minutes)
4. Benching and resting (10 to 15 minutes)
5. Makeup (20 minutes)
6. Proofing (15 to 20 minutes)
7. Baking (rolls, 20 minutes; loaves, 60 minutes)
8. Cooling
9. Storing

Thus, the total preparation time in the straight-dough method is from 2 to 4 hours. The actual work time is about 45 minutes for 100 portions or 11 lb of dough.

A sponge yeast is made by mixing all of the ingredients except for about 20 to 40 percent of the flour. The dough is fermented, and then the rest of the flour added. Best-quality bread results from this method.

A no-time yeast bread is made by the straight-dough method, with an increase in the amount of sugar and yeast and a higher baking temperature. The bread often remains in the mixer for its rapid fermentation rise. Makeup and proof times are shortened, and baking times are normal. The bread is quick to produce, but it lacks something in quality, being slightly yeasty and staling quickly.

Weigh all ingredients before starting. Most operations use an upright or vertical mixer with a dough hook; small batches can be made by hand. Rehydrate the yeast in a small bit of liquid, and then add the remaining liquid and fat at proper temperature for good yeast growth. Add dry ingredients and mix until a soft, sticky ball is formed and the gluten is well-developed. When properly mixed, the dough should stretch out into a thin sheet, indicating good gluten development. Doughs should come from the mixer at 74 to 84°F (23 to 29°C). A relative humidity of 75

percent is desirable. If the humidity is lower, cover the dough with a damp cloth or brush the top of it with oil or shortening. When the dough doubles in volume or when fingers can be inserted easily about 3 inches into the dough, with it reacting by puckering or receding, the fermentation period is over. During fermentation, the dough ripens and develops into a soft, smooth, and silky mass with good elasticity. Extended fermentation or too high temperatures cause the dough to develop too much acid, giving a sour bread of poor texture when baked. The most desirable fermentation temperature is 81°F (27°C).

Punching is a technique of folding a dough over from its sides into the middle, and then turning the dough completely over in its container. This remixes food that the yeast needs so that it can start working again. Some strong doughs may be allowed to ferment a second time. Benching and resting are done to let the dough recover from punching. Rich doughs and doughs containing a lot of yeast need a shorter rest. If the makeup time is extended, the rest is shortened. Makeup involves shaping the dough into the form needed for baking. Dinner rolls are portioned at about 1 lb per dozen; breakfast rolls are made into 1 oz (strong) if two are served, and 2 oz if only one is served. Baking loss is about 12 percent. If 1 lb per pan is desired, the scaling weight at makeup is 18 oz. After being weighed (scaled), the dough is formed into smooth, round shapes (see Fig. 16-3). A rough surface allow carbon dioxide to escape and gives a poor-looking product. A short bench rest, called *bench proofing*, of 8 to 12 minutes allows the dough to recover from the effects of rounding. This helps to relax the dough and let it shape up better in the pan. Similar shaping techniques are used with bread loaves.

Proofing is the final stage of conditioning before baking. The relative humidity for proofing is from 80 to 85 percent or more, at a temperature of from 90 to 100°F (32 to 38°C). This gives a soft, delicate product with maximum increase in volume. Too much proofing results in an open-grained, gray-colored bread, as well as a loss of flavor and lower volume. Some breads are docked just after proofing and before baking. Docking is gashing the top to allow the bread to rise without breaking or bursting. This procedure is done on hard-crusted rye and other breads.

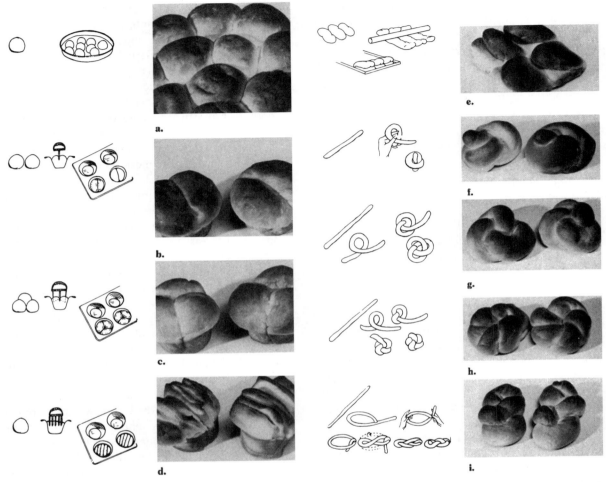

FIGURE 16-3 Rolls and techniques for making them: (*a*) pan rolls; (*b*) twin rolls; (*c*) cloverleaf rolls; (*d*) fantans or butter rolls; (*e*) Parker House rolls; (*f*) single or bow-knot rolls; (*g*) double or rosette rolls; (*h*) kaiser-knot rolls; (*i*) triple or braided rolls.

When proper procedures are followed, baking yields a product with maximum volume, good texture and flavor, and good appearance. Bake loaves at 425°F (218°C) for 15 minutes and then at 375°F (191°C) for 45 minutes. Larger loaves may need slightly lower temperatures and longer baking. Rolls that are separated should be baked for 15 to 30 minutes at 425°F (218°C); pan rolls and other rolls that touch must be baked nearly as long as loaf breads. Rich doughs are baked at lower temperatures because they tend to burn

or scorch at bread-baking temperatures (see Table 16-4). Bread rises rapidly during the first period of baking, and a slight tan develops on the top. This rise is called *oven spring*. A well-baked bread gives a hollow sound when it is tapped. Appendix G lists reasons that should be checked to ascertain why bread fails.

After baking, loaves, large rolls, and other products that touch in the pans are dumped for cooling. This allows the steam and alcohol to escape, resulting in a less soggy bread. Breads should not be placed in

TABLE 16-4   Baking Temperatures for Various Bakery Items

Product	Pan Size	Units per Pan	Scaling Weight	Temperature °F	Temperature °C	Baking Time (min)
Bread, rye	1 lb	1	18 oz	400	204.5	45
white	1¼ lb	1	26½ oz	400	204.5	60
Cinnamon rolls	8 in.[a]	8–12	1–2 oz	375–385	190–196	25
	17 × 25 in.	72–100		375	190	30
Coffee cakes, fancy	8 in.[a]	1	10–12 oz each with 2 oz filling	375	190	30
	17 × 25 in.	3–5				
topped	8 in.[a]	1	10–12 oz each with 2 oz topping	375	190	30
	17 × 25 in.	8				
Rolls, small	8 in.[a]	12	1–1⅓ oz	400–425	204.5–218.0	15–20
	17 × 25 in.	72–100				
medium	8 in.[a]	8–12	1½–1¾ oz	400–425	204.5–218.0	15–20
	17 × 25 in.	60–85				
large	8 in.[a]	8	2–2⅓ oz	390–420	198.8–215.5	20–25
Sweet rolls	17 × 25 in.	24	2 oz	380–400	193–204.5	25
Tea rings	8 in.[a]	1	10–12 oz each with 2 oz filling	375	190	30
	17 × 25 in.	4				

[a] Round or square pan.

drafts because these might cause them to crack or shrink. Hard-crusted bread and some others are not wrapped or stored in tight containers; instead they are put into open boxes and left at room temperature. Other breads may be stored in good containers or put into moisture- and vapor-proof wraps. Hard-crusted breads should be used as soon as possible. Carry-over bread can be made into crumbs, croutons, or stuffings. Specialty breads may be used as decorations (see Fig. 16-4).

Quick Breads

Quick breads are made from batters and doughs. They are prepared much more rapidly than yeast breads, since they can be baked as soon as mixed. Leavening may be obtained from chemicals, air, steam, or yeast. Slight variation in method, ingredients, or timing can yield a completely different product, even though the same procedures are followed (see Table 16-5).

Biscuits are the closest things to yeast bread. Shortening is worked into the flour, leavening agent,

FIGURE 16-4   Bread made into special shapes may enhance a buffet line, as this bread does on St. Patrick's Day.

and salt until the mixture looks like cornmeal. Liquid is then added. The dough is worked on a floured board for several minutes until smooth. It is then rolled out and cut into biscuits. Flakier biscuits are made by using slightly more shortening and cutting it

TABLE 16-5 Comparative Analysis of Quick Breads

I. Approximate Ingredient Percentages

Ingredient	Biscuits (%)	Muffins (%)	Cornbread (%)	Griddle-cakes (%)	Waffles (%)
Baking powder, double-acting	6	6	8½	6–10	6–8
Eggs	0–10	20–25	20–25	15–35	20–65
Flour, bread pastry	0–50 50–100	100[a]	100[b]	100	100[a]
Liquid	60–70[c]	70–80	80–90	125–200	130–180
Milk, dry, nonfat	7	8	8	10–15	10–15
Salt	1–2	1–2	1–2	1–2	1–2
Shortening	20–30	20–40	6–20	5–15	50–40
Sugar	0–2	10–65	5–25	2–10	15–30

[a]For greater tenderness, cake flour may be used.
[b]About 50% of the flour is cornmeal.
[c]For shortcakes, reduce liquid to 50 to 60% and increase sugar to 15 to 20%

II. Scaling Weights, Baking Temperatures, and Baking Times

Product	Pan Size	Units per Pan	Scaling Weight	Baking Temperature °F	Baking Temperature °C	Time (min)
Biscuit, 2¼ in.	17 × 25 in.	88	1¼ oz	425	218	15–20
Breads, date and nut	7½ × 3½ × 2¼ in.	1	22 oz	350	176.5	60
quick	4 × 9 × 4 in.	1	1¾ lb	350	176.5	60–75
	7½ × 3½ × 2¼ in.	1	18 oz	350	176.5	60
Brown bread, steamed	4 × 9 × 4 in.	1	1½ lb	Steam	Steam	60–90
Cornbread	17 × 25 in.	1	6 lb	425	218	30
Dumpling	8 in.[a]	12	1½ oz	Steam	Steam	20–30
Griddlecakes			4 oz	330–375	176.5–190	3–5
Muffins, bran			1½ oz	425	218	20–25
plain			1⅓–1½ oz	425	218	15–20

[a]Round or square pan.

only to the size of a small pea. Good kneading is also required; however, overmixing results in a tough biscuit. A good biscuit has an attractive shape, with straight sides and a level top (see Table 16-6).

Dropped biscuits are not shaped like other biscuits because they are not kneaded. The dough is simply dropped onto greased pans after mixing. Beaten biscuits are like baking powder biscuits without the baking powder; they are popular in the South. Scones are a type of baking powder biscuit. Shortcakes are made from a rich biscuit dough containing extra sugar and shortening. Dumplings are a variation of biscuits and may contain eggs. All biscuit items are baked at 425°F (218°C) for 15 to 20 minutes. Dumplings are steamed or cooked over boiling liquid with a tight cover on the container.

Muffins are made from a rather thick batter that is produced by blending flour, salt, sugar, and baking powder with a liquid (eggs, milk, melted fat, or oil). Very little mixing is given to the batter, which is kept cold to prevent gluten from forming. In a mixer, the batter is mixed for only 15 to 20 seconds, at low speed. The flour mixture should just disappear, and the mixture should seem pebbly. Muffin pans should be conditioned and only lightly greased. A light dusting of flour over the greased tins reduces sticking. Muffins should be panned and baked immediately after mixing; they are not as good if allowed to stand before being baked. Portioning into tins for baking should be done by taking batter from the outside edge in, to eliminate working the batter. Normally, 6 to 7 lb of flour (14 to 16 lb of batter) yields 100 portions (150 muffins).

Overmixing, mixing too slowly, putting too much flour or liquid into the batter, using insufficient leavening, or setting too low an oven temperature gives a poor muffin. A good-quality muffin is large for its weight. The crust is crisp, shiny, pebbly, and golden brown. The top is rounded, free from knobs, and quite rough. A moist, light, and tender crumb is evident on breaking. No tunneling should be evident. The flavor should not be bready or too sweet. In taste, a muffin is midway between a cake and bread. Loaf breads, quick coffee cakes, and other variety breads are based on muffin mixtures.

Griddle cakes, pancakes, and hotcakes are different names for the same food item, made with the same batter. Waffle batter is richer and slightly sweeter and thicker. Both batters are mixed very much as muffins are, but more mixing can be given because the additional liquid helps separate the gluten particles and lessens the danger of gluten development. Thick waffle batters, however, should not be mixed too much. For 100 portions of hotcakes (two cakes per portion) or waffles, about 13 lb of flour or 30 lb (4 gal) of batter is required.

Hotcakes are baked on a well-greased griddle, allowing about 2 oz for a large cake. The cake cooks on the bottom and bubbles appear on top. When a slight dryness appears at the edge, the cake is turned and allowed to cook on the other side. The less turning, the better. If the top is pitted after baking, the

TABLE 16-6	Biscuit Defects and Their Causes
Defect	*Possible Causes*
Heavy or compact crumb	(1) Overmixed or over kneaded dough; (2) insufficient baking powder or shortening; (3) too much liquid or flour; (4) oven not hot enough
Pale crust	(1) Oven not hot enough; (2) baked in too deep a pan; (3) too much flour
Poor volume	(1) Oven not hot enough; (2) baked in too deep a pan; (3) too much flour
Light but not flaky	(1) Shortening cut too finely into flour; (2) insufficient kneading
Poor shape	(1) Dough too slack; (2) uneven rolling; (3) twisting the cutter; (4) careless cutting or panning

cake was cooked too long on the first side or the heat was too high. A hotcake that has a raised middle may not be completely cooked. Good hot-cakes have an even, good, brown color, and a good rounded shape. They should be tender and moist. The outside should be slightly crisp, and the flavor pleasing and slightly bready.

A waffle grid should be conditioned well. The batter should contain enough fat to make greasing unnecessary, about 4 oz per lb of flour (the same proportions work well for hotcakes). If the griddle sticks even though the recipe is correct, the grid may need conditioning or the temperature may be too low or too high. Put enough batter in so that when the top is lowered, the batter completely spreads over the iron, giving a waffle with an even top and bottom. A large waffle needs about 4 oz of batter. A waffle is considered done when it stops steaming in the grid.

Good waffles are evenly browned with a crisp outer surface and only slight moistness inside. The grid markings are distinct. Flavor is sweet, nutty, and pleasant.

FIGURE 16-5 This special dessert, known as a *croquembouche*, is made of tiny cream puffs stacked upon one another.

■
DESSERTS

Desserts are usually sweet and are used to conclude a meal. They should blend with the meal and complement it. Desserts are supposed to end the appetite, and for this reason they are sweet. They can be used to heighten meal interest and to add something extra to a meal (see Fig. 16-5).

Cakes

Cakes are low in cost, easy to prepare, and can be carried over without too much loss in quality. Cakes made from mixes hold less well than cakes made from scratch. High-quality ingredients, exact measurement, good equipment, and skill are needed to make good cakes.

Three types of cakes exist: butter, pound, and foam. Butter cakes (the most popular type) are made from shortening (in the past, butter), sugar, eggs, liquid, flour, salt, leavening, and flavoring. Pound cakes contain, proportionally, 1 lb each of shortening, eggs,

flour, and sugar. The shortening, eggs, and sugar are creamed and then folded into the flour. Fruit cakes, nut cakes, and steamed puddings are made in this way. Angel foods, sponge cakes, and chiffon cakes are foam cakes. The leavening is contributed by air in the egg foam, although some sponges and chiffons may also contain some baking powder. Foam cakes are difficult to prepare in quantities greater than 100 portions per batch because they are so delicate.

Mixing methods vary, depending on the cake type. The prescribed method for making a butter, pound, or foam cake should be followed carefully.

Butter cakes can be made by creaming the shortening, eggs, and sugar until a light mass is achieved. Then a mixture of flour, salt, and baking powder is added alternately with the liquid to form a thick batter. Many butter cakes, however, are made by the blending method, which gives a tighter grain, a more velvety texture, and a better-keeping cake. The flour and shortening are mixed with a mixer paddle on low speed for 3 to 5 minutes, during which time the bowl is scraped frequently. If chocolate is to be used, it is added in melted form at this stage. Sugar, salt, baking powder, dry milk solids, and cocoa (if used) are sifted and added with one-third to one-fourth of the liquid in two parts to give a smooth batter. Spices are also added at this time. Mixing continues for another 3 to 5 minutes, again with frequent scraping down. Then the remaining liquid, flavoring, and eggs are added in about three parts, and mixing proceeds for 3 to 5 minutes at low speed with frequent scraping down. This method enables the fat to coat the flour, yielding a tenderer, more even-grained cake and reducing moisture loss. The batter is thin, but this is not a defect.

A third mixing method is the muffin method in which all liquid ingredients (including oil or liquefied shortening) are blended into the dry ingredients. A cake mixed in this way must be served soon, since it keeps poorly. A conventional sponge cake is used in very hot climates. Because the shortening may be too soft to cream, half of the sugar is mixed into the shortening, after which dry ingredients are sifted together and then blended in, followed by liquids. Warm eggs beaten to a fluffy sponge are then folded into the mixture with the remaining sugar.

Pound cakes are made by creaming the shortening and then adding sugar and creaming these together well. The eggs are added and creaming is continued until the mass looks like thick whipped cream. Finally, flour is blended in carefully. Some pound cakes call for a bit of liquid and some baking powder. A good pound cake begins with having the ingredients at 75°F (24°C).

Foam cakes depend on egg foam for their lightness and texture. For angel cakes, egg whites at 100 to 125°F (44 to 52°C) are beaten until they reach a soft peak, at which time one-third to one-half of the sugar is beaten in. The remaining sugar is sifted into the flour to help blend the egg whites more easily. The egg whites are folded into the flour and salt with a down-and-under motion, with care being taken not to break the foam. Angel cakes are usually baked in regular tubular ungreased pans; greasing gives a lower volume because the batter cannot climb up the sides of the pan. Angel cakes can be baked in loaf pans or in sheets. A sponge cake is made of warm whole eggs or yolks (if only yolks are used, some other moisture is added). After a stable foam is made with salt, lemon

juice, or cream of tartar, sugar is added; the flour is then folded in. A delicate hot butter or hot milk sponge cake is difficult to make. The butter must be melted and hot; the milk is usually added at 140°F (60°C). Chiffon cakes are made somewhat like sponges, except that they contain oil. The grain of chiffon cakes is tight but light, and they keep well.

Panning is important to cake quality. Butter and pound cake batter should evenly fill pans one-half to two-thirds full. Straight-sided, ungreased sides give greater volume. Greased bottom paper liners are often used. Foam cake batter is placed in nongreased pans, filling them about three-fourths full. Since black metal absorbs heat, black pans are preferred. Butter cake batter fills pans at about 0.2 oz per square inch of pan. An 18 × 26 in. pan holds about 6 lb of batter. The scaling weights of cakes are usually specified (see Table 16-7).

Cake baking occurs in four stages. Preheated ovens should be used. The first stage consists of a slight batter thinning accompanied by a rapid rise. Rising starts at the sides and moves in toward the center. In the second stage, the center starts to rise slightly higher than the sides. Coagulation and gelatinization are taking place. A slight darkening is seen on the surface of the batter, with some bubbles. The cake is in a critical state; any jarring will cause the cake to fall or lose volume. The process is now one of capturing maximum rise through structural firming. No further rise occurs in the third stage; but the crust now takes on color and the structure sets. At the end of this stage, the cake structure is firm enough to allow the pans to be moved gently. In the fourth stage, the cake shrinks slightly and moves away from the pan sides, and the aroma of baked cake is evident. Good bakers never look at a cake to see if it is done. Instead, they judge doneness by aroma. The cake will be firm to the touch, springing back and leaving no imprint; however, foam cakes and some rich butter cakes such as German chocolate cake may lack this spring. When a wire tester or toothpick that is inserted into the cake comes out clean, the cake is done.

Pans that touch the oven sides or each other do not produce good cakes as consistently as do pans that the air is free to circulate around. When baking is done in a convection oven, the fan should not be used during the first two stages, because the air flow will

TABLE 16-7   Baking Procedures for Various Cakes[a]

Type of Cake	Scaling Weight	Baking Temperature °F	Baking Temperature °C	Time (min)
Layer				
Butter or pound (½ to 2 in. deep)				
6 in. diameter	6–8 oz	375	191	18
7 in. diameter	9–11 oz	375	191	20
8 in. diameter	12–14 oz	375	191	25
10 in. diameter	20 oz	360	182	35
12 in. diameter	1½–2 lb	360	182	
14 in. diameter	2¼–3 lb	360	182	
Foam				
6 in. diameter	4–5 oz	375	191	20
8 in. diameter	9–10 oz	375	191	
10 in. diameter	16–18 oz	360	182	
12 in. diameter	1½–1¾ lb	360	182	
Loaf				
Angel, 3¼ × 3½ × 8 in.	7–10 oz	365	185	
tube, small	8–10 oz			
tube, 10 in.	1½–2 lb	360	182	50
Fruit, 3¼ × 3½ × 8 in.	1½ lb	315	157	90
Pound, 3¼ × 3½ × 8 in.	1 lb	355	179	50
3¼ × 6 × 11 in.	3 lb	325	163	100
Sheet				
Butter, 1 × 18 × 26 in.	6–7 lb	360	182	35
3 × 18 × 26 in.	8–10 lb	350	177	
Sponge, 1 × 18 × 26 in.	3 lb	360	182	25

[a]Weights, times, and temperatures are average only; adjust for each cake. Butter cakes made by the blended method may be baked at slightly higher temperatures than those used for conventionally made cakes; temperatures given here are for the blended type.

move the batter and cause a rippled effect on the top; volume will also be less. As with meat, small products are baked at higher temperatures than large ones, and thin products are baked at higher temperatures than thick ones. Some cakes require more than one temperature over the course of baking.

The final steps in cake production are cooling, storing, and makeup. Cool cakes in the pan for at least 15 minutes; loaf and pound cakes may take 30 minutes. Foam and pound cakes should cool upside down. Foam cakes are easier to remove from pans while they are still slightly warm. Depan the cakes, and store them when completely cool in a moisture- and vapor-proof wrap. Makeup includes cutting, shaping, filling, frosting, and decorating (see Fig. 16-6). Cakes can be made up before being stored.

a.

b.

FIGURE 16-6 (*a*) An elegantly decorated cake makes any occasion special when it is served. (*b*) Cakes and desserts cut and decorated into small, bite-sized individual servings known a *petits fours* will dress up any buffet or dessert presentation.

Cookies

Cookies use virtually the same ingredients as cakes, and some are made in exactly the same way as cakes. Cookies are usually richer than cakes, however, and contain the same ratio of flour to shortening as pie doughs do. Sugar, icebox, and similar cookies are made by the creaming method. Foam-type cookies such as ladyfingers and meringues are related to the foam cake family. A chewy cookie is made with syrups containing a large amount of invert sugar (such as molasses, honey, or corn syrup), plus more eggs and less fat to flour. Crisp cookies contain high ratios of fat and sugar to liquid and are baked at a higher temperature. Soft cookies contain a lot of liquid and are baked at a lower temperature. Thickness and size affect crispness and softness. Better spread is obtained if soda or ammonia is used as the leavener than if baking powder is used. Spread is also greater with cookies that are high in sugar; the coarser the sugar, the greater the spread. Cool cookies well and store them in airtight containers (see Fig. 16-7).

FIGURE 16-7 Cookies are always well received as a dessert item.

Egg-Thickened Desserts

Desserts prepared from eggs are quite delicate and can fail unless properly done. Such desserts include custards, soufflés, and meringues (see Fig. 16-8).

A good baked custard is firm and soft. When cut, it shows an even, glossy, cream-colored surface. The

FIGURE 16-8 Another egg-thickened dessert is *crème brûlée*.

FIGURE 16-9 Meringues being prepared for baking into hard shells.

top should be even, not cheesy in texture, and golden brown. Doneness is indicated when a pointed knife inserted near the edge comes out clean. Another test is to shake the custard and note its consistency. It can be removed when slightly underdone in the center since baking continues after oven removal. A stirred or soft custard is made in the same way as baked custards, except that the mixture is constantly stirred over simmering water.

Meringues are blends of whipped egg white foam and sugar. Pies are topped with soft meringues; hard meringues are the base for meringue shells, which are filled with whipped cream or act as the base for a Schaum torte. Whip the whites and add the sugar gradually during whipping. Soft foams are made in a 1:1 ratio of egg whites to sugar (by weight). Hard meringues have a ratio of 1:1.75 or 1:2. Per pound of whites, use ¾ tsp of cream of tartar and a bit of salt. Hard meringues sometimes contain 1 oz of liquid per pound of whites. If the liquid is lemon juice, the cream of tartar is omitted.

Soft meringues are baked for from 12 to 18 minutes at 350°F (177°C); hard meringues are baked at 275°F (132°C) for 1½ hours or for 50 minutes at 325°F (163°C). Hard meringues are scaled at 1¼ oz each or 15 to 18 oz per dozen (see Fig. 16-9). A 9-in. pie needs 4 to 5 oz of soft meringue to cover it well.

Fried Desserts

Doughnuts are popular for breakfast, snacks, or desserts. Cake doughnuts are made from a rich muffin dough that is handled as muffin dough would be—kept cold and not overworked. The doughnuts are cut from the rolled dough and dropped into hot fat. Yeast doughnuts are made from a rich (and usually sweet) yeast dough. They are cut from the dough, proofed, and deep fried. Some bakers cut simple circles instead of doughnuts with holes. After proofing, they lift the circle of dough up carefully and put their forefinger though it, making a hole. The doughnut is then dropped from the fingers into the hot fat. Frying temperatures for doughnuts may be as high as 385°F (196°C). Doughnuts should be drained well on absorbent paper after frying. They can be coated with various frostings or glazes.

An unusual dessert is fried ice cream (see Fig. 6-10). Two other popular fried desserts are fritters and crêpes. Fritters, which usually are deep-fried, are of two kinds: fruit that is dipped into batter and fried; and fruits or other items mixed into a muffin batter and deep-fried. Most fritters are served with syrup or a sweet sauce. Thin pancakes are called crêpes, *blini*, or some other name, depending on the type and on the culture from which they come. They are also popular filled with meat, poultry, or fish mixtures and

FIGURE 16-10 Fried ice cream is usually a scooped ball of vanilla ice cream that has been breaded and quickly deep fried, browning the crust while the ice cream remains frozen.

FIGURE 16-11 Sherbet, sorbet, or granite may be served as a dessert but are sometimes served as an intermezzo, to cleanse the palate between courses.

served as entrées. The crêpe is a mixture of egg, a small bit of liquid, a bit of salt, and flour. The thin batter is poured into well-conditioned buttered pans, and moved quickly around by twisting the pan to distribute the batter evenly in all areas. It is browned well on one side, flipped over, and browned on the other. Special crêpe pans have been designed for producing crêpes. Crêpes Suzette contain a rich, sweet orange filling and are doused with liqueurs and flamed. Crêpes can be made ahead of time, then refrigerated or frozen, and made up for service later.

Frozen Desserts

Frozen desserts include ice cream, sherbet, ices, sorbets, mousses, mallows, glacés, biscuits, parfaits, junkets, puddings, and whips. A slush, granite, or frappé is a coarse, granular product that is only partially frozen (see Fig. 16-11). Most frozen desserts are purchased ready-made. They usually are a low-cost product and require little labor.

Ice cream and sherbet contain a lot of air, which gives them volume and texture and enhances their flavor. Therefore, in dishing or scooping, it is important that the ice cream or sherbet be given maximum volume and not be packed firmly. Figure 16-12 shows a good way to scoop. These frozen items should be

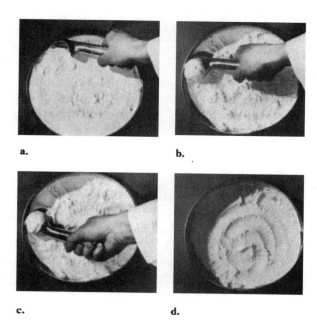

FIGURE 16-12 Scooping ice cream: (*a*) insert the scoop about ½ inch deep into the frozen dessert, starting at the outer edge or wherever the last scoop left off; (*b*) draw the scoop lightly and evenly across the surface, rolling the dessert into a ball; (*c*) when the scoop is filled, turn it up with a twist of the wrist; (*d*) keep the surface smooth, working evenly across the top.

		Number Scoops per
Item	Scoop Size	Portion
A la carte portion	No. 10	1
Banana split	No. 30	3
Bowl of ice cream	No. 30	4
Ice cream soda, malt or milkshake	No. 24	2
Parfait	No. 30	3
Pie, cake, or pudding à la mode	No. 20	1
Sundae	No. 20	2
Sundae, meal portion	No. 12	1
Table d'hôte, plain	No. 16	1

TABLE 16-8   Portion Size of Various Ice Cream Desserts

FIGURE 16-13   Fresh fruit makes an excellent dessert.

served at 8 to 15°F (–13.2 to –9.4°C). Table 16-8 gives the portion size in scoops. Store frozen desserts at temperatures of from 8 to 15°F (–13.2 to –9.4°C).

Fruit Desserts

Fresh fruit makes one of the best and simplest desserts. Fruits are low in calories and high in nutrients (see Fig. 16-13). Serve them cold. A crisp, cold, and sweet yet tart apple, a mellow peach or pear, a flavorful bunch of grapes, a spicy orange, delicious slices of pineapple, and a bowl of succulent berries all vie with the finest bakery creations for credit as the most popular of desserts. Many are served plain; others may be served with sugar and cream. Acidic or antioxidant solutions can be used to prevent tarnishing of freshly cut fruit.

Frozen fruits have a better texture if they are served while still slightly frozen. A 5-lb carton of frozen fruit at 0°F (–18°C) takes 2 to 3 hours to thaw at room temperature and 4 to 5 hours to thaw under refrigeration. A portion of such fruit is usually ½ cup, or 4 oz.

Fruits can also be cooked and served. Some dried fruits need no sugar. Others can be served in a syrup. Fruits in extra-heavy syrup are called preserves; fruits

in heavy syrup are called *compotes*, and fruits in light or medium syrup are called *stewed fruits* or *fruit sauces*. About 15 lb EP of fruit cooked in 1½ gal of light or medium syrup yields 100 ½-cup portions of sauce; four No. 10 cans of fruit usually provide enough fruit to serve 100, but the count or size of the fruit must be appropriate for this purpose.

A crisp is fruit mixture topped with a blend of flour, sugar, fat, and seasonings, and then baked. Oatmeal can be used in place of flour to give a crisper and more nut-flavored top. Betties are layers of fruit alternating with layers of bread or cake crumbs, sugar, and butter or margarine; after being properly layered, the betty is baked. Shortcakes are rich biscuits filled and topped with fruit and whipped cream. Cobblers are fruit dishes with biscuit dough or pie dough on top.

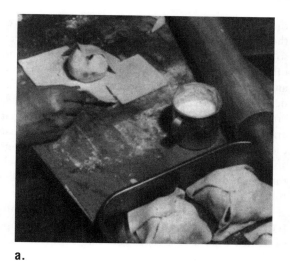

a.

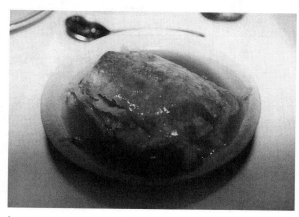

b.

FIGURE 16-14 Apple dumplings make a tasty dessert: (*a*) preparing apple dumplings; (*b*) the finished product served.

Apple or fruit dumplings are popular dessert items. They consist of rich biscuit dough rolled ⅛ in. thick and filled with fruit, which is then sealed inside the dough (see Fig. 16-14). The dumplings are baked in sweetened water, which thickens slightly during baking and is served as a sauce over the dumplings.

Gelatin Desserts

Gelatin desserts include plain gelatin mixtures, gelatin mixtures with fruit, whipped gelatin mixtures, bavarians, and Spanish creams. They are popular and usually fairly inexpensive to make. They lend themselves well to merchandising. Many fruits, fruit juices, and other products carried over can be used in gelatin desserts.

Pies and Pastries

Pies and pastries are popular, but they require considerable labor and expensive ingredients. Many operations purchase ready-made items because they do not have the specialized labor necessary to make them.

Single-crust pies have only a bottom crust. They can be prebaked, filled with various fillings, and topped with meringues or whipped cream. Unbaked single crusts are filled with custard, pumpkin filling, or

a pecan filling. Double-crust pies are usually filled with fruit and covered with a top crust that is sealed around the edge to the bottom crust. Baked pie shells can be filled with ice cream, gelatin mixtures, whips, and other products. Tarts, dumplings, and other desserts are closely related to pies.

Pie Crust

Pie dough is usually made from flour, shortening, salt, and water. The fat shortens the dough and makes a tender crust. Flakiness is achieved by having fat particles separate layers of dough. The fat melts, creating a pool; then steam forms here and raises the dough, and (when coagulation and gelatinization occur) solid, flaky layers of crust are formed. Salt adds flavor but toughens the gluten, so minimal amounts of it are added. Working a dough also develops gluten. The ingredients and their amounts (in percentages based on 100 percent for flour) for different pie doughs are shown in Table 16-9.

Bread flour is used in making puff paste, giving a firmer dough so that more flakiness results. About 10 percent gluten flour is used for the other crusts—although, to make a good flaky crust, some bread flour may be used. To give sufficient shortness when bread flour is used, the shortening is increased. Flaky and

TABLE 16-9   Relative Percentages of Ingredients in Various Pie Doughs (Flour = 100 Percent)

Ingredient	Mealy Crust	Semi-flaky Crust	Flaky Crust	Puff Paste
Flour, bread				100
pastry	100	100	100	
Salt	2–3	2–3	2–3	1
Shortening	50–60	75	100	100
Water	25–30	30–35	35–40	35–40

puff paste doughs contain as much shortening as they do flour. A weak flour can be mixed more and can contain less shortening than a stronger flour.

Specialty crusts are made from ground pecans, almonds, filberts, or other products, replacing 25 percent of the flour. Adding these ingredients weakens the flour and produces a shorter crust. Adding soy flour does the same. Graham cracker crusts are made by weight: 100 percent cracker crumbs, 30 percent sugar and 55 percent butter or margarine. Cinnamon, almond flour, powdered sugar, or other ingredients may be added (see Fig. 16-15). Gingersnaps, vanilla wafers, chocolate wafers, or other dried crumbs can be used for variety crusts. About 2 oz of the crumb mixture packed well on the bottom and sides are sufficient for a 9-in. pie. The crust can be baked in a 350°F (177°C) oven for 10 minutes. Some foodservices prefer to refrigerate such crusts overnight without baking them.

Production Methods

Mealy, semiflaky, and flaky crusts are made by varying the ingredients and method of mixing. Rubbing the shortening thoroughly into the flour produces a mealy crust. Mixing it in only to the size of peas gives a semiflaky crust, and leaving the grains the size of walnuts produces a flaky crust. The materials should be cold to reduce gluten formation. Mixing should be limited to the amount needed to spread water around. A pie dough is better if it is left standing for at least 15 minutes after mixing. When this is done,

the dough is easier to roll and the baking shrinkage of the dough is less. Mealy crusts soak less than others, so they are used for custard pies, pumpkin pies, and other pies in which an uncooked filling is put into an unbaked crust. Semiflaky crusts have good tenderness and flakiness and make a good all-purpose crust. A flaky crust is used for baked pie shells that are to be filled later and for tops of double-crust pies. Since a flaky crust soaks easily, cream fillings and others that would soak are not put into the shell until just before needed. Puff paste is made by the rolled dough method.

Regular pie crust doughs are usually made from 6 oz of dough for bottoms and 5 oz of dough for tops of 9-inch pies. The crust is rolled to about ⅛ inch thick. Machine rolling is used for quantity production. Hand-rolling with a rolling pin or roller that is bout 1½ inches in diameter and about 18 inches long is common. A smooth surface, rolling cloth, or board is used, with only a small amount of rolling flour on it. Additional flour toughens dough. The piece of dough should be well-rounded and flat. The roller pushes out the dough gently, allowing the dough to move on its own and not be stretched; stretching causes shrinkage in baking. The gluten strands need to be relaxed in rolling. The dough is lifted and turned as often as is necessary to get a perfectly round dough. An experienced baker can do this and get the exact shape needed so that no trim is required. The pie crust

FIGURE 16-15   Many fine cheese cakes feature graham cracker crusts containing cinnamon, sugar, and other ingredients.

is folded and laid gently into the pan without stretching. It is then shaped to the pan. Crusts may be baked docked (pierced with a fork all over) to keep the crust from blistering or bubbling on its surface. Double-panning—putting another pan tightly over the crust—also prevents blistering. The extra pan is removed shortly before the crust finishes baking so that the crust can take on a more even color. Many bakers turn the first pan upside down, put the dough over it, and then put the other pan down over this.

Single crusts should be baked at temperatures of from 425 to 450°F (218 to 232°C) for about 15 minutes. Double-crust pies should be baked at temperatures of from 425 to 450°F (218 to 232°C) for 10 minutes; then the temperature should be dropped to between 350 and 375°F (177 and 191°C) for the remainder of baking. A tough crust results from too low a temperature. High heat causes overbrowning and, in double-crust pies, an unbaked filling. Putting hot fillings into double-crust pies reduces their baking time.

Pie Fillings

Many different fillings are used for pies. Single-crust pies can be topped with meringues, whipped cream, toasted coconut, crumbs, nuts, or other products. Unbaked single crusts are filled with uncooked fillings, such as custards, chiffons, soufflés, or pecan custard. A 9-in. pie needs 3 to 4 cups of filling. A meringue or whipped cream topping takes up some space, so less filling is used. Management should require that pie markers be used, to ensure that all crusts are the same. Some pies are of better quality if slightly warm at service.

Pastries

Tarts, tartlets, turnovers, fried pies, and other items may be made from pie doughs or puff pastes (see Fig. 16-16). Tartlet tins are filled, and then dry beans or rice are poured in to keep the crust from blistering during baking. Double-panning can also be used when baking tartlet crusts. Docking after the dough is put on the outside of the small pan is also possible. For tartlets, bakers often use a very rich pastry dough that is more a cookie dough than a pie dough (see Fig. 16-17). Rings or hoops can be placed on greased baking sheets and

FIGURE 16-16 Puff pastry makes this dessert extra special.

filled in with dough for large tarts. The dough shapes are given a border and then docked and baked. Dumplings and turnovers can also be made in fancy shapes and baked. If they are deep-fried, the dough should be leaner; the results of this method are called *fried pies.*

Puddings

Most puddings can be made quickly in quantity at reduced labor cost. Material cost is often low. Every menu should offer some of the wide variety of puddings. An old-fashioned rice custard pudding or bread

FIGURE 16-17 Display trays of various finished puff pastries.

custard pudding commonly sells out before other desserts.

Steamed (Boiled) Puddings

Steamed puddings are usually heavy products served warm with a sauce. They are usually either steamed or boiled, although some cooks modify this and bake them. They often use suet as the shortening. Bread crumbs, eggs, and strong bread flour are used to hold heavy quantities of fruit and nuts. Some contain no added moisture, since the fruit furnishes all that is needed; this is the case with carrot pudding or apple pudding. The ratio of batter to fruit and nuts is the same one used in fruit cakes—from 1:1 to 1:2.5. Some steamed puddings are made from pound cake batters. Soda is usually the leavening agent, which results in a darkened product because the fruit, nuts, spices, and some other ingredients darken in the presence of alkalies. The texture is porous. A good steamed pudding is rich, delicate in texture, and flavorful. The puddings are steamed in large or individual-portion containers, which are greased and floured well. The batter swells and so the containers are filled to about two-thirds full; then they are covered. Steaming takes from 2½ to 3 hours. Baking is done with the puddings standing in water in a 325°F (163°C) oven. Steamed puddings usually keep well and can be reheated.

Starch-Thickened Puddings

Starch thickens many fine puddings. Cornstarch, tapioca, or arrowroot may be used as the thickening agent. A blancmange is made by adding cornstarch, sugar, and vanilla to milk and cooking the mixture in a double boiler, steam-jacketed kettle, or steamer. A steamer is not the best piece of equipment to use, since the product should be stirred as it cooks.

The amount of cornstarch needed varies from 6 to 12 oz per gal of liquid. Molded puddings require more cornstarch, while cream puddings require less. A cream pudding is a blancmange with eggs added; it needs less cornstarch because the eggs act as thickeners (see Table 16-10). Some margarine or butter may be added. Pearl, granular, or quick-cooking tapioca can also be used to thicken puddings. Indian pudding is thickened with cornmeal and contains molasses. A portion is ½ cup; about 3⅓ gal per 100 portions are needed.

■

SUMMARY

Bakery products can do much to enhance a meal and to bring in added income in commercial operations.

Flour is a starchlike product used to give structure and body to bakery products. Hard-wheat or bread flour is used for yeast breads and other items needing

TABLE 16-10 Thickeners Needed for 1 Gallon of Pudding

Starch Thickener	Amount of Thickener [a]	Whole Eggs
Cake or bread crumbs	1–2 lb (1–2 qt) [b]	
Cornmeal, farina, or other	1 lb (1 qt))	
Cornstarch	4–5 oz (1 cup)	1½ lb (1½ pt or 16 eggs)
Flour	8 oz (1 pt)	1½ lb (1½ pt or 16 eggs)
Rice, cooked	1 to 1½ gallon	
uncooked	9–14 oz (1–1½ pt)	
Tapioca, granular	6–9 oz (1½–2 cups)	1½ lb (1½ pt or 16 eggs)
pearl	12 oz (3 cups)	1½ lb (1½ pt or 16 eggs)

[a] Increase thickener if eggs are omitted.
[b] Quantity depends on moistness of crumbs.

strong structures. Pastry or soft-wheat flour is used for pastry, quick breads, cakes, and cookies. The gluten or protein in flour gives products their basic structure. It is elastic. If overworked, it can make some products tough. A yeast bread needs a strong structure, so the dough is worked considerably; pie dough is not, because a tender product is wanted.

Shortening adds richness, color, sheen, grain, and texture to bakery goods. It makes items tender by lubricating gluten.

Sugar adds sweetness and also tenderizes and gives color, texture, and moistness to products. Some sugars draw moisture and thus make products moister. Sugars are also used frosting, sanding, and coating materials.

Leavening agents aerate, or make bakery goods light. They help to give volume, shape, and texture. Air, which is a good leavener, is brought into bakery goods by creaming, by beating, or by being incorporated into egg foams. Steam, another good leavener, expands 1600 times in volume in changing from water to vapor. Steam is the only leavener in items such as cream puff shells and is also important in leavening cakes, breads, and other products. Baking soda, one of the oldest chemical leaveners, gives off carbon dioxide gas, which leavens the product. An acid reacts with soda to produce the gas; heat with moisture also produces the gas. Baking powders combine dry acid ingredients with soda. When they are moistened and heat is applied, they react to form carbon dioxide gas. Single-acting baking powders react completely when moistened in the cold stage. Heat is not necessary to produce carbon dioxide. Double-acting baking powders react partially when cold and more fully in the hot stage. Double-acting baking powders are usually used in foodservices. Baking ammonia, which gives off carbon dioxide and ammonia gas when moistened and heated, is used in cookies, cream puff shells, and other thin-shelled products.

Yeast is also a leavener. It feeds on carbohydrates, producing alcohol and carbon dioxide gas. Yeast is a plant and must be nurtured by favorable conditions to make good products.

Eggs are added to bakery products to add structure, tenderness, volume, nutritional value, flavor, color, and moisture. They coagulate and thus help to firm up bakery products. Spices and flavorings are used to give flavor, richness, and color to bakery goods.

Breads are either yeast or quick-type and can be made from scratch or from mixes. A straight yeast dough is made by mixing all ingredients at one time; a sponge yeast dough is made by withholding part of the flour and adding it after the first fermentation. A no-time dough is made in the same way as a straight dough but is richer in sugar and contains more yeast. This method is used to produce yeast breads in as short a time as possible. A rich or sweet yeast dough is used for Danish pastries, coffee cakes, rich breakfast rolls, and other products. Leaner yeast doughs are used for regular loaf bread, sandwich bread, rolls, and hard-crusted breads.

After mixing, a yeast dough is left to ferment in a warm area. After the dough reaches a stage in which it will pucker when a hand is put into it, it is punched. Punching distributes the food so that the yeast can get new energy-giving materials. The dough is then benched (given a rest) and afterward made up into the desired shapes. In the proofing stage, temperatures and relative humidity are higher than for fermentation, in order to get a fast rising action before baking. The bread is then baked, cooled, and stored. Hard-crusted breads are not wrapped.

Quick breads are leavened by chemical leaveners—usually baking powder. Biscuits are made from a fairly stiff dough and resemble bread. Muffins are made from a more slack dough; care must be taken to ensure that too much gluten does not develop and make them tough. Muffin dough is kept cold to reduce the chance for gluten development. The batter for griddlecakes and waffles is thinner than for muffins. They often are served for breakfast.

Desserts on a menu should be selected for their flavor, richness, color, variety, texture, form, appearance, and ability to lend themselves to quantity production. They need not be expensive and can do much to improve a meal.

Cakes are not difficult to prepare. Butter cakes (the most popular) are made by creaming shortening, sugar, and eggs, and then adding sifted dry ingredients and liquid; or they may be made by the blending method, which calls for blending first the fat and flour and then the other ingredients. A muffin method and a sponge method also are used occasionally to make butter cakes. Pound cakes are made by creaming equal quantities by weight of shortening, sugar, and eggs, and then folding in the same amount of flour. Foam cakes (chiffon, angel, and sponge) are leavened largely by the air in egg foams. These delicate cakes are difficult to make. Cookies are made with the same ingredients, mixing techniques, and other methods used for cakes.

Custards are liquids thickened with eggs. They can be baked or stirred over heat. Custards contain 1½ to 2 lb of whole eggs to 1 gal of liquid. Soufflés are very light baked desserts thickened and leavened by eggs. Meringues are made of egg white and sugar, with cream of tartar and salt added. Soft meringue tops pies and desserts, and hard meringue is a base for ice creams or tortes.

Fried desserts include doughnuts, fritters, and crêpes. Cake doughnuts are made from a rich muffin batter-type dough. Yeast doughnuts are made from a sweet (rich) yeast dough. Fritters consist of fruit that is either batter-dipped and deep-fried or incorporated into a muffin batter and then deep-fried. Fritters usually are served hot with a syrup or sweet sauce. Crêpes are very thin pancakes that usually contain a sweet filling and often are flamed.

Frozen desserts, such as ice cream and sherbets, and fruits served as desserts are very popular in today's foodservice operations.

Pies and pastries can be made from various kinds of pie doughs and then filled. Pie dough is made from flour, shortening, salt, and water. The amount of fat and how it is incorporated are varied to produce a mealy crust, a semiflaky crust, or a flaky crust. Puff paste is made by a special method using pie dough ingredients. A mealy crust is used for unbaked single-crust pies that are filled with an uncooked filling, such as custard. A semiflaky, all-purpose crust can be used for all pies. A flaky crust is used for baked pie shells or for topping double-crust pies. Single-crust pies are frequently topped with meringue or whipped cream. Double-crust pies are usually filled with fruit.

Pie dough should not be worked too much and should not be stretched as it is being rolled. Baked shells are docked or double-panned for baking. Pastries made from pie doughs or puff pastes include tarts, tartlets, turnovers, and fried pies.

Most puddings are low in cost and can easily be made in quantity. Steamed (boiled) puddings are rich products containing fruit, nuts, spices, and other ingredients. The shortening is often suet. They can be boiled, steamed, or baked and are served warm with a sauce. Starch-thickened puddings include blancmanges, cream puddings, arrowroot puddings, and tapioca puddings.

■

CHAPTER REVIEW QUESTIONS

1. Describe high-quality muffin. Does it matter if the batter is overmixed?

2. What two kinds of wheat flours are used in bakery goods? Why is each used? Why is a good patent flour added to whole-wheat flour or rye flour in making bread?

3. What is gluten: what does it do in flour and in bakery goods? What procedures should be followed to develop or retard development of gluten? Name some products that need a strong gluten and some products that need a weak gluten.

4. List some of the other more commonly used bakery ingredients, and list the job each does.

5. Name the four major types of leavening agents and the subtypes. How does each work?

6. What methods are used to make yeast breads? Give the steps used to make bread by the straight-dough method. Explain what is done in each step. List the proper temperatures that should be used in the various steps. How are hard-crusted breads handled in storage versus soft-crusted breads?

7. List the various kinds of quick breads and indicate how they are made. Give some of the standards that should be looked for in these products.

8. List the three major classifications of cakes. How are butter cakes made by various methods? What kind of foam cakes are common? Indicate

the properties of each type of cake in the major classifications.

9. How do cookies differ from cakes in ingredients and methods of making?

10. Describe the making of a meringue. State the proportions of sugar and eggs in soft and hard meringues.

11. How are fruits used as desserts? List the ways fruits can be served cooked.

12. What is a crisp? a betty? a shortcake? a cobbler?

13. (a) How are steamed puddings made and cooked? (b) How are starch-thickened puddings made?

CASE STUDIES

1. A large foodservice corporation with many multiple units has made a rather good merchandising program by serving its own homemade breads, pastries, and desserts; but now it is having difficulty finding good bakers. Management is also wondering whether running its own bakery in individual units pays off. It calls in a consultant to review the operation's entire baking and dessert program. It wants to know if it can purchase adequate breads and dessert products to satisfy menu needs. Would it pay to have some units produce their own items or should the foodservice purchase entirely from outside, or should it set up a central commissary in its most heavily concentrated sales areas to provide the products for those areas and let the individual units outside make their own? In other words, which way should the operation go? With a limited background on the company and on relevant conditions, set up the pros and cons that you as a consultant should consider in assessing the situation. Why should a foodservice today make its own breads, pastries, desserts, and other bakery products? Why should it not? Why should a large foodservice chain depend on products from the outside? There are many other potential factors you should add to your summary to cover the problem thoroughly. Set up a summary of your ideas in outline form, since you are to present this account verbally before a meeting of executives of the company.

2. Convenience baked products have been available for years. Foodservice operator Louis Pruse has decided to stop baking his own bread and move toward a convenience product. His steak house serves either a basket of rolls or a loaf of bread with whipped butter on request. In researching the Web site at *www.richs.com*, you are to recommend to Louis a bread loaf that he might serve as well as a dinner roll. Explain why you have decided on the products you are recommending to Louis.

VOCABULARY

Gluten

Ash

Patent

Strong flour

Hard-wheat flour

Soft flour

Weak flour

Middlings

Clears

Cake flour

Pastry flour

Bread flour

Graham flour

Shortening

10X

Leavening

Carbon dioxide

Single-acting

Double-acting

Baking powder

Baking ammonia

Yeast

Natural flavoring

Pumpernickel

Sourdough bread

Dough cutter

Rolling pin

Lean dough

Sweet dough

Fermentation

Punch

Bench

Rest

Makeup

Proof

Docking

Biscuit

Beaten biscuit

Dropped biscuit

Dumpling (steamed)

Muffin

Griddlecake

Waffle

Hotcake

Pancake

Waffle griddle

Butter cake

Creaming method

Blending method

Foam cake

Angel cake

Sponge cake

Blistering

Single crust

Double crust

Fried pie

Chiffon cake

Pound cake

Conventional sponge method

Muffin method

Panning

Scaling

Spread

Meringue (soft)

Meringue (hard)

Cake doughnut

Yeast doughnut

Fritter

Crêpe

Crêpes Suzette

Stewed fruit

Compote

Preserve

Crisp

Shortcake

Betty

Cobbler

Tartlets

Mealiness

Flakiness

Blancmange

ANNOTATED REFERENCES

Amendola, J. 1993. *The Baker's Manual*, 4th ed. New York: John Wiley & Sons, Inc. (Total baking procedures are reviewed in this classic text.)

Berl, C. 1998. *The Classic Art of Viennese Pastry*. New York: John Wiley & Sons, Inc. (Pastry is the topic of this text with recipes.)

Boyle, T., and T. Moriartz. 1997. *Grand Finales: The Art of Plated Desserts*. New York: John Wiley & Sons, Inc. (Dessert presentation is pictured and discussed in this text.)

Boyle, T., and T. Moriartz. 1998. *Grand Finales: A Modern View of Plated Desserts*. New York: John Wiley & Sons, Inc. (Dessert presentation is pictured and discussed in this book.)

Donovan, M. (ed.). 1996. *The New Professional Chef*, 6th ed. New York: John Wiley & Sons, Inc., pp. 375–418 and 1025–1128. (Baking and pastry as well as baked products are considered in these chapters.)

Donovan, M. 1997. *Cooking Essentials for the New Professional Chef*. New York: John Wiley & Sons, Inc., pp. 425–470 and 709–759. (Baking and pastry as well as baked products are considered in these chapters.)

Friberg, B. 1996. *The Professional Pastry Chef*, 3rd ed. New York: John Wiley & Sons, Inc. (Baked products and their accompaniments are discussed in this chapter.)

Gielisse, V., M. E. Kimbrough, and K. G. Gielisse. 1999. *In Good Taste*. Upper Saddle River, N.J.: Prentice Hall, pp. 327–370. (Baked products are discussed in this chapter.)

Gisslen, W. 1995. *Professional Cooking*, 3rd ed. New York: John Wiley & Sons, Inc., pp. 653–793. (Breads, cakes, icings, cookies, pies, pastries, creams, custards, puddings, frozen desserts, and sauces are considered in these chapters.)

Janericco, T. 1994. *The Book of Great Desserts*. New York: John Wiley & Sons, Inc. (Dessert recipes and presentation are the basis for this book.)

Kapoor, S. 1996. *Healthy and Delicious: 400 Professional Recipes*. New York: John Wiley & Sons, Inc., pp. 365–467. (Healthy recipes for baked products are presented in this chapter.)

Karousos, G., B. J. Ware, and T. H. Karousos. 1994. *The Patissier's Art*. New York: John Wiley & Sons, Inc., pp. 63–170. (Baked products are considered in this text.)

Labensky, S. R., and A. M. Hause. 1999. *On Cooking: A Textbook of Culinary Fundamentals*, 2nd ed. Upper Saddle River, N.J.: Prentice Hall, pp. 861–1080. (Breads, pies, pastries, cookies, cakes, frostings, custards, creams, frozen desserts, and sauces are presented in these chapters.)

Pauli, P. 1997. *Classical Cooking the Modern Way*, 3rd ed. New York: John Wiley & Sons, Inc., pp. 339–400. (Baked products are reviewed in this chapter.)

Stogo, M. 1998. *Ice Cream and Frozen Desserts*. New York: John Wiley & Sons, Inc. (Frozen desserts are focus of this book.)

CHAPTER 17

Dairy Products and Eggs

■

LEARNING OBJECTIVES

By the end of this chapter, the reader will:

1. Learn about the nutrition, quality, and grading of eggs.

2. Recognize the difference between the various cooking methods used in foodservice operations to cook and present eggs.

3. Know the types of dairy products, including milk and cheese, and how they are used and served in foodservice operations.

■

SELECTED WEB SITE REFERENCES

American Egg Board
www.aeb.org/

California Egg Commission
www.eggcom.com/

Belgioioso Cheese, Inc.
www.belgioioso.com/

Commercial Creamery Company
cheesepowder.com/

Cabot Creamery
www.cabotcheese.com/

Land O'Lakes Incorporated
landolakes.com/Main.cfm

As noted in Chapter 15, heat, especially if high and prolonged, toughens proteins of the type found in high protein animal products. The proteins in dairy products and eggs are more delicate and respond to heat more drastically than in flesh foods, so they require special care in cooking.

■

DAIRY PRODUCTS

Milk and its products are among our most nutritious foods. They are an excellent source of calcium and phosphorus, and they also provide a limited amount of iron. They are a good source of some B vitamins, especially riboflavin. The protein in milk is complete and very digestible. Milk fat is fairly saturated. Table 17-1 gives the percentage makeup of different milks. Available grades of milks, creams, butter, and cheese include the following (from best to worst): AA, A, B, and C.

Milk and Cream

Coffee or table cream is 18 percent milkfat; light whipping cream is 30 to 34 percent milkfat; heavy cream is 34 to 36 percent milkfat; half-and-half is about 10.5 percent milkfat; and regular milk is 3¼ to 4 percent

TABLE 17-1	Approximate Composition of Different Milks (Percent)			
Type of Milk	*Fat (Butterfat)*	*Nonfat Solids (Protein, Milk, Sugar, Minerals)*	*Water*	*Sucrose (Sugar)*
Dry whole	27	70	3	
Evaporated nonfat	Trace	20–30	70–80	
Evaporated whole	8	18	74	
Fresh nonfat	Trace	9	91	
Fresh whole	3.5	8.5	88	
Nonfat dry	Trace	96	4	
Sweetened condensed whole	8	20	30	42

milkfat. The milkfat of yogurt, buttermilk, and other dairy products varies. Sour cream is 18 percent milkfat. Evaporated and condensed milk are fairly high in milkfat, since they are concentrated from regular milk.

Milk and its products are used in many foods as a liquid base in which other ingredients are brought together. It contributes flavor, color, texture, and richness. A cake baked with milk keeps better. Dry milk is satisfactory for much cooking and baking. Butter is made from cream and is 80 percent fat; the remainder is salt and liquid. Margarines are also 80 percent fat. Most contain milk solids for flavoring. Low-saturated-fat margarines are available. The fat in butter is fairly saturated. Butter may be made from either sour or sweet cream. Sweet cream butter is very mild in flavor and can be ordered unsalted. Grades for butter are the same as for other dairy products, except that it can also be ordered by grade score: AA is 93 score, A is 92 score, B is 89 score, and C is 88 score. Foodservices normally order AA (93 score) or A (92 score). Cooking butter may be A or B. Little or no grade C butter is used. Butter is usually available in pats, in wrapped quarters, half-pounds, or pounds, or in cubes of pound bricks stacked 4 × 4 × 4 or 64 pounds. Cooking butter is often ordered in cubes. Pats come usually in 5-lb boxes, layered between sheets of parchment, wax, or polyethylene paper, with seventy-two to ninety pats per pound.

Principles

Milk is a delicate substance that can curdle and scorch easily. Fresh milk is almost neutral, but bacteria that produce lactic acid can sour it. With a pH of about 4.6, milk becomes unstable and curdles, separating into casein and whey. Wine, lemon juice, tomato juice, vinegar, or other acids easily curdle milk. Curdling is a problem since these are often blended with milk in cooking. Milk can also be curdled by salts such as nitrites or nitrates in cured meat, by tannins in potatoes, asparagus, and other products, by very hard water, and by coffee. Heat and salt speed curdling. Evaporated milk is the most stable form of milk, with fresh milk next and dry milk last.

To prevent curdling, cook at low heat and withhold salt until just before service. Any acid should be withheld until it must be added. Adding soda to milk makes it more alkaline and helps it resist curdling, but it also destroys vitamin C and thiamine and has a bad effect on flavor and (sometimes) color. Curdling substances, such as a tomato velouté for tomato soup, must be handled in a special manner. First pour a small bit of the curdling product into the warmed milk; stir vigorously to blend, and then add this tempered mixture into the tomato sauce.

Some milk proteins coagulate when heated. As a result, a scum forms over the top or milk solids collect on the bottom, which leads to scorching. Putting a bit of fat over the milk top prevents the scum from forming. To prevent scorching, heat milk in a double boiler, a steam-jacketed kettle, or a steamer. It also can be heated over a hot surface in a very heavy pan, if it is stirred well. Milk heated for a long time acquires a darker color and a flatter, caramelized flavor.

Pasteurization and Homogenization

Dairy products are pasteurized to destroy pathogens. Milk is pasteurized by being heated for 30 minutes at 143°F (62°C). It can also be flash-pasteurized by being heated for 15 seconds at 160°F (71°C); even higher temperatures and shorter times are now common. Ultrapasteurized products are subjected to the highest temperatures of all. Homogenization divides fat globules so that they remain suspended in the product instead of rising to the top. Milk to be homogenized is forced under very high pressure through tiny holes; this breaks up the fat into a fine emulsion. Whipping cream is usually not homogenized, since homogenization reduces its whipping ability.

Foams

Milk and its products have the ability to foam because they contain protein and fat. The more fat present, the stabler the foam.

Whipping cream and the container in which it is whipped should both be cold. Warmth destroys whipping ability. Using a deep bowl helps develop foam. The increase in volume is about one to two times; recipes usually state the amount of unwhipped cream to use. Using powdered sugar as a sweetener instead of granulated gives a more stable foam. Whipping cream turns into sweet butter if overwhipped.

To whip evaporated milk, scald it first and then cool it down. (Cans may be put into boiling water for 5 to 10 minutes). Adding a bit of hydrated gelatin (½ tsp unhydrated gelatin per pt) aids in giving a more stable foam. When the milk is at 40°F (4.5°C), it can be whipped. When the foam has almost developed, add 3 Tbsp of lemon juice per pt (with sugar and flavorings) and finish beating. Then refrigerate.

Although it is difficult to get a stable foam with a product that is less than 30 percent milkfat, whole or nonfat dry milk can be used to make a low-calorie whipped product. The increase is to four times the original volume, but the foam is not stable. Use 1 qt of instant dry milk, or ¾ qt of regular dry milk per qt of ice water to make a foam. Whip this until a soft peak forms, and then add a cup of lemon juice. The resulting foam has a lemony flavor. Whip until the product is stiff, and add 2 cups of sugar and other flavorings. Then refrigerate.

Uses of Milk Products

Dry milk is milk evaporated until only the solids remain. It becomes like regular milk when this moisture is replaced; 1 lb of dry milk per gal of water makes a slightly more concentrated milk than natural milk (see Table 17-2). Dry milk made a day before use tastes better than newly reconstituted milk, since the lactose takes time to go into solution. Dry milks may curdle with hard or extended boiling. Add 5 oz of butter or other fat per gal of nonfat milk to give the milk a fat content equivalent to that of regular milk.

Adding extra milk solids to baked products and frozen desserts improves their quality, texture, and flavor. They also keep better. Frozen desserts with extra dry milk solids are smoother. Some recipe adjustment may be necessary.

Instant dry milk has been processed to go into solution easily. Regular dry milk does not, so it must be mixed more carefully. In baking, add dry milk with the sifted dry ingredients, and add the extra liquid with the liquid used in the recipe. Weigh dry milk, rather than measuring it. This allows easy interchange of instant and regular dry milk. The best way to blend regular dry milk into water is to sift it over the top and use a wire whip to blend it in.

TABLE 17-2 Processed Milk Equivalents of 1 Gallon of Milk

For 1 gal whole liquid milk use:	For 1 gal nonfat liquid milk use:
18 oz dry whole milk	13 oz dry nonfat milk
7½ lb or 3¾ qt water or 13 oz nonfat dry dry milk 5 oz melted fat	7¾ or 3 qt 2½ cups water or 4½ lb nonfat evaporated milk
7½ lb or 3¾ qt water or 4¼ lb evaporated whole milk	4¼ lb or 2 qt ½ cup water or 3 lb condensed nonfat milk
4½ lb or 2¼ qt water or 4 lb condensed whole milk	7 lb or 3½ qt water[a]
6¼ lb or 3¼ qt water[b]	

[a]Will also contain 1 lb 5 oz sugar.
[b]Will also contain 1 lb 10 oz sugar.

Evaporated milk is milk that has had a lot (about 55 percent) of its water extracted. If the evaporated milk is from whole milk, multiply the quantity of evaporated milk by 2.2 to get the equivalent amount of whole milk. Thus, 1 pt of evaporated milk (16 oz) equals 35.2 oz of whole milk (2.2 × 16 = 35.2). The regular No. 1 can holds 14.5 oz, which is equivalent to 1 qt of whole milk (2.2 × 14.5 = 31.9). Sweetened condensed milk is evaporated milk that contains 45 percent added sugar. It is seldom used in foodservices.

Cream is made from milk. It is lighter than milk and so can be centrifuged out of it. Customers add cream to coffee until they see a desired color. The higher the milkfat content, the better the color. About 1 Tbsp (½ oz) is a portion for 6 oz of coffee. Cream is often portioned. Homogenized cream has a smoother flavor in coffee than nonhomogenized. Many operations use synthetic products rather than true cream.

Cheese

Cheese is primarily composed of milkfat and casein. It is separated from milk by adding bacteria that develop lactic acid, by adding acid, or by adding rennet (an enzyme that solidifies milk into a clabber so that it can be separated into curds, which are solids, and whey, which is liquid). There are many kinds of cheese, and each type is made by a slightly different technique. Most cheese is made from cows' milk; others are made from the milk of goats, sheep, reindeer, or buffalo.

The aging of cheese is divided into three periods, called *current, medium*, and *aged* or *ripened;* American current is aged up to 30 days; medium, 30 to 90 days; and aged or ripened, over 90 days. Swiss current is aged up to 60 days; medium, 60 days to 6 months; and aged, over 6 months. Most foodservices purchase their American cheese in daisies weighing about 25 lb, but they can get cheese in other size units as well. Swiss cheese comes in different-sized units, too. Large wheels of Swiss can weigh 50 to 100 lb. Processed cheese usually comes in 5-lb bricks. Stages of cheese processing are identified in Figure 17-1.

Aged (ripened) cheese has a relatively tart, sharp flavor and is softer and more mellow than fresh cheese. Fresh cheese is rubbery and more difficult to blend into sauces. Moisten dry cheeses such as Parmesan before trying to blend them with other ingredients.

Cheese is a delicate product. High heat hardens it. In a sauce or rarebit, high heat causes curdling. Cheese is blended into sauces at temperatures below the boiling point. Cheese can curdle milk, as sometimes happens in cheese sauces. To avoid this, first thicken the milk with starch, binding the casein in, and then add the cheese.

Enzymes and bacteria aid in cheese development. Limburger is ripened by having bacteria work their way into the cheese, making it softer and giving it a ripe, mellow flavor. Brie, Camembert, brick, and other cheeses ripen in the same way. Others, such as Swiss, ripen by having bacteria work from within them; for this to happen, bacteria must be added to the cheese as it is being made. While the cheese is aging, these bacteria are at work; the large bubbles or eyes in Swiss cheese come from bacteria-developed gas in the cheese. Other cheeses are ripened with molds, including Roquefort, blue, Gorgonzola, and Stilton.

Some cheeses are quite dry because the temperature was raised while they were being made, driving out moisture and hardening the curd. Other cheeses, such as mozzarella used for pizza, are tacky and stringy. These are pulled and worked so that the casein joins together into a sticky mass. Some cheeses become very smooth as they age; their flavor gets richer and also more tart. Selecting the right cheese for a particular cooking need is not difficult if a proper study has been made of cheese and its properties.

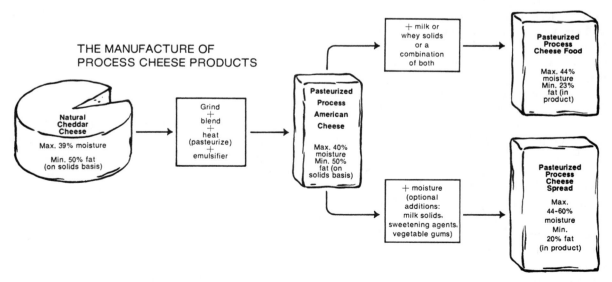

FIGURE 17-1 Manufacturing cheese and processed cheese products.

Natural cheese is a product made by coagulating milk and then separating the solid curd from the whey. Some natural cheeses are ripened to develop flavor and textures, while others are used unripened. Ripened cheeses often are labeled as to their degree of ripening. Cheddar cheese, for example, may be labeled "mild," "medium," or "aged (ripened)." Some connoisseurs prefer natural cheese to others because of its flavor and texture. Cheese flavors range from bland (cottage cheese) to tangy (blue) to pungent (Limburger). Textures vary from the smooth creaminess of cream cheese to the firm elasticity of Swiss.

A processed cheese does not age because, after being brought to the proper flavor and texture, it is pasteurized (which prevents further changes). Processed cheese is 98 percent cheese. It can be flavored with other items. It goes into solution very easily in sauces because it contains an emulsifier that helps blend it in. It does not have a rind, so there is little waste. Processed cheese is available in blocks, slices, or spreads.

Pasteurized cheese food is a blend of fresh and aged natural cheese, processed cheese, and other substances such as dry whey, dry milk solids, and fillers. It may lack the milkfat content that is required to be in regular cheese. It is usually pasteurized, and it is often made into spreads.

■

EGGS

Eggs are a versatile food. They can be a complete dish or entrée, or they can be served for breakfast or in a luncheon omelet. They can be stuffed and used as appetizers or chopped and used for sandwiches or in salads. They thicken and bind some products and cover others. When whipped to a foam, they bring air into items such as cakes, soufflés, and foamy omelets. Eggs, especially yolks, can emulsify, as in making mayonnaise or hollandaise sauce. They can clarify soups or give stability and texture to frozen desserts and candies. In fact, eggs are one of the kitchen's workhorses. They need special consideration in each different use.

Nutrition

Eggs contain the most digestible of proteins and supply good quantities of vitamins A and D, thiamine,

riboflavin, niacin, fat, minerals, and calories. Eggs' iron is well used in the body. One large egg contains over 200 mg of cholesterol, so people on low-cholesterol diets must restrict the number of eggs in their diet. The American Medical Association recommends that the total diet contain not more than 300 mg of cholesterol per day.

Quality

Eggs are at their highest quality immediately after laying. Unless chilled and held at low temperatures, they deteriorate rapidly. They are neutral or slightly acidic when laid but gradually become alkaline. Moisture is lost and the air sac gets larger as they age. A fresh egg has a high yolk and a white that clings to the yolk; there is little thin liquid. As aging occurs, the yolk's height drops, the white becomes thinner, and the yolk breaks easily when the egg is shelled. Older eggs whip to a better foam than fresh eggs do.

An egg's flavor changes as it ages. The white contains sulfur, and the yolk contains iron. When the iron and sulfur join, they make ferrous sulfide—an offensively strong-flavored and odoriferous product. Rotten eggs contain a lot of iron sulfide. Because iron sulfide is black, it also discolors eggs. Some off-flavors bake out in old eggs used in bakery goods, but this does not occur in breakfast eggs. Refrigerate eggs at 40°F (4.5°C). Remove only those that are to be used soon.

Processed eggs can be used in the bakeshop or in other production (see Table 17-3). A high-quality frozen whole egg is available that can be used for French toast, omelets, and custards. Frozen whole eggs, frozen yolks, frozen whites, a sugared yolk (10 percent sugar), and whole eggs containing added yolks (proprietary eggs) are used. All processed eggs must be pasteurized. Never use bakeshop-quality eggs for breakfast eggs or for regular cooked items.

Grades

Eggs are graded for quality on the following scale (from highest to lowest grade): AA, A, B, and C. C grade cannot be sold on the market, but instead must go into processed eggs. Few grade B eggs are used. Shell color is not a grade factor. Eggs are sized as

TABLE 17-3   Processed Dried Eggs and Their Fresh Egg Equivalents

I. Processed to Fresh

Processed Product	Egg to Water	Fresh Equivalent EP (Large Eggs)
No. 10 can (3 lb) dried whole		100 whole eggs
1 lb dry whole and 2½ pt water	1:2½	3½ lb whole eggs
1 lb dry yolks and 1½ pt water	1:1½	2½ lb yolks
6 oz dry whole and 1⅞ cups water	3:7	dozen whole eggs
1 lb dried whole and 2½ pt water	1:2½	3 doz whole eggs (3½ lb)
1 lb dried whites and 5 pt water	1:5	100 whites (6 lb)
1 lb dried yolks and 1¾ cups water	8:7	47 yolks

II. Fresh to Processed

Fresh Eggs (Number to Give 1 lb)	Quantity for 1-lb Equivalent			
	Dried Egg		Water	
	Ounces	Measure	Ounces	Measure
Whole (9)	4½	1 cup 2 Tbsp	11¼	1 cup 7 Tbsp
White (17–19)	2¼	½ cup	13¾	1¾ cups
Yolk (20–24)[a]	7¼	1 cup	8¾	1 cup 1 Tbsp
(25–27)[b]	7½	1 cup	8½	1 cup 1 Tbsp
(16–20)[c]	6½	1 cup	8	1 cup

[a]43% solids.
[b]45% solids.
[c]Add 1.6 oz of sugar to give yolks with 10% sugar as used in the bakeshop for frozen sugared yolks.

shown in Figure 17-2. Foodservices usually use large eggs, and recipes calling for eggs are based on the large size.

Principles

When heat strikes eggs, the protein changes color, becoming firmer as it does in meats, fish, and poultry. The following coagulation temperatures are useful to know: whole eggs, 156°F (69°C); yolks, 144 to 158°F (62 to 70°C); whites, 140 to 149°F (60 to 65°C). When eggs are mixed with sugar or liquids, the co-agulation temperature rises to around 175 to 185°F

FIGURE 17-2   Egg sizes and weights per dozen.

(79 to 85°C). If the temperature is raised quickly, a higher coagulation temperature is needed than if the temperature is raised slowly. Coagulation can also occur when eggs are mixed with strong acids or alcoholic drinks. High temperatures cause egg protein to

toughen and develop a strong flavor. Adding an acid such as cream of tartar, lemon juice, or tomato juice makes them tenderer, retards undesirable flavor development, lowers the coagulation temperature, and increases the eggs' thickening power. Eggs beat to a foam better when they are slightly acidic.

Uncooked egg protein can combine with moisture. When heat strikes this mixture, the water binds tightly to the protein, causing the firmness seen in custards. Too much moisture or too high a heat can cause the protein to lose moisture—a process called *syneresis*. Gentle temperatures and a slightly acidic condition help to avoid it.

Beating air into eggs develops foam in whole eggs, yolks, or whites. Egg white does not foam if fat is present. Thus, whole eggs are harder to whip to a foam because of the fat in the yolk. Yolks whip well alone, however, older eggs foam better than fresh ones do. Dried or frozen egg products also whip well into foams. Eggs that have warmed to about 60°F (16°C) separate best into yolks and whites; they whip best if they are warmed to 75 to 110°F (24 to 43°C) and if they are slightly acidic. Some cooks think adding salt also helps. Acid also lightens the eggs. The amount of acid to use varies; about ¼ oz of cream of tartar to 1 lb of eggs is normal. Sugar gives egg foam stability. Eggs containing sugar can be whipped more without danger of being overwhipped. Some sugar is usually added soon after a good foam is obtained. Table 17-4 identifies the four different foam stages.

Cooking

Eggs for breakfast can be prepared in quantity, but it is difficult to do so. Custards and soufflés, however, pose no problem in quantity.

TABLE 17-4	Stages of Egg Foams			
	First Stage	*Second Stage*	*Third Stage*	*Fourth Stage*
Foam appearance	Liquid, but well blended; foam is in large bubbles	Medium-sized air cells throughout mass; foam is shiny, moist, and fluid; tips fold over into rounded peaks; liquid separates out in standing	Stiff foam; small air cells; no longer fluid, especially whites; still moist, smooth, and glossy; points stand when peaked	Dry, dull, brittle foam; flakes off and can be cut into rigid parts; curds may appear indicating coagulation; it is difficult to beat whole eggs or yolks to this stage
Foam use	Clarifying soups; French toast; coating foods; blending into mixtures as liquid	Sponge or angel cakes; soufflés; foamy omelets	Cooked frostings; divinity; soft or hard meringues; tortes; sponge cakes	Has no use in food work; eggs are so overextended that they will not extend further in baking, causing a failure in the product

Boiled Eggs

Table 17-5 gives cooking times for soft-, medium-, and hard-boiled eggs. The texture ranges from a soft, almost runny white and only warmed yolk for soft-boiled eggs to a completely coagulated egg for hard-boiled eggs. The latter should have no dark ring around the yolk; such a ring contains iron sulfide, which gives poor appearance and flavor. Good eggs cook to clear, shiny colors; poor ones are dull and have off-colors. Overcooking turns a good egg into a bad one. Overcooked eggs are tough, rubbery, or puffy.

Eggs to be stuffed or sliced should be boiled fairly hard, to give them enough toughness to handle. Simmer-cooking gives a tender egg that is delicate to eat but breaks in handling. Very cold eggs added to hot water can crack. Therefore, eggs may be pre-warmed or added to cooler water for cooking. To boil eggs in quantity, put them into a perforated basket, lower this into tepid water, and bring the water quickly to a boil. After cooking, plunge the eggs into cold water and send them immediately—as hot as possible—to service. The quick blanch stops the cooking. Eggs can also be steamed in the shell, but it takes skill to tell time and to get correct doneness.

To peel eggs, plunge the hot eggs into cold water and let them stand for 5 to 10 minutes. Then crack well and start to peel at the large end, where the air sac is. If they peel with difficulty, peel them under cold running water. The water helps lift the shell from the egg. If chopped eggs are needed, break eggs into lightly greased pans to an egg depth of about 2 in.,

and steam them until hard-cooked. Then turn them out of the pan, and chop.

Coddled eggs are eggs put into boiling water—1 pt per egg. After they have been added, the pot is covered, and the eggs are allowed to cook, without further heat, to desired doneness.

Fried Eggs

Eggs can be fried *over easy, sunny-side-up*, or *basted*. Over easy is frying until the egg is almost done, and then turning it over in the hot fat only to coagulate the top. A sunny-side-up egg is fried without being turned. Basted eggs are also called *country-style* eggs. They are fried with a tight cover on the pan so that steam cooks the tops. Alternatively, they can be basted with hot fat during frying. Some cooks fry country-style eggs sunny-side-up and then add a bit of water and cover tightly until steam forms a white coating over them.

To fry, heat about ⅛ in. of fat in a pan to a temperature of from 350 to 375°F (182 to 190°C). Slide the egg into the hot fat, and quickly lower the heat; the hot fat helps keep the egg from spreading. Too hot a fat toughens the bottom of the egg, browning it and giving it a poorer flavor. Egg pans should be conditioned for frying. Eggs can also be fried on a grill. They are not as attractive as pan-cooked eggs, but more can be produced. Use good-quality fat or clarified butter. Set the grill's thermostat at from 300 to 325°F (149 to 163°C). Start some eggs at one place on the grill, and add more while the first batch is frying.

	Simmering		
TABLE 17-5	**Cooking Times for Boiled Eggs**		
Boiling Style	212°F[a] (100°C)	190–195°F[a] (88–91°C)	Steam Pressure (7 psi)
Soft-cooked	3 min	6 min	1 min 25 sec
Medium-cooked	4 min	8 min	
Hard-cooked	12–15 min	20–25 min	3 min 10 sec

[a]These times vary at elevations above 3000 ft.

Add more grease as necessary. Remove the eggs in the order they have been put onto the grill.

To baste, cover the eggs with a small pan. Often eggs are only partially fried before being put into pans and sent to the service area, where the pans are placed in the steam table and the eggs finish cooking.

Omelets

Omelets are made from whole eggs. Cook omelets at a higher temperature than most egg products, giving them a brown or delicate tan outside. They should be moist, tender, and delicate in flavor. Omelets should be rolled neatly and have a good shape. A ragged, separated omelet lacks proper appearance. A French omelet is made by beating whole eggs to a stage-one foam and adding them to a pan covered with about ⅛ in. of hot fat. Stir lightly to lift cooked portions and allow uncooked portions to flow underneath, but stop before the eggs become dry, so a *whole* mass can be obtained. Cheese, creamed fillings, marmalade, or other items can be placed at the center, and the omelet is then folded over. This is done while some moist egg remains on the interior, and this then cooks to seal the filling inside. The omelet can be flipped to cook on the other side, if desired. To unpan, tip the pan and let the omelet slide down. Place a plate under the pan, and let the omelet fall gently onto it. Omelets with sweetened centers can be "burned" as shown in Figure 17-3. Omelets can be made on a grill, but this is a difficult procedure.

Poached Eggs

Shelled eggs cooked in water are called *poached*. The water should be 2 to 2½ in. deep in a 4-in.-deep pan. Use 1 Tbsp of salt and 2 Tbsp of vinegar per gallon of water. Have the water gently boiling, and slide the eggs (already broken) from a plate or platter into the water. Sliding rather than dropping the eggs helps reduce flattening the egg. Poach 8 to 16 eggs per gallon of water. The same water can be used for 3 to 4 batches before it must be discarded. Cooking time varies from 3 to 5 minutes.

Scrambled Eggs

Scrambled eggs are the easiest type to prepare in quantity. Stage-one mixed whole eggs (salted lightly) can

a.   b.

c.   d.

FIGURE 17-3 Making an omelet: (*a*) when the fat is fairly hot, pour the beaten whole eggs into the pan; (*b*) lift until the entire mass is well coagulated, then fill and fold; (*c*) shape, if desired; (*d*) if the filing is sweet, dust with powdered sugar, and burn with a poker.

be added to a steam-table pan and scrambled there. Finished scrambled eggs should be in fairly large segments, so stirring should be limited while they cook. Let the eggs cook on the bottom, and then lift them up and allow the liquid eggs to flow under to cook. Using too much heat during holding or holding for too long can cause a dark, tough bottom and a poor flavor. The eggs should be tender, moist, and delicate in flavor. Liquid such as milk or cream can be added in a ratio of 1:4 to eggs. Medium cream or white sauce can also be used. Scrambled eggs hold well for about 30 minutes. In the kitchen, they are cooked in large skillets, steamers, double boilers, steam-jacketed kettles, grills, or in bain-maries. Scrambled eggs to order are cooked in regular egg pans.

Shirred Eggs

Shirring is a method of baking an egg. Eggs are put into shallow ceramic dishes that are liberally buttered. These are placed over heat until a trace of coagulation appears on the bottom. They are then finished under a broiler or in an oven. The whites should be cooked, but not the yolks. Excessive heat makes them rubbery. Good shirred eggs have a bright, solid appearance;

poorly made ones are dull and puffy. Sometimes ham or several cooked strips of bacon are placed on the dish bottom, and the eggs are put over them and shirred. Cream and cheese can be put over the eggs before they go into the oven. Creamed chicken, seafood, Newburg, chicken livers, tomato quarters, mushrooms, diced ham, or other mixtures are placed over shirred eggs after cooking.

Soufflés

A soufflé can be an entrée or a dessert. It is a mixture of well-beaten egg whites folded into a starch base containing the yolks and a seasoning mixture. Thus, the paste for a cheese soufflé contains milk, roux, and seasonings, along with cheese. This paste is cooked, and the yolks are added. Then the stage-three beaten egg whites are folded in, and the mixture is baked in a greased casserole or other dish. The oven is set at about 300°F (149°C), or at 375°F (190°C) if the baking dish is placed in a pan of water.

Soufflés are delicate and can fall when they are removed from the oven. Some cooks pull soufflés to the edge of the oven when they are done and leave the door open for a few minutes to allow them to adjust. A good soufflé should have a slightly rounded top that is puffy and slightly cracked. The crust should be a delicate brown (see Fig. 17-4). Chopped meat, seafood, poultry, and other foods can be added to the paste. As a dessert, soufflés may be flavored with chocolate, strawberry, vanilla, or other flavors.

FIGURE 17-4 Freshly baked soufflé.

Custards

A custard is a combination of milk and eggs. Custards can be sweet and used as a dessert, or they can be unsweetened and combined with vegetables for service as a timbale. A fondue (different from a Swiss fondue) is a light custard-like entrée resembling a soufflé. It combines milk and eggs and may be baked with cheese, poultry, or seafood. The fondue is then served with a sauce. Soft bread cubes are usually added to give it a light texture. Swiss fondue is a melted mixture of white wine, garlic, Swiss (Emmentaler) and Gruyère cheese that is served in a chafing dish. People at the table dip cubes of hard-crusted bread into the mixture. The Swiss also have a fondue that uses hot oil in a chafing dish to deep-fry meat items.

A stirred custard is a sweetened mixture of milk, eggs, and seasoning that is cooked while being stirred over hot water, until it coats a spoon. Constant, thorough stirring and gentle heat must be used to prevent curdling. Stirred custard is used as a sauce over desserts. Sabayon (French) and zabaglione (Italian) are stirred custards composed of eggs and a wine such as Marsala; they are cooked until slightly thickened and then served over desserts. Often only egg yolks are used because they are less likely to curdle than whole eggs.

The proportion of eggs to liquid for a custard is 1½ to 2 lb of whole eggs to each 1 gal of liquid. If yolks are used or if cereals are combined with the custard, the quantity of eggs may be reduced. Thus, fewer eggs are used to make a bread or rice custard pudding. A custard bakes better if the milk is already heated to about 140°F (60°C) when it is added to the eggs. Custards often are baked in pans of water, to protect them from the heat. Too much heat causes syneresis. Baking temperatures usually range from 325 to 340°F (164 to 171°C), with the custard baked in a pan of water.

■

SUMMARY

Milk, which is an excellent food nutritionally, is used in cooking to blend substances together. Acids, tannins, salts, and some other substances cause milk to curdle, as will heat. Curdling is the separation of the casein in the milk from the liquid whey. Milk has proteins that will coagulate, as seen in the scum that forms on heated milk or collects on the side of the pan. The collected protein on the bottom of a pan causes milk to scorch easily, so milk usually is cooked over water, in a steamer, or in a steam-jacketed kettle.

Milk and its products are pasteurized to destroy harmful bacteria. They may also be homogenized to separate the milkfat into tiny globules so that it does not rise to the top. Milk can be made into foams. Dry milks will foam if properly handled, as will evaporated milk. Whipping cream comes from a cream that contains more than 30 percent milkfat. To whip properly, the milk or cream and the utensils used in whipping should be very cold.

Milk products used in production are creams, half-and-half, soured products, buttermilk, yogurt, dry whole or nonfat milk, evaporated milk, and condensed milk.

Cheese is made from milk by separating the casein and milkfat from the whey. Ripened cheese is sharper and stronger in flavor than unripened cheese. It is also softer and goes into solution in sauces and other foods more easily. A processed cheese is a natural cheese that has been pasteurized to stop ripening. It contains an emulsifier, which makes it go into solution more easily. Fruits, vegetables, meats, and other products are added to processed cheese. Some processed cheese is made into spreads.

Eggs and milk products are high-protein foods that require similar treatment in cooking and baking.

Eggs contain excellent protein and other important nutrients. They add color, richness, and flavor to food.

Egg quantity is important. A very fresh egg with a high yolk and a firm white is desired for breakfast purposes. High-quality frozen processed eggs also can be used for breakfast. All processed eggs are pasteurized to destroy harmful bacteria. Processed eggs include dried whole, dried whites, dried yolks, frozen whole, frozen whites, frozen yolks, and other specialty frozen products. They are used primarily in baking.

Eggs coagulate when heat strikes them. The protein changes color and firms. If the egg has a lot of liquid added to it and the heat is too high, the liquid can separate out, causing a curdling called *syneresis*. Eggs will foam. Eggs separate best at around 60°F (16°C) and whip best at about 110°F (43°C). Whites do not whip well if any grease is present.

Egg yolks contain iron, and the whites contain sulfur. When iron joins with sulfur, it makes ferrous sulfide, which is an offensively odorous, bad-tasting, dark product. Overcooked and old eggs easily develop this substance. As fresh eggs age, they become more alkaline, which favors the development of ferrous sulfide. Adding acid ingredients to eggs discourages the formation of this chemical.

Eggs can be boiled, coddled, fried, poached, shirred, scrambled, and made into omelets and soufflés. Eggs also are used to make baked or stirred custards. The grades of eggs used in foodservices are AA, A, and B. The sizes are jumbo, extra large, large, medium, small, and peewee.

■

CHAPTER REVIEW QUESTIONS

1. A cook has curdled a batch of cream of mushroom soup. Set up an outline of what a chef should tell the cook about preparing such products.

2. You want to present a lecture to a group of trainees on how to cook eggs. Set up an outline on what should be covered.

3. Name five ways eggs are used in food production, and indicate the procedure used. For instance, if you say that eggs are used to make emulsions, state how the egg is used to form an emulsion.

4. Why are eggs and milk products nutritious foods?

5. Describe a fresh egg, and indicate what happens when an egg ages. Are fresh eggs *always* best to use in cooking and baking? Explain.

6. What is coagulation in an egg? What is ferrous sulfide, and how does it form in eggs? What does heat do in the formation of ferrous sulfide?

7. Name the types of processed eggs. Why must they now be pasteurized? What are the grades and sizes of eggs? How are eggs sized?

8. What would you tell an employee about how to make a good egg farm?

9. Describe how eggs are cooked. Prepare some of these products in the laboratory, and critique them for quality.

10. List some of the products that come from milk, and identify how they can be used in food production.

11. What makes milk curdle? How can milk's tendency to curdle be reduced?

12. Certain phenomena occur when milk is used in cooking. Why do coagulation, scum formation, and scorching happen? What are pasteurization and homogenization, as applied to milk and its products?

13. What kinds of foams can be made from milk products? Describe how each foam should be made.

14. How should the various milk products be handled in food production?

15. What is cheese? processed cheese? cheese food? How do they differ? State several principles that should be observed in handling cheese in food preparation.

■

CASE STUDIES

1. You're thinking of opening a new foodservice operation in California that will feature the serving and presentation of cheese after the meal. As one interested in learning more about this dairy product, you research the Internet beginning at the site *www.calif-dairy.com*. Prepare a report on cheese containing information about its history, production, types, service, and nutrition.

2. In your pursuit of happiness, you have decided to open a new foodservice operation on campus, serving breakfast. You have decided that the name of the facility will be The Egg Head. In order to serve the finest of products, you research the Web site at *www.aeb.org* for information on egg safety. Your answer should include a discussion on the grading, purchasing, safe handling, and nutrition of eggs. As a final comment to your report, discuss *Salmonella* as it relates to egg safety.

■

VOCABULARY

Fresh egg
Ferrous sulfide
Proprietary eggs
Pasteurization
Coagulation
Grade AA
Curdling
Hard-cooked

Soft-cooked
Sunny-side-up
Over easy
Basted
Country-style
Omelet
Poached egg
Shirred egg

Soufflé

Half-and-half

Coffee cream

Light whipping cream

Heavy whipping cream

Dry nonfat milk

Evaporated milk

Condensed milk

Butter

Homogenization

Cheese

Processed cheese

Cheese food

Ripened cheese

Syneresis

■

ANNOTATED REFERENCES

Gisslen, W. 1995. *Professional Cooking*, 3rd ed. New York: John Wiley & Sons, Inc., pp. 551–576. (Breakfast preparation and dairy products are considered in this chapter.)

Kapoor, S. 1996. *Healthy and Delicious: 400 Professional Recipes*. New York: John Wiley & Sons, Inc., pp. 321–336. (Healthy recipes for eggs and cheese are presented in this chapter.)

Labensky, S. R., and A. M. Hause. 1999. *On Cooking: A Textbook of Culinary Fundamentals*, 2nd ed. Upper Saddle River, N.J.: Prentice Hall, pp. 534–557. (Eggs are presented in this chapter.)

Pauli, P. 1997. *Classical Cooking the Modern Way*, 3rd ed. New York: John Wiley & Sons, Inc., pp. 149–156. (A discussion on eggs and cheese is offered here.)

Appendices

Menu Analysis

Some of the most commonly used methods of analyzing menus are discussed here. Evaluations of each of these methods follow. As aids to understanding how methods differ, data for seven dessert items are provided in Tables A-1 and A-2.

■

SUBJECTIVE MENU EVALUATION

Menus are frequently evaluated on the basis of a cursory visual examination. This is a highly subjective method,

and wide variation can be found in judgments. The comments of viewers of the winning menus in the annual National Restaurant Association's menu contest offer ample proof of this. The advantage of such a method is that it allows evaluation of a wider range of factors than other analytical methods assess. The method is also highly flexible. A scoring system can be used to identify factors and assign them numerical values. These numerical scores can then be used to compare menus. If this is done by a knowledgeable person, the method can be a good way of evaluating many aspects of menus.

	TABLE A-1 Menu Count and Associated Popularity Index for Seven Dessert Items	
Menu Item	*Menu Count*	*Popularity Index (%)*
Strawberry cake	⊬⊬ ⊬⊬ ⊬⊬ /// (18)	13
Lemon pie	⊬⊬ ⊬⊬ ⊬⊬ ⊬⊬ // (22)	15
Fudge sundae	⊬⊬ ⊬⊬ // (12)	8
Caramel custard	⊬⊬ ⊬⊬ ⊬⊬ ⊬⊬ ⊬⊬ / (26)	18
Cherry torte	⊬⊬ ⊬⊬ ⊬⊬ ⊬⊬ ⊬⊬ /// (28)	20
Key lime pie	⊬⊬ ⊬⊬ ⊬⊬ ⊬⊬ ⊬⊬ //// (29)	20
Brownie	⊬⊬ /// (8)	6
Total sold 143		100

TABLE A-2 Sample Data for Menu Analysis of Seven Dessert Items

Menu Item	A (Count) Number Sold	B Selling Price	C Item Food Cost	D (A × C) Total Item Food Cost	E (A × B) Item Revenue	F (B − C) Gross Profit	G (A × F) Total Item Gross Profit	H (C/B) Item % Food Cost	I (A/K) Item % of Total Sold
Strawberry cake	18	$1.50	$0.28	$ 5.04	$27.00	$1.22	$21.96	19	13
Lemon pie	22	1.65	0.44	9.68	36.30	1.21	26.62	27	15
Fudge sundae	12	1.30	0.26	3.12	15.60	1.04	12.48	20	8
Caramel custard	26	1.00	0.22	5.72	26.00	0.78	20.28	22	18
Cherry torte	28	1.60	0.50	14.00	44.80	1.10	30.80	31	20
Key lime pie	29	1.75	0.56	16.24	50.75	1.19	34.51	32	20
Brownie	8	1.00	0.20	1.60	8.00	0.80	6.40	20	6

Data derived from the figures above:

J = number of items in the study = 7
K = total sold in column A = 143
L = total food cost (total of column D) = $55.40
M = total revenues (total of column E) = $208.45
N = total gross profit (total of column G) = $153.05

O = average food cost percent (L/M) = 27%
P = average gross profit (N/K) = $1.07
Q = average number sold (K/J) = 20.43
R = average selling price (M/K) = $1.46

■
MENU COUNTS (POPULARITY INDEXES)

Using a menu count or calculating a popularity index (the percentage of an item sold out of all items sold) is easy to do and gives highly informative data. Table A-1 shows a count and a popularity index. Volume is an important factor in successful operation; if other factors are favorable, large volume can indicate a high level of patronage satisfaction and a profitable operation. If records are maintained, managers can use them as a historical file of successful and unsuccessful menu items.

Unfortunately, it is often difficult to compare values obtained from different times when the underlying indexes are based on a different number of menu items studied. Thus, if five items are studied one time and eight are studied the next, the item with four "competitors" has a better chance of having a high index rating than the one with seven. Thus, if an item among eight has an index of 18 percent, it is likely to have a score of better than 1.0, while an item among five that has an index of 18 percent is likely to have a score of less than 1.0.

■
HURST'S MENU SCORING

Hurst's menu scoring is a good method for developing an overall numerical evaluation of how well a menu will do. It tests for the combined effect of items such as volume, selling price, food cost, and gross profit. Specifically, the method involves obtaining a score by multiplying the percentage of patrons who select a menu item belonging to an arbitrarily established group of items (such as the seven desserts identified in Tables A-1 and A-2) by the items' average gross profit. Thus, if 143 out of 340 customers on a given day selected one of the seven desserts under study, the percentage would be equivalent to 143/340. If we also knew (as Table A-2 indicates) that the average gross profit of the seven desserts was $1.07, we could then calculate the menu score for these seven items as $143/340 \times 1.07 = 0.45$.

Because it is highly flexible and sensitive even to slight changes in a single factor, menu scoring lends itself to simulation procedures that check ahead for possible beneficial or undesirable effects of changes in price, food cost, or other variable. Menu scoring is not difficult to do and relies on data that are readily available. The effects of changes on individual menu items cannot be evaluated by this method, but their effect on the whole menu can—an important consideration.

■ MENU FACTOR ANALYSIS

Kotschevar's menu factor analysis tests individual items, giving them numerical values that indicate how well each item measures up to management's expectations. It can test for variations in food cost, gross profit, dollar sales, volume, and other factors. It is well-suited to simulations that involve trying out changes before running a menu item.

Understanding what the factors are is a prerequisite for drawing good information from them. The standard of 1.0 by which items are judged desirable or undesirable is like an average (mean) that marks the middle of the value range, with some items below

it and others above it. If numerically significant changes are made in one item, some other item (or items) consequently will change with respect to the average (mean). For example, if one item increases in number sold but the total number sold does not change, its percentage sold of total sold increases and the percentages sold of other items decrease as illustrated in Table A-3. In the first instance, item C has a factor of 1.08 (18/16.7), while in the second it has a factor of 0.96 (16/16.7); item A has gone from a factor of 0.65 (11/16.7) in the first instance to a factor of 1.08 (18/16.7) in the second. Even though it is below 1.0 in the second instance, item C may be a good item. It may just be that some items are better. Management must use menu factor values not as dictators of policy but as guides for decisions. Managers must look behind the index factor and weigh other variables before making a decision.

■ MATRIX ANALYSIS

Matrix analysis can be used to test individual menu items for their suitability on a menu, but it is limited to testing the effect of two factors and not the effect of a number of factors. It allows for the testing of different factors such as volume, food cost, gross profit, and

	Instance I			Instance II		
Item	Number Sold	Index (%)	Index Factor	Number Sold	Index (%)	Index Factor
A	15	11	0.66	25	18	1.08
B	20	14	0.84	18	13	0.78
C	25	18	1.08	22	16	0.96
D	18	13	0.78	18	13	0.78
E	22	16	0.96	17	12	0.72
F	40	29		40	29	
Total	140	100		140	100	
Average (mean)	23.3	16.7%		23.3	16.7%	

TABLE A-3   Changes in Index Factor Caused by Changes in Number Sold

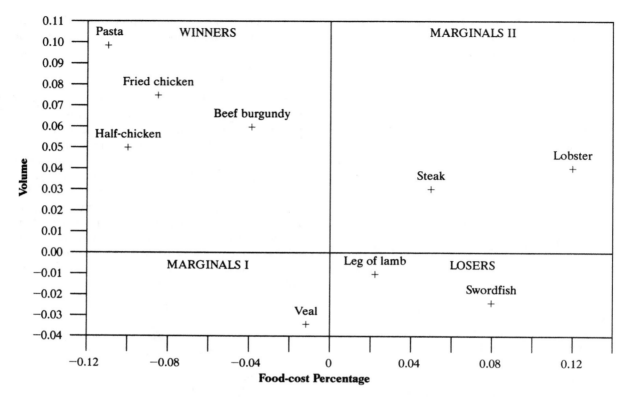

FIGURE A-1 Miller's menu-item analysis.

TABLE A-4	Methods of Calculating Matrix Values		
Item Value	*Standard Value*	*Item Value*	*Standard Value*
Miller		Pavesic	
Food cost: item food cost	Average food cost	*Food cost:* item food cost	Average food cost
Volume: number sold of an item	Average sold of all items	*Weighted gross profit:* item gross profit	Total gross profit/number items being studied
Smith–Kasavana		Pavesic–Kotschevar	
Gross profit: item's gross profit	Average gross profit	*Food cost:* item food cost	Average food cost
Volume: percent sold of item	([1/number of items studied] × 0.7 × 100)	*Weighted gross profit:* item gross profit	Total gross profit/(number of items sold/number of items being studied)

dollar sales, and factors can be weighted, as is done by Pavesic for gross profit and by Smith–Kasavana for volume. Because matrix analysis uses a mean or average as the standard for judging items, however, it has the same defect discussed previously relating to the standard of 1.0. The ideal, of course, would be to have all items exactly match the value of the average (mean). Any changes made in one item reposition the other items with respect to the mean. Matrix analysis lends itself to simulation study.

Matrix analysis has been treated by Miller, Smith–Kasavana, Pavesic, and Kotschevar. As Figure A-1 indicates, Miller evaluates menu items on the basis of food cost and volume (Miller 1980). Smith and Kasavana use volume percentage and gross profit as their criteria; they call their method *menu engineering* (Kasavana and Smith 1982). Pavesic uses food cost and a weighted gross profit (Pavesic 1985). Kotschevar's method is modeled on Pavesic's, except that Kotschevar's weighted gross profit margin is calculated differently. His other standard is food cost. Table A-4 describes how each matrix method compares individual menu item values with an established standard. In all four methods, the standard is subtracted from the item value to establish the item's matrix value. Thus, on the basis of data from Tables A-2 and A-4, strawberry cake has the matrix values given in Table A-5.

The matrices described in Table A-5 each possess four possible combinations of values. These are given different names by the four developers, as indicated in Figure A-2. Table A-6 gives the values for all four different matrices along with their calculations. Table A-7 summarizes these calculations, listing only the end values and their combination names.

■

GOAL ANALYSIS

The Hayes–Huffman goal analysis or numerical target method tests for the suitability of individual menu items and does not give an overall test for a menu. In this it resembles menu factor analysis and matrix analysis, but it considers more factors: gross profit

TABLE A-5   Strawberry Cake Values[a]

Miller
  Food cost: $(H - O)$ Volume: $(A - Q)$
          $(19 - 27) = -8$ $(18 - 20.43) = -2.43$

Smith–Kasavana
  Gross profit: $(F - P)$ Volume: $[I - (1/J \times 0.7)] \times 100$
          $(1.22 - 1.07) = +0.15$ $(13 - 10) = +3$

Pavesic
  Food cost: $(H - O)$ Weighted gross profit: $[G - (N/J)]$
          $(19 - 27) = -8$ $[21.96 - (153.05/7)] = +0.07$

Pavesic–Kotschevar
  Food cost: $(H - O)$ Weighted gross profit: $[G/(A/J)]/(N/Q)$
          $(19 - 27) = -8$ $[21.96/(18/7)]/(153.05/20.43) = +1.05$

[a] The value of 10 is the Smith–Kasavana standard for volume obtained using the model $1/J \times 0.7 \times 100$, where J = number of items in the study group or 7. The calculation is then $1/7 \times 0.7 \times 100 = .10$. (The letters in parentheses refer to the values in Table A-4.)

**Matrix
Layout**

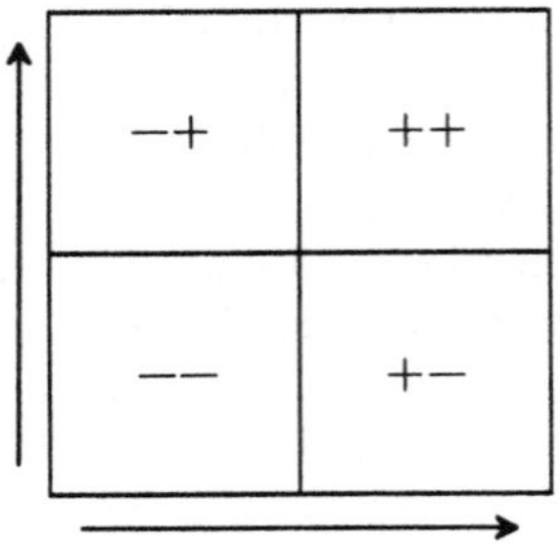

Possible Combination	Combination Name			
	Miller	*Smith- Kasavana*	*Pavesic*	*Pavesic- Kotschevar*
− +	Winner	Plowhorse	Prime	Gold
− −	Marginal I	Dog	Sleeper	Copper
+ +	Marginal II	Star	Standard	Silver
+ −	Loser	Puzzle	Problem	Tin

FIGURE A-2   Layout of matrix values, and names assigned to each combination of values.

Table A-6   Four Matrix Calculations for Seven Menu Items

	Miller		*Smith–Kasavana*	
Menu Item	*Food Cost*	*Volume*	*Gross Profit*	*Volume*
Strawberry cake	19 − 27 = −8	18 − 20.43 = −2.43	1.22 − 1.07 = +0.15	13 − 10 = +3
Lemon pie	27 − 27 = 0	22 − 20.43 = +1.57	1.21 − 1.07 = +0.14	15 − 10 = +5
Fudge sundae	20 − 27 = −7	12 − 20.43 = −8.43	1.04 − 1.07 = −0.03	8 − 10 = −2
Caramel custard	22 − 27 = −5	26 − 20.43 = +5.57	0.78 − 1.07 = −0.29	18 − 10 = +8
Cherry torte	31 − 27 = +4	28 − 20.43 = +7.57	1.10 − 1.07 = +0.03	20 − 10 = +10
Key lime pie	32 − 27 = +5	29 − 20.43 = +8.57	1.19 − 1.07 = +0.12	20 − 10 = +10
Brownie	20 − 27 = −7	8 − 20.43 = −12.43	0.80 − 1.07 = −0.27	6 − 10 = −4

	Pavesic		*Pavesic–Kotschevar*	
	Food Cost	*Weighted Gross Profit*	*Food Cost*	*Weighted Gross Profit*
Strawberry cake	19 − 27 = −8	21.96 − 21.89 = +0.07	19 − 27 = −8	21.96 − (18/7) − 7.49 = +1.05
Lemon pie	27 − 27 = 0	26.62 − 21.89 = +4.73	27 − 27 = 0	26.62 − (22/7) − 7.49 = +0.99
Fudge sundae	20 − 27 = −7	12.48 − 21.89 = −9.41	20 − 27 = −7	12.48 − (12/7) − 7.49 = −0.19
Caramel custard	22 − 27 = −5	20.48 − 21.89 = −1.61	22 − 27 = −5	20.28 − (26/7) − 7.49 = −2.02
Cherry torte	31 − 27 = +4	30.80 − 21.89 = +8.91	31 − 27 = +4	30.80 − (28/7) − 7.49 = +0.11
Key lime pie	31 − 27 = +5	34.51 − 21.89 = +12.62	32 − 27 = +5	34.51 − (32/7) − 7.49 = +0.85
Brownie	20 − 27 = −7	6.40 − 21.89 = −15.49	20 − 27 = −7	6.40 − (8/7) − 7.49 = 1.09

(1 − food cost % = gross profit), selling price, volume, variable cost, and fixed cost (1 − [variable cost % + fixed cost %] = fixed cost). It is probably the best method yet developed for testing individual menu items. It is a very sensitive test and lends itself well to simulation. Because its standards are averages (means) and because some items are always going to be above and others below the standard, manipulation of items merely reorganizes the order.

The Hayes–Huffman method of analysis has only one standard. The method involves developing a numerical target or goal for the standard and then

TABLE A-7   Results of Four Types of Matrix Analysis

Menu Item	Miller			Smith–Kasavana			Pavesic			Pavesic–Kotschevar		
	Food Cost	Volume	Name	Gross Profit	Volume	Name	Food Cost	Weighted Gross Profit	Name	Food Cost	Weighted Gross Profit	Name
Strawberry cake	−8	−2.43	Marginal I	+0.15	+3	Star	−8	+0.07	Prime	−8	+1.05	Gold
Lemon pie	±0	+1.57	Winner or Marginal II	+0.14	+5	Star	±0	+4.73	Prime or Standard	±0	+0.99	Gold or Silver
Fudge sundae	−7	−8.43	Marginal I	−0.03	−2	Dog	−7	−9.41	Sleeper	−7	−0.19	Copper
Caramel custard	−5	+5.57	Winner	−0.29	+8	Plowhorse	−5	−1.61	Sleeper	−5	−2.02	Copper
Cherry torte	+4	+7.57	Marginal II	+0.03	+10	Star	+4	+8.91	Standard	+4	+0.11	Silver
Key lime pie	+5	+8.57	Marginal II	+0.12	+10	Star	+5	+12.62	Standard	+5	+0.85	Silver
Brownie	−7	−12.43	Marginal I	−0.27	−4	Dog	−7	−15.49	Sleeper	−7	−1.09	Copper

TABLE A-8 Hayes–Huffman Goal Analysis of Seven Desserts

Item	(H)	(A)	(B)		(H)	Item Value – Standard
Strawberry cake	$(1 - 0.19) \times$	$18 \times$	$1.50 \times$	$[1 - (0.32 + 0.19)]$ =	10.72	$- 8.93 = +1.79$
Lemon pie	$(1 - 0.27) \times$	$22 \times$	$1.65 \times$	$[1 - (0.32 + 0.27)]$ =	10.86	$- 8.93 = +1.93$
Fudge sundae	$(1 - 0.20) \times$	$20 \times$	$1.30 \times$	$[1 - (0.32 + 0.20)]$ =	5.99	$- 8.93 = -2.94$
Caramel custard	$(1 - 0.22) \times$	$26 \times$	$1.00 \times$	$[1 - (0.32 + 0.22)]$ =	9.33	$- 8.93 = +0.40$
Cherry torte	$(1 - 0.31) \times$	$28 \times$	$1.60 \times$	$[1 - (0.32 + 0.31)]$ =	11.44	$- 8.93 = +2.51$
Key lime pie	$(1 - 0.32) \times$	$29 \times$	$1.75 \times$	$[1 - (0.32 + 0.32)]$ =	12.42	$- 8.93 = +3.49$
Brownie	$(1 - 0.20) \times$	$8 \times$	$1.00 \times$	$[1 - (0.32 + 0.20)]$ =	3.07	$- 8.93 = -5.86$

calculating a similar value for each specific menu item to see how well it compares with the standard. The Hayes–Huffman method uses a mathematical formula $A \times B \times C \times D$ to derive the standard value and item values, where

A = 1 – food cost %
B = number sold
C = selling price
D = 1 – (variable cost % + food cost %)

Using data in Table A-2 and a variable cost of 32 percent, we may calculate the goal or numerical target as follows:

$$(1 - 0.27) \times 20.43 \times 1.46 \times [1 - (0.32 + 0.27)] = 8.93$$
$$(O) \quad\quad (Q) \quad\quad (R) \quad\quad\quad\quad\quad (O)$$

The calculation of the data shown for individual items in Table A-8 according to the Hayes–Huffman model adopts the standard 8.93 as the means of determining whether individual items meet the goal or not. According to Table A-8, only two desserts—fudge sundae (at –2.94) and brownie (at –5.86)—fail to meet the goal. Of those that meet the goal, custard (at +0.40) is least successful and key lime pie (at +3.49) is most successful.

BREAK-EVEN ANALYSIS

Break-even analysis is a good method to use to find out what must be achieved overall by a menu in dollar sales, volume, patron count, or some other category. It gives no information about how successful individual items will be or how they will react with others in the menu.

CORRELATION OF METHODS ANALYZING INDIVIDUAL ITEMS

Aside from the Pavesic–Kotschevar matrix method, we have discussed five methods that can be used to test the suitability of individual menu items: Kotschevar's menu factor analysis. Miller's matrix method, Smith–Kasavana's matrix method, Pavesic's matrix method, and the Hayes–Huffman goal method. How well do these six correlate on what is or is not a desirable menu item? Even a cursory visual examination of the data in Table A-9 indicates that the methods varied considerably. Smith–Kasavana term one menu item a Plowhorse that Miller calls a Winner. Smith–Kasavana call two other items Stars that no one else gives a top rating. The others also had their disagreements.

Kendall's method for testing for coefficient of concordance was used to obtain a value to indicate whether or not any correlation existed between these five methods as a whole. Spearman's rank correlation and a test by Freidman were also run to check against Kendall's. They both agreed with the Kendall finding, which indicated no significant correlation existed (w = 0.04).

TABLE A-9 Hayes–Huffman Goal Analysis of Ten Entrées

(A) Item	(B) Number Sold (MM)	(C) Menu Mix (%)	(D) Item Food Cost	(E) Item Selling Price	(F) Item CM (E − D)	(G) Menu Costs (D × B)	(H) Menu Revenues (E × B)	(L) Menu CM (F × B)	(P) FC% Category	(R) MM% Category	(S) Menu Item Classification
Half-chicken	24	12.0	1.74	6.95	5.21	41.76	166.80	125.04	−0.10	0.05	Winner
Steak	20	10.0	4.78	11.95	7.17	95.60	239.00	143.40	0.05	0.03	Marginal II
Shrimp	16	8.0	3.17	8.50	5.33	50.72	136.00	85.28	0.02	0.01	Marginal II
Veal	7	3.5	3.01	8.95	5.94	21.07	62.65	41.58	−0.01	−0.03	Marginal III
Pasta	35	17.5	1.28	5.50	4.22	44.80	192.50	147.70	−0.12	0.11	Winner
Swordfish	9	4.5	4.65	10.95	6.30	41.85	98.55	56.70	0.07	−0.02	Loser
Lobster	22	11.0	6.85	14.50	7.65	150.70	319.00	168.30	0.12	0.04	Marginal II
Beef burgundy	26	13.0	2.17	6.95	4.78	56.42	180.70	124.28	−0.04	0.06	Winner
Fried chicken	29	14.5	1.78	6.75	4.97	51.62	195.75	144.13	−0.09	0.08	Winner
Leg of lamb	12	6.0	3.33	8.95	5.62	39.96	107.40	67.44	0.02	−0.01	Loser
Total sold	(N) 200					(I) 594.50	(J) 1698.35	(M) 1103.85			
Number of items	10										

$K = I/J$
35.0%

$O = M/N$
5.52

$Q = (100\%/\text{items}) \times (70\%)$
7.0%

$R = D/E − K$

433

Paired correlation tests were also run. The r's for these follow:

Smith–Kasavana and Hayes–Huffman	0.90
Miller and Pavesic	0.75
Kotschevar and Hayes–Huffman	0.72
Kotschevar and Smith–Kasavana	0.53
Kotschevar and Pavesic	0.43
Miller and Smith–Kasavana	0.43
Miller and Hayes–Huffman	0.26
Pavesic and Hayes–Huffman	0.26
Smith–Kasavana and Pavesic	0.25
Kotschevar and Miller	0.15

Out of the twenty possible chances each author had for agreement with other authors, the following percentages of agreement occurred:

Hayes–Huffman	80%
Smith–Kasavana	75%
Pavesic	70%
Kotschevar	65%
Miller	40%

A check was also made to see how these five authors agreed on their evaluations of the various desserts. The greatest degree of agreement was found with respect to strawberry cake, lemon pie, and caramel custard.

■

RESULTS OF COMPARATIVE ANALYSIS

The most common analytical methods for analyzing menus can yield a wide variety of information; the kind depends on which method is used. There is probably no best method because each gives specific and different information from the others. The choice of which method to use should be made on the basis of the conditions and needs of the user.

Menu analysis is a way to focus management's attention on getting information about exactly what menus or menu items do or should do. At the very least, they are better than nothing, and too many operations do nothing—losing efficiency and profitability thereby. Methods of menu analysis force management to scrutinize, study, and evaluate menus or items. They develop a numerical value that facilitates objective comparison. They are helpful in pretesting or in simulations that do not require actually running the menu.

Probably the best way to perform menu analysis is to enlist a combination of methods. Certainly any menu greatly benefits from scrutiny by subjective methods because this enables management to consider factors that are not raised in any other way. Using a scoring method yields a numerical factor that can be used to compare other menus that are given subjective evaluation. This can be followed by a test of the menu, using Hurst's menu scoring method, to place a numerical value on the effects various factors have on menu acceptance and profitability. Next, individual menu items should be tested for suitability. For this purpose, the Hayes–Huffman goal method may well be the best of those discussed for obtaining information on individual menu items. The last method used could be a check, using break-even analysis, to see how well the menu must perform overall in order to make a profit.

Food Composition

Linoleic (g)	Saturated (g)	Fat (×9 g)	Carbohydrate (×4 g)	Protein (×4 g)	Food Energy (calories)	Measure	Food Item	Weight (g)	Calcium (mg)	Iron (mg)	A (I.U.)	C (mg)	Thiamine (mg)	Riboflavin (mg)
							BREAD AND CEREAL							
Trace	1	6	16	3	129	1 biscuit, 2" diam.	Biscuit, enriched	35	42	.6	—	—	.07	.07
—	—		1	23	3	101	¾ cup	Bran flakes (40% bran) with thiamine added	28	16	1.3	—	—	.10	.05
							Bread							
—	—	—	8	1	43	2 tablespoons	Bread crumbs, enriched, grated	11	13	.4	—	—	.03	.03
—	—	1	20	2	93	1 slice, 3" × ¾"	Boston brown bread	44	39	.8	—	—	.05	.03
—	—	—	14	1	62	⅔ cup	Cornflakes, enriched, plain or pre-sweetened	16	3	.2	—	—	.07	.01
—	—	—	13	1	62	½ cup	Corn grits, enriched, cooked	121	1	.4	—	—	.05	.04
—	—	—	13	1	62	½ cup	unenriched, cooked	121	1	.1	—	—	.02	.01
—	—	—	26	2	119	1 cup	Cornmeal, enriched, cooked	238	2	1.0	142	—	.14	.10
—	—	—	26	2	119	1 cup	unenriched, cooked	138	2	.5	142	—	.05	.02
Trace	2	5	23	3	151	1 muffin, 2¾" diam.	Corn muffin, enriched	48	50	.8	144	—	.10	.11
—			12	2	60	1 slice	Cracked wheat bread, 20 slices per pound	23	20	.3	—	—	.03	.02
—		1	10	1	60	2 medium	Crackers, graham	14	3	.3	—	—	.04	.02
—		1	6	1	35	2 crackers	saltine	8	2	.1	—	—	—	—
—		—	20	4	98	¾ cup	Farina, enriched, instant cooking	179	138	11.5	—	—	.02	.02
Trace	Trace	1	13	2	65	1 slice	French, Vienna, or Italian bread, 20 slices per pound, enriched	23	10	.6	—	—	.06	.05
—	—	1	13	2	65	1 slice	unenriched	23	10	.2	—	—	.02	.02
1	14	25	44	18	475	1 cup	Macaroni (enriched) and cheese	220	394	2.0	970	—	.22	.46
—	—	—	32	5	155	1 cup	Macaroni, enriched, cooked (tender)	140	11	1.3	—	—	.20	.11
—	—	—	32	5	155	1 cup	Macaroni, unenriched, cooked (tender)	140	11	.6	—	—	.01	.01
Trace	1	4	17	3	118	1 muffin, 2¾" diam.	Muffin, enriched	40	42	.6	40	—	.07	.09
Trace	1	1	19	4	100	½ cup	Noodles, enriched, cooked	80	8	.7	61	—	.12	.07
1	1	2	15	4	98	⅔ cup	Oatmeal or rolled oats, cooked	156	14	1.1	—	—	.15	.03
Trace	Trace	2	8	2	60	1 cake, 4" diam.	Pancake, enriched	27	34	.3	30	—	.05	.06
Trace	3	6	23	8	180	1 piece, ⅛ of 14" pie	Pizza pie, cheese	75	157	.7	570	8	.03	.09
—	—	1	11	2	55	1 cup	Popcorn, popped	14	2	.4	—	—	.05	.02
—	—	—	4	—	20	5 sticks	Pretzels, small stick	5	1	—	—	—	—	—
—	—	1	12	2	60	1 slice	Raisin bread, 20 slices per pound	23	16	.3	—	—	.01	.02
—	—	—	22	2	100	½ cup	Rice, white cooked	84	7	.3	—	—	.01	.01
—	—	—	12	1	55	1 cup	puffed, enriched	14	2	.3	—	—	.06	.01
—	—	—	12	2	55	1 slice	Rye bread, American, light, 20 slices per pound	23	17	.4	—	—	.04	.02
—	—	—	16	3	78	½ cup	Spaghetti, enriched, cooked (tender)	70	6	.7	—	—	.10	.06
Trace	5	18	33	11	339	1 cup	with meat sauce	250	23	1.8	771	20	.10	.10
Trace	2	7	37	9	261	1 cup	with tomato sauce, cheese	250	80	2.2	1,079	12	.27	.17
Trace	3	7.8	27	6.6	206	1 waffle, 5½" diam.	Waffle, enriched	75	179	1.0	173	—	.11	.17
—	—	—	17	2	74	¾ cup	Wheat flakes, enriched	21	9	.9	—	—	.13	.03
—	—	1	18	2	81	1 biscuit	shredded	22	9	.7	—	—	.02	.03
—	—	—	11	2	51	1 cup	Wheat, puffed, enriched	14	4	.6	—	—	.08	.03
—	—	—	11	2	51	2 tablespoons	Wheat flour, enriched, sifted	14	2	.3	—	—	.06	.04
—	—	—	11	2	51	2 tablespoons	Wheat flour, unenriched, sifted	14	2	.1	—	—	.01	.01
—	—	1	12	2	63	1 slice	White bread, enriched, 20 slices per pound	23	20	.6	—	—	.06	.04
—	—	1	9	1	47	1 slice	26 slices per pound	17	16	.5	—	—	.04	.03
—	—	1	12	2	63	1 slice	Unenriched, 20 slices per pound	23	16	.2	—	—	.02	.02
—	—	1	9	1	47	1 slice	26 slices per pound	17	16	.1	—	—	.01	.01
—	—	1	11	2	55	1 slice	Whole-wheat bread, 20 slices per pound	23	23	.5	—	—	.06	.03
							Rolls							
Trace	1	2	24	4	125	1 bun	Bun, enriched	46	42	1.2	—	—	.12	.09
Trace	1	2	31	5	162	1 roll	Hard, round, enriched	52	24	1.4	—	—	.14	.12
Trace	1	2	20	3	115	1 roll	Plain, enriched	38	28	1.0	—	—	.11	.07
Trace	1	4	21	4	136	1 roll	Sweet, pan	43	37	.3	30	—	.03	.06
							MEAT, POULTRY, FISH, ALTERNATES							
							Meat							
							Beef, cooked							
Trace	5	10	15	15	209	1 cup	Beef and vegetable stew	235	28	2.8	2,303	16.5	.14	.16
Trace	2	4	—	19	114	4 thin slices, 4" × 5"	Dried or chipped beef	56	12	2.8	—	—	.04	.18
Trace	5	10	—	23	185	3 ounces, lean	Hamburger, broiled	83	10	3.0	20	—	.08	.20

Fatty Acids from Fat		Calorie Carriers						Minerals			Vitamins			
Linoleic (g)	Saturated (g)	Fat (×9 g)	Carbo-hydrate (×4 g)	Protein (×4 g)	Food Energy (calories)	Measure	Food Item	Weight (g)	Calcium (mg)	Iron (mg)	A (I.U.)	C (mg)	Thia-mine (mg)	Ribo-flavin (mg)
Trace	2	9	—	26	185	2 slices, 4″ × 1¼″ × ½″	Pot roast, rump, lean only	80	7	3.8	—	—	.08	.18
2	7	23	37	17	436	1 pie, 4¼″ diam.	Potpie, baked	227	23	2.3	931	—	.07	.13
—	—	7	—	28	302	2 slices, 4″ × 2½″ × ½″	Rib roast, lean, some fat	106	8	3.5	—	—	.06	.22
Trace	2	2.6	—	11	70	2 slices, trimmed	lean only	41	4	1.4	—	—	.03	.09
Trace	7	13	—	25	222	1 slice, 4″ × 3″ × ⅜″	Round, broiled, lean and fat	85	0	2.6	—	—	.07	.19
Trace	1	4	—	22	134	2 pieces, 4″ × 1″ × ¼″	broiled, lean only	71	9	2.6	—	—	.06	.17
1	13	27	—	20	329	3-ounce piece, 4″ × 2¼″ × 1″	Steak, sirloin, broiled, lean and fat	85	8	2.5	50	—	.05	.16
Trace	3	6	—	27	175	3-ounce piece, 4″ × 2¼″ × 1″	lean only	85	11	3.3	15	—	.08	.21
							Cured meat							
1	2	14	2	5	147	3 slices	Bacon, broiled or fried crisp	23	4	.6	—	—	.14	.07
1	7	16	1	7	173	2 slices, 4½″ × ⅛″	Bologna	57	4	1.0	—	—	.09	.12
1	13	26	—	20	316	3 slices, 3″ × 2¼″ × ¼″	Corned beef, medium fat	85	8	2.5	—	—	.02	.15
1	6	14	1	6	155	1 frankfurter	Frankfurter, cooked	51	3	.8	—	—	.08	.10
2	7	19	—	18	246	3 ounces	Ham, smoked, lean and fat	85	8	2.2	—	—	.40	.15
1	5	14	—	9	168	2 ounces	Luncheon meat, canned	57	5	1.3	—	—	.18	.12
							Lamb							
1	18	33	—	25	402	1 thick chop	Chop, broiled, lean and fat	112	10	3.1	—	—	.14	.25
Trace	3	6	—	21	139	1 thick chop	lean only	74	9	2.5	—	—	.11	.20
Trace	9	16	—	22	237	2 slices, 4″ × 3¼″ × ⅛″	Leg, roasted, lean and fat	85	9	2.8	—	—	.18	.23
1	13	23	—	18	287	2 slices, 4″ × 3½″ × ⅛″	Shoulder, roasted, lean and fat	85	8	2.4	—	—	.11	.20
Trace	3	6	—	17	131	2 slices, 3″ × 2¾″ × ⅛″	lean only	64	8	2.2	—	—	.10	.18
							Liver							
Trace	2	3	6	15	131	1 slice, 3″ × 2¼″ × ⅜″	Beef, fried	57	6	5.0	30,438	15	.14	2.38
							Pork							
3	12	27	—	26	359	1 thick chop	Chop, broiled, lean, low fat	100	12	3.5	—	—	1.00	.29
1	2	12	—	23	203	1 thick chop	lean only	75	10	2.9	—	—	.85	.25
1	4	11	—	20	173	2 slices, 3″ × 2½″ × ¼″	Roast, lean only	68	9	2.6	—	—	.83	.21
5	18	50	—	21	538	2 patties, 2¼″ diam.	Sausage, bulk	113	8	2.7	—	—	.89	.38
Trace	4	9	—	23	183	3 ounces	Veal cutlet, broiled, lean only	85	9	2.7	—	—	.06	.21
Trace	7	14	—	23	229	2 slices, 3″ × 2″ × ⅛″	Veal roast, lean and fat	85	10	2.9	—	—	.11	.26
							Poultry							
2	3	3	—	20	116	¼ bird, boned	Chicken, broiled	85	8	1.4	68	—	.04	.16
2	3	8	—	25	185	½ breast, boned	fried	79	9	1.2	103	—	.04	.21
2	4	12	—	27	234	Thigh and drumstick, boned		89	11	1.8	187	—	.06	.40
2	2	5	—	25	152	2 slices, 3½″ × 2⅝″ × ⅛″	Turkey, roasted	80	6	1.4	—	—	.04	.14
							Fish							
—	—	7	6	18	167	1 piece, 3″ × 3″ × ⅜″	Bass, average, oven-fried	85	71	.9	72	—	.05	.12
—	—	—	—	14	82	4 large or 9 small	Clams, raw meat only	100	—	3.4	—	—	—	—
—	—	5	—	24	145	3 ounces	Cod steak, broiled	85	26	.9	—	—	.04	.08
2	4	9	7	17	176	4–5 sticks	Fishsticks, breaded, cooked	100	11	.4	—	—	.04	.07
—	—	6	6	20	165	1 fillet, 3″ × 3″ × ½″	Haddock, fried	100	40	1.2	—	—	.04	.07
—	—	9	—	32	214	5 ounces	Halibut, broiled	125	20	1.0	850	—	.06	.09
—	—	3	1	20	108	1 small	Lobster, boiled	306	74	.7	—	—	.11	.06
—	—	2	3	8	66	5–8 medium	Oysters, raw	100	94	5.5	310	—	.14	.18
—	—	11	6	16	195	1 medium fillet	Perch, ocean, breaded, fried	85	14	1.3	50	—	.09	.10
—	—	7	—	27	182	1 small serving	Salmon, cooked, broiled or baked	100	—	1.2	160	—	.16	.06
Trace	3	8	—	17	145	3 ounces or ½ cup	red, canned	85	220	1.1	200	—	.03	.14
4	2	9	1	22	180	6 sardines	Sardines, Atlantic, canned in oil	85	367	2.5	190	—	.02	.14
—	—	1	1	16	87	6–8 medium	Shrimp, raw	85	54	1.4	—	—	.02	.03
—	—	23	—	21	290	3 ounces	Trout, lake, broiled	88	27	1.1	460	—	.07	.20
4	2	7	—	25	170	½ cup	Tuna, canned	85	7	1.2	70	—	.04	.10
							Eggs							
1	2	6	—	7	88	1 egg, 24 oz./doz.	Egg, whole, large	54	26	1.2	550	—	.06	.16
—	—	—	—	4	17	1 white	white	33	3	—	—	—	—	.09
Trace	2	5	—	3	60	1 yolk	yolk	17	24	.9	550	—	.04	.07
							Nuts, peanuts, peanut butter							
4	2	19	7	7	215	¼ cup	Amonds, shelled	36	83	1.7	—	—	.09	.33
—	3	3	4	—	43	2 tablespoons	Coconut, shredded, sweetened	8	2	.2	—	—	.01	—

Linoleic (g)	Saturated (g)	Fat (×9 g)	Carbohydrate (×4 g)	Protein (×4 g)	Food Energy (calories)	Measure	Food Item	Weight (g)	Calcium (mg)	Iron (mg)	A (I.U.)	C (mg)	Thiamine (mg)	Riboflavin (mg)
3	4	18	7	10	210	4 tablespoons	Peanuts, roasted, shelled	36	26	.8	—	—	.12	.05
2	2	8	3	4	93	1 tablespoon	Peanut butter	16	12	.4	—	—	.02	.02
6	Trace	10	2	2	104	8–12 halves	Walnuts	16	16	.4	—	—	.06	.02
							Dry beans and peas							
—	—	1	24	8	132	½ cup	Lima beans, cooked	96	28	2.8	—	—	.13	.06
—	—	—	21	8	115	½ cup	Red beans	128	37	2.3	—	—	.06	.06
—	—	1	26	10	148	½ cup	Split peas, cooked	125	14	2.1	60	—	.18	.11
—	—	1	60	16	308	1 cup	White beans, cooked without pork	261	183	5.2	140	5	.13	.10

FRUITS AND VEGETABLES
Fruits

Linoleic (g)	Saturated (g)	Fat (×9 g)	Carbohydrate (×4 g)	Protein (×4 g)	Food Energy (calories)	Measure	Food Item	Weight (g)	Calcium (mg)	Iron (mg)	A (I.U.)	C (mg)	Thiamine (mg)	Riboflavin (mg)
—	—	1	33	—	133	1 apple, 3" diam.	Apple, raw	220	16	.7	200	9	.07	.05
—	—	—	30	—	126	½ cup	Applesauce, sweetened	127	5	.6	51	1	.03	.02
—	—	—	27	1	105	4 halves, 2 tablespoons syrup	Apricots, syrup packed	122	13	.4	2,130	5	.02	.03
2	4	16	6	2	167	½ avocado, 3¼"×4"	Avocado, peeled	100	10	.6	290	14	.11	.20
—	—	—	22	1	85	1 banana, 6"	Banana, small	100	8	.7	190	10	.50	.60
—	—	1	15	2	64	¾ cup	Blackberries, raw	108	35	1.0	218	23	.04	.05
—	—	1	11	1	43	½ cup	Blueberries, raw	70	11	.7	70	10	.02	.04
—	—	—	14	1	60	½ melon, 5" diam.	Cantaloupe	200	28	.8	6,800	66	.08	.06
—	—	—	11	1	40	½ cup or 9 large	Cherries, sweet, raw	57	13	.2	63	6	.03	.03
—	—	—	26	—	101	¼ cup	Cranberry sauce, sweetened	69	4	1.4	14	1	.01	.01
—	—	—	34	1	123	5–6 dates	Dates, pitted	45	26	1.4	25	—	.04	.04
—	—	1	25	1	98	½ cup	Fruit cocktail, syrup pack	128	12	.5	180	3	.02	.02
—	—	—	10	—	40	½ grapefruit, 4" diam.	Grapefruit, medium	100	16	.4	440	36	.04	.02
—	—	1	20	1	75	1 cup	Grapefruit sections, raw	194	31	.8	20	72	.07	.03
—	—	—	23	1	95	1 cup	Grapefruit juice, fresh	246	22	.5	20	92	.09	.04
—	—	1	8	1	35	½ cup or 11–12 grapes	Grapes, raw, American type	77	7	.2	50	2	.03	.02
—	—	—	21	1	83	½ cup	Grape juice	127	14	.4	—	—	.05	.03
—	—	—	1	—	5	1 tablespoon	Lemon or lime juice	15	1	—	—	7	—	—
—	—	—	18	1	70	1 orange, 3" diam.	Orange	150	63	.3	290	66	.12	.03
—	—	—	10	1	35	1 peach, 2" diam.	Peaches, raw	114	9	.5	1,320	7	.02	.05
—	—	—	17	—	66	¼ of 12-ounce carton	sliced, frozen	85	5	.4	443	25	.01	.03
—	—	1	25	1	100	1 medium pear	Pears, raw	182	13	.5	30	7	.04	.07
—	—	—	23	—	90	2 halves, 2 tablespoons syrup	syrup pack	117	6	.2	—	2	.01	.02
—	—	—	10	—	40	2 halves, 2 tablespoons juice	water pack	122	6	.3	—	2	.01	.03
—	—	—	10	1	44	½ cup	Pineapple, raw, diced	84	13	.2	108	20	.07	.02
—	—	—	26	—	95	2 small slices, 2 tablespoons syrup	syrup pack	122	35	.7	100	11	.09	.02
—	—	—	15	1	55	2 small slices, 2 tablespoons juice	juice pack	100	29	.6	80	9	.07	.03
—	—	—	16	1	60	½ cup	Pineapple juice	125	19	.6	100	11	.07	.02
—	—	—	7	—	30	1 plum, 2" diam.	Plums, raw	60	10	.3	200	3	.04	.02
—	—	—	19	1	70	4 prunes	Prunes, dried	32	14	1.0	430	1	.02	.05
—	—	—	31	1	115	4 tablespoons	Raisins, dried	40	25	1.4	8	1	.05	.03
—	—	—	18	1	70	¼ of 10-ounce carton	Raspberries, frozen	71	20	.4	55	11	.01	.03
—	—	1	13	1	53	¾ cup	red, raw	92	20	.8	120	23	.02	.07
—	—	—	49	1	193	½ cup	Rhubarb, cooked, sugar added	136	—	.6	35	9	.01	—
—	—	—	19	—	75	¼ of 10-ounce carton	Strawberries, frozen	71	16	.4	30	29	.01	.04
—	—	—	7	—	28	½ cup	raw, capped	75	15	.8	45	44	.02	.05
—	—	—	10	1	40	1 medium	Tangerine	114	34	.3	360	26	.05	.01
—	—	1	29	2	120	1 wedge, 4"×8"	Watermelon	925	30	.9	2,530	26	.20	.22

Vegetables

Linoleic (g)	Saturated (g)	Fat (×9 g)	Carbohydrate (×4 g)	Protein (×4 g)	Food Energy (calories)	Measure	Food Item	Weight (g)	Calcium (mg)	Iron (mg)	A (I.U.)	C (mg)	Thiamine (mg)	Riboflavin (mg)
—	—	—	3	2	20	6 spears	Asparagus, cooked	96	18	1.8	770	17	.06	.08
—	—	—	3	1	13	½ cup	Beans, green, cooked in small amount water, short time	63	23	.5	415	9	.05	.06
—	—	1	15	4	75	½ cup	lima, immature, cooked	80	23	1.4	230	12	.11	.07
—	—	—	8	1	35	½ cup	Beets, cooked, diced	83	18	.6	15	6	.02	.04

Linoleic (g)	Saturated (g)	Fat (×9 g)	Carbohydrate (×4 g)	Protein (×4 g)	Food Energy (calories)	Measure	Food Item	Weight (g)	Calcium (mg)	Iron (mg)	A (I.U.)	C (mg)	Thiamine (mg)	Riboflavin (mg)
—	—	—	5	3	26	1 large spear	Broccoli, cooked	100	88	.8	2,500	90	.09	.20
—	—	—	6	3	30	5–6 sprouts	Brussels sprouts, cooked	65	22	.9	260	31	.03	.08
—	—	—	5	1	20	½ cup	Cabbage, cooked in small amount water, short time	85	39	.4	75	27	.04	.04
—	—	—	3	1	13	½ cup	raw, finely shredded	50	23	.3	40	25	.03	.03
—	—	1	5	1	23	½ cup	Carrots, cooked, diced	73	19	.5	9,065	3	.04	.04
—	—	—	5	1	20	1 carrot, 5½" × 1"	raw	50	20	.4	6,000	3	.03	.03
—	—	—	5	2	23	¾ cup	Cauliflower, cooked	90	20	1.0	83	26	.05	.08
—	—	—	2	1	10	½ cup	Celery, raw, diced	50	25	.3	—	4	.03	.02
—	—	1	7	4	38	½ cup	Collards, cooked	95	237	1.5	7,250	42	.08	.23
—	—	1	21	3	85	½ cup	Corn, canned, solids and liquid	128	5	.7	260	7	.04	.07
—	—	1	16	2	65	From 1 ear, 5" × 1¾"	sweet, yellow, cooked	85	4	.5	300	6	.09	.08
—	—	1	13	6	75	½ cup	Cowpeas, immature, cooked	80	30	2.0	310	16	.23	.07
—	—	—	1	—	5	6 slices	Cucumber, raw, pared	50	5	.2	—	4	.02	.02
—	—	1	8	3	40	½ cup	Dandelion greens, cooked	90	169	2.8	13,655	15	.12	.11
—	—	—	2	1	10	10 leaves	Endive, curly (including escarole)	57	45	1.0	1,700	6	.04	.07
—	—	1	8	4	45	1 cup	Kale, cooked	110	248	2.4	9,220	56	.08	.25
—	—	—	1	1	5	2 large or 4 small	Lettuce leaves	50	11	.2	270	4	.02	.04
—	—	—	5	2	15	½ cup	Mushrooms, canned	122	9	1.0	—	2	.02	.30
—	—	—	3	2	15	½ cup	Mustard greens, cooked	70	154	2.1	5,025	32	.04	.13
—	—	—	3	1	15	4 pods	Okra, cooked	43	35	.3	315	9	.03	.03
—	—	—	11	2	50	1 medium	Onion, raw	110	35	.6	60	10	.04	.04
—	—	—	5	—	25	6 small onions	young, green	50	68	.4	30	12	.02	.02
—	—	—	—	—	1	1 tablespoon	Parsley, raw, chopped	4	7	.2	290	7	—	.01
—	—	1	10	4	55	½ cup	Peas, green, cooked	80	18	1.5	575	12	.20	.11
—	—	1	9	2	50	1 tablespoon	Peppers, hot, red, dried; ground chili pepper	15	20	1.2	11,520	2	.03	.20
—	—	—	3	1	15	1 pepper	sweet, green	62	6	.4	260	79	.05	.05
—	—	—	21	3	90	1 potato, 2½" diam.	Potatoes, baked, peeled after baking	99	9	.7	—	20	.10	.04
—	—	—	23	3	105	1 potato, 2⅜" diam.	boiled, peeled after boiling	136	10	.8	—	22	.13	.05
4	2	7	20	2	155	10 pieces	french-fried	57	9	.7	—	8	.06	.04
—	—	1	15	2	73	½ cup	mashed, milk added	98	24	.5	25	9	.06	.06
4	2	7	10	1	110	10 chips	Potato chips	20	6	.4	—	2	.04	.02
—	—	—	2	—	10	4 radishes	Radishes, raw	40	15	.4	10	10	.01	.01
—	—	—	4	1	15	½ cup	Sauerkraut, canned, drained	75	27	.4	30	12	.03	.05
—	—	1	3	3	23	½ cup	Spinach, cooked	90	—	1.8	10,600	27	.07	.18
—	—	—	4	1	18	½ cup	Squash, summer, cooked	105	16	.4	275	12	.04	.08
—	—	1	12	2	48	½ cup	winter, baked	103	25	.8	6,345	7	.05	.16
—	—	1	36	2	155	1 medium	Sweet potato, baked	110	44	1.0	8,970	24	.10	.07
1	2	8	60	2	295	1 potato, 3½" × 2¼"	candied	175	65	1.6	11,030	17	.10	.08
—	—	—	5	1	23	½ cup	Tomatoes, canned	121	14	.8	1,270	20	.07	.04
—	—	—	6	2	30	1 medium	raw	150	16	.9	1,640	35	.08	.06
—	—	—	5	1	25	½ cup	Tomato juice	121	9	.5	1,270	19	.06	.04
—	—	—	5	1	20	½ cup	Turnips, white, cooked	78	31	.4	—	14	.03	.05
—	—	—	6	1	26	½ cup	yellow, cooked	80	44	.3	280	9	.04	.06
—	—	1	4	2	23	½ cup	Turnip greens, cooked	77	188	1.8	7,685	44	.05	.30

MILK AND CHEESE
Milk

Linoleic (g)	Saturated (g)	Fat (×9 g)	Carbohydrate (×4 g)	Protein (×4 g)	Food Energy (calories)	Measure	Food Item	Weight (g)	Calcium (mg)	Iron (mg)	A (I.U.)	C (mg)	Thiamine (mg)	Riboflavin (mg)
—	—	—	13	9	90	1 cup	Buttermilk	246	298	.1	10	2	.10	.44
Trace	3	6	27	8	190	1 cup	Chocolate-flavored milk drink	250	270	.4	210	2	.09	.41
—	—	18	58	11	421	1 regular, 8 ounces milk	Chocolate milk shake	345	363	.9	687	4	.12	.55
Trace	6	11	26	9	235	1 cup	Cocoa	242	286	.9	390	2	.09	.45
—	—	—	11	7	73	4 tablespoons	Dry, nonfat	20	260	.1	5	2	.07	.36
1	11	20	24	18	345	1 cup	Evaporated, undiluted	252	635	.3	820	3	.10	.84
—	—	—	13	9	90	1 cup	Fluid, nonfat (skim)	246	298	.1	10	2	.10	.44
Trace	6	10	12	9	165	1 cup	whole	244	285	.1	390	2	.08	.42
Trace	1	2	1	—	20	1 tablespoon	Half-and-half	15	16	—	70	—	—	.02
Trace	1	2	7	4	60	½ cup	Yogurt	123	148	.05	85	1	.05	.22

Cheese

Linoleic (g)	Saturated (g)	Fat (×9 g)	Carbohydrate (×4 g)	Protein (×4 g)	Food Energy (calories)	Measure	Food Item	Weight (g)	Calcium (mg)	Iron (mg)	A (I.U.)	C (mg)	Thiamine (mg)	Riboflavin (mg)
Trace	5	9	—	6	105	1 ounce	Blue mold (Roquefort-type)	28	122	.2	350	—	.01	.17
Trace	1	2	—	2	30	1 tablespoon	Cheddar, grated	7	55	.1	90	—	.02	.03

439

| Fatty Acids from Fat | | Calorie Carriers | | | | | | Minerals | | Vitamins | | | |
Linoleic (g)	Saturated (g)	Fat (×9 g)	Carbo-hydrate (×4 g)	Protein (×4 g)	Food Energy (calories)	Measure	Food Item	Weight (g)	Calcium (mg)	Iron (mg)	A (I.U.)	C (mg)	Thia-mine (mg)	Ribo-flavin (mg)
Trace	5	9	—	7	105	1 ounce or 1 slice	processed	28	214	.2	350	—	—	.12
Trace	1	1	1	4	30	2 tablespoons	Cottage cheese, creamed	28	25	.1	50	—	.01	.08
—	—	—	1	5	25	2 tablespoons	uncreamed	28	26	.1	—	—	.01	.08
Trace	3	6	—	1	55	1 tablespoon	Cream cheese	15	9	—	230	—	—	.04
Trace	4	8	1	7	105	1 ounce	Swiss	28	271	.3	320	—	.01	.06

FATS AND OILS

Trace	3	6	—	—	55	1 tablespoon before whipping	Cream, heavy, whipping	15	10	—	240	—	—	.02
Trace	2	3	1	—	35	1 tablespoon	light	15	15	—	130	—	—	.02

Fats, Cooking

1	5	14	—	—	135	1 tablespoon	Lard	14	—	—	—	—	—	—
1	3	12	—	—	110	1 tablespoon	Vegetable fats, solid	13	—	—	—	—	—	—

Oils, Salad or Cooking

7	3	14	—	—	125	1 tablespoon	Corn, cottonseed, or soybean	14	—	—	—	—	—	—
1	2	14	—	—	125	1 tablespoon	Olive	14	—	—	—	—	—	—

Salad Dressings

5	2	10	1	1	90	1 tablespoon	Blue cheese	16	11	—	30	—	—	.02
3	1	6	2	—	60	1 tablespoon	Commercial, mayonnaise-type	15	2	—	30	—	—	—
3	1	6	2	—	60	1 tablespoon	French	15	3	.1	—	—	—	—
6	2	12	—	—	110	1 tablespoon	Mayonnaise	15	2	.1	40	—	—	—
4	1	8	1	—	75	1 tablespoon	Thousand Island	15	2	.1	60	2	—	—

Spreads

Trace	6	11	—	—	100	1 tablespoon	Butter	14	3	—	460	—	—	—
1	3	11	—	—	100	1 tablespoon	Margarine	14	3	—	460	—	—	—

SUGARS, SYRUPS, CANDIES

Trace	2	3	22	1	120	3 caramels	Caramels, plain	28	36	.7	50	—	.01	.04
Trace	5	9	16	2	145	1-ounce bar	Chocolate, sweetened, milk	28	61	.3	40	—	.03	.11
—	—	—	11	—	40	1 tablespoon	Chocolate syrup	20	3	.3	—	—	—	—
Trace	2	3	23	—	115	1¼" square	Fudge, plain	28	14	.1	60	—	—	.02
—	—	—	23	1	90	1-ounce bar	Marshmallow	28	—	—	—	—	—	—
—	—	—	9	—	37	2 squares	Hard candy	10	—	—	—	—	—	—
—	—	—	17	—	60	1 tablespoon	Honey	21	1	.2	—	1	—	.01
—	—	—	14	—	55	1 tablespoon	James, jellies, preserves	20	2	.1	—	1	—	—
—	—	—	6	—	23	1 marshmallow	Marshmallow, plain	7	—	—	—	—	—	—
—	—	—	13	—	50	1 tablespoon	Molasses	20	33	.9	—	—	.01	.01
—	—	—	13	—	50	1 tablespoon	Sugar, brown, firmly packed	14	10	.4	—	—	—	—
—	—	—	4	—	17	1 teaspoon	granulated, cane or beet	4	—	—	—	—	—	—
—	—	—	8	—	30	1 tablespoon	powdered	8	—	—	—	—	—	—
—	—	—	15	—	55	1 tablespoon	Syrup, table blends	20	9	.8	—	—	—	—

DESSERTS
Cakes

1	—	—	23	3	110	1/12 of 8" cake	Angel food cake	40	2	.1	—	—	—	.05
1	5	14	70	5	420	1/16 of 10" cake	Chocolate layer cake, fudge icing	120	118	.5	520	—	.03	.10
Trace	1	6	24	2	152	1 piece, 3" × 3" × ½"	Fruitcake, enriched, dark	40	29	1.0	48	—	.05	.06
Trace	2	10	27	2	206	1 piece, 2" × 2" × 2"	Gingerbread	57	63	1.4	69	—	.07	.05
Trace	1	5	31	4	180	1 piece, 3" × 2" × 1½"	Plain cake, no icing	55	85	.2	200	—	.02	.05
Trace	2	6	62	5	320	1/16 of 10" cake	with icing	100	117	.4	280	—	.02	.07
Trace	1	3	23	3	130	1 cupcake	Plain cupcake, no icing	40	62	.2	150	—	.01	.03
Trace	1	3	31	3	160	1 cupcake	with icing	50	58	.2	140	—	.01	.04
1	2	7	15	2	130	1 slice, 2¾" × 3" × ⅝"	Pound cake	30	16	.5	300	—	.04	.05
Trace	1	2	22	3	115	1/12 of 8" cake	Sponge cake	40	11	.6	210	—	.02	.06

Cookies

Trace	1	3	19	2	110	1 cookie, 3" diam.	Assorted	25	6	.2	—	—	.01	.01
—	—	6	15	1	116	1 cookie	Butterscotch, refrigerator	24	11	.5	22	—	.04	.03

Fatty Acids from Fat		Calorie Carriers							Minerals		Vitamins			
Linoleic (g)	*Saturated (g)*	*Fat (×9 g)*	*Carbohydrate (×4 g)*	*Protein (×4 g)*	*Food Energy (calories)*	*Measure*	*Food Item*	*Weight (g)*	*Calcium (mg)*	*Iron (mg)*	*A (I.U.)*	*C (mg)*	*Thiamine (mg)*	*Riboflavin (mg)*
—	—	4	3	—	17	1 small cookie	Chocolate or gingersnap	4	3	.1	3	—	.02	.02
—	—	1	12	1	55	1 small bar	Fig bar	16	11	.2	—	—	—	.01
—	—	2	10	9	63	1 cookie	Oatmeal with raisins	14	3	.4	7	—	.15	.11
—	—	2	8	—	17	1 small cookie	Vanilla wafer	4	4	—	14	—	.02	.08
							Ice Cream and Frozen Custard							
Trace	3	5	21	5	143	⅛ quart or ½ cup	Ice milk	94	146	.1	195	1	.05	.21
Trace	5	8	15	3	137	⅛ quart or ½ cup	Regular, 10% fat	71	104	—	312	—	.03	.15
Trace	5	11	13	2	158	⅛ quart or ½ cup	Rich, 16% fat	71	55	—	469	—	.01	.08
							Pies							
1	4	15	51	3	346	⅐ of 9″ pie	Apple	135	11	.4	40	1	.03	.03
1	4	15	52	4	352	⅐ of 9″ pie	Cherry	135	19	.4	594	—	.03	.03
1	4	15	32	8	320	⅐ of 9″ pie	Custard	135	130	.8	311	—	.07	.22
1	4	11	45	4	306	⅐ of 9″ pie	Lemon meringue	120	17	.6	204	4	.04	.10
1	2	16	56	3	366	⅐ of 9″ pie	Mince	135	38	1.4	—	1	.09	.05
1	5	15	33	5	274	⅐ of 9″ pie	Pumpkin	130	66	.7	3,211	—	.04	.14
							Other							
Trace	3	5	24	4	152	½ cup	Cornstarch pudding, vanilla	125	144	.1	195	—	.04	.21
Trace	5	9	23	9	205	1 custard	Custard, baked (4 to 1 pt. milk)	157	163	1.1	607	—	.08	.32
Trace	1	6	16	2	125	1 average doughnut	Doughnut, cake-type	32	13	.4	26	—	.05	.05
—	—	1	30	1	140	½ cup	Sherbet	97	15	—	58	2	.01	.03
							MISCELLANEOUS							
—	—	—	18	—	71	1 cup	Beverage, carbonated, ginger ale	230	—	—	—	—	—	—
—	—	—	23	—	90	1 cup	cola-type	230	—	—	—	—	—	—
—	—	—	—	—	5	1 cube	Bouillon cube	4	—	—	—	—	—	—
—	—	—	4	—	18	1 tablespoon	Chili sauce	17	3	.1	238	3	.02	.01
—	—	14	8	3	141	1 ounce square	Chocolate, unsweetened	28	22	1.9	17	—	.01	.07
—	—	—	—	9	34	1 tablespoon	Gelatin, dry, plain	10	—	—	—	—	—	—
—	—	—	17	2	71	½ cup	Gelatin dessert, ready-to-eat, plain	120	—	—	—	—	—	—
—	—	—	20	2	80	½ cup	with fruit	120	—	—	4	—	—	—
Trace	Trace	2	—	—	35	4 extra large	Olives, green	22	14	.4	66	—	—	—
Trace	Trace	3	1	—	28	4 extra large	ripe	22	15	.3	13	—	—	—
—	—	—	3	1	15	1 pickle, 4″ × 1¾″	Pickle, dill	135	34	1.6	420	8	—	.09
—	—	—	7	—	29	1 pickle, 2¾″ × ¾″	sweet	20	2	.2	18	1	—	—
							Soups, canned, ready-to-serve							
Trace	2	6	22	8	168	1 cup	Bean, with pork	250	63	2.3	650	3	.13	.08
Trace	1	1	5	2	38	¾ cup	Chicken, with rice	188	6	.2	113	—	—	.18
1	Trace	3	13	2	84	1 cup	Clam chowder, Manhattan-type with water	255	36	1.0	918	—	.03	.03
4	4	11	13	5	168	¾ cup	Mushroom, cream of, with milk	191	149	.4	191	—	.03	.27
1	Trace	3	21	9	148	1 cup	Pea, split	250	30	1.5	450	—	.25	.15
1	1	2	12	2	66	¾ cup	Tomato (water)	184	11	.6	754	9	.04	.04
Trace	1	2	10	5	80	1 cup	Vegetable, beef	250	13	.8	2,750	—	.05	.05
—	—	—	7	—	30	1 tablespoon	Starch, corn, arrowroot, etc.	8	—	—	—	—	—	—
1	5	8	6	3	107	¼ cup	White sauce, medium	66	74	.2	328	—	.02	.01

Ounce and Pound Amounts for Multiple Portions

Ounces: 25 to 1000 Portions

25	50	75	100	200	300	400	500	600	700	800	900	1000
—	—	⅛ oz	⅛ oz	¼ oz	⅜ oz	½ oz	⅝ oz	¾ oz	⅞ oz	1 oz	1⅛ oz	1¼ oz
—	—	⅙ oz	⅙ oz	⅓ oz	½ oz	⅔ oz	⅚ oz	1 oz	1⅙ oz	1⅓ oz	1½ oz	1⅔ oz
—	—	⅕ oz	⅕ oz	⅖ oz	⅗ oz	⅘ oz	1 oz	1⅕ oz	1⅖ oz	1⅗ oz	1⅘ oz	2 oz
—	—	¼ oz	¼ oz	½ oz	¾ oz	1 oz	1¼ oz	1½ oz	1¾ oz	2 oz	2¼ oz	2½ oz
—	—	⅓ oz	⅓ oz	⅔ oz	1 oz	1⅓ oz	1⅔ oz	2 oz	2⅓ oz	2⅔ oz	3 oz	3⅓ oz
—	¼ oz	⅜ oz	½ oz	1 oz	1½ oz	2 oz	2½ oz	3 oz	3½ oz	4 oz	4½ oz	5 oz
—	⅓ oz	½ oz	⅔ oz	1⅓ oz	2 oz	2⅔ oz	3⅓ oz	4 oz	4⅔ oz	5⅓ oz	6 oz	6⅔ oz
—	⅜ oz	⅝ oz	¾ oz	1½ oz	2¼ oz	3 oz	3¾ oz	4½ oz	5¼ oz	6 oz	6¾ oz	7½ oz
¼ oz	½ oz	¾ oz	1 oz	2 oz	3 oz	4 oz	5 oz	6 oz	7 oz	8 oz	9 oz	10 oz
½ oz	1 oz	1½ oz	2 oz	4 oz	6 oz	8 oz	10 oz	12 oz	14 oz	1 lb	1 lb 2 oz	1 lb 4 oz
¾ oz	1½ oz	2¼ oz	3 oz	6 oz	9 oz	12 oz	15 oz	1 lb 2 oz	1 lb 5 oz	1 lb 8 oz	1 lb 11 oz	1 lb 14 oz
1 oz	2 oz	3 oz	4 oz	8 oz	12 oz	1 lb	1 lb 4 oz	1 lb 8 oz	1 lb 12 oz	2 lb	2 lb 4 oz	2 lb 8 oz
1¼ oz	2½ oz	3¾ oz	5 oz	10 oz	15 oz	1 lb 4 oz	1 lb 9 oz	1 lb 14 oz	2 lb 3 oz	2 lb 8 oz	2 lb 13 oz	3 lb 2 oz
1½ oz	3 oz	4½ oz	6 oz	12 oz	1 lb 2 oz	1 lb 8 oz	1 lb 14 oz	2 lb 4 oz	2 lb 10 oz	3 lb	3 lb 6 oz	3 lb 12 oz
1¾ oz	3½ oz	5¼ oz	7 oz	14 oz	1 lb 5 oz	1 lb 12 oz	2 lb 3 oz	2 lb 10 oz	3 lb 1 oz	3 lb 8 oz	3 lb 15 oz	4 lb 6 oz
2 oz	4 oz	6 oz	8 oz	1 lb	1 lb 8 oz	2 lb	2 lb 8 oz	3 lb	3 lb 8 oz	4 lb	4 lb 8 oz	5 lb
2¼ oz	4½ oz	6¾ oz	9 oz	1 lb 2 oz	1 lb 11 oz	2 lb 4 oz	2 lb 13 oz	3 lb 6 oz	3 lb 15 oz	4 lb 8 oz	5 lb 1 oz	5 lb 10 oz
2½ oz	5 oz	7½ oz	10 oz	1 lb 4 oz	1 lb 14 oz	2 lb 8 oz	3 lb 2 oz	3 lb 12 oz	4 lb 6 oz	5 lb	5 lb 10 oz	6 lb 4 oz
2¾ oz	5½ oz	8¼ oz	11 oz	1 lb 6 oz	2 lb 1 oz	2 lb 12 oz	3 lb 7 oz	4 lb 2 oz	4 lb 13 oz	5 lb 8 oz	6 lb 3 oz	6 lb 14 oz
3 oz	6 oz	9 oz	12 oz	1 lb 8 oz	2 lb 4 oz	3 lb	3 lb 12 oz	4 lb 8 oz	5 lb 4 oz	6 lb	6 lb 12 oz	7 lb 8 oz
3¼ oz	6½ oz	9¾ oz	13 oz	1 lb 10 oz	2 lb 7 oz	3 lb 4 oz	4 lb 1 oz	4 lb 14 oz	5 lb 11 oz	6 lb 8 oz	7 lb 5 oz	8 lb 2 oz
3½ oz	7 oz	10½ oz	14 oz	1 lb 12 oz	2 lb 10 oz	3 lb 8 oz	4 lb 6 oz	5 lb 4 oz	6 lb 2 oz	7 lb	7 lb 14 oz	8 lb 12 oz
3¾ oz	7½ oz	11¼ oz	15 oz	1 lb 14 oz	2 lb 13 oz	3 lb 12 oz	4 lb 11 oz	5 lb 10 oz	6 lb 9 oz	7 lb 8 oz	8 lb 7 oz	9 lb 6 oz

Pounds: 25 to 1000 Portions

25	50	75	100	200	300	400	500	600	700	800	900	1000
4 oz	8 oz	12 oz	1 lb	2 lb	3 lb	4 lb	5 lb	6 lb	7 lb	8 lb	9 lb	10 lb
5 oz	10 oz	15 oz	1 lb 4 oz	2 lb 8 oz	3 lb 12 oz	5 lb	6 lb 4 oz	7 lb 8 oz	8 lb 12 oz	10 lb	11 lb 4 oz	12 lb 8 oz
6 oz	12 oz	1 lb 2 oz	1 lb 8 oz	3 lb	4 lb 8 oz	6 lb	7 lb 8 oz	9 lb	10 lb 8 oz	12 lb	13 lb 8 oz	15 lb
7 oz	14 oz	1 lb 5 oz	1 lb 12 oz	3 lb 8 oz	5 lb 4 oz	7 lb	8 lb 12 oz	10 lb 8 oz	12 lb 4 oz	14 lb	15 lb 12 oz	17 lb 8 oz
8 oz	1 lb	1 lb 8 oz	2 lb	4 lb	6 lb	8 lb	10 lb	12 lb	14 lb	16 lb	18 lb	20 lb
9 oz	1 lb 2 oz	1 lb 11 oz	2 lb 4 oz	4 lb 8 oz	6 lb 12 oz	9 lb	11 lb 4 oz	13 lb 8 oz	15 lb 12 oz	18 lb	20 lb 4 oz	22 lb 8 oz
10 oz	1 lb 4 oz	1 lb 14 oz	2 lb 8 oz	5 lb	7 lb 8 oz	10 lb	12 lb 8 oz	15 lb	17 lb 8 oz	20 lb	22 lb 8 oz	25 lb
11 oz	1 lb 6 oz	2 lb 1 oz	2 lb 12 oz	5 lb 8 oz	8 lb 4 oz	11 lb	13 lb 12 oz	16 lb 8 oz	19 lb 4 oz	22 lb	24 lb 12 oz	27 lb 8 oz
12 oz	1 lb 8 oz	2 lb 4 oz	3 lb	6 lb	9 lb	12 lb	15 lb	18 lb	21 lb	24 lb	27 lb	30 lb
13 oz	1 lb 10 oz	2 lb 7 oz	3 lb 4 oz	6 lb 8 oz	9 lb 12 oz	13 lb	16 lb 4 oz	19 lb 8 oz	22 lb 12 oz	26 lb	29 lb 4 oz	32 lb 8 oz
14 oz	1 lb 12 oz	2 lb 10 oz	3 lb 8 oz	7 lb	10 lb 8 oz	14 lb	17 lb 8 oz	21 lb	24 lb 8 oz	28 lb	31 lb 8 oz	35 lb
15 oz	1 lb 14 oz	2 lb 13 oz	3 lb 12 oz	7 lb 8 oz	11 lb 4 oz	15 lb	18 lb 12 oz	22 lb 8 oz	26 lb 4 oz	30 lb	33 lb 12 oz	37 lb 8 oz
1 lb	2 lb	3 lb	4 lb	8 lb	12 lb	16 lb	20 lb	24 lb	28 lb	32 lb	36 lb	40 lb
1 lb 1 oz	2 lb 2 oz	3 lb 3 oz	4 lb 4 oz	8 lb 8 oz	12 lb 12 oz	17 lb	21 lb 4 oz	25 lb 8 oz	29 lb 12 oz	34 lb	38 lb 4 oz	42 lb 8 oz
1 lb 2 oz	2 lb 4 oz	3 lb 6 oz	4 lb 8 oz	9 lb	13 lb 8 oz	18 lb	22 lb 8 oz	27 lb	31 lb 8 oz	36 lb	40 lb 8 oz	45 lb
1 lb 3 oz	2 lb 6 oz	3 lb 9 oz	4 lb 12 oz	9 lb 8 oz	14 lb 4 oz	19 lb	23 lb 12 oz	28 lb 8 oz	33 lb 4 oz	38 lb	42 lb 12 oz	47 lb 8 oz
1 lb 4 oz	2 lb 8 oz	3 lb 12 oz	5 lb	10 lb	15 lb	20 lb	25 lb	30 lb	35 lb	40 lb	45 lb	50 lb
1 lb 5 oz	2 lb 10 oz	3 lb 15 oz	5 lb 4 oz	10 lb 8 oz	15 lb 12 oz	21 lb	26 lb 4 oz	31 lb 8 oz	36 lb 12 oz	42 lb	47 lb 4 oz	52 lb 8 oz
1 lb 6 oz	2 lb 12 oz	4 lb 2 oz	5 lb 8 oz	11 lb	16 lb 8 oz	22 lb	27 lb 8 oz	33 lb	38 lb 8 oz	44 lb	49 lb 8 oz	55 lb
1 lb 7 oz	2 lb 14 oz	4 lb 5 oz	5 lb 12 oz	11 lb 8 oz	17 lb 4 oz	23 lb	28 lb 12 oz	34 lb 8 oz	40 lb 4 oz	46 lb	51 lb 12 oz	57 lb 8 oz
1 lb 8 oz	3 lb	4 lb 8 oz	6 lb	12 lb	18 lb	24 lb	30 lb	36 lb	42 lb	48 lb	54 lb	60 lb
1 lb 12 oz	3 lb 8 oz	5 lb 4 oz	7 lb	14 lb	21 lb	28 lb	35 lb	42 lb	49 lb	56 lb	63 lb	70 lb
2 lb	4 lb	6 lb	8 lb	16 lb	24 lb	32 lb	40 lb	48 lb	56 lb	64 lb	72 lb	80 lb
2 lb 4 oz	4 lb 8 oz	6 lb 12 oz	9 lb	18 lb	27 lb	36 lb	45 lb	54 lb	63 lb	72 lb	81 lb	90 lb
2 lb 8 oz	5 lb	7 lb 8 oz	10 lb	20 lb	30 lb	40 lb	50 lb	60 lb	70 lb	80 lb	90 lb	100 lb
2 lb 12 oz	5 lb 8 oz	8 lb 4 oz	11 lb	22 lb	33 lb	44 lb	55 lb	66 lb	77 lb	88 lb	99 lb	110 lb
3 lb	6 lb	9 lb	12 lb	24 lb	36 lb	48 lb	60 lb	72 lb	84 lb	96 lb	108 lb	120 lb
3 lb 12 oz	7 lb 8 oz	11 lb 4 oz	15 lb	30 lb	45 lb	60 lb	75 lb	90 lb	105 lb	120 lb	135 lb	150 lb
4 lb 4 oz	8 lb 8 oz	12 lb 12 oz	17 lb	34 lb	51 lb	68 lb	85 lb	102 lb	119 lb	136 lb	153 lb	170 lb
4 lb 8 oz	9 lb	13 lb 8 oz	18 lb	36 lb	54 lb	72 lb	90 lb	108 lb	126 lb	144 lb	162 lb	180 lb
5 lb	10 lb	15 lb	20 lb	40 lb	60 lb	80 lb	100 lb	120 lb	140 lb	160 lb	180 lb	200 lb
5 lb 12 oz	11 lb 8 oz	17 lb 4 oz	23 lb	46 lb	69 lb	92 lb	115 lb	138 lb	161 lb	184 lb	207 lb	230 lb
6 lb 4 oz	12 lb 8 oz	18 lb 12 oz	25 lb	50 lb	75 lb	100 lb	125 lb	150 lb	175 lb	200 lb	225 lb	250 lb

Metric Conversion Tables

Metric Equivalents

Linear Measure

1 centimeter = 0.3937 in.	1 in. = 254 centimeters
1 decimeter = 3.937 in. = 0.328 ft	1 ft = 3.048 decimeters
1 meter = 39.37 in. = 1.0936 yards	1 yard = 0.9144 meter
1 dekameter = 1.9884 rods	1 rod = 0.5029 dekameter
1 kilometer = 0.62137 mile	1 mile = 1.6093 kilometers

Square Measure

1 sq. centimeter = 0.1550 sq. in.	1 sq. in. = 6.452 sq. centimeters
1 sq. decimeter = 0.1076 sq. ft	1 sq. ft = 9.2903 sq. decimeters
1 sq. meter = 1.196 sq. yd	1 sq. yd = 0.8361 sq. meter
1 are = 3.954 sq. rods	1 sq. rod = 0.2529 are
1 hectare = 2.47 acres	1 acre = 0.4047 hectare
1 sq. kilometer = 0.386 sq. m.	1 sq. m. = 259 sq. kilometers

Measure of Volume

1 cu. centimeter = 0.061 cu. in.	1 cu. in. = 16.39 cu. centimeters
1 cu. decimeter = 0.353 cu. ft	1 cu. ft = 28.317 cu. decimeters
1 cu. meter = 1.308 cu. yd	1 cu. yd = 0.7646 cu. meter
1 stere = 0.2759 cord	1 cord = 3.642 steres
1 liter = 0.908 qt dry	1 qt dry = 1.101 liters
1 liter = 1.0567 qt liquid	1 qt liquid = 0.9463 liter
1 dekaliter = 2.6417 gal	1 gal = 0.3785 dekaliter
1 dekaliter = 0.135 peck	1 peck = 0.881 dekaliter
1 hectoliter = 2.8375 bushels	1 bushel = 0.3524 hectoliter

Weights

1 gram = 0.03547 oz	1 oz = 28.35 grams
1 kilogram = 2.2046 lb	1 lb = 0.4536 kilogram
1 metric ton = 1.1023 English ton	1 English ton = 0.9072 metric ton
	1 kilogram = 1000 grams

Approximate Metric Equivalents

1 decimeter = 4 in.	1 metric ton = 2200 lb
1 meter = 1.1 yards	1 liter = 1.06 qt liquid
1 kilometer = ⅝ mile	1 liter = 0.9 qt dry
1 hectare = 2½ acres	1 hectoliter = 2⅝ bushels
1 stere or cu. meter = ¼ of a cord	1 kilogram = 2⅕ lb

Weights

The *kilo*, or kilogram (kg), has 1000 grams and is the standard metric weight unit. It is the *exact* equivalent of 2.205 American pounds. Conversions for recipes are:

Grams		Ounces	Ounces		Grams
1	=	0.035	1	=	28.35
2	=	0.07	2	=	56.70
3	=	0.11	3	=	85.05
4	=	0.14	4	=	113.40
5	=	0.18	5	=	141.75
6	=	0.21	6	=	170.10
7	=	0.25	7	=	198.45
8	=	0.28	8	=	226.80
9	=	0.32	9	=	255.15
10	=	0.35	10	=	283.50
20	=	0.70	11	=	311.85
30	=	1.05	12	=	340.20
40	=	1.40	13	=	368.55
50	=	1.75	14	=	396.90
			15	=	425.25
			16	=	453.59

Metrics is a system of weights and measures in which the gram is the basic unit of weight, the meter is the unit of length, and the liter is the measure of capacity or volume.

Metric measurements are made according to the decimal system, in units of 10. The meter, then, which is the approximate equivalent of 1 yard, contains 100 centimeters. One liter, which is approximately 1 quart liquid measure, contains 100 centiliters.

The basic vocabulary of the metric system is:

- *milli:* 1/1000 of, as a millimeter equals 1/1000 of a meter.
- *centi:* 1/100 of, as a centimeter equals 1/100 of a meter.
- *deci:* 1/10 of, as a decimeter equals 1/10 of a meter.
- *kilo:* 1000 times, as a kilometer equals 1000 meters.

Liquid Measure

The *liter* (L) has 1000 milliliters (mL) and is the standard metric unit for liquid volume. The liter is commonly divided into tenths (deciliters or dL) and hundreths (centiliters or cL). One liter is the *exact* equivalent of 1.057 American quarts. Conversions for recipes are:

Fluid Ounces		Centiliters	Centiliters		Fluid Ounces
1	=	2.96	1	=	0.34
2	=	5.92	2	=	0.7
3	=	8.87	3	=	1.0
4	=	11.83	4	=	1.4
5	=	14.79	5	=	1.7
6	=	17.74	6	=	2.0
7	=	20.70	7	=	2.4
8	=	23.66	8	=	2.7
9	=	26.62	9	=	3.0
10	=	29.57	10	=	3.4
11	=	32.53	20	=	7.0
12	=	35.48	30	=	10.0
13	=	38.44	40	=	14.0
14	=	41.40	50	=	17.0
15	=	44.36	60	=	20.0
16	=	47.32	70	=	24.0
			80	=	27.0
			90	=	30.0
			100 (liter)	=	34.0

A major advantage of the metric system is that liquid volume and weights are related. One liter of water equals 1 kilo or 1000 grams.

Portion Measurement Data

Ounce and Cup Equivalents

Food Item	Ounces per Cup (Approximate)	Cups per Pound (Approximate)	Food Item	Ounces per Cup (Approximate)	Cups per Pound (Approximate)
Eggs, fresh, hard cooked, diced	6	2⅔	Mincemeat	8	2
fresh or frozen, white	8⅔	2	Molasses, cane	11½	1⅓
fresh or frozen, yolks	8⅔	2	Noodles, 1-in. pieces, dry	2⅔	6
Egg solids, pan dried, powdered	4⅔	3½	Nut meats, chopped	4	4
whites, spray dry, sifted	3	5	Oats, rolled (quick)	3	5⅔
whole, packed	4	4	Oils, all	8	2
whole, sifted	3	5	Onions, dried	3½	4½
yolks, packed	3½	4¾	Peanuts	5	3¼
yolks, sifted	3	5⅔	Peanut butter	8	2
Figs, cut fine	6	2⅔	Peas, split	7	2¼
Filberts, whole	4¾	3⅓	Pecan halves	3¾	4¼
Flour, corn	4	4	Pimiento	7	2¼
gluten	5	3¼	Potato starch, sifted	5	3¼
graham	5	3¼	Prunes, cooked, drained	8	2
rice	4½	3½	pitted, dried, raw	7	2¼
rye	3	5⅔	Raisins, seeded, chopped	6½	2½
soy, full fat	2	7½	seeded, whole	5	3¼
soy, low fat	3	5½	seedless, chopped	8	2
white, all-purpose	4	4	seedless, whole	5¾	2¾
white, bread	4	4	Rice, white, long grain	6½	2½
white, cake	3⅓	4¾	white, medium grain	6¾	2⅓
white, pastry	3½	4½	white, short grain	7	2¼
white, self-rising	4	4	wild	5½	3
whole-wheat	7	2¼	Rice flour, sifted	4½	3½
Gelatin, flavored	7	2⅓	Rye flour	3	5⅔
granulated	5⅓	3	Salt, cooking	8	2
Ginger, crystalized, diced	6	2⅔	free running	10¼	1½
Grapenuts	4	4	Sorghum	11⅔	1⅓
Hominy, grits	5½	3	Soybeans	7½	2¼
whole	6½	2½	Soy flour, full fat	2	7½
Honey	12	1⅓	low fat	3	5½
Hydrogenated fat	6⅔	2½	Soy grits	5	3⅓
Lard	8	2	Spaghetti, 2-in. pieces	3⅓	4¾
Lentils	7	2¼	Suet, chopped	4¼	3¾
Macaroni, 1-in. pieces, dry	4	4	Sugar, cane or beet		
shell, dry	4	4	confectioners	4½	3½
Maple syrup	11	1½	granulated	7	2¼
Margarine	8	2	superfine	7	2¼
Marshmallows	4	4	Tapioca, quick cooking	5½	3
Milk, buttermilk	8⅔	2	Tea	2½	6⅓
condensed	10¾	1½	Vermicelli, 2-in. pieces	3	5⅔
evaporated	9	1¾	Walnuts, English, half	3½	4½
nonfat, dry, crystal	2¾	5⅔	Wheat germ	2½	6⅔
nonfat, dry, powder	4	4	Yeast, dry	4¾	3⅓
skimmed	8⅔	2			
whole	8⅔	2			
whole, dry	3¾	4¼			

Common Can and Jar Sizes

The labels of cans or jars of identical size may show a net weight for one product that differs slightly from the net weight on the label of another product, due to the density of the food. An example would be lima beans (1 lb), and blueberries (14 oz), in the same-size can.

Can Size (Industry Term)	Average Net Weight or Fluid Measure per Can (Check Label)	Average Cups per Can	Cans per Case
No. 10	6 lb 8 oz (104 oz) to 7 lb 5 oz (117 oz)	12–13	6
No. 3 Cyl. or 46 fl. oz	3 lb 3 oz (51 oz) or 1 qt 14 fl. oz (46 fl. oz)	5¾	12
No. 2½	1 lb 13 oz (29 oz)	3½	24
No. 2	1 lb 4 oz (20 oz) or 1 pt 2 fl. oz (18 fl. oz)	2½	24
No. 303	16 to 17 oz	2	24
No. 300	14 to 16 oz	1¾	24
No. 1 Picnic	10½ to 12 oz	1¼	48
8 oz	8 oz	1	48 or 72

Meats, fish and seafoods are known and sold by weight of contents of can.

Measures for Portion Control

The most dependable method to use in measuring serving sizes or portions is to serve the food with dippers, scoops, ladles, or spoons of standard sizes.

Dippers or Scoops

The number of the scoop indicates the number of scoopfuls it takes to make 1 quart. The following table shows the level measures of each scoop in cups or tablespoons:

Dipper or Scoop No.	Level Measure
6	⅔ cup
8	½ cup
10	⅖ cup
12	⅓ cup
16	¼ cup
20	3⅕ tablespoons
24	2⅔ tablespoons
30	2⅕ tablespoons
40	1⅗ tablespoons

Scoops may be used for portioning such items as drop cookies, muffins, meat patties, and some vegetables and salads.

Ladles

Ladles may be used in serving soups, stews, creamed dishes, sauces, gravies, and similar products.

¼ cup	¾ cup
½ cup	1 cup

Common Food Measures

3 teaspoons	=	1 tablespoon
2 tablespoons	=	⅛ cup or 1 fluid ounce
4 tablespoons	=	¼ cup
8 tablespoons	=	½ cup
12 tablespoons	=	¾ cup
16 tablespoons	=	1 cup
2 cups	=	1 pint
2 pints	=	1 quart
4 quarts	=	1 gallon
8 quarts	=	1 peck
4 pecks	=	1 bushel

Abbreviations

AP	as purchased
EP	edible portion
Cyl	cylinder
tsp	teaspoon
T or tbsp	tablespoon
lb	pound
c	cup
pt	pint
qt	quart
gal	gallon
oz	ounce
fl. oz	fluid ounce
No.	number
wt	weight
incl	including
excl	excluding

Ounce Equivalents in Decimal Parts of One Pound

(1) Number of Ounces	(2) +0 Ounce	(3) +$\frac{1}{4}$ Ounce	(4) +$\frac{1}{2}$ Ounce	(5) +$\frac{3}{4}$ Ounce
	Pound	Pound	Pound	Pound
0	—	0.016	0.031	0.047
1	0.062	0.078	0.094	0.109
2	0.125	0.141	0.156	0.172
3	0.188	0.203	0.219	0.234
4	0.250	0.266	0.281	0.297
5	0.312	0.328	0.344	0.359
6	0.375	0.391	0.406	0.422
7	0.438	0.453	0.469	0.484
8	0.500	0.516	0.531	0.547
9	0.562	0.578	0.594	0.609
10	0.625	0.641	0.656	0.672
11	0.688	0.703	0.719	0.734
12	0.750	0.766	0.781	0.797
13	0.812	0.828	0.844	0.859
14	0.875	0.891	0.906	0.922
15	0.938	0.953	0.969	0.984

To convert 10½ ounces to a decimal part of a pound, find 10 in column 1, then follow this line across to column 4, which shows that 0.656 pound corresponds to 10½ ounces.

To convert a decimal part of a pound such as 0.531 to ounces, find 0.531 in the decimal pound readings—then refer to column 1 on the same line and find 8, the number of whole ounces. At the top of the column in which 0.531 is located, the +½ ounce should be added to the 8 ounces. Thus 0.531 pound corresponds to 8½ ounces.

Food Equivalence Data

TABLE F-1 Approximate Equivalents in Quality Food Products

Ingredient	Substitute	Measure	Weight
Butter (1 pt, 1 lb)	Margarine	1 pt	1 lb
Butter or margarine (1 pt, 1 lb)	Fat or oil	1⅞ c plus ½ t salt, 2 T water	14 oz, ½ t salt, 1 oz water
Cake flour (1 qt, 1 lb)	Hard or all-purpose flour	3¾ c and ¼ c cornstarch	13½ oz and 2½ oz cornstarch
Chocolate, bitter (2 c, 1 lb)	Cocoa	3½ c, ¾ c shortening	12½ oz, 3½ oz shortening
Cream, coffee, 18% (1 qt, 2 lb)	Milk, nonfat, and butter	3½ c milk, ¾ c butter	26 oz milk, 6 oz butter
whipping, 40% (1 qt, 2 lb)	Butter	Butter	40% of cream weight
	Milk, nonfat, and butter	3 c milk, 1⅓ c butter	20 oz milk, 12½ oz butter
Eggs, whites (1 pt, 1 lb)	Dried whites	1⅛ c, 1¾ c water	3 oz, 14 oz water
	Frozen whites	1 pt (18 whites)	1 lb
whole (1 pt, 1 lb)	Dried whole eggs	1¼ c, 1¾ c water	5 oz, 12½ oz water
	Frozen whole	1 pt (10 eggs)	1 lb
yolks (1 pt, 1 lb)	Dried yolks	1⅛ c, 1¾ c water	6 oz, 11 oz water
	Frozen yolks	1 pt (24 yolks)	1 lb
Flour (2 c, 8 oz)	Cornstarch	1 c	
Leavening agents			
SAS phosphate baking powder (⅜ c, 2 oz)	Phosphate baking powder	½ c plus 1 T	3 oz
	Soda plus liquid	3 T plus 1 c sour milk, or 1 c buttermilk, or 1 c milk, 1 T vinegar or lemon juice, or ½ to 1 c molasses	
Tartrate baking powder (¾ c, 4 oz)	SAS phosphate baking powder	⅜ c	2 oz
	Soda and cream of tartar	3 and ⅔ c	
Milk, nonfat (1 qt, 2 lb)	Dry milk, nonfat[a]	¾ c, 37/8 c water	3½ oz, 1⅞ lb water
Milk, whole (1 qt, 2 lb)	Dry milk, nonfat[a]	¾ c, 3½ c water, 2 T shortening	3½ oz, 1¾ lb water, 1 oz shortening
	Dry milk, whole[a]	7/8 c, 37/8 c water	3¾ oz, 1⅞ lb water
	Evaporated milk	No. 1 tall can plus water to equal qt	14½ oz plus water to equal 1 qt
Syrups, honey, etc. (1 pt, 1 lb 6 oz)	Sugar	2½ c sugar, ½ c water, ⅛ t cream of tartar	1¼ lb sugar, 4 oz water, ⅛ t cream of tartar

[a]Measure equivalent is for regular dry milk; for instant use 1¹/₃ c for every ³/₄ of regular.

TABLE F-2   Food Equivalents: Weights and Measures

Food	Weight	Approximate Measure
Beverages		
Cocoa	1 lb	4½ c
Coffee, urn grind	1 lb	4½ c
instant	1 oz	½ c
Tea	1 lb	1½ qt
Cereals and cereal products		
Barley, pearl	1 lb	2½ c
Bran, all-bran	1 lb	2 qt
Bran flakes	1 lb	3 qt
Bread, crumbs, dry	1 lb	1¼ qt
crumbs, dry sifted	1 lb	1 qt
crumbs, fresh	1 lb	2½ qt
slices, ⅝ in.	1 lb	16 slices
soft, broken, ¾-in. cubes	1 lb	2¼ qt (packed 2 qt)
Cake crumbs, soft	1 lb	1¼ qt
Cracked wheat	1 lb	3½ c (5 to 6 c cooked)
Crackers, crumbled	1 lb	2 to 2½ qt
crumbs	1 lb	1¼ qt
graham	1 lb	40 crackers
large, soda	1 lb	56 crackers
small, square saltines	1 lb	108 crackers
Cornflakes	1 lb	1¼ gal
Cornmeal	1 lb	3½ c (3 qt cooked)
Cornstarch, stirred	1 lb	3½ c (1 c = 4¾ oz, 1 oz 3½ T)
Farina	1 lb	2⅔ c
Flour, cake, sifted	1 lb	1 qt
cake, unsifted	1 lb	3¾ c
graham or whole-wheat	1 lb	3½ c
rye, straight grade, sifted	1 lb	1¼ qt
white, bread, sifted	1 lb	1 qt
white, bread, unsifted	1 lb	3¾ c
Hominy grits	1 lb	2½ to 3 c (6½ lb or 3¼ qt cooked)
Macaroni, 1-in. pieces	1 lb	3½ to 4 c (3¾ lb or 2½ qt cooked)
Noodles	1 lb	6 to 8½ c (3¾ lb or 2¼ qt cooked)
Oats, rolled	1 lb	4¾ c (2¼ qt cooked)
Rice	1 lb	2⅛ c (2½ lb or 2½ qt cooked)
Soya flour	1 lb	1¼ to 1½ qt
Spaghetti, 2-in. pieces	1 lb	1¼ qt (4 lb or 2½ qt cooked)
Tapioca, pearl	1 lb	2¾ c (soaked and cooked 7½ c)
quick cooking	1 lb	2⅔ c (7½ c cooked)
Wheat, shredded	1 lb	20 small biscuits
Wheat cereals	1 lb	2⅞ c (cooked 6 c)

(continues)

TABLE F-2   Continued

Food	Weight	Approximate Measure
Dairy products		
Butter or margarine (see also fats)	1 lb	2 c
Cheese, cottage	1 lb	2¼ c
cubed	1 lb 1 oz	1 qt
grated or ground	1 lb	2¼ c tight pack, 3¾ c loose pack
Philadelphia cream	1 lb 9 oz	3 c
Cream, 18%	8¾ oz	1 c
30 to 40%, whipping	1 lb	1 pt (doubles volume in whipping)
Milk, condensed, sweetened	11 oz	1 c
dry, instant	1 lb	5¾ c
dry, nonfat, regular	1 lb	1 qt
dry, whole, regular	1 lb	3¾ c
evaporated	1 lb	1⅞ c
fresh, liquid	8½ oz	1 c
Eggs, large[a]		
Eggs, dry, whole, packed	1 oz	¼ c
hardcooked, chopped	1 lb	2½ c (1 doz = 3½ c)
in shell	1½ lb	1 doz
whites	1 lb	1 pt (17 to 20)
whites, dry	1 lb	2 qt (¾ c [2 doz] and 1½ c water = 1 doz whites)
whole	1 lb	1 pt (9 to 11)
whole, dry	1 lb	1¼ qt (1½ c [6 oz] and 1⅞ c water = 1 doz eggs)
yolks	1 lb	1 pt (19 to 23)
yolks, dry	1 lb	4¼ c (1⅛ c [4 doz] and ⅝ c water = 1 doz yolks)
Meringue	6 oz	1 c
Fats and oils		
Bacon fat	15 oz	1 lb (1 lb = 2⅛ c)
Butter or margarine	14 oz	1 pt
Creamed fat	1 lb	2½ c
Hydrogenated shortening	14½ oz	1 pt
Oil	1 lb	2¼ c
Suet, chopped	1 lb	3¾ c
Fruits		
Apple nuggets	1 lb	6⅔ c
Apples, canned, solid pack	1 lb	1 pt
diced, ½ in.	1 lb	1 qt
fresh	1 lb	3 size 113 (3 c pared, diced, or sliced)
sliced	1 lb	4 to 4½ c
Applesauce	1 lb	1⅞ c
Apricots, canned, halves, no juice	1 lb	1 pt (21 halves)
canned, heavy pack	1 lb	1 pt
dried	1 lb	3¼ c (1¾ lb or 5 c cooked)
fresh	1 lb	8 medium

TABLE F-2 Continued

Food	Weight	Approximate Measure
Avocados, Calavos, medium size	1 lb	2 to 3
Bananas, AP, medium size	1 lb	3 (peeled 10 oz)
peeled	1 lb	2½ c diced, 1 medium banana = 30⅛ in. slices, or ¾ c, or ⅓ to ½ c mashed
Blackberries, fresh	1 lb	1 qt
water pack, drained	1 lb	3 c
Blueberries, fresh	1 lb	3 c
Cantaloupe	1 lb	1 melon 4 in. in diameter
Cherries, candied	1 lb	3 c or 120 cherries
Maraschino	1 qt	60 to 70 cherries
red, heavy pack, drained	1 lb	3 c
Royal Anne, drained	1 lb	3 c
Citron, chopped	1 lb	2½ c
Cranberries, dehydrated	1 lb	8½ c
fresh	1 lb	1 qt (1 lb AP = 3¼ c sauce)
whole	2½ lb raw	1 qt cooked
Currants	1 lb	3½ c (1 c = 4½ oz)
Dates, pitted	1 lb	2¾ c (1 c = 6 oz, 1 c = 8¼ oz if packed)
unpitted	1 lb	2½ c (1¾ c pitted)
Figs, dry	1 lb	3 c (1 c = 5 oz)
Grapefruit, 32s	1 lb	12 sections, 1¼ c juice
Grapes, cut	1 lb	2⅔ c
whole, stemmed	1 lb	1 qt
Oranges, 88s, diced with juice	1 lb	2¼ c (1 orange = ½ c diced or ⅓ c juice)
88s, Florida	1 doz	1 qt juice
rind, grated (also lemon)	1⅔ lb	¼ c (1 t = ⅛ oz)
	6½ oz	1 c
Peaches, canned, sliced with juice	1 lb	⅞ c
dry, loose pack	1 lb	1 qt
fresh	1 lb	3 to 5 peaches
Pears, canned, drained, diced	1 lb	2½ c
Pineapple, slices	1 lb	8 to 12 slices (2½ c)
Prunes, cooked, pitted, with juice	1 lb	2¼ c
dried, size 30 to 40, uncooked	1 lb	3 c (2½ lb or 5 to 6 c cooked)
Pumpkin	1 lb	2½ c
Raisins	1 lb	3 c (1 c = 5¼ oz, 1 lb cooked = 1 lb 9½ oz or 1 qt)
Raspberries	1 lb	3½ c (2¼ c cooked)
Rhubarb, raw, 1-in. pieces	1 lb	1 qt (cooked 1⅜ lb or 2½ c)
Strawberries, fresh	1 lb	3¼ c
Meats		
Bacon, cooked	1 lb	85 to 95 slices
diced, packed	1 lb	2¼ c
raw, sliced	1 lb	15 to 25 slices

(continues)

TABLE F-2   Continued

Food	Weight	Approximate Measure
Meats (*cont.*)		
Beef, cooked, diced	1 lb	3 c
dried, solid pack	1 lb	1 qt, scant
ground, raw	1 lb	1 pt
Chicken, cooked, cubed	1 lb	2½ c
ready-to-cook	5 lb	5 c cooked, diced meat (40% yield)
Crabmeat, flaked	1 lb	3 c
Ham, cooked, diced	1 lb	3¼ c
cooked, ground, packed	1 lb	1 pt
raw, AP	1 lb	1 c fine diced cooked
Meats, chopped, cooked, moist, packed	1 lb	1 pt (loose pack = 1 qt)
Oysters, 1 qt, Eastern	2 lb	40 large, 60 small
Salmon, canned	1 lb	1 pt
Sardines, canned	1 lb	48, 3 in. long
Sausage, link	1 lb	16
Sausage meat	1 lb	1 pt
Shrimp, 2 lb AP	1 lb EP	3¼ c (5 lb in shell = gal)
Tuna fish	1 lb	1 pt
Turkey, ready-to-cook	30 lb	15 lb clear meat
Wieners	1 lb	10 (frankfurters 6 to 7)
Miscellaneous		
Chocolate (*see* Spices)		
Gelatin, granulated, unflavored	1 lb	3½ c (1 oz = 3½ T)
prepared, flavored	1 lb	2⅓ c (1 oz = ¼ c)
Marshmallows (1¼ in.)	1 lb	80
Yeast, compressed	½ oz	1 cake
	8½ oz	1 c
dry active	1 lb	2½ lb compressed
Nuts		
Almonds, blanched	1 lb	3 c
shelled	1 lb	3½ c (¼ lb shelled)
Coconut, ground or fine shred	2⅜ oz	1 c
shredded	1 lb	4½ c to 7 c depending on type shred and tightness of pack
shredded, medium	1 oz	7 T
Filberts	1 lb	3⅓ c (½ lb shelled)
Nut meats, ground	4¼ oz	1 c
Peanut butter	1 lb	1⅞ c
Peanuts, chopped	1 lb	1 qt (⅔ lb shelled)
Pecans	1 lb	4¼ c (⅓ lb shelled)
Walnut meats, chopped	1 lb	1 qt
whole	1 lb	4¾ c (½ lb shelled)

TABLE F-2   Continued

Food	Weight	Approximate Measure
Salad dressings and condiments		
Catsup or chili sauce	9 oz	1 c
Cooked salad dressing	1 lb	1 pt
French dressing	1 lb	2⅛ c
Horseradish, ground	1 lb	2¼ c
Mayonnaise	1 lb	2⅛ c
Olives, small	1 lb	3½ c or 135 olives, 1 No. 10 = 4½ lb drained weight or 350 large olives
Pickles, chopped	1 lb	2½ c
small	1 gal	80 (about 225 gherkins or 25 large per gal)
Spices, seasonings, and leavenings		
Allspice, ground	1 lb	4½ c (1 oz = 4½ T)
Baking powder	1 lb	2½ c (1 T = $^7/_{16}$ oz, 1 oz = 2½ T)
Celery seed	1 lb	1 qt (1 oz = ¼ c)
Chili or curry powder	1 oz	3 T
Chocolate, grated	1 lb	1 qt (1 lb = 16 squares)
melted	1 lb	1⅞ c (1 oz = 2 T)
	1 oz	5 T (1 c = 3¾ oz)
Cinnamon, ground	1 lb	1 qt (1 oz = ¼ c)
Cloves, ground	1 lb	3¾ c (1 oz = 3¾ T)
whole	1 oz	5 T
Cream of tartar	1 lb	3 c (1 oz = 3 T)
Flavoring extracts	⅜ oz	1 T (⅛ oz = 1 t)
Ginger, ground	1 lb	4¾ c (1 oz = 4¾ T)
Mustard, ground	1 lb	5 c (1 oz = 5 T)
Nutmeg, ground	1 lb	3½ oz (1 oz = 3½ T)
Paprika	3¼ oz	1 c (1 T = ⅜ doz)
Pepper	1 oz	¼ c
Sage, ground	1 oz	½ c
Salt	1 lb	1⅔ c (1 oz = 1 T, 2 t)
Soda	1 lb	2½ c (1 oz = 2 T, 6½ t = 1 oz)
Vinegar	1 lb	2 c (1 oz = 2 T)
Worcestershire sauce	9½ oz	1 c
Sugars and syrups		
Corn syrup	11 oz	1 c
Honey	12 oz	1 c
Jam or jelly	1½ lb	1 pt
Molasses	11 oz	1 c
Sugar, brown	1 lb	3 c (packed 2¼ c)
cocktail cube	1 cubelet	½ t (small cube = 1 t, tablet = 1½ t, 96 cubes, medium, per lb)
confectioners, 4X, sifted	1 lb	4½ c (unsifted 2¾ c)
confectioners, stirred	1 lb	3½ c
granulated	1 lb	2¼ c (superfine 2 c)

(continues)

TABLE F-2   Continued

Food	Weight	Approximate Measure
Vegetables		
Asparagus, canned, cuts, drained	1 lb	2½ c
canned, tips, drained	1 lb	19 stalks
fresh	1 lb	20 stalks
Bean sprouts	1 lb	1 qt
Beans, baked	1 lb	1⅞ c
kidney, dry, AP	1 lb	2⅓ c (2¼ lb or 1½ qt cooked)
lima, drained, cooked, fresh or canned	1 lb	2⅔ c (1½ lb = 1 qt)
lima, dried, small, AP	1 lb	2⅓ c (2½ lb or 1½ qt cooked)
lima, fresh, shelled	1 lb	2¼ c
lima, fresh, unshelled	1 lb	⅔ c shelled
navy, dry, AP	1 lb	2⅓ c (2½ lb or 1¾ c cooked)
string, cut, uncooked, EP	12 oz	1 qt
Beets, cooked, diced, drained	1 lb	2¼ c (3 to 4 medium whole)
cooked, sliced, drained	1½ lb	1 qt (1 lb = 2¾ c)
Brussels sprouts, AP	1 lb	1 qt
Cabbage, shredded, EP	12 oz	1 qt (1 lb = 5½ c or 1 lb = 7 c loose pack)
shredded, cooked, drained, AP	1 lb	3½ c
Carrots, AP	1 lb	4 medium, 6 small
diced, cooked, drained	1 lb	2½ to 3 c
ground, raw, EP	1 lb	3¼ c
half-inch cube, raw	1 lb	3¼ c
Cauliflower, crate	12½ lb, net, EP	10 qt
head, medium	12 oz	4 to 5 portions
Celery, dehydrated	1 lb	9½ c
diced, EP	1¼ lb	1 qt (1 lb = 3¼ c)
Corn, cream style	1 lb	1⅞ c
whole kernel, drained	1 lb	2⅓ c
Cucumbers, diced	1 lb	2½ c
Eggplant, diced, half-inch cubes	1 lb	4½ c
sliced, 4 in. diameter, ½ in. thick	1 lb	8 slices
Garlic, crushed	1 oz	6 to 9 cloves
Lettuce, average head	1 lb	10 to 12 leaf cups
leaf	1 lb	30 salad garnishes
shredded	1 lb	8 c (packed 5 c)
Mushrooms, fresh	1 lb AP	1⅓ c cooked
Onions, AP	1 lb	4 to 5 medium
chopped	1 lb	2½ to 3 c
dehydrated	1 lb	9½ c
grated or minced	5 oz	1 c

TABLE F-2   Continued

Food	Weight	Approximate Measure
Parsley	1 lb	3 bunches (6 c chopped)
Parsnips, AP	1 lb	3 to 4 medium
diced, cooked	1 lb	2½ c
diced, raw	1 lb	2½ to 3 c
mashed	1 lb	1 pt
Peas, canned, dried	1 lb	2¼ c
dried, split	1 lb	2⅓ c (2½ lb or 5½ c cooked)
fresh, 2½ lb AP	1 lb EP	1 pt scant, 5 portions
Peppers, green	1 lb	5 to 6 medium
green, chopped	1 lb	3½ c
Pimientos, chopped	8 oz	1 c
Potatoes, sweet	1 lb	3 medium
Potatoes, white, cooked, diced, ½-in. cubes	1 lb	3 c
white, dehydrated, cube	1 lb	4¾ c
white, dehydrated, flake	3½ oz	1 c
white, dehydrated, granule	7 oz	1 c
white, medium, AP	1 lb	3 to 4 (¾ lb pared, 1 pt mashed)
Potato chips	1 lb	5 qt (20 1-c portions ¾ oz)
Pumpkin, cooked	1 lb	1 pt
Radishes, whole, topped and cleaned	1 lb	1 qt
Rutabagas, cubed, cooked	1 lb	3 c
raw, cubed, EP	1 lb	3⅓ c
Sauerkraut, uncooked	1 lb	3 c
Spinach, canned, drained	1 lb	1 pt
Spinach, raw	1 lb	5 qt, loose pack
1 lb raw, AP, cooked	13 oz EP	1½ c cooked, 3 portions
Squash, Hubbard, cooked, mashed	1 lb	2⅛ c
summer, AP	1½ lb	1 3 in. diameter
Tomatoes, canned	1 lb	1 pt
dried	1 lb	3½ c
fresh	1 lb	3 to 4 medium
fresh, diced	1 lb	2¼ to 2¾ c
Turnips, AP	1 lb	4 to 5 medium
raw, diced	1 lb	3½ c
Watercress	1 lb	5 bunches

[a]Eggs lose approximately 11 to 12% weight in shelling. Medium eggs are about 10% less in weight than large.

Troubleshooting Bread Failures

TABLE G-1 Troubleshooting Bread Failures

Fault	Possible Causes	Possible Remedies
Excessive volume	Too much yeast	Reduce yeast to 2 to 3%; check weighing procedures.
	Too little salt	Maintain from 2 to 2½%; check weighing procedures.
	Excess dough	Reduce scaling weights.
	Overproofed	Reduce proofing time; keep between 70–30 min. or 80–20 min. fermentation time for sponges.
	Too cool oven	Increase temperature.
Poor volume	Weak flour	Blend strong flour into flour or use a stronger flour; give less mixing, shorter fermentation, and less proofing time.
	Flour too old or too new	Use aged flour; check age of flour.
	Water too soft or too alkaline	Use a conditioner; additional salt improves too soft a water.
	Lack of leavening	Use good yeast and handle it properly; have dough at proper temperature; reduce quantity of salt.
	Undermixing	Increase mixing times until gluten in dough is properly developed; proper volume of dough to mixer is also a factor to check.
	Overfermented dough	Reduce fermentation time.
	Overmixing	Reduce mixing.
	Improper proofing	Proof between 90 and 100°F and 80 to 85% relative humidity; watch proofing time and maintain proper ratio between fermentation and proofing procedures.
	Too much or too little steam in oven	Open or close oven dampers; if steam is introduced into oven, establish better controls.
	Too hot oven	Reduce temperature.

TABLE G-1   Continued

Fault	Possible Causes	Possible Remedies
Too dark crust	Excess sugar or milk	Reduce; check diastatic action of flour; it may be breaking down too much starch into sugars.
	Overmixing	Reduce mixing.
	Dough too young	Increase fermentation and proof periods.
	Too hot oven	Correct oven temperatures.
	Too long baking	Reduce baking time.
	Too dry oven	Close oven damper during part of baking, or use steam.
Too pale or dull color on crust	Wrong proportion or lack of right ingredients	Check ratios of sugar, salt, or milk, and diastatic action of flour; increase ingredients to proper ratios; add diastase syrup.
	Soft water	Increase salt or add conditioner.
	Overfermentation	Reduce temperature or time of fermentation.
	Excessive dusting flour	Cover bench only with bare minimum.
	Too high proof temperature	Reduce temperature.
	Cool oven	Increase temperature.
	Improper use of steam	Avoid excessive steam; open dampers to increase oven moisture.
Spotted crust	Improper mixing	Follow correct mixing procedures and sequence of adding ingredients.
	Excess dusting flour	Reduce dusting flour.
	Excess humidity in proofing	Reduce relative humidity to between 80 and 85%.
	Water in oven or excessive moisture in steam	Check steam pipes and ovens; open dampers.
Hard crust or blisters	Lack of sugar or diastatic action	Increase sugars or check diastatic action of flour; check weighing of ingredients.
	Slack dough	Reduce liquid; check mixing.
	Improper mixing	Check mixing procedures and sequence of ingredient addition.
	Old or young dough	Correct the fermentation time.
	Improper molding or makeup	Correct procedures.
	Cool oven or too much top heat	Check damper handling procedures and oven temperatures; check heating elements and heat source to see if functioning properly; check oven circulation.
	Cooling too rapidly	Cool more slowly; keep out of drafts.
	Too much fat on product	Reduce brushing of fat after makeup.
Poor shape	Improper makeup or panning	Correct procedures.
	Overproofing	Reduce.
Flat top or sharp corner	New flour	Age flour six to eight months under proper conditions.
	Low salt	Increase; check weighing procedures.
	Slack dough	Reduce liquid; check mixing.
	Young dough	Increase fermentation time.
	Excessive humidity in proofing	Reduce humidity.

(continues)

TABLE G-1   Continued

Fault	Possible Causes	Possible Remedies
Excessive break on side	Overmixing	Reduce mixing.
	Improper molding	Check molding, especially seam folds; place seam folds down on bottom of pan.
	Young dough	Correct; check proofing time.
	Hot oven	Reduce temperatures.
Thick crust	Low shortening, sugar, or milk	Increase; check scaling procedures.
	Low diastase	Check diastatic action of flour; add malt syrup or diastase compound.
	Mixing improper	Correct mixing procedures.
	Improper proofing	Correct temperature, relative humidity, or time of proofing; check for wet crusts after proofing.
	Old dough	Correct fermentation and/or proofing time.
	Improper baking	Correct temperatures and times; reduce steam and check for excessive or insufficient moisture in ovens.
Tough crust	Old or young dough	Check fermentation times.
	Improper mixing	Correct.
	Excess proof or wrong proof conditions	Correct.
	Oven cold or excess steam	Correct.
Lack of break or shred	Excess diastase	Decrease amount; use nondiastatic malt; check diastatic diastatic action of flour.
	Soft water	Increase salt or use conditioner.
	Slack dough	Reduce water substantially; check mixing.
	Improper fermentation or proof time	Correct.
	Too hot or too dry an oven	Correct temperatures by damper control; introduce steam.
Ragged scaling or shelling on top	Green or overly old flour	Use flour properly aged.
	Old or young dough	Check fermentation or proof times and conditions.
	Stiff dough	Reduce flour or increase liquid.
	Crusting during proofing	Increase relative humidity; brush lightly with shortening.
	Excess salt	Reduce.
	Underproofing	Increase proofing time.
	Excessive top heat in oven	Check heat circulation and heat source in oven; check damper control.
	Cold dough	Add warmer liquid; check mixing, fermentation, and proofing temperatures.
	Lack of salt or milk	Check recipe and weighing procedures.
Too close grain	Low yeast	Increase; check weighing procedures.
	Underproofing	Correct.
	Excess dough in pan	Check scaling procedures.

TABLE G-1   Continued

Fault	Possible Causes	Possible Remedies
Too coarse or open grain	Hard or alkaline water	Add vinegar or conditioner.
	Old dough	Excess yeast; reduce fermentation time.
	Slack dough	Reduce liquid; check mixing times.
	Improper molding	Correct.
	Overproofing	Reduce time or check temperatures.
	Improper pan size	Check.
	Cold oven	Increase oven temperature.
	Excessive greasing	Check oiling or greasing of dough.
Gray crumb	High diastatic action	Reduce.
	High dough temperature or overfermentation	Check mixing, fermentation, and proof temperatures and times.
	Cold oven	Check temperatures and conditions of baking.
	Pans greasy	Check greasing.
Streaked crumb	Improper mixing	Check ingredient sequence of adding in mixing.
	Too slack or stiff dough	Check liquid or flour quantities; check to see if proper mixing times given.
	Excessive oil, grease, or dusting flour used	Correct.
	High relative humidity	Reduce relative humidity in fermentation or proofing.
	Crusting of dough in fermentation	Increase relative humidity or brush with fat; cover to prevent moisture loss.
Poor texture	Alkaline or very hard water	Use conditioner or vinegar.
	Too slack or too stiff dough	Reduce or increase ingredients to correct ratios; check mixing.
	High sugar or excess yeast	Check ingredient ratios; check diastase, and decrease or increase as required.
	Lack of shortening	Increase.
	High dough temperature	Reduce liquid temperature or temperatures during fermentation or proofing.
	Overfermentation or proofing	Reduce.

Herb, Spice, and Seasoning Data

	Appetizers	Soups	Salads and Dressings	Vegetables	Eggs and Cheese	Meats and Sauces	Poultry and Fish	Desserts and Baked Goods
Allspice	Pickles, relishes, cocktail meatballs, pickled beets, fruit compote	Green pea, vegetable, beef, minestrone, asparagus, tomato	Cottage cheese, fruit salad, cheese dressing	Eggplant, spinach, beets, squash, turnips, red cabbage, carrots	Egg casserole, cream cheese	Beef stew, meat loaf, hamburgers, baked ham, roast lamb, pot roast, cranberry sauce, meat gravies, tomato sauce	Boiled fish, oyster stew	Mincemeat, tapioca and chocolate puddings, spice cake, fruit cake, baked bananas, cookies, pies
Bay leaves	Hot tomato juice, pickles, pickled beets	Bouillon, bouillabaisse, fish chowders, lobster bisque, vegetable, minestrone, oxtail, potato, tomato, turtle	Seafood salad, tomato aspic, chicken aspic, beet salad, French dressing	Potatoes, artichokes, carrots, beets, eggplant, lentils, onions, rice, squash, zucchini	Eggs Creole	Beef stew, meat pie, corned beef, pot roast, roast beef, veal, meat sauces, lamb, spare ribs, gravies	Capon, boiled chicken, à la king, fricassée, turkey roast), boiled shrimp and lobster, fish stews, baked salmon	
Caraway	Soft cheese spreads, pickles	Cream soups, clam chowder, borsch, vegetable	Potato salad, sour cream dressing, spiced vinegar, coleslaw	Cabbage, cauliflower, potatoes, carrots, onions, turnips, broccoli, brussels sprouts, sauerkraut	Cottage cheese, cream cheese	Sauerbraten, roast pork, beef à la mode, liver, kidney stew	Tunafish casserole, roast goose	Rye bread, muffins, rolls, coffee cakes, cookies, loaf cake
Cayenne	Deviled eggs, seafood sauces, cottage and cream cheese dips and spreads, avocado dip	Clam and oyster stews, fish chowder, cream soups, shrimp gumbo, vegetable	Tuna, shrimp, chicken, macaroni, seafood, mayonnaise, thousand island dressing, sour cream	Green beans, lima beans, cauliflower, cut corn, kale, broccoli	Welsh rarebit, egg dishes, cheese soufflé, cottage and cream cheese	Pork chops, veal stew, ham croquettes, barbecued beef, sandwich fillings, meat sauces	Creamed chicken and croquettes, oysters, shrimp, poached salmon, tuna salad	
Celery seed (salt, flakes, seeds)	Deviled eggs, ham spread, tomato juice, kraut juice, cream cheese spread, pickles	Cream of celery or tomato, fish chowders and bisques, vegetable, bean, potato, bouillon	Coleslaw, potato, egg, tuna, vegetable, kidney beans, salad dressings, sour cream dressing	Cabbage, stewed tomatoes, potatoes, cauliflower, turnips, braised lettuce	Welsh rabbit, boiled and fried eggs, cheese casserole, omelets, cheese sauce, deviled eggs	Meat loaf, pot roast, meat stews, short ribs of beef, braised lamb	Chicken croquettes, fish stews, chicken pie, oyster stew, stuffings	Rolls, biscuits, salty breads
Chili Powder	Avocado and cheese dips, seafood cocktail sauce	Corn soup, pepperpot, fish and clam chowders, tomato, bean, shrimp, gumbo, vegetable, chili soup	French dressing, kidney bean salad, thousand island dressing, chili sauce	Vegetable relishes, green peas, eggplant, rice, tomatoes, corn Mexicali, green beans, lima beans	Omelets, soufflés, casseroles, boiled and scrambled eggs, cheese sauces, rarebits	Chili con carne, arroz con pollo, tamales, meat loaf, hamburgers, stews, sauces	Creamed seafood, shrimp, chicken and rice, chicken pie	
Cinnamon (ground and stick)	Cinnamon toast, sweet gherkins, hot spiced beverages, pickled fruits		Fruit salad, dressings for fruit salads	Sweet potatoes, squash, pumpkins, spinach, turnips, green beans, beets, parsnips		Pork chops, ham, sauce for pork and lamb	Boiled chicken (stick cinnamon), boiled fish, special chicken and fish recipes	Chocolate and rice pudding, stewed fruits, apple desserts, buns, coffee cake, muffins, spice cake, molasses cookies
Cloves (whole and ground)	Sweet gherkins, pickled fruits, hot spiced wines, fruit punch	Beef, bean, cream of tomato, cream of pea, mulligatawny	Topping for fruit salad	Beets, baked beans, candied sweet potatoes, squash		(Whole) Ham and pork roast, stews, gravies, sausage; boiled tongue	Baked fish, chicken à la king, roast or smothered chicken, chicken croquettes	Preserved and stewed fruits, apple, mince, and pumpkin pies, chocolate, rice, and tapioca pudding, stewed pears
Curry powder	Tomato juice; sauce for dips, sweet pickles, deviled eggs, salted nuts	Clam and fish chowders, tomato soup, cream of mushroom, oyster stew	Fruit and meat salads, mayonnaise, French dressing	Rice, creamed onions, creamed potatoes, scalloped tomatoes, carrots, corn, celery, lima bean	Sauce for eggs, deviled eggs, cottage cheese, cream cheese, cheese sauce	Curried lamb, veal croquettes, stews	Chicken croquettes, chicken hash, curried chicken and turkey, fish croquettes, shrimp	

	Appetizers	Soups	Salads and Dressings	Vegetables	Eggs and Cheese	Meats and Sauces	Poultry and Fish	Desserts and Baked Goods
Dill	Cottage cheese, anchovy spread, cheddar cheese spread, pickles, sour cream dips, stuffed eggs	Split pea soup, cream of tomato, navy bean, borsch, chicken, lobster bisque, turkey, fish chowder	Coleslaw, cucumber, green bean, lettuce, mixed green, potato, seafood, mayonnaise, sour cream, French dressing	Carrots, beets, cabbage, lima beans, green beans, turnips, eggplant, cauliflower, zucchini, squash	Deviled eggs, omelets, scrambled eggs, cottage cheese, cream cheese, macaroni	Beef: pot roast, corned, stew, barbecued, hamburger; lamb chops or stew, roast pork	Chicken pie, creamed chicken, baked halibut, mackerel, salmon, creamed lobster, boiled and creamed shrimp	
Ginger (ground)	Pickles, broiled grapefruit, chutney	Bean soup, onion, potato	Ginger pears, French dressing	Beets, carrots, squash, baked beans	Cheese dishes	Broiled beef, lamb, veal, pot roast, stews, chopped beef	Roast chicken, squash, cornish hen, sautéed chicken	Gingerbread, cakes, cookies, pumpkin pie, custards, baked, stewed, preserved fruits, Indian pudding
Marjoram	Fruit punch, cream and cottage cheese dips, cheddar cheese spreads, pickles	Onion soup, clam, oyster, Boston clam chowder, minestrone, oxtail, spinach	Mixed green salad, asparagus, chicken, fruit, seafood	Carrots, eggplant, peas, spinach, string beans, onions, summer squash, tomatoes, celery, broccoli, brussels sprouts	Omelets, soufflés, creamed eggs, scrambled, cheese sauce, cheese soufflé, rarebits, cheese straws	Roast beef, pork, veal, stews, meat pies, loaf, pot roast, short ribs, spare ribs	Chicken croquettes, duck, goose, pheasant, guinea hen, codfish balls, halibut, salmon loaf, shad roe	
Mustard (ground)	Pickles, pickled onions, ham spreads, Chinese hot sauce, hot English mustard, deviled eggs	Lobster bisque, bean, onion	Egg salad, shrimp, lobster, potato, fruit, salad dressings	Asparagus, beets, broccoli, brussels sprouts, cabbage, onions, green beans, potatoes, baked beans	Deviled eggs, casseroles, cheese sauces, cream cheese	Baked ham, kidneys, pickled meat, sauces	Shrimp, creamed and stewed oysters, boiled fish, fish sauces	Molasses cookies, ginger-bread
Oregano	Cheese spreads, pizza, vegetable juice, avocado dip, creamed and cottage cheese spreads	Bean, beef, vegetable, tomato, lentil, minestrone, navy bean, onion, spinach	Salad dressings, seafood, avocado, green bean, mixed green, potato, tomato, tomato aspic	Peas, onions, potatoes, spinach, green beans, stewed tomatoes, mushrooms	Creamed eggs, omelets, scrambled eggs, cheese sauce, soufflé, cottage, cream, straws, rarebits	Swiss steak, beef stew, broiled and roast lamb, meat loaf, sauces, gravies, spare ribs, veal scalloppini	Chicken: cacciatore, sauté, roast, guinea hen, pheasant, stuffed fish, boiled shrimp, clams	
Paprika	Canapé, deviled eggs, cream cheese spreads, stuffed celery, creamed seafood, seafood cocktails	Cream soups, chicken soup, chowders	Coleslaw, potato salad, mayonnaise, French dressing	Cauliflower, potatoes, celery, creamed vegetables	Deviled eggs, scrambled eggs, Welsh rarebit, cottage cheese, cheese and egg dishes	Hungarian goulash, ham, gravies	Poultry and seafood dishes, shellfish, fried chicken	
Poppy seed	Cheese spreads, cottage cheese, cheese dips	Onion soup	Green salads, salad dressings	Peas, potatoes, rutabaga, sweet potatoes, carrots, zucchini	Fried and scrambled eggs, omelets, cottage cheese	Noodle dishes		Coffee cake, cookies, pie crusts, bread, rolls, pastries
Rosemary	Deviled eggs, pickles, sour cream dips	Mock turtle, chicken, lentil, minestrone, split pea, spinach, chowders	Meat salad, fruit salad	Peas, potatoes, mushrooms, onions, celery, lima beans, green beans, broccoli, cucumbers	Deviled eggs, omelets, soufflés	Roast and broiled lamb, beef, pork, veal, beef stew, pie, pot roast, swiss steak, spare ribs	Capon, chicken fricassee, sauté, roast pheasant, partridge, quail, salmon, baked halibut, baked sole	
Saffron		Chicken, bouillabaisse, lobster bisque, turkey	Seafood salads	Rice	Scrambled eggs	Gravy for roast chicken, roast turkey, roast veal, Spanish sauce, rabbit	Arroz con pollo, bouillabaisse, chicken stew, chicken fricassée, creamed lobster, baked halibut, sole	Rolls, breads, buns, cake, frostings and icings
Sage	Cheese spreads	Consommé, fish and corn chowders, cream soups, asparagus, cream of tomato, minestrone, turkey	Salad greens, salad dressings	Brussels sprouts, onions, lima beans, peas, tomatoes, carrots, eggplant, winter squash, turnips	Creamed eggs, soufflés, cheese sauce, rarebits, egg and cheese casseroles, cottage and cream cheese	Beef barbecue, stew, pie, roast, pot roast; barbecued lamb, roast lamb, pork, veal stew	Capon, chicken stuffing, goose, duck	
Sesame seed	Soft cheeses	Most soups	Coleslaw, salad dressings	Asparagus, green beans, tomatoes, spinach, noodle and vegetable casseroles, potatoes, rice	Cream cheese	Meat pies, Hawaiian ham steak	Fried chicken, chicken casseroles, fish	Top dressings on pies, cookies, coffee cake, rolls, breads, buns, crumpets
Tarragon	Vegetable juice cocktail, liver paté, herb butters, cheese spreads, seafood cocktails, stuffed eggs, pickles	Bean, chicken, consommé, seafood chowders and bisques, mushroom, pea, tomato, turtle	Asparagus, celery, chicken, coleslaw, cucumber, egg, green bean, kidney bean, mixed green, tomato	Asparagus, beans, broccoli, cabbage, cauliflower, celery root, mushrooms, potatoes, spinach, tomatoes	Deviled eggs, omelets, scrambled eggs, cottage cheese	Meat marinades; broiled steak, pot roast, braised lamb, lamb stew, veal stew, béarnaise sauce, brown gravies	Chicken, chicken sauté, broiled chicken, turkey, duck, broiled halibut, baked salmon, trout, tuna, broiled lobster	

Standard Portions

TABLE I-1 Standard Portions

Food	Portion and Serving Method
Breads	
Biscuits	2 to 3 1 oz each[b]
Bran rolls	2 1 oz each
Cinnamon rolls	2 1½ oz each
Cornbread, coffee cake, etc.	1 piece 2 oz, cut 18 × 26 in. baking sheet 6 by 8
Griddle cakes	3 3 to 4 oz each
Hot rolls	2 1 oz each
Muffins	2 2½ oz each
Potato doughnuts	2 2 oz each
Sweet dough items, breakfast	1 3 oz each
White or other bread, sliced	1 to 2 slices, 1 oz each
Dressings and sauces	
Cranberry sauce, applesauce, etc.	2 to 2½ oz, scant serving spoon or No. 16 scoop
French or other liquid	1 to 2 T; portion depends on salad size
Mayonnaise, boiled, etc.	1 to 2 T; portion depends on salad size
Fish	
Creamed fish dishes	4-oz ladle rounded; 1 slice toast, 1 biscuit, or No. 16 scoop of rice
Croquettes	No. 10 scoop for 1; No. 20 scoop for 2; 1½ oz sauce
Fillet, baked or fried	3 to pound before cooking; 4 oz if breaded
Fish and noodles	Serving spoon rounded, 5 to 6 oz
Loaf	4-oz slice; in 17 × 25 pan, cut 5 by 9, in 12 × 20 pan, cut 4 by 6: bake in these 1 in. deep
Scalloped salmon, tuna, etc.	1 4-oz ladle rounded, 5 to 6 oz; if thick, rounded serving spoon
Shrimp, deep-fried, fantail	4 to 5, tongs
Shrimp wiggle	4-oz ladle rounded; slice of toast or biscuit or No. 16 scoop of rice
Soufflé	Cut 17 × 25 pan, 5 × 9, 12 × 20 pan 4 by 6
Steak	3 to pound unless wasteful in eating, then 6 oz
Strips, breaded and deep-fried	1 oz each, serve 3 tongs, about 35% breading
Tunafish, potato chip dish	Serving spoon rounded, 5 to 6 oz
Luncheon entrées[a]	
American noodles	5 oz, serving spoon well-rounded
Baked beans	6-oz ladle or two serving spoons or one heaped
Baked eggs, Creole	4-oz ladle, rounded
Baked lima beans	6 oz, serving spoon heaped
Beef biscuit roll	1, 4 in. diameter, 2-oz ladle gravy
Buttered apples with sausage	3 apple halves, 2 sausages
Cheese fondue	4 oz, 1 oz sauce; cut 12 × 20 pan 4 by 6, spoon and ladle
Cheeseburgers	1 No. 16 scoop, 2 each, slice cheese ¾ to 1 oz each
Creole spaghetti	1 serving spoon well-rounded (6 oz)
Eggs à la king or creamed eggs	2 halves egg on half slice of toast; 2 oz sauce
Goulash	6-oz ladle
Italian delight	4-oz ladle
Italian spaghetti	1 heaped serving spoon spaghetti, 4-oz ladle sauce
Macaroni and cheese	1 heaped serving spoon, 5 to 6 oz
Macaroni hoe	1 heaped serving spoon, 5 to 6 oz

TABLE I-1   Continued

Food	Portion and Serving Method
Meat soufflé	1 heaped serving spoon, 1½ oz sauce
Omelet	4 oz, spoon; if cut, use spatula
Pizza pie	Cut 18 × 20 in baking sheet 4 by 5, use spatula
Rice and cheese baked	5 oz (⅔ c), well-rounded serving spoon
Scalloped ham and potatoes	1 heaped serving spoon, 5 to 6 oz
Scalloped meat dishes	1 heaped serving spoon, 5 to 6 oz
Scrapple	4 oz, 2 slices
Spanish rice	1 well-rounded serving spoon, 5 oz
Stuffed cabbage	1 or 2
Swedish meatballs	2 2 oz each after cooking; portion with rounded No. 20 scoop
Tamale pie	1 heaped serving spoon, 5 to 6 oz
Welsh rarebit	½ c, 4-oz ladle, on toast, biscuit, or No. 16 scoop rice
Meats	
American chop suey	4-oz ladle, rounded
with corn soya	2 T, No. 32 scoop
with rice	No. 16 scoop
Baked hash, beef or corned beef	No. 10 scoop or heaped serving spoon, 5 to 6 oz
Beef or other meat and noodles	No. 8 scoop rounded or 6-oz ladle rounded, about 7 oz (¾ c)
Beef or other meat stew	6-oz (¾ c), ladle
Beef patty	No. 8 scoop before cooking, use tongs
Cabbage rolls	2 rolls, 3 oz each (use 2 oz meat filling), use spoon
Chili con carne	6-oz ladle rounded to give 8 oz
Cold cuts	3 oz, tongs or spatula
Corned beef and cabbage	3 or 4 oz sliced beef, tongs; 3 to 4 oz cabbage, spoon
Creamed meats	½ c, 4-oz ladle; use 8 to 10 lb cooked meat per 100; serve over toast, biscuits, or No. 16 scoop rice
Croquettes	No. 10 or 12 scoop for 1; No. 20 scoop for 2; 1½ oz sauce
Frankfurters, 6 to 7 lb	2, tongs
with sauerkraut	1 rounded serving spoon, 3 oz
Fritters	2, tongs; portion with No. 20 scoop; 2 strips bacon
Ham, baked, boned	5 to 6 oz before cooking, tongs
baked, slices	3 to 4 oz after cooking, tongs
fried	6 oz before cooking, tongs
Ham à la king	4-oz ladle rounded
Hamburgers	2, portion with No. 20 dipper, tongs
Liver, braised	3 to 4 oz before cooking, 2 strips bacon, tongs
Meatballs and spaghetti	No. 20 scoop, 2 balls, 2 serving spoons spaghetti and sauce
Meatloaf	4 to 5 oz cooked, slice and use spatula or tongs
Meat pie	2 serving spoons, rounded 8 oz
with biscuit	2½ in. diameter, serve 1 with 6 oz stew
with pie crust	Cut 17 × 25 baking pans 5 by 9, 45 portions
Meat sandwich, hot	2 oz meat, 1 or 2 slices bread, tongs; 2-oz ladle gravy
Meat turnover	2 oz meat, No. 16 scoop; serve with 2 oz gravy, ladle
Mock drum sticks	5 oz before cooking, serve 1
New England boiled dinner	6 oz before cooking, 3 to 4 oz after, tongs, 5 oz vegetables and 5 oz potatoes

(continues)

TABLE I-1 Continued

Food	*Portion and Serving Method*
Meats (*cont.*)	
Pork chop	3 to lb before cooking, serve 1; 6 to lb serve 2
with dressing	2 to 3 oz; use serving spoon or No. 16 scoop
with pocket	3 to lb, 1½ oz stuffing; tongs
Roasts, meat or poultry	3 oz cooked, tongs
with dressing	2 to 3 oz meat cooked; 4 oz (½ c), No. 10 scoop rounded of dressing
Sausage, bulk	3 oz before cooking, tongs
link, 14 to 16 per lb	2, tongs
Spareribs	8 to 12 oz before cooking, tongs
Steak, braised, swiss, etc.	6 to 7 raw, 4 to 5 oz cooked, spoon
dinner, dry-heat type	8 oz AP, no bone, tongs; size may vary with institution
ground	3 to pound, No. 8 scoop rounded; 4 oz cooked, tongs
stuffed	5 to 6 oz before cooking; 1½ oz dressing, tongs
Stew	No. 8 scoop rounded or 6-oz ladle rounded (¾ c)
Veal, birds	5 oz before cooking; 1½ oz dressing; spoon
chop	5 oz
cutlet	4 oz before breading; 5 oz breaded
Wieners, 10 to lb	2, tongs
Poultry	
Chicken, creamed	6 oz (¾ c); about 2 oz cooked chicken meat per portion
fried	2 pieces or half (12 oz before cooking)
Chicken fricassée, unboned	12 oz raw meat, spoon
Chicken or turkey, roast	2 to 3 oz with dressing, 4 oz without, 2 oz gravy
Duck or geese	12 oz to 1 lb before cooking
Salads	
Brown bean	4 to 5 oz
Coleslaw	3 oz, serving spoon
Cottage cheese	No. 20 scoop
Deviled egg	2 halves
Fish or meat salad, entrée-type	5 to 6 oz, 1 c
Gelatin	12 × 20 pan, 1 in. deep, cut 5 by 10, 50 portions
Head lettuce, 1 lb average	⅛ head, 2 oz serving
Mixed fruit	1 rounded serving spoon, No. 2 scoop
Mixed vegetable	1 rounded serving spoon
Potato, cold or hot	1 No. 10 or No. 12 scoop (4 to 5 oz)
Sliced tomato	2 large or 3 medium slices
Waldorf	1 rounded serving spoon, 3 oz
Soup	
Bowl	8 oz, 1 c
Cup	6 oz, ¾ c
Tureen	10 to 12 oz, 1¼ to 1½ c
Vegetables	
Most canned vegetables	3 oz (½ c), 1 rounded serving spoon
Apples, buttered	½ c, 3 to 4 pieces, serving spoon
Asparagus tips	3 to 5 canned, 4 to 6 fresh

TABLE I-1   Continued

Food	Portion and Serving Method
Beans, navy, lima, or other	4 to 5 oz, serving spoon
Beet greens, other greens	3 oz ($\frac{1}{2}$ c), tongs or serving spoon
Beets, Harvard	$\frac{1}{2}$ c, serving spoon rounded
Broccoli, buttered	2 to 3 pieces, 3 to 4 oz, tongs
Cabbage, steamed, fried, etc.	3 oz ($\frac{1}{2}$ c), serving spoon
Onions, creamed	2 to 3 small onions, serving spoon
Potato, au gratin, creamed, etc.	4 to 5 oz, serving spoon
baked	5 to 6 oz, tongs
browned, steamed, etc.	5 oz, serving spoon
french-fried	4 oz, 8 to 10 pieces, tongs or spoon
hash brown, etc.	4 to 5 oz, serving spoon
mashed	1 No. 10 scoop or serving spoon, 4 oz
Potato cakes	4 oz, serving spoon
Potato puff	5 oz, $\frac{2}{3}$ c, spoon
Rice, steamed	No. 10 scoop rounded ($\frac{2}{3}$ c)
Scalloped sweet potatoes and apples	4 oz, serving spoon
Squash, acorn, baked or steamed	$\frac{1}{3}$ or $\frac{1}{2}$ squash
hubbard	6- to 7-oz piece before baking
mashed	4 oz, rounded serving spoon
Sweet potatoes, baked	5 to 6 oz, tongs
candied or glazed	2 slices, 4 oz
Tomatoes, escalloped or stewed	4-oz ladle
Vegetable pie	5 oz, well-rounded serving spoon
Vegetables, creamed	3 to 4 oz

Desserts	Pan Size	Portion
Cakes, butter		
Round, 2 layer	8 in. diameter	Cut 12
Sheet, 1 layer	18 × 26 in.	Cut 6 by 8, 48 portions
	13$\frac{1}{2}$ × 22$\frac{7}{8}$ in.	Cut 5 by 9 or 6 by 8, 45 or 48 portions
	12$\frac{3}{4}$ × 23 in.	Cut 5 by 9 or 6 by 8, 45 or 48 portions
2 layer	18 × 26 in.	Cut 12 by 5, 60 portions
	13$\frac{1}{2}$ × 22$\frac{7}{8}$ in.	Cut 3 by 20, 60 portions
	12$\frac{3}{4}$ × 23 in.	Cut 3 by 20, 60 portions
Square, 1 layer	9$\frac{1}{2}$ × 9$\frac{1}{2}$ in.	Cut 3 by 4, 12 portions
2 layer	9$\frac{1}{2}$ × 9$\frac{1}{2}$ in.	Cut 3 by 7, 21 portions
Eight 8-in. round layer cakes will serve 96		
Six 9 × 13 in. sheets will serve 96		
One 9 × 13 in. sheet, two layer, will serve 30		
Four 12-in. round layer cakes will serve 120		(See also portioning information in
Three 14-in. round layer cakes will serve 120		chapter on cakes and cookies)
Angel food	16 oz	Cut 16
Chocolate roll, jelly, etc.	18 × 26 in. rolled	34 to 36 portions
Cupcakes	1 No. 16 scoop	1 each
Doughnuts, cake	1 oz.	2

(continues)

TABLE I-1 Continued

Desserts	Pan Size	Portion
Cookies		
Brownies, date bars, etc.	18 × 26 in.	Cut 54, serve one
	13½ × 22⅞ in.	Cut 5 × 9, 45 portions
Ice cream, etc.		
Brick	quart	Cut 8
Bulk		No. 12 scoop
Sundae		No. 16, 2 oz sauce
(See also portioning information in section on frozen desserts)		
Miscellaneous		
Graham cracker roll, etc.	Loaf 9⅝ × 5½ × 3¼ in.	Cut 16
Meringues		⅓ c, 2-oz ladle syrup or sauce, 2 T whipped cream
Pineapple delicious		½ c, No. 10 scoop rounded
Sauces for topping		3 T, vary with richness
Shortcake		2½-in.-diameter biscuit, ⅓ c fruit, 2 T whipped cream
Steamed pudding	1 qt mold	Cut into 12 (3½ oz); 2 oz sauce
	12 × 20 in.	Cut 6 by 10
Pies		
Crust, double	9 in.	12 oz
single	9 in	6½ oz
Filling, cream	9 in.	1½ to 2 pt (1½ to 2 lb)
custard-type	9 in.	1½ pt (1½ lb)
fruit	9 in.	1½ pt (1½ lb)
One or two crust	10 in.	Cut 8
(use marker)	9 in.	Cut 7
	8 in.	Cut 6
Puddings		
Apple crisp, brown betty, etc.		4 oz
Apricot whip		¾ c
Bread pudding	13½ × 22⅞ in.	½ c, cut 5 by 9
Cobblers, etc.	13½ × 22⅞ in.	½ c, cut 5 by 9
	12 × 20 in.	½ c, cut 6 by 8
Cream puff or éclair, batter		1 oz (small), 2 oz (large)
filling		1½ oz, No. 20 scoop
Cream, rice, tapioca, etc.		½ c, No. 10 scoop
Icebox cake	12¾ × 23 in.	Cut 5 by 10
Icebox pudding		No. 20 scoop
Jello	12¾ × 23 in.	Cut 5 by 9
	12 × 20 in.	Cut 6 by 8
Whipped cream topping		2 T; 2 qt whipped topping per 100 portions

[a]A 12 × 20 in. baking pan 4 in. deep with food (16 to 18 lb of food) may be cut 5 by 8 to give 40 6- to 7-oz portions. (Use 6-in.-deep pan.)

[b]Weight is calculated from raw weight before baking.

4qt = 1gL
16oz = 1lb
512Tbsp = 1gL

Conversion Scales

Weights and Liquid Measures

Weights					
½ oz	=	14 g	9 oz	=	254 g
¾ oz	=	21 g	10 oz	=	283 g
1 oz	=	28 g	11 oz	=	311 g
1½ oz	=	43 g	12 oz	=	340 g
1¾ oz	=	50 g	13 oz	=	368 g
2 oz	=	57 g	14 oz	=	396 g
2½ oz	=	71 g	15 oz	=	425 g
2¾ oz	=	78 g			
3 oz	=	85 g	1 lb	=	453 g
3½ oz	=	99 g	1¼ lb	=	566 g
4 oz	=	114 g	1½ lb	=	679 g
5 oz	=	142 g	1¾ lb	=	792 g
6 oz	=	170 g	2 lb	=	905 g
7 oz	=	199 g	2¼ lb	=	1018 g
8 oz	=	226 g			

Liquid Measures					
1 teaspoon	=	0.005 L	1½ cups	=	0.36 L
1 tablespoon	=	0.015 L	1¾ cups	=	0.42 L
2 tablespoons	=	0.03 L			
			1 U.S. pt	=	0.47 L
¼ cup	=	0.06 L	1¼ U.S. pt	=	0.60 L
½ cup	=	0.12 L	1½ U.S. pt	=	0.72 L
¾ cup	=	0.18 L	1¾ U.S. pt	=	0.83 L
1 cup	=	0.24 L	1 U.S. qt	=	0.94 L
1¼ cups	=	0.30 L			

Note: Throughout this volume, quantities quoted in pints and quarts refer to the U.S. measures.

Fahrenheit/Celsius

−40°F = −40°C	120°F = 49°C	320°F = 160°C
−30°F = −34°C	125°F = 52°C	330°F = 166°C
−20°F = −29°C	130°F = 54°C	340°F = 171°C
−10°F = −23°C	135°F = 57°C	350°F = 177°C
−5°F = −21°C	140°F = 60°C	360°F = 182°C
0°F = −18°C	145°F = 63°C	370°F = 188°C
5°F = −15°C	150°F = 66°C	380°F = 193°C
10°F = −12°C	155°F = 68°C	390°F = 199°C
15°F = −9°C	160°F = 71°C	400°F = 204°C
20°F = −6°C	165°F = 74°C	410°F = 210°C
25°F = −4°C	170°F = 77°C	420°F = 216°C
30°F = −1°C	175°F = 79°C	430°F = 221°C
32°F = 0°C	180°F = 82°C	440°F = 227°C
35°F = 2°C	185°F = 85°C	450°F = 232°C
40°F = 4°C	190°F = 88°C	460°F = 238°C
45°F = 7°C	195°F = 91°C	470°F = 243°C
50°F = 10°C	200°F = 93°C	480°F = 249°C
55°F = 13°C	205°F = 96°C	490°F = 254°C
60°F = 16°C	210°F = 99°C	500°F = 260°C
65°F = 18°C	212°F = 100°C	510°F = 266°C
70°F = 21°C	220°F = 104°C	520°F = 271°C
75°F = 24°C	230°F = 110°C	530°F = 277°C
80°F = 27°C	240°F = 116°C	540°F = 282°C
85°F = 29°C	250°F = 121°C	550°F = 288°C
90°F = 32°C	260°F = 127°C	560°F = 293°C
95°F = 35°C	270°F = 132°C	570°F = 299°C
100°F = 38°C	280°F = 137°C	580°F = 304°C
105°F = 41°C	290°F = 143°C	590°F = 310°C
110°F = 43°C	300°F = 149°C	600°F = 316°C
115°F = 46°C	310°F = 154°C	

Glossary

Acceptance buying—Method used by large-volume buyers in which the vendor submits products to a federal inspector, who ensures that they meet government specifications before final purchase is authorized.

Accompaniment salad—Type of salad served to complement the meal, as compared to a main course salad, dessert salad, and so on.

Acrolein—Acrid substance produced by the glycerol component of fat when fats are overheated.

AGA (American Gas Association)—National association that, among other things, inspects equipment using natural gas as fuel (e.g., stoves, heaters) and issues its seal of approval only if the equipment meets national safety standards.

A la carte—French phrase which means that each menu item is charged for separately (i.e., there are no package deals for a complete meal).

American service—Method of service in which food is cooked in the kitchen, portioned onto individual plates by the cook, and served by the waiter.

Amoeba—Single-cell protozoan. Certain kinds, if contracted from impure drinking water of food, can cause amoebic dysentery. It is usually transmitted from fecal matter on the unwashed hands of an infected person.

Appetizer—Food or drink served to whet the appetite before the main course (*see* Hors d'oeuvres).

Appetizer salad—Type of salad served at the beginning of the meal, as compared to a main course salad, dessert salad, and so on.

Arabica—Colombian coffee plant producing fine-quality beans. First discovered in Arabia, it is today considered the first in quality of the two varieties of Arabica and Robusta.

Arrowroot—Food derived from the tubers of the plant *Maranta arundinacea*. Composed almost entirely of starch, it is useful in thickening soups and sauces.

Assessment—Process whereby management's decision-making abilities are tested and employees can be evaluated for their managerial potential.

Au jus—Unthickened browned meat juice made by deglazing a pan in which meat has cooked.

Automatic brewer—Machine that brews coffee automatically by spraying a regulated amount of hot water over a filter basket holding ground coffee.

Bacteria—One-cell microscopic organisms of the Monera kingdom (neither plant nor animal). Some species spoil food or cause disease; others aid in food production or digestion.

Baking—Cooking in an oven with indirect heat. Confusingly, also called *roasting* when applied to meats, nuts, and certain fruits and vegetables.

Baking ammonia—Leavening agent that emits carbon dioxide and ammonia when moistened and then heated.

Balance sheet—Financial statement produced at the end of an accounting period that shows the assets and liabilities of an establishment at that time. The categories of the statement balance out, so that total assets equal total liabilities plus owner's equity.

Banquet service—American, French, or Russian styles of service, or a combination of these services, organized so that every table is served at the same time.

Barbecuing—Method of cooking under or over direct heat, like broiling. It differs from broiling in that a highly seasoned sauce with a tomato base is used to baste the food or is served with it.

Barrow—Young male hog that has been castrated.

Base or underliner, salad—Foundation of a salad.

Batch cooking—Quantity cooking of vegetables at one time in order to serve patrons for 20 minutes.

Bay scallop—*See* Shellfish.

Beaten biscuit—Like baking powder biscuits but without the baking powder.

Béchamel sauce—Basic sauce made from a rich veal or chicken stock, thickened with roux, to which milk or cream is added.

Beef—Meat from steers, heifers, cows, stags, and bulls that are over one year old. The best beef comes from steers that are castrated when young and fattened on grain. Good-quality beef is bright red in color with firm, fine-grained, well-marbled flesh. Grades of beef in descending order of quality are Prime, Choice, Select, Standard, and Commercial. Prime is rarely sold in consumer retail stores. Wholesale cuts of beef are the rump, loin, flank, short loin, plate, rib, brisket, chuck, and shank. A whole of beef is divided into four quarters: a left and right forequarter and a left and right hindquarter. The forequarter contains the chuck, rib, short plate, brisket, and foreshank. The hindquarter includes the short loin, sirloin, round, tip, and flank. Following is a description of the major wholesale and more common cuts of beef.

Baron of beef: Extralarge cut that includes part of the ribs and both sirloins. It is traditionally served at Christmas.

Brisket: Cut from the breast, including parts of five ribs and part of the shoulder. It is an economical cut that should be cooked a long time to make it tender.

Chuck: Portion of the forequarter remaining after the removal of the foreshank, brisket, short plate, and rib. Chuck also includes clod, the beef just above the blade bone that is suitable only for stewing.

Flank: Fleshy part between the ribs and the hip. The flank steak, a flat muscle embedded in the inside of the udder end of the flank, is obtained from this cut.

Loin: Portion of the hindquarter remaining after removal of the round, flank, hanging tender, kidney knob, and excess fat. Sirloin steak comes from the loin. T-bone steak, porterhouse steak, shell steak, club steak, and top loin steak come from the tenderloin.

Rib: Portion of the forequarter remaining after removal of the crosscut chuck and short plate. It contains part of the seven ribs, sixth to twelfth inclusive, the section of the backbone attached to the ribs, and the tip of the blade bone. The standing rib roast, rib steak, boneless rib steak, Delmonico steak, and rib-eye roast come from this cut.

Round: Portion of the hindquarter remaining after removal of the untrimmed loin. Tender, expensive cuts come from the round, including the top round, side of round, round steak, and bottom round.

Rump: Portion between the loin and the round. This is an economical cut.

Shank: Portion from the upper or sometimes lower part of the leg. It is an economical cut.

Short loin: Anterior portion of the full loin where the loin eye and tenderloin are located. Filet mignon comes from the short loin.

Short plate: Portion of the forequarter immediately below the primal rib, yielding both the short ribs and boiling or ground meat.

Sirloin tip: Part of the hindquarter, also known as the knuckle.

Beginning inventory—Inventory on hand at the start of a certain time period.

Behavioral training—Involves the employee in learning proper behavior on the job as compared to just the task to be accomplished.

Belt conveyor—Flight-type dishwashing unit with a continuous belt allowing dishes, pots, pans, and other large units to be placed directly on the belt, thus eliminating the need for racks. After exiting the final rinse section, the conveyor travels for several feet in the open to allow time for drying before the dishes are removed. Glasses and silver are placed on racks to be sent through on the conveyor.

Bench—Placing dough to rest after the punch.

Benefit packages—Along with compensation and incentives, additional rewards for performance are given mostly in insurance-related areas or retirement payments.

Benefits—Part of a compensation package that would include incentives as well.

Betty—Layers of fruit alternating with layers of bread or cake crumbs, sugar, and butter or margarine. After being properly layered, the betty is baked.

Beurre manie—Butter and flour kneaded together and used to thicken sauces.

Beurre meuniere—Butter that has been melted with a bit of lemon juice and chopped parsley and cooked to a light brown color, just short of becoming beurre noisette.

Beurre noir—Butter that has been melted with a bit of lemon juice and chopped parsley and cooked to almost black in color, but not burned.

Beurre noisette—Butter that has been melted with a bit of lemon juice and chopped parsley and cooked to a nut-brown color.

Beverage cost—Cost to an establishment of a beverage that is in turn sold to a customer.

Bibb lettuce—Dark, crisp leaves with a delicate nutty flavor.

Biotin—One of the B-complex vitamins, which is sometimes called vitamin H. Good food sources include vegetables, cereals, nuts, and organ meats. Biotin is important in metabolic processes such as fat synthesis. The avidin in raw egg whites renders this vitamin useless.

Biscuit—Quick bread that is the closest thing to yeast bread.

Bisque—Soup, either thick or thin, having a base of shellfish, such as lobster or seafood bisque.

Black tea—Tea that is fully fermented.

Blanching—To plunge a food item into *boiling* water for a short time. Usually, a step preliminary to further cooking by another method.

Blancmange—Cream pudding to which eggs have been added.

Blank check—To order an item without checking the cost. Since the establishment is obligated to pay whatever the vendor charges, only reputable vendors should be relied on in this way.

Blind receiving—Receiving goods when the quantity, quality, weight, and price are omitted from the invoice. The goods must be checked with the packing slip on delivery.

Blond—French for a method of braising in which the meat is not browned first.

Blue crab—Crab that comes from north Atlantic and Gulf waters.

Bluepoint—Type of oyster from Maryland and Chesapeake Bay.

Boar—Mature sexually active male hog.

Boiled dressing—Cooked dressings thickened with starch and perhaps eggs containing less oil than other dressings.

Boiling—Method of cooking by immersion of food in water or some liquid brought to 212°F or 100°C.

Boston chowder—Similar to New England chowder; has clams and the addition of cream to its base of thickened fish stock. *See also* Chowder.

Boston lettuce—Also known as butter lettuce, this salad green has a loose head with delicate, sweet leaves.

Botulism—Rare but extremely dangerous form of food poisoning caused by the anaerobic bacterium *Clostridium botulinam*. Since the spores are difficult to kill and thrive without oxygen, infected canned goods present the greatest potential threat.

Bouillon—Soup made from stock, extra meat, and seasonings.

Bouquet garni—French for herbs and seasonings, including parsley, garlic, clove, peppercorn, thyme, and bay leaf, tied together between two celery stalks or bundled in knotted cheesecloth, and used to flavor sauces and soups.

Braising—Method of cooking in which vegetables or meat are prepared by first browning or searing in a small amount of fat, then cooking in a small amount of liquid, such as water, stock, juice, or meat drippings, that produces steam and keeps the meat soft. A thickening agent is often added when vegetables are braised.

Brand—Product(s) of one manufacturer that can be identified and distinguished from similar goods produced by its competitors.

Break-even analysis—Analysis to determine the break-even point or that sales volume at which revenues and expenses exactly equal each other. Above this point is increasing profit; below it is increasing loss.

Break-even point—Sales volume at which revenues and expenses exactly equal each other. Above this point is increasing profit; below it is increasing loss.

Breakfast cocoa—Special type of hot chocolate that has 22 percent cocoa butter.

Brisk flavor—Taste with a zestful, stimulating quality.

Broiler—(1) Piece of equipment that cooks food mainly by radiant heat supplied by charcoal, coke, heated ceramic, gas lamp, or electric heating element. It usually cooks one side at a time, although some models cook both sides at once. (2) *See* Poultry.

Broiling—Method of cooking whereby the food is exposed to intense direct heat, as under a broiler flame or over a charcoal fire.

Brown sauce—Most commonly used sauce made from thickening a brown stock with added tomato purée, seasonings, and spices.

Brown stock—Thin liquid rendered from simmering roasted beef and veal bones in water with vegetables and seasonings for six to eight hours.

Brun—French for a method of braising in which the meat is first browned by searing.

Btu—Standard unit of heat; 1 British thermal unit (Btu) equals the heat required to raise the temperature of 1 pound of water by 1°F.

Budget—Estimate of future expenses and revenues that helps an establishment plan ahead and control future operations.

Buffalo chopper—Chopping machine that operates with the combined action of a revolving metal bowl and rapidly spinning blades. Food is put into the bowl, which revolves in the opposite direction from the spinning blades; the food is chopped as it passes through the blades.

Bullock—Young uncastrated bull.

Business entity—Concept of separation of the business from the person who provides the assets.

Butter—Made from cream and is 80 percent fat with the remainder being salt and liquid.

Butterfly—These fillets of fish are the two sides of the fish cut lengthwise away from the backbone and held together by the uncut flesh and skin of the belly. May apply to meat and poultry as well.

Butter sauce—Butter or margarine blended with lemon juice and a dash of cayenne pepper.

Buyer—Person responsible for making decisions regarding quality, amount, price, potential profit to be gained from commodities to be purchased, and ultimately customer satisfaction.

Cake doughnut—Made from a rich muffin dough that is handled as muffin dough would be, that is, cold and not overworked. The doughnuts are fried in deep fat.

Calcium—Essential mineral required for proper formation of bones and teeth. Sources include milk products, certain dark-green vegetables, including broccoli, and small-boned fish, such as sardines. Vitamin D aids in the body's absorption of calcium.

Calf—Young beef animal 14 to 52 weeks old.

Calorie—Unit expressing the amount of energy produced by food when it is used by the body (e.g., fat offers 9 calories per gram, protein and carbohydrates offer 4 calories per gram).

Canadian service—Variation of English service, in which no waiter is used. The host or hostess apportions the food and then passes a plate to each guest.

Canapé—Eye-appealing appetizers made of crackers or shaped slices of toast spread with a flavorful paste.

Capon—*See* Poultry.

Carbohydrate—Compound within the body that is a source of energy. Carbohydrates are manufactured by plants and are obtained for use from the diet. All break down in the body to the simple sugar glucose. Excess carbohydrates are converted in the liver to fat.

Catalyst—Any substance that accelerates the chemical reaction occurring in other substances without itself undergoing any permanent change.

Celsius—Scale for measuring temperature. The freezing point is arbitrarily set at 0°; the boiling point at 100°. The distance between these two points is divided into 100 equal degrees. Degrees F = 9/5 (degrees C) + 32 degrees.

Centralized service—Trays are prepared in the serving section of the main kitchen and dispatched from there.

Chapon—Dried piece of bread flavored with garlic.

Chaudfroid—Special sauce made to decorate cold meats, especially on the buffet.

Cheese—Dairy product made by separating the milk curds from the whey. Cheese was originally made to preserve excess milk but is now a popular and nourishing protein food produced in virtually all countries of the world. Any list of cheese cannot be exhaustive. What follows are some of the better-known cheeses of the world.

Altenburger: Soft goat's milk cheese from central Germany. Altenburger has a very strong flavor.

Banon: Goat's milk cheese made in the Marseilles area of France.

Bel Paese: Very mild cow's milk cheese from Italy.

Bierkase: In Germany, a term for a small, round cheese dropped into beer.

Bleu d'Auvergne: Cow's milk blue cheese from central France.

Bleu de Bresse: Small, rich cheese, similar to Gorgonzola; made near Lyons in France.

Boursin: French cheese flavored with garlic and chopped herbs.

Brick: Semisoft American-made cheese with a mild flavor.

Brie: Soft cow's milk cheese with a white rind; made in Melun, Coulommiers, and Meaux in France.

Cabecou: Round, flat goat's milk cheese made in the Landes district south of Bordeaux in France.

Caerphilly: White, crumbly, easily digested cheese made in the Clamorgan area of England.

Camembert: Soft, whole-milk cheese from Camembert, France.

Chabichou: Goat's milk cheese made near Poitiers in France.

Cheddar: Hard whole-milk yellow-orange cheese originally from Somerset, England, but now produced in Scotland, Canada, Australia, New Zealand, and the United States.

Cheshire: Hard, salty cow's milk cheese first produced in Cheshire and said to be the oldest of English cheeses.

Chevreton: Strong goat's milk cheese from France. Chevreton has a hard rind and a soft, runny interior.

Colby: American-made cheese similar to cheddar but more moist.

Coon: American-made, crumbly, dark-brown cheddar with a tangy flavor. Coon is cured in a different manner than regular cheddar.

Cottage cheese: Bland unripened cheese with white soft curds.

Cream cheese: White, soft, rich, and creamy uncured cheese made of cream and whole milk.

Danablu: Creamy, salty, blue-veined cheese from Denmark.

Danbo: Samsoe family cheese with a nutty, sweet flavor; made in Denmark.

Derby: Mild, pale cheddar similar to double Gloucester; made in England.

Edam: Mild, wax-covered cheese made in northern Holland.

Elbo: Mild Danish cheese of the samsoe family.

Emmentaler: Hard swiss cheese named after the Emme Valley. Emmentaler is made from whole or part-skim milk.

Feta: Soft, spicy cheese from Greece; made from ewe's milk.

Fontainbleau: Triple cream cheese made in the Ile-de-France.

Fontina: Yellow, semisoft ewe's milk cheese made in the Piedmont region of Italy.

Gammelost: Potent sour-milk cheese from Norway.

Géromé: Munsterlike cheese originally made in the Strasbourg region of France.

Gjetost: Norwegian cheese made from cow and goat milk. The cheese has a moldy flavor that is an acquired taste.

Gloucester: Cheddarlike mellow cheese from the Gloucester area of England.

Gorgonzola: Salty, strong, crumbly ewe's milk cheese with blue veins; made in the Milan area of Italy.

Gouda: Mild, Edamlike cheese made in southern Holland.

Gruyère: Creamy, firm cheese from the Gruyère valley in Switzerland.

Havarti: Mild, bland cheese from Denmark.

Herve: Soft limburger type of cheese from Belgium.

Kasseri: Hard ewe's milk cheese from Greece.

Le Cantral: Hard, yellow, cow's milk cheese from France.

Limburger: Very strong semisoft cheese made in Liège Province of Belgium and in Germany.

Livarot: Soft, pasty cheese from a small town in the Calvados region of France.

Mascarpone: Soft, fresh cream cheese made in the Lombardy region of Italy.

Minnesota Blue: American copy of Roquefort cheese made from cow's milk.

Mont d'Or: Munsterlike cheese made east of Lyons near Switzerland.

Monterey Jack: Cheddar made without coloring in Monterey County, California.

Morbier: Hard cheese made in the French Alps.

Mozzarella: Mild buffalo's milk cheese made in Italy.

Munster: Fatty, semihard cow's milk cheese made in the Alsace region of France.

Neufchâtel: Very white, tube-shaped skim or whole-milk cheese made in France.

Oka: Mild-flavored cheese similar to Port Salut. It is made by trappist monks in Quebec, Canada.

Parmesan: Hard, yellow grating cheese from the Parma region in Italy.

Pecorino: A hard, aromatic grating cheese made in Italy from ewe's milk.

Petit Suisse: Soft, fluffy dessert cheese made in France.

Picodou: Soft goat's milk cheese made in the Haute Savoie region of France.

Pineapple: Hard cheddar cheese molded into a pineapple shape; originally made in Connecticut.

Pont l'Evêque: Semihard fermented cheese from the Calvados region of France.

Port Salut: Creamy, yellow whole-milk cheese originally made by trappist monks near the Laval region of France.

Provolone: Smoky, semihard cheese from Italy.

Ricotta: Soft white cheese similar to cottage cheese; made in Italy.

Roquefort: Pungent blue-veined cheese made from ewe's milk in the Saint-Affrique district of France.

Saint-Marcellin: Soft, slightly salty cheese made in the Savoie district of France.

Sainte-Maure: Soft, creamy goat's milk cheese made in the Touraine region of France.

Samsoe: Cheddarlike, nutty cheese from Denmark.

Schabzigei: Pungent grating cheese also known as sapsago; made in Switzerland.

Stilton: Strong blue-veined cheese from England.

Tilsiter: Port Salut type of cheese with small holes; made on the borders of Lithuania in Germany.

Tybo: Very mild cheese of the samsoe family; made in Denmark.

Vacherin: Soft, runny cheese made in France and Switzerland.

Vendôme: Hard ewe's milk cheese made in the Loire Valley in France.

Cheese food—Similar to processed cheese but contains less cheese and fat and more milk and whey solids.

Cheese, processed—Blend of shredded cheeses, pasteurized and processed with added emulsifiers.

Cheese, ripened—Cheese developed over time through the action of enzymes and bacteria naturally present.

Chef de rang—French for chef of rank or the principal waiter in the French restaurant system.

Chef des parties—French for the heads of production in a kitchen, such as chef de saucier.

Cherrystone—Type of clam. *See also* Shellfish.

Chicken stock—Thin liquid rendered from simmering rinsed chicken bones in water with vegetables and seasonings for up to no more than three hours.

Children's menu—Special menu designed and printed to present to patrons typically under the age of 12.

Chincoteague—Type of oyster from Maryland and Chesapeake Bay.

Chlorine—Extremely pungent poisonous element used in small amounts to kill bacteria and to purify water. It can damage lungs and skin and should be handled with extreme care.

Choice—*See* Beef.

Cholesterol—Steroid alcohol found in high concentration in organ meats, egg yolks, and saturated fats. It appears as a waxy or soapy crystalline substance. Although it is found naturally in the body and is necessary for life, ingesting too much has been linked with heart ailments and circulatory problems.

Choline—Essential nutrient required by the body to produce acetylcholine. It is important in building and maintaining cell structure.

Chowder—Soup characterized by a main ingredient of seafood, corn, or mushrooms with diced potato, chopped onion, and bacon fat or salt pork.

Chromium—Mineral found in brewer's yeast, flesh, nuts, whole grains, and American cheese and thought to be useful in changing carbohydrate to energy.

Chuck—*See* Beef.

Clean—Lack of soil or dirt on an item.

Coagulation—Conversion of proteins from a relatively fluid state to a more solid one, as happens when eggs are cooked.

Cobalt—Mineral that appears in trace amounts in various nutrients (e.g., in vitamin B_{12} molecules). In excess, it can be poisonous.

Cobbler—Fruit dish with biscuit dough or pie dough on top.

Cocci—Bacteria that are spherical in shape.

Cocktail—Alcoholic beverage usually served iced, often with a garnish, in small glasses. Also, nonalcoholic food item such as shrimp, crab, oysters, clams, fruit cup, fish, fruit or vegetable juice. Both types may be served alone or, most often, preceding a meal.

Codes—Set of rules established to set a standard, including building, construction, fire, and safety.

Coenzyme—Nonprotein compound that chemically assists enzymes during catalysis. Since most coenzyme molecules contain one of several different vitamins, they depend on appropriate dietary intake for their proper functioning.

Coliform—Beneficial aerobic bacteria that are normally found in the human gastrointestinal tract, where among other things they ferment the sugar lactose. Outside the intestinal tract, as when in food, milk, or water, they are an indication of contamination and are themselves a contaminant.

Collagen—Solid protein found in bone, skin, and tendons that is transformed by boiling into gelatin. Thus, by cooking meats, this connective tissue is softened.

Combo-oven—Oven that cooks both with dry heat and with steam heat, making it possible to have it act as a dry heat oven, an oven that has enough steam to bake hard-crusted breads or to act as a steamer.

Commis de rang—French for assistant waiter.

Commodity—Goods or tangible items of value.

Compactor—Machine that reduces the volume of waste by compressing it to a small fraction of its original size.

Compensation—Well-planned package that includes not only remuneration but fair evaluations, participation in decision making, and fair treatment through integrity by management.

Competitive bid—Purchasing method in which sellers are invited to submit their prices in writing. Normally, the supplier who offers the lowest-priced goods that meet the purchaser's specifications is chosen.

Complementary flavors—Blending flavors in foods so that they enhance one another, such as serving roast beef with a horseradish sauce.

Complete protein—Proteins containing all the essential amino acids.

Compote—Fruit served in a heavy syrup.

Computer-assisted artwork—Menu artwork that is generated by the computer

Concept of consistency in accounting—Theoretical basis for using the same methods for all accounting transactions.

Conduction—Movement of heat through solid matter, as it spreads from one molecule to the next.

Coney Island chowder—Similar to Manhattan, Philadelphia, or Long Island chowder, which have unthickened fish stock with tomatoes and clams. *See also* Chowder.

Connective tissue—Collagen and elastin that hold muscle fibers together. The more connective tissue, the tougher the meat.

Conservatism—Concept based on the theory that the values of transactions should not be overstated.

Consommé—Rich, delicate, clear soup made by clarifying a rich stock or broth after enriching it by simmering it with meat and other flavoring ingredients.

Construction documents—Detailed drawings that identify all aspects of a facility exactly as it is to be built.

Construction schedule—Outline of the times by which various elements of construction will be completed.

Contribution margin—Net difference between sales income and direct costs. Also called *marginal income*.

Control points—Consistent level in the quality of menu items is achieved by establishing a standard for the product and then setting up checks at critical control points to see that the desired quality is achieved. Examples include correct ingredients and their measurement, preparation techniques, times and temperatures, correct tools, equipment and utensils, and so on.

Convection—Movement of heat in a liquid or gas based on the principle that hotter portions of a substance expand and rise through colder ones.

Convection oven—Oven in which heated air is circulated evenly by a fan throughout the cavity, thus allowing faster cooking of greater quantities of food than with a conventional oven. Also called *forced convection oven*.

Convection steamer—Equipment that cooks food with steam at normal pressure. The steam is circulated within the cooking compartment so that cooking is rapid and even. The compartment can hold three 12 in. × 20 in. × 2.5 in. pans.

Convenience food—Any food that has been prepared before packaging to save later time and/or effort.

Convenience service—Uses ready-prepared and ready-to-eat foods to which the consumer gives final production.

Convenience stock—Manufactured base for stock, often used when needed bones or necessary staff is not available to prepare a stock from scratch.

Cook chill—Foodservice system in which foods are mass produced in the individual unit, stored, and then later taken out for service in partial lots.

Copper—Metallic element, traces of which are present in the human body in certain enzymes and in the blood. Some copper in the diet is necessary, but toxic levels can be reached quickly. The metal itself is also used in the manufacture of some cooking utensils because it is a good conductor and spreads heat evenly and rapidly throughout the vessel walls.

Copy—Printed words used to name and describe items on a menu. Good menu copy helps sell.

Cost factor—Ratio of the user's cost for meat (U) to the supplier's price (S). The user's cost is that involved in meat preparation (e.g., butchering). When the supplier's price changes, the cost factor (C) can be used to estimate the user's new cost. Thus, $C = U/S$ and $U = C \times S$.

Cost of goods—Total cost of goods that have been acquired, including purchase price, transportation, storage, and other factors.

Cost plus—System in which an establishment contracts to buy all of certain items (e.g., food) from a given supplier and agrees to pay the supplier's costs plus a set percentage as markup.

CO_2 extinguisher—Extinguishes fire by discharging a foam or powder containing carbon dioxide.

Counter dishwasher—Smallest type of dishwashing unit available. It comes in either 24- or 20-in. models, both of which fit on top of the counter.

Counter service—Method of service in which guests sit at a counter, behind which the waitress or waiter stands. Service is usually faster than at tables.

Court bouillon—Rich fish stock in which tender seafood is sometimes poached.

Cow—Older female beef animal that has had a calf.

Cream, coffee—Table cream having 18 percent milkfat. *See also* Half-and-half.

Cream, heavy whipping—Cream having 34 to 36 percent milkfat.

Cream, light whipping—Cream having 30 to 34 percent milkfat.

Crepe—Thin pancake filled with main course food items, such as chicken or meat, or prepared tableside as a dessert.

Crepes Suzette—Dessert crepe filled with a rich, sweet orange mixture and flamed tableside after being doused with liqueurs.

Crisp—Fruit mixture topped with a blend of flour, sugar, fat, and seasonings, and then baked.

Critical incidents—Most important aspects of the job to be done.

Crustacean—Shellfish with crustlike shell, including crab, crayfish (or crawfish), lobster, and shrimp.

Cultural diversity—Encompasses differences among workers in age, affectional orientation or sexual preference, economic status, educational background, ethnicity, gender, geographic location, income, marital status, lifestyle, religion, and so on.

Curb appeal—Appearance of the operation from the external environment, from parking lot to landscape and entrance into the foodservice facility.

Curb cuts—Design of the curbs upon entering an operation should allow for easy car access.

Curdling—Process whereby eggs, sweet cream, or milk separate into protein and liquid fractions when combined with acids such as those found in sour cream or lemons.

Curly endive—Salad green similar to chicory which has curly leaves and a bitter taste.

CVAP (controlled vapor technology)—Cooking, holding, and thermalizing technology using vapor pressure to maintain food temperature and moisture content.

Cycle—Menu that list different meals each day for a given period (e.g., 30 days). After that time, the same menu list is used again for the next period.

Daily receiving report—Form summarizing all the food deliveries to an establishment for that day; it includes the supply source, the description of the items, and the cost.

Danger zone—Temperature range most favorable for bacteria (i.e., about 40 to 140°F).

Database—Organized grouping of computer data in a central place to facilitate their retrieval by management.

Dead-bolt lock—Type of lock on door providing additional security.

Decentralized service—Service method in which food prepared in a central kitchen is sent in bulk to service pantries, where it is placed onto dishes and trays and dispatched.

Decision making—Process of using good judgment, practical knowledge, and astute thinking to take action and bring plans to fruition.

Deep-frying—Cooking food by immersing it in boiling fat or oil.

Demiglaze—Reduction of a stock by one-half of its volume.

Design development phase—Characterized by detailed drawings that include specific aspects of the mechanical, plumbing, equipment, structural, electrical, and fire protection systems.

Dessert salad—Type of salad served at the end of the meal as compared to an accompaniment salad, main course salad, and so on.

Deveined—Shrimp are deveined when purchased as PDQ (peeled, deveined, and quick-frozen) market form.

Direct deliveries—Food or supplies taken directly to their destination (e.g., the kitchen) without being placed in storage.

Directs—Food received that goes right into production in the kitchen rather than being stored.

Direct seller—One who sells directly from manufacturers or packers. A direct seller may provide some savings to a buyer because of the elimination of marketing agents and services.

Docking—Gashing the top of a dough product such as bread to allow it to rise without breaking or bursting.

Dollies—Pieces of mobile equipment on which food items or supplies are placed to facilitate their movement from one place to another.

Dough cutter—Rectangular metal or plastic blade used to cut and section dough and scrape benchtops.

Drained weight—Weight of canned food remaining when the liquid is drained off. Two different brands of the same item may have the same total weight, but one may contain more actual food.

Dropped biscuit—Dough is dropped onto greased pans after mixing, with no kneading taking place.

Dry storage—Room for the storage of food that does not have to be refrigerated (e.g., in cans and cartons).

Duck—*See* Poultry.

Du jour—French for the menu of the day. The term refers to a menu composed of special dishes offered on that day.

Dumpling (steamed)—Variation on biscuit; it is steamed or cooked over boiling liquid with a tight cover on the container.

Dungeness—*See* Shellfish.

Edema—Excess retention of water by the body or part of the body that causes swelling. The various causes include prolonged protein deficiency and kidney disease.

Egg, basted—Egg is covered with a small pan to finish cooking on a steam table prior to service. Alternatively, the egg may be basted with hot fat during frying. Also known as *country style*.

Egg, country-style—*See* Egg, basted.

Egg, fresh—Egg is at its highest quality immediately after laying.

Egg, Grade AA—Highest level of egg grade.

Egg, hard-cooked—Completely coagulated egg.

Egg, omelet—Made by beating whole eggs together and adding them to a pan covered about ⅛ in. of hot fat. They should be rolled neatly with good shape and be moist, tender, and delicate in flavor.

Egg, over easy—Flipping an egg to cook both sides when frying, yet keep the yolk runny.

Egg, poached—Cooked in water out of shell.

Eggs, proprietary—Whole eggs containing added yolk can be frozen and used in the bakeshop or other production area.

Egg, shirred—Method of baking an egg.

Egg, soft-cooked—Soft, almost runny white and only warmed yolk.

Egg, souffle—Mixture of well-beaten egg whites folded into a starch base containing the yolks and a seasoning mixture. The mixture is baked and result in a delicate product.

Egg, sunny-side up—Egg is not turned when fried such that the yolk remains yellow in the white.

Elastin—One of the proteins in meat responsible for holding muscle fibers together. It is insoluble and not altered by cooking; the more of it present, the tougher the meat.

Employee selection—Selection that involves not only short-term need goals but long term as well. The proper type of recruitment material must be available at all times for successful employee selection.

Emulsion, permanent (stable)—Emulsion in which the two liquids do not readily separate out.

Emulsion, temporary—Unstable mixture of two fluids (e.g., oil and water) that tend to separate. They will meld temporarily if shaken or stirred.

Ending inventory—Inventory at the close of a time period.

English service—Style of service in which a host helps to apportion food onto plates that are then distributed to other guests by the waiter. Such service is usually done only at private parties or other special events.

Entrée—French term for main course of the meal. It is usually the most expensive item of the meal.

Equal Pay Act of 1963—Legislation governing pay to employees referencing same pay for same work.

Equity—Owner's equity portion of the balance sheet reflects the difference between total assets and total liabilities. If the assets outweigh the liabilities, the equity is a positive figure.

Escarole—Plant of the chicory family with broad, flat leaves and a slightly bitter taste.

Escoffier—Great French chef given credit for having established the organization of the modern kitchen.

Espagnole sauce—Same as brown sauce.

Espresso—Italian for strong, heavy coffee made by forcing steam through a finely ground dark roast and traditionally served after a meal.

Essence—Rich stocks simmered with wine, vegetables, and herbs. After simmering, they are strained and reduced to the desired consistency, usually that of a demiglaze. They are used to flavor and enrich items.

Essential amino acid—Eight amino acids that cannot be synthesized within the human body. Sufficient quantities must be obtained from the diet.

Estimated versus actual food costs—Actual food cost is the amount spent for food during a certain period, regardless of how much was originally budgeted, or estimated.

Ethylene—Gas that accelerates the ripening of fruit.

Evaporated milk—Milk having had about 55 percent of its water extracted.

Eviscerated—Fish cleaned by taking the insides out.

Ewe—Mature female sheep.

Extended cost—In yield testing, the total cost of the material tested.

External environment—Area of the operation which the customer first sees upon arrival, including parking, lighting, signage, landscape, and accessibility.

External security system—Measures taken to ensure the operation is secure, including lighting, locks, and alarm systems.

Fahrenheit—Method measuring heat in which 32 degrees is the freezing point of water and 212 degrees the boiling point. It is 100/180 of a *Celsius* degree. Degrees C = 5/9 (F degrees −32 degrees).

Fair Labor Standards Act of 1938—Legislation governing how employees should be treated.

Fat fish—Fish containing more fat than lean ones.

Fat-soluble vitamin—*See* Vitamins A, D, E, and K.

Feasibility study—Intensive and detailed analysis of the market, performed to plan site location and judge type and location of competition, size of operation, cost, financing, and budgeting. The study determines the likelihood of the operation's success.

Fermentation—Breakdown of a compound that does not require the use of oxygen. Such chemical changes can alter the makeup of food.

Ferrous sulfide—Movement of sulfur from the egg's white into the iron of the yolk results in this chemical combination that appears green if an egg is overcooked.

Fettuccine—Small ribbon-shaped pasta.

Fiber—Food tissue that cannot be digested or absorbed. It is believed to be beneficial in many illnesses; recent research suggests that it may help prevent certain forms of cancer. Foods high in fiber content include wholemeal cereals and flour, root vegetables, nuts, and fruit. Fiber is also called *roughage*.

FIFO—Principle that goods purchased first should be used first. This helps reduce food spoilage.

Fillet—Sides of a fish that have been cut away from the backbone and are ready to cook.

Finger bowl—Small silver or glass bowl placed on a doily and silver underliner, filled one-third with warm water and lemon juice or fragrance. It is used by a guest to clean his or her fingers after eating messy foods such as lobster or steamed clams.

Fish—Fish is high in protein and low in fat. There are numerous varieties. Below is a list of some of the more popular fish in the United States.

Albacore: One of the Pacific tuna, with the lightest meat. Canned, it is labeled white meat.

Anchovy: Fish of the herring family, found on both the Atlantic and Pacific coasts. Not usually eaten whole but pickled and salted.

Blackback: Fish with a white, firm flesh; abundant in the Northeast in the fall and winter months. It is a type of flounder.

Bluefish: Fish with dark meat and oily flesh; abundant in the Northeast from May to November and on the Gulf coast in the winter months.

Butterfish: Fish with nonoily flesh; abundant in the Northeast in summer and on the Gulf coast in winter.

Carp: Lean, firm fish from the Great Lakes; abundant in spring.

Catfish: Southern freshwater fish with delicate white flesh.

Cod, Atlantic: One of the most abundant fish in the United States, particularly in the Northeast during the spring and summer months. It has a flaky, nonoily flesh.

Cod, Pacific: Fish with a flaky, nonoily flesh; abundant on the Pacific coast all year long.

Croaker: Lean fish abundant in the mid-Atlantic states from March to October.

Cusk: Delicate, lean fish; abundant in the Northeastern states in spring and summer.

Dolphin: Also called *mahimahi* in Hawaii, where it is abundant. It has a firm flesh.

Eel: Fresh- or saltwater snakelike fish without scales. Conger is the most popular variety.

Flounder: Group of fish including fluke, yellowtail, and blackback. Flounder is available all year in the Atlantic, Gulf, and Pacific coast states. It has a flaky, delicate flesh.

Flounder, starry: Flounder found in the Pacific Northwest.

Flounder, yellowtail: Flounder found in the Gulf and mid-Atlantic states from summer to winter.

Fluke: Flounder abundant in the Northeast and mid-Atlantic states all year long.

Grouper: Fish with a delicate, lean flesh; abundant on southern and Gulf coasts from April to December.

Haddock: Fish with lean, nonoily flesh; abundant in the North Atlantic from March to November.

Hake: Fish with a lean, delicate flesh; found on the Gulf and Pacific coasts. Most abundant in August and September.

Halibut: Fish with delicate, white flesh abundant in the Northwest and Northeast during spring and summer.

Lake sturgeon: Firm, delicate freshwater fish; usually smoked. A major source of caviar.

Lake trout: Firm, oily freshwater fish; abundant in the Great Lakes from May to October.

Lingcod: Fish with a delicate, lean flesh; most abundant in the Pacific Northwest from October to May.

Mackerel, Atlantic: Fish with a dark, oily flesh; most abundant in the New England states in summer.

Mackerel, King: Fish with an oily flesh; most abundant in the Gulf states from November to March.

Mackerel, Spanish: Fish with lighter flesh than the other mackerels. Available all along the Atlantic coast and parts of the Pacific coast.

Mullet: Fish with firm, medium-oily flesh; most abundant in the Gulf coast states and Florida from April to November

Northern pike: Lean, flaky freshwater fish. No longer fished commercially.

Ocean perch: Also known as *redfish* and *rockfish*. Fish with flaky, white flesh; found on both coasts.

Pollack: Fish with white, nonoily flesh; found in the Northeast and mid-Atlantic states. Most abundant from October to December

Pompano: Fish with firm, white flesh; abundant in Florida from March to May.

Porgy: Fish with delicate, white flesh; abundant on the Atlantic coast.

Rainbow trout: Delicate white-fleshed freshwater fish; available throughout the United States.

Redfish: Fish with firm, light flesh; abundant in the Gulf states from November to February.

Red snapper: Fish with delicate, white flesh; abundant in the Atlantic and Gulf coast states during the summer months.

Rockfish: Group of fish of over 50 varieties abundant in the Pacific Northwest. They have a firm, white flesh.

Salmon: Species of fish including chinook, coho, sockeye, pink, chum, and Atlantic salmon. Salmon have firm, fatty flesh and are one of the most prized of American fish.

Salmon, Atlantic: Salmon found in Atlantic waters; usually sold smoked.

Salmon, chinook: Salmon found in the Pacific Northwest from May to October.

Salmon, chum: Salmon found in the Pacific Northwest from August to October. It is of poor quality and is usually canned.

Salmon, coho: Salmon found in the Pacific Northwest from June to September.

Salmon, pink: Salmon found on the Pacific coast from June to November. It has a pink, oily flesh and is usually sold canned.

Salmon, sockeye: Salmon found in the Pacific Northwest; abundant in the summer months. It has a pink, oily flesh.

Sardine: Fish with dark, oily flesh, found on both coasts. Most sardines in the United States are canned.

Sea bass, black: Bass with white, flaky flesh; abundant in the mid-Atlantic states in spring.

Sea bass, white: Bass with white, flaky flesh; found on the Pacific coast from May to September.

Sea trout: Fish with delicate, lean flesh; abundant on the Gulf coast during spring and fall.

Shad: Fatty, bland fish found in northeastern states and Alaska. The roe is highly prized.

Smelt: Small, oily fish. Lake smelt are abundant in the Great Lakes during March and April. Sea smelts are found in the Gulf and mid-Atlantic states.

Sole, gray: Fish with firm, white flesh; found in the Gulf and Atlantic states.

Sole, lemon: Fish with firm, white flesh; found in the Gulf and Atlantic states.

Sole, petrale: Fish with firm, white flesh; found in the Pacific states.

Swordfish: Fish with firm, medium-oily flesh; found in the tropical waters of the Pacific and Atlantic.

Tilefish: Fish with firm, light flesh; abundant in the mid-Atlantic and northeastern states.

Tuna, albacore: Tuna with light flesh; abundant in Californian and Mexican waters.

Tuna, yellowfin: Tuna with light flesh, but not as light as albacore. It is abundant in California and waters south of California.

Whitefish: Flaky, white freshwater fish; abundant in the Great Lakes from May to August.

Whiting: Light, nonoily fish found in northeastern and mid-Atlantic waters. Most abundant in the summer months.

Yellow perch: Fish with firm, white flesh; abundant in fresh water areas from April to November.

Fish stock—Thin liquid rendered from simmering fish bones in water with vegetables and seasonings for about 45 minutes.

Fixed costs—Cost of running an establishment that is the same amount regardless of the quantity of production. The manager's salary, for example, normally remains the same day to day regardless of the sales volume.

Flakiness—Extremely crumbly texture of a finished baked product.

Flank—*See* Beef.

Flounder—*See* Fish.

Flour—Fine, soft powder resulting from sifting and grinding the meal of a grain, particularly wheat. The following are among the more common types.

All-purpose: Mixture of hard and soft wheat flour; used for all cookery except fine cakes.

Bread: Made from hard wheat flour; it has a high gluten content.

Cake flour: Made from soft wheat; contains 7 to 8.5 percent protein.

Cornmeal: Meal ground from yellow or white corn; used in breads and puddings.

Cracked wheat: Coarse flour made by cutting rather than grinding the whole wheat grain.

Family flour: Soft and hard wheats blended to make an all-purpose flour.

Graham: Made from unsifted whole wheat.

Hard wheat: Made from wheat with a strong, high-gluten quality; suitable for bread baking.

High-ratio: Very fine and powdery, this flour can be combined with up to twice as much sugar as other flours, as in cake baking.

Instant: Flour that has been processed so that it has a granular texture and dissolves easily in hot and cold liquids.

Middlings: Coarse flour made from the grain's outer cost.

Pastry or Cake: Made from soft wheat flour.

Potato: Made from potatoes. This flour is a good thickening agent and is used in baking combined with other flours.

Rye: Dark flour made from rye grains. It is used to make pumpernickel or is combined with wheat flour to make rye bread.

Self-rising: Contains salt and baking powder so that the dough rises without requiring yeast.

Semolina: Grainlike portions of wheat remaining in the bolting machine after fine flour has been sifted through. It is a hard wheat flour ideal for making pasta.

Soft wheat: Made from wheat with low gluten and a high starch content. It is suitable for use in making pastries and cakes.

Strong flour: Hard wheat or bread flour, so called because of its strong quality of gluten or protein.

Whole wheat: Dark flour that contains the whole grain, including the germ, endosperm, and bran. This flour is often stone-ground and unsifted. It is thought to be the most nutritious of flours.

Fluorine—Mineral that helps make and maintain strong teeth and bones. It is found in fluoridated water, flesh, and plant foods.

Focused groups—Meetings conducted with the potential patrons to identify specific likes and dislikes about an existing or planned foodservice.

Folacin—Also known as folic acid, this B vitamin is found in liver and dark-green vegetables. It is important in the synthesis of amino acids and the purines and pyrimidines that appear in nucleic acid.

Food cost—Cost to an establishment of the food it sells to the customers.

Food cost percentage—Ratio of cost of food to the revenue it produces upon resale, multiplied by 100 (i.e., food cost divided by food sales × 100%).

Foodservice industry—Dynamic, ever-growing industry, with over $300 billion in sales, that can best be understood by visiting the Web site *www.restaurant.org* (National Restaurant Association).

Forecast—Prediction of the amount of future sales volume and/or profit based on current trends.

Formal buying—Competitive buying in which several vendors are invited to submit bids in writing.

Four-tops—Freestanding tables in a dining room that seat four people.

Free fatty acid—Any fatty acid that is not linked to a glycerol molecule in a fat or oil. Such fat breakdown indicates deterioration in a fat.

Free-vent—Placement of food over steam to allow it to cook.

Freeze-dry—Item frozen and placed into a chamber where a vacuum is created. A small amount of heat is applied and the moisture goes from ice to water vapor without turning to water. *See also* Sublimation.

French service—Elaborate service in which the kitchen does only part of the work. The final preparation and apportioning is done at the table in the dining room.

Fricassee—French for a braised product.

Fritter—May be either fruit that is dipped into batter and fried or fruits and other items mixed into a muffin batter and deep-fried.

Front of the house—Customer's area in a foodservice establishment (i.e., the dining room).

Fryer—*See* Poultry.

Frying—Method of cooking in fat.

Fumet—Rich essence reduced to almost a glace and thinned with sherry or Madeira wine.

Functional flow design—Also known as *process method of design,* this system is used when many and various products are made to order. It assumes that numerous unrelated products are being prepared simultaneously in different areas of the work center.

Functional organization—Organizational structure set up on a functional basis such as one division for sales, another for planning, another for procurement, and so on.

Function sheet—List of specific details (type and number of entrées, wines to be served, etc.) regarding banquets to ensure that they run smoothly.

Functions of management—Management's job to establish goals, plan how to reach them, and organize, staff, lead, and control so as to reach them.

Fundamental accounting equation—All transactions relate to how assets, liabilities, and capital are recorded in accounting.

Futures-and-contract buying—Type of formal buying in which contracts are established in the present for goods to be delivered in the future. Since the price is set in the present, the establishment can predict related costs for the duration of the contract.

Garde-manger—French for the place in the kitchen where food is stored and basic preparatory work is done before the food's removal to the kitchen to be cooked. It is also used to designate the person in charge of this work.

Garnish, salad—Decoration or "icing on the cake" of an otherwise plain salad, offering contrast in color, flavor, and texture as well as harmony with the rest of the salad.

Gel—Short for *gelatin,* a protein derived by boiling beef bones, cartilage, hooves, tendons, and similar animal tissues. It is used to make jellylike dishes that are known as *gels* and combines well with other foods since it adds no flavor or color of its own; a term applied to a gelled mass.

Gender equity—Fair treatment of both male and female workers, recognizing that the job to be done can be accomplished successfully by either.

Genetic engineering—Scientific process of introducing genes from various organisms into crops and livestock to ensure the continued productivity of agriculture and forestry.

Gilt—Young female hog that has not had a litter.

Glace de poisson—Reduction of a fish stock to a paste consistency.

Glace de poulet—Reduction of a chicken stock to a paste consistency.

Glace de viande—Reduction of a brown or white stock to a paste consistency.

Glace de volaille—Reduction of a chicken stock to a paste consistency.

Glaze—Reduction of a stock by one-fourth of its volume.

Glucose—Most elementary chemical form of sugar. Glucose in the bloodstream is the primary bodily source of energy, hence also called *blood sugar.* In food products it is often called *dextrose.*

Going concern—Concept based on the assumption that a business will continue to operate for an indefinite period of time.

Goose—*See* Poultry.

Government regulation—Federal, state, and local mandates that influence decisions made in day-to-day operations.

Grade—Indication of the quality of meat or some other food product, used to differentiate quality levels in market products.

Grading, quality—Done on the basis of conformation or yield of the meat as judged by the size of the rib eye and other conformation factors, the amount of flesh marbling, the color of the lean, the texture and appearance of the flesh, age, and so on, for beef, veal, calf, lamb, and mutton.

Grading, yield—Done on the basis of flesh to bone and fat on beef, veal, and calf.

Green tea—Nonfermented tea.

Griddle-broiling—Cooking on a hot, dry surface and removing liquid fat from the food item as it accumulates.

Griddlecake—Another name for a pancake or hotcake.

Grilling—Cooking on a flat heated surface or griddle at moderate temperatures (i.e., 325–400°F).

Gueridon—In French service, a side table, usually on wheels, at which the waiter prepares food at tableside.

Gumbo—Heavy soup containing seafood or chicken, vegetables, rice, and okra, a vegetable of distinct flavor and some thickening power.

HACCP (hazard analysis critical control points)—Program to ensure food safety until the food is consumed by the patron.

Half-and-half—Commercial mixture of half 18 percent cream and half 3.5 percent milk.

Halon gas—System that draws away and consumes the oxygen in an area, thus eliminating fire.

Ham hock—*See* Pork.

Heart of the house—Employee's domain of an operation, revolving largely around the purchasing of food, its storage, and preparation. Also called *back of the house*.

Heifer—Young female cattle that has not had a calf.

Hepatitis—Inflammation of the liver. The type called *infectious hepatitis* can be caused by drinking or eating food (e.g., oysters) infected with the responsible virus or ameba.

Heuristic—Attacking a computer problem by the trial-and-error method.

Hollandaise—A basic sauce known as the *queen of sauces* for its richness of egg yolks blended with clarified butter and seasoned with lemon juice and a dash of cayenne pepper.

Home-meal replacement—Involves operations offering a high-quality meal that is eaten away from the restaurant or foodservice facility.

Homogenization—Process whereby the fat globules in milk are divided so that they remain suspended in the product instead of rising to the top.

Horizontal mixer—Large good mixer used in high-volume bakeries and kitchens. It consists of a horizontal cylinder in which a rotating shaft is driven by a heavy-duty motor.

Hors d'oeuvres—French for bite-sized morsels often picked up by the fingers and eaten as appetizers.

Human resources—Department within a foodservice operation which manages people.

Human resources cycle—Three steps of people development: selection, performance, and appraisal.

HVAC—Heating, ventilating, and air conditioning.

Hydrogenation—To add hydrogen to vegetable oils to make them more saturated and firm (i.e., solid), as in the production of margarine.

Iceberg lettuce—Pale green leaves in a tight head. Iceberg has a crisp texture and a bland flavor.

Image—Combination of the elements that make up a restaurant, including food, service, menu, location, price, market attractiveness, the personality and philosophy of management, and how this combination is viewed by those outside the establishment (i.e., potential and actual customers).

IMPS (institutional meat purchase specifications)—Number that identifies a cut of meat, its size, limits the amount of its surface fat, and so on.

Infection—Illness caused by the ingestion of a food containing the infection-causing organism.

Informal buying—Handling purchasing negotiations orally, either in person or over the phone.

Infrared—Electromagnetic energy with wavelengths just below those of visible red light. The radiant energy they supply can be used for cooking (e.g., in quartz broilers).

Ingredient file—List of all ingredients needed for recipes to be prepared.

Ingredient room—Method of planning production as compared to the production sheet method or laissez-faire method.

Internal recruitment—Process of selecting from those within the operation to fill open positions.

Internal security system—Measures taken by management to secure items of value within an operation including a safe, locks, and so on.

Internet—Electronic transmission of information available to the community via the personal computer.

Intoxication—Poisoning caused by the ingestion of a substance that is toxic (e.g., excessive alcohol, bacteria, or mercury in contaminated fish).

Inventory extension—Practice of calculating the value of inventory held by counting what is contained in inventory and multiplying that by the price paid for them.

Invoice—Form that lists and describes goods ordered and their cost. It should accompany their shipment.

Invoice receiving—Receiving technique in which actual goods received are compared with what the accompanying invoice says should be in the shipment.

Iodine—Element essential for proper thyroid function. Since many areas lack natural sources (which include seafood), iodine is frequently added to table salt during processing. Deficiency results in goiter.

Iron—Mineral essential to health. It plays a major role in the transfer of oxygen in the body, among other important functions. Too little can cause anemia, too much can cause hemochromatosis. Found in various foods such as liver and cereals.

Irradiation—Uses special light rays to destroy contaminating substances on meats and poultry.

Issues—Food that is placed into storage and later called from storage to production.

Issuing—Supplying food from the storerooms to the cooks and other production personnel, usually by requisition.

Japanese roulette—"Game" sometimes played in Japan of eating puffer fish, which is safe only if it is prepared correctly—otherwise, it is deadly poisonous.

Job descriptions—Statements of the tasks employees must perform in completing their jobs.

Job design—Involves not only the tasks to be accomplished but the successful scheduling and staffing of employees to accomplish what is to be done. The employee must participate in his or her job design.

Jowl—*See* Pork.

Jus lie—Very slightly thickened au jus, used in much the same way as a sauce over meat dishes.

King (Alaskan) crab—*See* Shellfish.

Kitchen exhaust—System installed in an operation to remove hot air, smoke, grease, and odors.

Labor cost—Total cost of all employees who comprise the workforce of an operation.

Labor schedule—Having the correct numbers of employees at work at a given time results in proper labor costs.

Lamb—Young sheep under a year old.

Lean dough—Less rich mixture used to make some rolls and breads.

Lean fish—Those containing less fat than others.

Legumes—Beans, lentils, peas, and other seeds of leguminous plants. These vegetables contain high-quality proteins that include all eight essential amino acids.

Liability—Responsibility that one person has to another that is enforceable in court.

Liaison—One part egg yolk to three parts cream, providing added flavor, smoothness, and body when added to soups and sauces.

Limited menu—Menu that offers only about six to twelve entrées, to save on labor costs and food waste. Usually only the most popular items are included.

Line organization—Lines of authority and responsibility flow from the head down through the various mid-management layers to the workers.

Lipid—All fats and related compounds (e.g., wax) that contain fatty acids. Lipids are found in all living cells.

Littleneck—Type of clam. *See also* Shellfish.

Lobster, chicken—Lobster that weighs about 1 lb.

Logo—Distinctive symbol that identifies a particular restaurant and may express its theme (e.g., a coat of arms or a photograph of some part of the interior).

Long-grain rice—Rice grains over 6 mm in length; stay light and fluffy when cooked. This is the best kind of rice to accompany stews and other meat dishes.

Low-temperature storage—Frozen-food storage at temperatures of 0°F (–18°C) or less.

Magnesium—Metallic element essential for the proper functioning of muscles, nerves, and certain enzymes. It is found in chlorophyll and is therefore present in green plants.

Main-dish salad—Type of salad served as the main course to a meal, as compared to an accompaniment salad, dessert salad, and so on.

Maître d'hôtel—French for the master of the house referring to the person who is in charge of the dining room in a food-service facility. Also known as the "maître d."

Maître d'hôtel butter—Butter that has been melted with a bit of lemon juice and chopped parsley.

Makeup—Shaping of bread or pastry dough into the form desired.

Manganese—Metallic element essential for the functioning of certain enzymes and the synthesis of protein, among other things. Best sources are whole grains, peas, and some leafy vegetables.

Marbling—Thin streaks of fat running through meat. Marbling increases tenderness as well as flavor.

Marinating—Process of soaking food in an acidic solution of vinegars and spices to tenderize meats, making them more palatable, moist, and flavorful.

Market—Region in which an establishment's products and those of its competitors are sold.

Marketing—Process of determining who the customers are and what they want.

Market research—Study or process of learning who the customers are, what they want, and how to meet their needs.

Mayonnaise—Permanent emulsion of acidic liquid (such as vinegar or lemon juice), oil, and egg yolk or whole eggs, with added seasonings.

Mealiness—Fine (and not flaky) texture of a finished baked product.

Mechanical oven—Gas or electric oven in which food is cooked on trays that move through the heated cavity on a belt or a wheel. Several round trips may be needed before the oven is ready to be unloaded. Mechanical ovens have a large capacity and are energy and labor efficient.

Menu analysis—Process of collecting data about menu sales in order to make decisions as to which items to keep and which to replace.

Menu clip-ons—Small ads attached to menus to entice patrons to buy.

Menu mix—Balance of items from different categories on a menu. Based on price, the menu mix might range from expensive to inexpensive, for example.

Menu programming—Use of a computer in menu planning, preparation, and management.

Merchandising—Presentation of factors that entice people to buy.

Meringue (hard)—Made in up to a 1:2 ratio of egg whites to sugar by weight.

Meringue (soft)—Made in a 1:1 ratio of egg whites to sugar by weight.

Mesophiles—Bacteria that thrive in moderate temperatures.

Microorganism—Bacteria, molds, viruses, protozoa, and other microscopic forms of life.

Microwave—High-frequency electromagnetic energy: very short waves of energy that when broadcast at food, make its molecules vibrate, creating friction and heat within the food that then cooks the food from the inside out.

Microwave oven—Oven in which food is cooked by exposure to high-frequency electromagnetic radiation.

Milk, dry nonfat—Nonfat milk evaporated until only the solids remain.

Milk, sweetened condensed—*Evaporated milk* that contains 45 percent added sugar.

Mirepoix—Seasoning for a stock consisting of 50 percent onions, 25 percent celery, and 25 percent carrots.

Mise en place—French for the orderly progression of work within a kitchen, such as organizing beforehand, getting all materials needed, and then keeping things organized, cleaned up, and moving toward the final goal.

Mollusk—Shellfish with soft bodies that are partially or wholly enclosed in hard shells of mineral composition, including clams, oysters, and scallops.

Molybdenum—Metallic element essential in small amounts for normal functioning because it interacts with the enzyme xanthine oxidase, which helps control metabolism. Too much molybdenum, however, is poisonous.

Money and cost concepts in accounting—All items are recorded at cost, based on the assumption that items recorded at their cost value are less subjectively valued than items recorded at a relative value such as a market appraisal value.

Muffin—Made from a rather thick batter that is produced by blending flour, salt, sugar, and baking powder with liquid.

Mulligatawny—National soup of India.

Multiplier method—Handy rule of thumb to calculate selling prices to be placed on a menu.

Mutton—Sheep over a year old, usually castrated.

Myoglobin—Red iron-containing protein in meat that turns brown when heated, thus accounting for the color change in meat during cooking.

Myoinositol—One of the B-complex vitamins which probably performs metabolic functions. It is found in organ meats, brains, whole-grain cereals, and yeast.

Natural flavoring—Flavoring derived from plant or animal products rather than from synthesized chemicals.

Needs analysis—Objective is to seek information about an operation and its personnel to determine discrepancies between desired and actual performance levels. The analysis results in human resource needs being identified and met.

Negligence—Act of not fulfilling one's part of the bargain in a situation. The result in business is a lawsuit against the operation or management, or both.

Negotiation—Discussion of price between buyer and potential seller in order to reach an agreed-upon selling price for an item or items.

Niacin—B vitamin important for the metabolism of carbohydrates. Insufficient amounts cause pellagra. Natural sources include liver, meat, and cereal germ.

Nutrition—Study of food in relation to dietary needs and normal bodily functions.

Olla podrida—National soup of Spain.

Olympia—Type of oyster

1,2,3 method—Preparation technique used to make ice tea.

Oolong tea—Type of tea that is partially fermented.

Open-face—Type of sandwich presented with no top slice of bread, meant to be eaten with a knife and fork.

Organization—Arrangement of people in jobs designed to accomplish the goals of the operation.

Orientation—Initial overview of the job, providing the employee with a solid understanding of what he or she must do.

OSHA (Occupational Safety and Health Act)—Federal law that established an agency to inspect U.S. businesses, including foodservice establishments, to assure healthful and safe working conditions for employees.

Ovenizing—Process of frying in the oven.

Oxymyoglobin—Combination of oxygen and myoglobin (muscle protein) that is responsible for the bright-red color of meat exposed to air. Meat not so exposed is a darker red.

Pan-broiling—Cooking on a hot, dry surface in which conduction, not radiation, is the process of heat transfer. Any fat present during cooking is discarded before it can accumulate.

Pantothenic acid—B vitamin needed to metabolize fats and carbohydrates. Since it is found in all plant and animal tissues, deficiency is unlikely. Especially rich sources include eggs, organ meats, certain cereals, and vegetables.

Papain—Enzyme derived from the papaya plant that breaks up the protein chains in meat and is used as a tenderizer.

Parasite—Harmful organism that dwells in or on a host, drawing nutrients from it and thus weakening it.

Parboiling—Cooking food in boiling water until it is partially done; similar to *blanching,* but the item is in the water for a longer period of time.

Par stock—Maximum amount of an item that should normally be on hand. When ordering new supplies, the amount ordered should be enough so that when it is added to existing inventory the total reaches the par level.

Partial-blind receiving—Combination of invoice receiving and blind receiving. The receiving clerk gets a partially filled-out invoice or purchase order with the names of items listed but the amounts missing. He or she must count or weigh the items and complete the form.

Passover—A Jewish holiday in which special foods are served.

Pasta—Paste made from hard wheat flour, rolled into various shapes and sizes. Popularized in Italy, pasta is high in carbohydrates and is often served with a sauce. The following are the most popular types.

Agnolotti: Turnover-shaped pasta that are filled with meat.

Ancini di pepe: Literally, peppercorns. Very small rectangular-shaped pasta used in soups.

Bocconcini: Literally, small mouthfuls. Tubular pasta ½ in. in diameter and 1½ in. long.

Bows: American egg pasta shaped like rounded bows.

Bucati: Literally, with a hole. Hollow pasta.

Bucatini: Hollow pasta similar to macaroni but thinner.

Cannelloni: Large, hollow pasta served stuffed with cheese or meat fillings.

Capelli d'angelo: Literally, angel's hair. Very fine spaghetti-shaped pasta.

Capellini: Very thin spaghetti-shaped pasta, most often served in soups.

Cappeletti: Literally, little hats. Hat-shaped pasta served stuffed.

Cavatelli: Short, curled noodles.

Diali: Small, thimble-shaped pasta.

Elbow: Semicircular, hollow pasta in various sizes.

Farfalle: Literally, butterflies. Similar to American bows but not made of egg pasta.

Fettuccine: Literally, small ribbons. Wide spaghetti-shaped pasta.

Fusilli: Spaghetti twisted into corkscrew shapes.

Gemelli: Short spaghetti strands twisted together.

Giant shells: Large, grooved, shell-shaped pasta.

Gnocchi: Dumplings made from pasta dough.

Lasagne: Wide, flat pasta with rippled edges.

Linguine: Literally, small tongues. Narrow, thick spaghetti.

Macaroni: General term for tubular pasta.

Manicotti: Literally, small muff. Large, tubular pasta often served stuffed with meat or cheese fillings.

Orzo: Small rice-shaped pasta.

Pappardelli: Broad noodles.

Penne: Tubular pasta cut on a diagonal to resemble quills.

Rigatoni: Large, grooved pasta tubes.

Rotini: Wheel-shaped pasta.

Spaghetti: Solid, round, long, and thin pasta.

Tagliatelli: Similar to fettuccine but wider.

Tubetti: Small tubes, often used in soups.

Vermicelli: Literally, little worms. Very thin spaghetti.

Ziti: Literally, bridegrooms. Large, tubular-shaped macaroni.

Pasteurization—Use of heat to kill most of the bacteria in a product.

PDQ—Peeled, deveined, and quick-frozen, referring to shrimp.

Peeler—Piece of equipment that peels hard root vegetables with a lightweight spinning abrasive disk.

Performance appraisals—Communication of what is expected of both management and the employee and whether or not that job is being accomplished. Compensation, promotion, and career development are all considered in the process to be successful.

Perpetual inventory—Continuously updated accounting of what is in storage without having to take a physical count.

pH—Abbreviation for potential hydrogen, a 15-point acid–alkalinity scale in which 0 = extremely acid, 7 = neutral, and 14 = extremely alkaline.

Phosphorus—Mineral found in all human tissues, especially bones and teeth. It is found in so many foods that deficiency is unlikely.

Physical inventory—Count of each item physically on the premises.

Planned profit method—Method of establishing selling prices by figuring in a desired level of profit as well as all costs, and then calculating the price accordingly.

Plate—*See* Beef.

Poaching—To cook food in lightly bubbling water to prevent overcooking.

Pork—Fresh meat of pigs. Pigs are usually slaughtered at five to six months. Good-quality pork has a soft gray-pink color, a fine grain, and white fat. The best pork cuts are from the loin, leg, and shoulder. Pork should be properly cooked to prevent trichinosis. Standard primal cuts are as follows:

Boston butt: Portion separated from the shoulder by a straight cut parallel to the breast side. The cut should have a slight amount of marbling.

Crown roast: Entire pork loin, usually served filled with mashed potatoes.

Ham hock: Ankle joint and its associated meat.

Jowl: Meat from the cheek and neck region; one source of bacon.

Pork fresh ham: Leg portion separated from the side by a straight cut perpendicular to the skin surface to a line parallel to the shank bones.

Pork loin: Back cut remaining after removal of the shoulder, ham, belly, and fat back. This cut yields the tenderloin, chops, cutlets, roasts, and ribs.

Pork shoulder: Portion separated from the side by a straight cut perpendicular to skin surface. The pork shoulder yields the pork shoulder, fresh shoulder, butt, boneless shoulder butt, picnic pork, and shoulder steaks.

Pork fresh ham—*See* Pork.

Pork loin—*See* Pork.

Portion—Measured serving size of a meat, fruit, or vegetable.

Portion cost—Cost per serving, derived by dividing total yield by portion size to determine the number of portions. The number of portions is then divided into the extended cost (total cost) to get the portion cost.

Positioning—Process of enticing patrons to buy one menu item over the other depending on its position on the menu.

Potassium—Mineral important for muscular and nervous function. It is found in most foods, although bananas and oranges are especially rich sources. Deficiency can directly affect the action of the heart.

Poultry—Domesticated species of birds, including chickens, ducks, geese, and turkeys, bred for their eggs or for the table.

Chicken: Chickens are high in protein and vitamin B. It is among the most versatile of meats with innumerable methods of preparation. They are graded by age and size as regulated by law.

Broiler: Young bird that is four to six weeks old and weighs up to $2\frac{1}{2}$ lb. The French word for broilers is *poussin.* They are usually cooked whole and have little flavor or flesh.

Capon: Young castrated cock that has been fattened and weighs between 6 and 10 lb.

Cock: Mature male chicken with dark, tough meat.

Fryer: Bird that weighs as much as $3\frac{1}{2}$ lb. with tender, tasty flesh. Fryers are also known as spring chickens.

Roaster: Bird that weighs $3\frac{1}{2}$ to 5 lb; roasters have the tastiest flesh.

Stewing hen: Female bird ranging in age from ten months to more than a year and a half old.

Duck: Swimming bird of which there are 60 to 70 species. Most domesticated ducks are descendants of the mallard or common wild duck of North America. The meat is darker and more tender than that of the chicken. In the United States, the Long Island duckling is the most favored.

Goose: Aquatic bird with dark, tender flesh and a gamey flavor. The goose is more fatty than chicken or turkey. The most common types are Embden, Toulouse, Chinese, and Canadian.

Turkey: Bird native to North America and one of our most valued table birds. The meat is lean and mildly flavored. The most popular breeds include the Bronze and the Bourbon Red. Turkeys average 12 to 14 lb.

Precosting—Using sales forecasts and ingredient costs to calculate food costs for a given menu item in advance of sales.

Preserve—Fruit served in an extraheavy syrup.

Preventive maintenance—System that stresses proper regular care and servicing of all aspects of the foodservice property.

Pricing factor method—Method for establishing selling prices based on separating foods that require little or no preparation from those that require a lot, and then allocating labor, energy, and other costs accordingly.

Primary market—Large central markets that set the price of a commodity and where supply and demand are largely controlled.

Prime—*See* Beef.

Processed foods—Foods prepared before packaging to save time in labor or preparation. Also known in the past as *convenience foods,* or *value-added* more recently.

Profit—Amount by which income exceeds costs.

Profit-and-loss statement—Financial statement for a given period that itemizes revenues, expenses, and the resulting profit or loss.

Profit-sharing—Act by management to return to employees a portion of the profit realized by an operation.

Proof—Holding the dough at a controlled temperature and humidity to promote the fermentation process prior to baking.

Protein—One of the three principal food components in addition to carbohydrates and fats. Each protein is composed of a long chain of amino acids, of which there are twenty-two. The exact sequence and number of amino acids determine the nature of the protein. Some proteins serve as the building blocks of all living cell structures; others function as enzymes.

Protozoa—Microorganisms composed of a single cell. Many of these one-celled creatures (e.g., amoebas) can cause disease.

Pulper—Machine that takes all kinds of waste material and breaks it down under water into small pieces. The water is then extracted and the particles are discharged into a container for easy removal. Nonpulpable material, such as cans and silver, is automatically rejected. This machine has the advantage of cleaning and deodorizing waste as well as reducing its volume.

Pumpernickel—Bread that uses all of the rye kernel and is somewhat heavy and compact because it lacks gluten.

Punch—Folding dough over from its sides into the middle and literally punching it down during leavening to redistribute the yeast.

Pupa—Stage in the metamorphosis of some insects between the larvae and adult stages; characterized by a cocoon holding an essentially motionless immature insect.

Purchase order—Form used to order supplies for the kitchen or bar. Copies go not only to the supplier but also to the receiving department and accounting departments.

Purchasing invoice stamp—Rubber stamp used to print on invoices a short form that can be filled in by the receiving clerk and others as needed. The form can include lines for the date received, the quantity received, the price, and the date paid.

Purée—Thick or thin soup of vegetables, legumes, or fruit, which are cooked, strained, and added to the desired base, either a plain or thickened stock. Neither milk nor cream are added to purées.

Purveyor—Seller or vendor.

Pyridoxine—Known as B_6, this vitamin is required for amino acid and nervous-system functioning. Deficiency, although rare, produces such symptoms as irritability and convulsions. Rich sources include meat, herring, whole grains, bananas, and potatoes. It is a water-soluble vitamin.

Quality control—All procedures designed to ensure that products for sale are of satisfactory quality.

Quotation and order sheet—Sheet on which items needed by an establishment are listed, along with the prices quoted from several leading suppliers. The sheet allows price comparison.

Rack conveyor—Dishes are moved on racks through the wash and rinse compartments by a pair of motor-driven chains.

Radiation—Movement of heat in waves through an available space (e.g., from white-hot broiler coils to meat cooking under the broiler).

Ram—Mature sexually active male sheep.

Rancidity—Chemical breakdown of fats caused by oxidation, resulting in spoilage.

Raw food cost—Total dollar cost of food to be prepared.

RDA (Recommended Dietary Allowance)—Guide to the amounts of foods needed daily to ensure the adequate intake of needed nutrients.

RDI (Reference Daily Intake)—Food and Drug Administration list used as the basis for judging a day's proper allowance.

Ready-to-cook—Referring to meat, poultry, and seafood totally dressed for cooking.

Rechaud—French for chafing dish mounted on a rolling cart, or *gueridon,* used to prepare special dishes and keep food warm.

Recipe file—Collection of standardized recipes whereby one may be pulled to support the next, as is the case in preparing French onion soup, when one must pull the recipe for a beef stock in order to make the soup.

Reconstituting oven—Equipment that can quickly defrost frozen foods without impairing their quality. Some models combine cycles of heat and refrigeration to reduce drying on the edges while gently thawing the center of the food. Other models, using infrared radiation, may reach temperatures as high as 850°F and are faster.

Recruitment—Process of selecting employees successfully.

Reel oven—Mechanical oven in which food is cooked on trays that revolve in Ferris-wheel fashion. Food is loaded and unloaded one tray at a time through the oven's door.

Relish—Typically, a combination of raw fruits and vegetables.

Request-for-credit memorandum—Receiving control form that lists shortages in quantity, nonconformance of the quality to specification, or any other discrepancy in order to request credit.

Requisitions—Forms that can be used by any department within an establishment to request supplies from the storeroom.

Rest—Allowing dough to relax as the yeast is allowed to work.

Retail—Markets that purchase from wholesalers or brokers and then sell to users.

Rib—*See* Beef.

Riboflavin—Known as B_2, this vitamin is required for the proper metabolism of protein, fat, and carbohydrates. Deficiency causes ariboflavinosis. Rich sources include whole grains, milk and milk products, green vegetables, and lean meat. It is a water-soluble vitamin weakened by sunlight.

Ripening (aging)—Holding meat for 10 to 14 days at specific temperatures and under controlled relative humidity to improve its quality.

Roaster—*See* Poultry.

Roasting—Dry-heat food preparation method whereby meats and vegetables are cooked in an uncovered pan in the oven in such a manner that the product is raised and will not cook in its own juices or fat. A trivet is frequently placed in the bottom of the pan, for example, to keep meats above their own juices.

Roasts—Larger cuts of meat, poultry, and fat fish that bake well in the oven.

Robusta—One of two varieties of coffee beans.

Rolling pin—Cylinder used in hand rolling of pastry dough.

Romaine—Also known as *cos,* this lettuce has dark green, oval leaves with a pungent flavor.

Rooster—Mature male chicken.

Rotini—Wheel-shaped pasta.

Round—*See* Beef.

Round (whole) fish—Cod, haddock, salmon trout, river trout, pike, salmon, and others.

Roux—Thickening agent for stocks, soups, and sauces consisting of equal parts of butter and flour by weight, cooked together. It may be white, blond, or brown, depending on the cooking time.

Roux, blond—*Roux* cooked for five to seven minutes until light in color.

Roux, brown—*Roux* cooked for ten minutes until nut in colored.

Roux, lean—*Roux* having more flour than fat by weight.

Roux, slack—*Roux* having more fat than flour by weight.

Roux, white—*Roux* cooked for three to five minutes.

RSVP—French for the responsibility for one having been invited to an affair to respond positively or negatively as to whether he or she will attend the event.

Russian service—Serving system in which the waiter or waitress picks up large platters of food in the kitchen and brings them to the table, where he or she divides the food among the diners.

Sachet—Small cloth bag of spices and seasonings (*see* Bouquet garni) that may be added to a stock.

Salad—Consists typically of chilled ingredients placed on an underliner and topped with a dressing and garnish.

Salad bar—Wide range of salad materials in containers, displayed on a refrigerated counter for self-service by the customer.

Salad body—Main ingredient of a salad, from which it usually gets its name, as compared to base, garnish, or dressing.

Salad dressing—Flavorful sauce applied to salads to enhance their appeal.

Salad greens—Following is a list of the most common salad greens.

Amaranth: Plant of over 50 varieties with a high protein content and a spinachlike flavor. It has rounded leaves with red splotches.

Arugula: Plant related to the mustard family. Arugula has small, dark-green notched leaves and a peppery flavor.

Belgium endive: Root of the chicory plant. Endive has pale yellow, boat-shaped leaves and a bitter flavor.

Bibb lettuce: Dark, crisp leaves with a delicate nutty flavor.

Borage: Plant with fuzzy leaves and a cucumber flavor. Borage grows wild in Europe.

Boston lettuce: Also known as *butter lettuce,* this salad green has a loose head with delicate, sweet leaves.

Chicory: Also known as *curly leaf endive,* this plant has curly leaves and a bitter taste.

Cress: Wild plant with green notched leaves and a peppery flavor. Also known as *watercress.*

Dandelion: Weed with leaves that have a tart flavor and are high in iron.

Endive: Salad green similar to chicory.

Escarole: Plant of the chicory family with broad, flat leaves and a slightly bitter taste.

Fiddlehead fern: Young shoots of ferns with a spiral shape and a nutty flavor.

Iceberg lettuce: Pale green leaves in a tight head. Iceberg has a crisp texture and a bland flavor.

Looseleaf: Group of lettuce including red leaf and green lettuce. Looseleaf lettuce has a mild, sweet flavor and tender leaves.

Mâche: Also known as *lamb's lettuce,* this is a wild green with small, round leaves that have a mild flavor.

Mustard greens: Plant of the mustard family with leaves that have a peppery flavor and a high vitamin A and B content.

Oyster plant: Gray-green leaves with a fresh oysterlike flavor. Also called *salsify.*

Radicchio: Plant of the chicory family with a compact head and red leaves. Radicchio has a tangy, bitter flavor.

Rape: Leaf plant with a mild flavor.

Romaine: Also known as *cos,* this lettuce has dark green, oval leaves with a pungent flavor.

Salad burnet: Plant with round, serrated leaves and a cucumber flavor.

Sorrel: Wild herb with a sour taste and a high vitamin C content. Types include mountain and french sorrel.

Spinach: Plant with bright green, medium-sized leaves and a mild, musky flavor.

Swiss chard: Plant of the beet family with leaves that have a mild flavor.

Salmon—*See* Fish.

Salmonella—Bacterium that commonly causes salmonellosis, a food poisoning. Cooking food long enough at high temperatures kills the bacteria.

Sanitary—Absence of all pathogens, poisons, and other harmful creatures or substances; thoroughly clean.

Saturated fats—Fatty acids in which each carbon atom in the chain is attached to the maximum number of hydrogen atoms. It is believed that such fats should be limited in the diet.

Saucier—French for the chef in charge of making sauces.

Sautéing—French for cooking in shallow fat from 335 to 425°F (168 to 223°C).

Scalding—Dipping a food into *boiling* water or allowing a hot water bath to strike the food for a very short time. Scalded milk is milk brought just to boiling and then taken off the flame to reduce its temperature rapidly.

Scaling—Process of weighing dough for *makeup.*

Schematic design phase—First phase of foodservice design, whereby the architect, designer, and owner develop the general scheme of the foodservice.

Scramble system—Cafeteria in which each different type of food is laid out on a separate counter. The customer walks from counter to counter, choosing what he or she wishes.

Seafood—Term that includes both fish and shellfish.

Sea scallop—*See* Shellfish.

Seasonality—Changes in income or expenses related to the time of year. For example, fresh fruit costs more in the winter; income in resort areas goes up during the tourist season.

Secondary market—Locations to which most of the products handled by the primary market will flow, and out of which they will be shipped to local markets.

Security—Prevention of access to valuables by those unauthorized to use or obtain them.

Select—*See* Beef.

Selenium—Mineral required by the body in trace amounts. It plays a role in several functions, including antioxidation, many enzyme systems, cell integrity, and antibody production. Excessive amounts can be toxic. Food sources include organ meats, seafood, whole grains, and food grown in regions where the soil contains an abundance of the mineral.

Selling prices—Guiding concepts used by managers to decide on prices. It may, for example, depend on a set markup over costs.

Semi-á la carte—Menu that offers the basic meal (e.g., an entrée with salad and vegetable) for a single price and lists extra charges for such things as appetizers and desserts.

Sequence—Order in which food items are listed on the menu.

Sexual harassment—Act by an employee offending another sexually.

Shank—*See* Beef.

Shellfish—Edible fish with shells; crustaceans or mollusks. Shellfish are high in protein, vitamin B, iodine, and mineral salts. The following are the more popular types.

Abalone: Marine snail found in the tropical waters of the world. In the United States, abalone is found on the Pacific Coast. The mussel has a rubbery texture.

Alaskan king crab: Crab related to the hermit crab, found in waters of the Bering Sea. They grow as large as 24 lb. Only the meat of the claws, legs, and shoulders is usually eaten.

American lobster: Also known as the *Maine lobster,* this shellfish is found from Canada to North Carolina. The average weight is 1 to 5 lb. Any over 3 lb is called a *jumbo lobster.* The lobster harvest has declined in recent years.

Blue crab: One of the most abundant crustaceans, found in the Atlantic from New Jersey to Florida. Blue crabs are sold in hard shell or soft shell.

Clam: Bivalve shells encase the edible body, consisting of a muscle and a neck. The most popular Atlantic species include hard clams or quahogs (also called cherrystone), surf clams, and softshell clams. The Pacific coast clams include razor clams, butter clams, littleneck clams, and geoducks.

Crayfish: Small, lobsterlike crustaceans particularly abundant in Louisiana. There are over two dozen species, with the red swamp crayfish being the most popular. Also called *crawfish.*

Dungeness crab: Crab with a flattened body and a reddish, spotted shell. This crab is most abundant on the Pacific coast.

Icelandic lobster: Imported species smaller than the American lobster. It has tender white flesh and a red shell that does not change color when cooked.

Mussel: Bivalve mollusk with dark blue shell. Mussels are abundant on the New England coast. They are high in protein and low in fat.

Oyster: Bivalve mollusk of the genus *crassostrea.* The most popular species include the Pacific oyster, the European oyster, the Atlantic oyster, and the Olympia oyster. Oysters are at their peak from October to May and are good sources of iron, iodine, and calcium.

Prawn: European term for large shrimp.

Rock shrimp: Member of the shrimp family with a rigid shell and flesh that is similar to lobster. Rock shrimp are abundant in southwest Florida.

Scallop: Mollusk with fluted shell and sweet-flavored muscles. Sea scallops are larger and less sweet than bay scallops. They are most abundant from April to October.

Shrimp: Small, 10-legged crustacean with a thin shell. The most popular varieties include the white shrimp, the pink shrimp, and the brown shrimp. The Gulf and Pacific coasts are the most abundant sources for shrimp, particularly from January to April.

Snow crab: Crustacean with long legs and a hard shell. They are abundant on the Pacific coast and have a sweet, delicate flavor.

Softshell crab: Crabs that are molting or have lost their shells and have not yet grown larger ones. Crabs in this temporary condition are considered a delicacy, for the soft shell may be eaten as well as the crab.

Stone crab: Crab abundant on the west coast of Florida, particularly from October to April. Stone crabs have very hard shells and pincers with black tips. Only the claws of the crab are eaten for their firm, sweet flesh.

Shield-grade stamp—Stamp in the shape of the federal shield used by the U.S. Department of Agriculture. It is used to certify the federal grade of a product.

Shortcake—Rich biscuits filled and topped with fruit and whipped cream.

Short loin—*See* Beef.

Shrinkage—Decrease in the size of a meat cut after cooking. The amount depends on the type of meat, cut, amount of fat, and cooking temperatures.

Simmering—To cook food immersed in a liquid at a temperature of 185 to 205°F (85 to 96°C), just below a full boil.

Simulation—Practice of one's job and position in training prior to being on the job for real.

Sirloin tip—*See* Beef.

Site analysis—Careful consideration of where the operation is to be located including traffic counts, visibility, accessibility, and so on.

Skills audits—These assess the abilities of management in a qualitative manner by bringing members of management together to discuss performance standards, skills, priorities, behaviors, and task resolutions.

Skills test—Periodic evaluations to determine if the employee has the ability to accomplish the job to be done.

Slope—Land on which the operation is to be built should have been considered for steepness of slope calculated by dividing the rise, or elevation, by the run, or distance.

Slurry—Blend of liquid and starch mixed together to provide thickening at the last minute.

Smoking point—Temperature at which cooking oils and fats begin to decompose and give off smoke.

S motion—Movement one makes when applying spread to sandwiches.

Sodium—Essential mineral required for proper functioning of the nervous system. One sodium atom is found in each molecule of table salt. Sodium, which is also present in other foods, such as shellfish, is believed to contribute to hypertension by increasing the amount of waste in the blood, thus adding to the blood's pressure in the artery. Since contemporary diets often include an abundance of salt and sodium in all forms, excess is far more common than deficiency.

Sodium sulfite—Additive used to bleach and preserve foods such as dried fruit and dried potatoes.

Sole—*See* Fish.

Sourdough bread—Made by adding a fermented starter or old dough to a new one.

Sow—Older female hog that has had a litter.

Spanish sauce—Same as a brown sauce.

Specification—Exact description of the quality and type of goods desired by an establishment. They are given to potential suppliers so that they may quote a price.

Spore—Quiescent state in which bacteria develop thick walls and suspend most life functions. They can then survive extremes of cold or heat, so that, for example, pasteurization does not succeed in killing them, and they can then germinate later under more favorable conditions.

Spread—Process of allowing the cake batter to be distributed evenly within a cake or sheet pan on which it will bake.

Staffing guide—Outline of basic personnel needs for a given period of time.

Staff organization—Support organizations established to render a service to the line organization.

Stag—Sexually mature male cattle that has been castrated.

Standard—*See* Beef.

Standardized recipe—List of each ingredient and the amount needed, along with the cooking procedures, to prepare a given item.

Staphylococcus—Cocci bacteria that are massed together in clusters. They exist in the air and can be found on most objects. Poisoning can result from a certain type that infects food and causes symptoms of nausea, cramps, diarrhea, and vomiting within two to four hours after eating. An absence of fever distinguishes this from many other kinds of food poisoning.

Staple food—Grocery item that can be kept in the storeroom as compared to perishables.

Steak—Slice of beef about ¾ to 1 in. thick. Also, the same for a cut of fish.

Steam cooker—Device that cooks food by the application of moist heat steam at 0 to 15 psi.

Steaming—Cooking with steam as in a steam cooker.

Steer—Young male cattle castrated before sexual maturity.

Stewed fruit—Fruit served in a light or medium syrup. Also called *fruit sauces*.

Stewing—Method of cooking in which small cuts or cubes of meat, poultry, and vegetables are cooked in a little water with seasonings in a pot over a burner.

Stewing hen—*See* Poultry.

Stir-fry—To cook food quickly in a pan at high heat with rapid tossing and stirring motions. The speed usually results in crisp textures and fresh tastes.

Stock—Thin, flavorful liquids derived from simmering meat, fish, or poultry bones with vegetables and seasonings.

Stock record card—Card used to record the quantity of stocks on hand, the amounts received, and the amounts issued. Each item goes on a separate card.

Stone crab—Crab that comes from the southern waters of the Atlantic.

Straight-line cafeteria—Cafeteria in which customers form a line and choose what they want as they walk past an array of foods.

Straight-line design—Products proceed in a line from one area of the kitchen to their final stop in another area.

Sublimation—Vaporization of a solid, when heated, directly into a gas without becoming liquid first. For example, solid carbon dioxide (dry ice) sublimates.

Substitution invoice—Form similar to an ordinary invoice used by the receiving clerk when goods arrive without an invoice.

Subsystem—As it applies to equipment implementation, an area within the kitchen where equipment is found (i.e., receiving, storing, preparation, cooking, serving, and cleanup).

Suggestive selling—Personnel trained to make suggestions to entice patrons to buy.

Sulfur—Element present in certain amino acid molecules such as cystine and certain vitamins such as B. It is essential for life.

Sweet dough—Mixture used to make coffee cakes, pastries, and baked products which might have shortening rolled into them, as is the case with puff paste.

Syneresis—Process that causes separation of the fluid portion from the solid.

Table d'hôte—Complete meal, usually including dessert and beverage, offered for a single fixed price. The opposite of à la carte.

Table tents—Small ads placed on top of the table to entice customers to buy something.

Tank—Name given to any of several sections that make up a dishwashing machine.

Tannin—Group of bitter phenol compounds found in such plant products as coffee, tea, and wine.

Target market—Group specifically identified as the segment of the population the foodservice wishes to attract for a special promotion.

Tartlets—Rich pastry dough cooked in decorative, fluted tins, and filled with fruit mixtures.

Temperature record book—Book or ledger in which the temperatures found in various parts of refrigerated areas can be recorded on a daily basis.

Thiamine—Known as B_1, this vitamin is required for proper carbohydrate metabolism, muscle coordination, and nerve tissue maintenance. Deficiency can cause beriberi. Rich sources include whole grains, cereals, and meat. It is a water-soluble vitamin.

Tilting skillet—Large flat-bottomed kettle, heated by gas or electricity, mounted in a sturdy framework with a mechanism that allows tilting of the kettle up to 90° for easy removal of the contents. It is a versatile piece of equipment and may be used for braising, boiling, griddling, thawing, deep and shallow frying, as well as for holding foods hot. Also called a *tilting fry pan*.

Tom—Male turkey marketed *young,* under a year old weighing 12 to 30 lb, and *mature,* over a year old.

Total quality management—Process used by management to eliminate quality problems before they occur rather than attempting to address them after they have happened.

Tray service—Serving method in which dishes are assembled onto trays so that a complete meal can be taken to a person at once. This method is typical of hospitals, for instance.

Trichinosis—Disease caused by eating undercooked pork or wild game that contains live cysts of the parasite *Trichinella spiralis.* Symptoms include abdominal pain and fever; stiff muscles occur because the worms burrow into muscle tissue. Adequate cooking kills the cysts and makes the meat safe.

Trivet—Platform or holder on which fish are simmered to ensure that they do not break up while cooking or when removed from the pan. Also, a stand or frame used to keep hot pots or dishes off a table surface.

Tureen—French for a deep, broad, covered dish of porcelain or silver in which soup is served. Tureens are usually very ornate and elegantly designed.

Turnover (fat)—Daily addition of new fat to the cooking supply without discarding the old, in order to keep it fresh and in good condition.

UL (Underwriters' Laboratories)—National laboratory that tests electrical equipment and awards its seal of approval only to those that meet its safety standards.

Unit cost—Cost of one unit of a given item. The unit cost is usually stated on the purchase invoice.

Unsaturated fats—Fatty acids in which each carbon atom in the chain has fewer hydrogen atoms attached than the maximum possible. They are believed to be more healthful in the diet than saturated fatty acids.

Upright mixer—Food mixer, floor or table mounted, with an overhead motor that drives a downward projecting shaft in a planetary motion, to which various types of mixing devices may be attached. The mixing bowl sits on a yoke that may be raised and lowered by hand or motor. There is usually a power takeoff on the motor that can be used to drive accessories such as grinders or choppers. Bowl capacities range from 5 to 140 qt, and motor horsepower from $\frac{1}{6}$ to 5 hp. Speeds vary from 55 to 318 rpm and may be controlled with a selector lever. Also called a *vertical mixer.*

Urn—Container used to brew and keep beverages warm such as coffee. It has a spigot at the bottom for serving. Urns vary in size.

U.S. Fancy or Extra Fancy—Grades above U.S. No. 1, representing better-than-average quality of canned and frozen fruits and vegetables.

U.S. Grade A—Grade designation to indicate high quality poultry, canned or frozen food, etc.

U.S. Grades A, B, and C—Consumer grades for fresh fruits and vegetables and may have to be used by foodservices when U.S. No. 1 and U.S. Fancy or Extra Fancy are not available.

U.S. No. 1—Grade for fresh fruits and vegetables of high quality suitable for use in foodservice.

Vacuum-dry—Procedure used on fruits and vegetables to reach 2 to 5 percent moisture content.

Value analysis—Analysis of what is needed compared to what is obtained, with value representing price divided into quality, or $V = Q/P$.

Variable costs—Expenses that change as the volume of production and sales change.

Variety meats—Organ meats, including sweetbreads, brains, tongue, and liver

VCM (vertical cutter/mixer)—Versatile high-speed food cutting and mixing machine consisting of a motor attached to the bottom of a mixing bowl that drives a set of cutting and mixing blades. Various types of interchangeable blades perform different functions, such as mixing, chopping, blending, emulsifying, grating, grinding, mashing, blending, or homogenizing. Capacities range from 25 to 130 qt. Blade speeds range from 1750 to 3500 rpm.

Veal—Meat of a calf slaughtered between the ages of six and fourteen weeks. The finest veal is milk fed and the flesh has a creamy white or pinkish hue. The best comes from France, Holland, and Italy. American veal is usually older and stringier than prime veal. Veal is often larded because it is so lean that it has a tendency to dry out. The better known cuts are as follows.

Breast: Portion of the foresaddle below the ribs.

Saddle cut: Back portions.

Veal hotel rack: Consists of seven ribs separated after the removal of the legs.

Veal leg, double: Portion of the hindsaddle remaining after the removal of the loin.

Veal loin, trimmed: Anterior portion of the hindsaddle remaining after removal of the legs.

Veal shoulder, unsplit: Portion including the large outside muscle system posterior to the elbow joint and ventral to the medial ridge of the blade bone.

Vector—Carrier of contamination.

Veloute—Stock thickened with roux.

Vending service—Use of machines to dispense food.

Vendor—Supplier; one who sells merchandise.

Ventilation—Proper movement of air within a foodservice facility to remove excessive heat, objectionable odors, and noxious gases.

Virus—Agent of infectious disease that is much smaller than a bacterium and is not even a complete cell. It is composed only of nucleic acids wrapped in a protein coat. Viruses can spread through food, water, or the air. Antibiotics are ineffective against them.

Vitamin—One of the main categories of nutrients contributing to the health and well-being of the body. Vitamins are received largely through diet.

Vitamin B₁₂—Required for red blood cell development, proper cell function, and maintenance of the nervous system. Deficiency affects almost all body tissues and may cause pernicious anemia and degeneration of the nervous system. Rich sources include organ meats, fish and shell-fish, milk, and eggs. It is a water-soluble vitamin and is also called *cobalamin*.

Vitamin C—Vitamin required for strong cell structure, healthy teeth and gums, and aids in the absorption of iron. Deficiency can cause easy bruising and scurvy. Rich sources include citrus fruits, potatoes, tomatoes, and green leafy vegetables. It is a water-soluble vitamin and is also called *ascorbic acid*.

Vitamins A, D, E, and K—Known as the fat-soluble vitamins and are stored in the body's fat tissues and thus persist in the body longer than water-soluble vitamins.

Waste compaction—Possible method for waste disposal whereby waste is crushed to reduce space in storage for pickup.

Waste incineration—Possible method for waste disposal whereby waste is burned to conserve.

Waste pulping—System that creates a liquefied waste by grinding foods with a recirculating water system.

Water-soluble vitamins—Vitamins of the B-complex and vitamin C. They are stored in body fluids (such as blood) and are more easily lost (e.g., in the urine) than are the fat-soluble vitamins. They are also easily leached out by cooking.

Waxy maize starch—Thickening agent, also called *converted starch,* used for fruit pie fillings since it has good clarity and is not too viscous when cool, giving a soft, pasty texture that is desirable in such pies.

Wet service—Method of making tea by covering it with boiling water and allowing it to steep from three to five minutes.

White stock—Thin liquid rendered from simmering rinsed veal (chicken) bones in water with vegetables and seasonings for four to six hours.

Wholesomeness—Clean and in good condition; suitable for human use or consumption.

Yearling—Young sheep about a year old.

Yeast—Class of fungi. Some are useful (e.g., for leavening bread) for fermenting sugar into alcohol, or as a rich protein source. Others produce contaminants.

Yeast doughnut—Made from a rich and usually sweet yeast dough. The doughnuts are fried in deep fat.

Yield test—Study to reveal the number of portions obtained from a quantity of food.

Zinc—Essential mineral required by several enzyme systems, and in the synthesis of nucleic acids and of protein. Best food sources include seafood, liver, eggs, and nuts. It is easily destroyed in cooking. Too much, however, can be toxic.

Ziti—Large, tubular-shaped macaroni.

Index